ELECTRONIC STRUCTURE

AND

CHEMICAL BINDING

ELECTRONIC STRUCTURE

AND

CHEMICAL BINDING

With Special Reference to Inorganic Chemistry

BY

OSCAR KNEFLER RICE

Kenan Professor of Chemistry
University of North Carolina

DOVER PUBLICATIONS, INC.

New York

Published in Canada by General Publishing Company, Ltd., 30 Lesmill Road, Don Mills, Toronto, Ontario.
Published in the United Kingdom by Constable and Company, Ltd., 10 Orange Street, London WC 2.

This Dover edition, first published in 1969, is an unabridged and slightly corrected republication of the work originally published in 1940 by McGraw-Hill Book Company, Inc., to which has been added a new preface by the author.

Standard Book Number: 486-62247-9
Library of Congress Catalog Card Number: 69-15363

Manufactured in the United States of America
Dover Publications, Inc.
180 Varick Street
New York, N. Y. 10014

PREFACE TO THE DOVER EDITION

Dover Publications has undertaken to reprint *Electronic Structure and Chemical Binding* in response to numerous requests that it again be made available. The subject has greatly changed, of course, since the book appeared in 1940, so that parts of it are now obsolete. Many of the data used in the book have since been redetermined, so for up-to-date data the reader is directed to the current literature. On the side of theory, also, much has changed, and a number of books have appeared in recent years which cover the new material.

On the other hand, the discussions of the experimental material, and of the relations between the facts and the physical ideas, remain in many essential respects unaltered. The Born-Haber cycle, which was much used in the latter part of the text, and even applied, I believe, in some novel cases, remains a most useful tool for correlating material which is of interest to the chemist. At the time of the writing of the book a number of the discussions possessed a certain degree of novelty, and some of these may still be of interest today. I hope that Chapters XIV and XVI may still be helpful summaries of some of the properties of crystals, and that the physical ideas discussed in Chapters XVII and XVIII may still offer the reader some degree of enlightenment, despite the great strides taken in these subjects in recent years. Finally, I believe that Chapter XIX offers an adequate summary of some of the grosser (and hence, in a sense, more important) features of aqueous solutions.

It is this material, as well as the exposition and point of view of *Electronic Structure and Chemical Binding*, which appear to have been of lasting value; for this reason it has seemed the wiser course to allow the work to be reprinted unchanged, rather than attempt a complete revision, which would amount in essence to a new book.

It seems desirable at this time to list—probably not exhaustively—the areas in which the text is now out-of-date, in the light of present knowledge.

Experimental data.—For the most part the more refined data now available do not alter any of my conclusions. One exception must be noted, however; the dissociation energy of F_2 is much lower than was believed at the time the book was written. The correct value now appears to be close to 37 kcal per mole, which will bring the electron affinity down to 82 kcal per mole. (The change cancels, however, in the calculation of the electronegativity from the thermochemical data.) There are also several changes in Table 7; for example, the dissociation energy of N_2 is about 225 kcal per mole, and those of the P_2, As_2, and Bi_2 are considerably larger than given. Also the value of the C-C and C-H bonds are now reasonably well known. For discussion of these and other bond energies, the reader may refer to T. L. Cottrell, *The Strengths of Chemical Bonds*, 2nd Edition (Butterworths, London, 1958). There are also various tabulations which are relevant, for example, "Selected Values of Chemical Thermodynamic Properties," *Circular* 500, *National Bureau of Standards* (U.S. Government Printing Office, Washington, 1952), and the *JANAF Thermochemical Tables* (issued by The Thermal Research Laboratory, Dow Chemical Company, Midland, Michigan).

Theory.—The treatment of the following topics might be misleading to the uninitated.

1. *Rotation in crystals.* While there are probably a few cases in which molecules or ions rotate in crystals, this phenomenon is less common than once thought, and many transitions formerly believed to be connected with onset of rotation are probably only transitions to a state in which a limited amount of randomness in orientation is involved.

2. *Atomic energy levels.* The discussion in Chapter VIII is misleading in some respects.

3. *The boranes.* The discussion of these compounds was written before the suggestion of three-center bonds was made, and hence is largely outdated. For a review, see W. N. Lipscomb, *J. Phys. Chem.* **61** 23 (1957).

4. *Hybridization and directed valence.* The discussion given in the first part of Chapter XV is incomplete, and too sketchy for present-day use.

5. *Hindered rotation about the C-C bond.* The view that the hindering potential can be ascribed to direct forces between the hydrogen atoms, implied on p. 273, today seems somewhat naive,

though it may perhaps be accepted as a very rough approximation. For recent references see *Annual Review of Physical Chemistry*, Vol. 18 (Annual Reviews, Palo Alto, California, 1967), p. 371.

6. *Boiling points of crystals*, discussed on pp. 367-8 and elsewhere. I would now stress the geometry and coordination number rather than the ionic or covalent character.

7. *Magnetic moment of the* O_2 *molecule*. Nowadays, of course, one would have recourse to a much more detailed discussion of molecular orbital theory.

O. K. Rice

Chapel Hill, N. C.
September, 1968

PREFACE TO THE FIRST EDITION

This book is the outgrowth of a course of lectures entitled "Advanced Inorganic Chemistry," which I gave for a number of years at Harvard University. The contents, however, bear little resemblance to those of the traditional works on inorganic chemistry; the object is, rather, to see what light can be thrown on typical phenomena in this field by the modern theory of the atom. In attempting a development of this point of view, the book has become essentially a discussion of the chemical bond as exemplified in inorganic compounds.

Since the book has been written primarily for students of chemistry, and since it has been anticipated that its principal use, at least in classes, will be by students of about the level of the first year of graduate work, its scope and character have been determined with the needs of such students in mind. A student who has had a good course in physical chemistry may be supposed to have a reasonably good knowledge of the properties of matter in bulk. On the other hand, it has been my experience that he is likely to have but a very superficial acquaintance with the properties of individual atoms. This is certainly less true today than it was a few years ago, but it is still a sufficiently good generalization to justify including a discussion of atomic theory. Such a discussion occupies about the first third of the text. It starts with a brief account of the development of the atomic theory in chemistry, and continues with the contribution of physics in uncovering the more detailed structure of matter. Carrying these considerations on through the more modern aspects of atomic theory, including the wave theory of matter and quantization, I have introduced applications to the hydrogen atom, the properties of the elements in general, and the nature of chemical binding. In this way the ground is broken for the more detailed applications of the later chapters.

This introductory material is presented in an elementary manner. In particular, I have made an effort to present the ideas lying behind the wave mechanical theories of covalent binding without giving all the mathematical detail. The treatment is based upon the fact that it is not necessary to consider the wave equation in order to study the wave phenomena associated with particles of constant velocity; thus the fundamental concepts can be elucidated, and applications to cases where the velocity of the particle is not constant can be looked upon as extensions of the theory, for which not all details need be given. The old quantum theory has been considered to be an approximation to the wave mechanics and has been applied where permissible.

The phenomena of inorganic chemistry are extremely complicated and involve such a variety of factors that any attempt at a complete wave mechanical analysis, without introduction of far-reaching simplifications, appears hopeless at the present time. Yet for the unraveling of the factors involved and the elucidation and classification of the phenomena in a more general manner, the semiquantitative method outlined in the first part of the book often yields surprisingly adequate results. It is not to be denied, of course, that anyone desiring to become an investigator in this field will wish to make a more thorough study of wave mechanics. It is hoped, however, that such a student will find a helpful approach to this more detailed study in the first part of the book, and that he will gain from it a knowledge of some of the physical implications, which may well precede the mastery of the mathematical details. For many students of chemistry, who do not have a mathematical turn of mind and who intend to specialize in other branches, the treatment given in the first part of the present volume may well be all that is required.

In writing the book I have tried to adhere as strictly as possible to topics that are directly concerned with the clarification of the nature of chemical binding. For example, I have freely used the results of crystallography but have limited discussion of the actual methods to a bare outline, only sufficient to indicate the general ideas involved. In cases of this kind references are given in which the details of the subject may be pursued further.

Similarly, little has been said about the actual measurement of spectral lines beyond indicating that that is the method by which energy levels are determined. And I have not stressed the details of atomic structure, which are of particular interest in spectroscopy. Such topics as the coupling of spin and orbital angular momentum, for example, are adequately treated in other works, and for the purpose of this book it was not necessary to lay much stress on the interaction of the electrons within an atom.

In discussing the chemical bond, I have generally made use of the concept of the localized bond rather than the molecular orbital. I fully realize the value of the latter method, but the localized-bond concept seems to lend itself more readily to an elementary treatment. Since the chemist is most interested in the lowest state of a molecule, no attempt has been made to treat the excited energy levels in any detail.

Phenomena that depend primarily upon rates of reaction, rather than upon the state attained in equilibrium, have also been excluded. Since the state of systems in equilibrium is our primary concern, great stress has been laid on the calculation of the energies of various systems under various conditions. In the last chapter, which deals with aqueous solutions, entropies also have been considered. However, in the earlier chapters calculation of entropies seemed less important, and it was felt to be advantageous to keep the book as elementary as possible by excluding detailed applications of the second law of thermodynamics. The emphasis was, therefore, laid on the energy relationships.

Within the self-imposed limitations outlined above, it has still been possible to discuss and correlate a considerable mass of material that is of interest to the inorganic chemist. Of course, no exhaustive treatment of the vast body of known facts could be attempted. I have tried to select typical applications with the aim of giving the student a "feeling" for the phenomena of inorganic chemistry that will make it possible for him to correlate other facts as these facts appear. Nevertheless, a considerable amount of material has been brought together in the last chapters and treated in a manner more or less novel, at least as far as the textbook literature is concerned. It is my hope that this latter part of the book will have some appeal to research workers, as well as to the students who seek an introduction to the field. In general, the more recent references have been given, since from these the earlier literature may be traced; the references are not designed to give a historical perspective. Even with a list confined to the more recent papers, it would be impractical to attempt to make it exhaustive, and I am well aware that many important references have been omitted; the fact that some particular piece of work has not been mentioned is not to be construed as meaning that it is not significant.

It is my pleasure to express my gratitude to many who have at one time or another been of assistance in the preparation of the text. In the early stages I received much help and encouragement from Dr. Victor Guillemin, Jr., who read and criticized all the earlier chapters. These chapters were also the subject of extremely helpful comments of Professor A. E. Ruark. Professor K. Fajans gave freely of his time and effort, and a number of the later chapters owe much to his friendly criticism. Several portions were painstakingly read and criticized by Professor Hertha Sponer, and a rather lengthy chapter was the subject of detailed comments by Professor Joseph E. Mayer. Colleagues who were kind enough to read special sections of the book and communicate with me on special topics are Professors J. C. Bailar, Jr., A. B. Burg, E. D. Eastman, C. J. Gorter, Ralph Halford, David Harker, W. M. Latimer, Edward Mack, Jr., A. R. Olsen, Linus Pauling, G. K. Rollefson, W. C. Root, W. H. Zachariasen, and others. Mr. W. L. Haden, Jr., assisted in the preparation of Fig. 91, and Mr. C. V. Cannon assisted in the proofreading.

I am indebted to a number of authors and publishers for permission to use cuts and quotations. Sources of these are acknowledged at the appropriate places in the text.

I should not close without mention of my former teachers and colleagues at the University of California and the California Institute of Technology. All of them through their influence on the early development of my ideas in the field of chemistry, and atomic chemistry in particular, are in some measure responsible for whatever merit this book may possess.

Finally, it is a pleasure to express gratitude for the aid rendered by my mother in the preparation of the manuscript. Without her encouragement and assistance, the completion of the book would scarcely have been possible.

O. K. RICE.

Chapel Hill, N. C..
December, 1939.

CONTENTS

TABLES

ELECTRONIC STRUCTURE

AND

CHEMICAL BINDING

ELECTRONIC STRUCTURE AND CHEMICAL BINDING

CHAPTER I

DEVELOPMENT OF THE ATOMIC THEORY IN CHEMISTRY

1.1. Early Development of Chemistry.—It is the ideal of the theoretical physicist ultimately to explain all the facts of chemistry in terms of the properties of the electrons and positively charged nuclei of which all atoms are composed. Actually, of course, he is yet far from accomplishing this ideal, and historically the development of the subject has necessarily been from the empirical to the theoretical rather than the reverse; hypotheses have usually followed facts and have developed from the specific and special to the general and inclusive.

Thus the alchemists labored for many years to convert base metals into gold and, though failing in this attempt, laid a foundation of empirical facts upon which theoretical chemistry could build. They found that substances could be transformed in various and manifold ways, but that it was not possible, at least with the means at their command, to continue this process indefinitely and change any substance into any other. This led to the fundamental concept of chemistry, the notion of elementary substances. As early as the seventeenth century, it was stated by Jungius and by Boyle that there were certain elementary substances which could not be changed into each other and could not be further decomposed, and that all other substances were composed of combinations of these elements.

This concept was crystallized by Lavoisier, who showed, for example, in his classical experiment on the decomposition of HgO that one could account for all the material of the HgO in the products, Hg and O_2, and that by proper treatment it was possible to get the HgO back again from the products. The importance of this work of Lavoisier was due to his use of the weight of the material as the measure of its amount. He could thus show that within his experimental error (which was, to be sure, rather large) there was just as much substance in the products, Hg and O_2, as in the original HgO from which the former were produced. It is thus possible to follow quantitatively the decomposition of a substance into two or more substances, all with different properties, such that the total weight of all the products is equal to the weight of the original substance; and finally it is possible to learn which substances cannot be decomposed and are thus elements. By the use of weight relationships, one can trace a reaction through in all its details, and can make sure that nothing is lost, and that no extraneous material (oxygen or other gases of the atmosphere, for example) becomes involved. By aid of his own and other work, Lavoisier was able to make a list of elementary substances which was reasonably accurate, considering the difficulties involved. This list was corrected and greatly added to during the first part of the last century.

1.2. Laws of Chemical Combination.—The truth of the first law of chemical combination, the law of definite proportions, was at least indicated by the work of Lavoisier. The establishment of the laws of multiple and reciprocal proportions followed early in the nineteenth century. Though these laws are well known to all chemists of the present day, a statement of them may not be out of place.

1. Law of definite proportions: The chemical elements combine always in a definite weight ratio when forming a given compound.

2. Law of multiple proportions: If an element A combines with an element B to form different compounds in which different weights, say α, β, γ, . . . , of A are combined with unit weight of B, then $\alpha:\beta:\gamma: \ldots = a:b:c \ldots$, where a, b, and c are integers (usually small integers, at least in inorganic compounds).

3. Law of reciprocal proportions: Let α be the weight of A and β the weight of B which combine with unit weight of C. Then the weight of A which combines with a weight β of B will be either α or $\alpha(q/r)$, where q and r are integers (usually small integers).

These laws have been combined and stated as one in the following way by Muir:[1]

> To every homogeneous substance can be assigned a certain number, expressing a definite mass of the substance, which may be called its combining weight, or its reacting weight, all chemical reactions between elements and compounds occur between masses of them which can be expressed by the numbers in question, or by whole multiples of these numbers.

Final proof of the exact and practically universal validity of these laws and of the fact that the combining weights of elements do not, in general, depend on the source or the method of preparation was given by Stas in a series of memoirs published between 1840 and 1882. His conclusions have been somewhat modified in exceptional cases by the discovery of isotopes, but this does not change the essence of the ideas we wish to develop (see footnote 2).

1.3. The Atomic Theory.—The laws of chemical combination readily follow if it is assumed (1) that each element is composed of ultimate indivisible particles, *atoms*, all the atoms of a given element being exactly alike, and in particular having the same weight;[2] (2) that in forming a compound a number (usually a small number) of atoms of various kinds combine in a definite way to form a *molecule;* and (3) that a compound consists of an aggregation of such molecules, the composition of all molecules of a given compound being exactly alike. The atomic hypothesis of the constitution of matter, however, did not wait for the discovery of the laws by which elements combine, for some of the

[1] Muir, "History of Chemical Theories and Laws," p. 99, John Wiley & Sons, Inc., 1906 (permanently out of print). This book contains an interesting account of the development of chemical theory.

[2] The existence of isotopes necessitates some modification of this statement. Isotopic atoms are practically identical in all respects except in mass. The laws of chemical combination will follow from the atomic hypothesis, however, if all the isotopes of a given element are always present in constant proportion, whether the element is free or combined, so that there is a definite *average* atomic mass for the element. The condition that the isotopes be present in constant proportion is practically always fulfilled to an extremely high degree of approximation in manipulations involving ordinary chemical processes, though recently there have been successful attempts to separate isotopes. It is also true that radioactive lead has a different isotopic constitution and hence a different atomic weight from ordinary lead. This results in an exception to the conclusion of Stas.

ancients held that matter was composed of atoms. Still, they can hardly be said to have formulated a true scientific theory; the modern atomic theory may be considered to date from the deduction by Dalton of the laws of chemical combination and his description of chemical compounds and reactions in terms of atoms.

1.4. The Determination of Atomic Weights.—If, now, the laws of chemical combination are established, and if we have a fairly good idea as to which substances are elements and which are compounds, a new problem at once presents itself. We wish to determine combining weights. And once the atomic theory is accepted in the form stated above, this resolves itself into a more special problem, the determination of atomic weights. This, of course, involves the determination not only of the relative weights of elements that will combine with each other, but of the chemical formulas of the compounds formed.

The pioneer in the determination of combining and atomic weights was Berzelius, who analyzed some two thousand substances in the course of his lifetime. These analyses were made with remarkable accuracy, considering the equipment which he had at hand, and the chief difficulty turned out to be the determination of the formulas of the compounds. Here, it was often necessary to make arbitrary assumptions. For example, if the formula Cr_2O_3 were chosen for chromous oxide, then the formula for chromic oxide would have to be CrO_3; but since it is not possible to know how many atoms are in combination in a molecule, how might one be sure that chromous oxide ought not to have the formula CrO_3 and chromic oxide, therefore, the formula CrO_6? Berzelius frequently used what appeared to him to be the simplest set of formulas. This often gave him correct, sometimes incorrect, results. But without some other criterion, it is obvious that atomic weights determined by arbitrary assignment of chemical formulas might easily be in error by such a factor as 2 or 3.

There is a very useful physical principle which is of great service in the determination of atomic weights; it is, namely, Avogadro's law, which states that equal volumes of gas at standard conditions of temperature and pressure contain equal numbers of molecules, regardless of the nature of the gas. Berzelius, however, did not recognize the validity of this law; he assumed, instead, what he considered to be simpler, that equal volumes of elementary gases contained equal numbers of atoms and that a different and unknown law held for compound substances. Thus from the fact that two volumes of hydrogen combined with one of oxygen in the formation of water, he concluded that two atoms of hydrogen combined with one of oxygen, and he wrote

$$2H + O \rightarrow H_2O$$

arriving at the correct formula for water, in spite of his incorrect premise. The fact that two volumes of water are formed from two of hydrogen and one of oxygen, he did not attempt to interpret. By this method, simply because so many of the elementary gases were diatomic, Berzelius arrived at the correct results for the formulas of many compounds. These results were then extended to other compounds by analogy. Thus if the formula for ammonia is NH_3, the formula for the chemically similar phosphine would be assumed to be PH_3.

Two kinds of physical analogy also played an important role in the early determination of atomic weights. One of these is indicated by the law of Dulong and Petit. These workers found that for a great many elements the specific heat per gram atom per degree centigrade was approximately 6 cal.; there is now known to be a statistical mechanical basis for this law,[1] but at first it was found empirically. It will be observed that it depended upon knowledge of the atomic weights of the elements investigated; otherwise, the amount of material constituting a gram atom would not be known. However, once it was established for a series of elements whose atomic weights were known, at least provisionally, it could be extrapolated, so to speak, to

[1] Modern developments have shown that Dulong and Petit's law is an asymptotic approximation, valid only at high temperatures (and then exactly true only in the absence of other complicating factors), and the limiting asymptotic value of the specific heat has been shown to be $3R$ = 5.96 cal. per gram atom per deg. (R is the gas constant). (This is for constant volume, however, whereas usually specific heats are measured at constant pressure.) Low values of the specific heat, found for particular elements at room temperature, are due to the fact that room temperature is not a sufficiently high temperature for these substances, and raising the temperature brings an approach to the limiting high temperature value. For a more detailed account, the student is referred to Taylor, "A Treatise on Physical Chemistry," 2d ed., vol. 1, pp. 276*ff*., and vol. 2, pp. 1393*ff*., D. Van Nostrand Company, Inc., 1931.

other elements. It was then possible to say that the amount of any element which required six calories to heat it up one degree centigrade was one gram atom. Though it was never supposed that the law of Dulong and Petit would give an exact value for the specific heat per gram atom, the atomic weight could thus be roughly found, and then more exactly determined by analyses; since this law would almost always give results much closer than the factor of two or three which was in doubt in the determination of the atomic weight, this combination of methods was helpful in many cases.

Another type of physical analogy is expressed by Mitscherlich's law of isomorphy. Mitscherlich made a study of the properties, especially the crystalline form, of a series of analogous compounds of arsenic and phosphorus; he also investigated a number of groups of oxides and salts of analogous composition. This study led him to enunciate the following law:[1] "Equal numbers of atoms, combined in the same manner, produce the same crystalline forms; identity of crystalline form is independent of the chemical nature of the atoms and is determined only by their number and relative positions." This law was applied in many cases to determine the formulas of unknown compounds from those whose formulas were believed to be known. Thus, Berzelius, having concluded that chromous oxide had the formula Cr_2O_3, noted that ferric oxide and aluminum oxide had the same crystal form, and therefore wrote Fe_2O_3 and Al_2O_3.

1.5. The Combining Volumes of Gases and Avogadro's Law.—We shall now return to a consideration of the combining volumes of gases and the proper use of Avogadro's law. Despite the pioneer work of Avogadro himself, of Gay-Lussac, and of Dumas, confusion hovered over this subject until the middle of the last century, when the difficulties were cleared up by the work of Laurent, of Gerhardt, and particularly of Cannizzaro, and Avogadro's law came to furnish the most powerful method for the determination of the formulas of compounds.

It is obvious that if Berzelius had interpreted the relation

2 volumes of hydrogen + 1 volume of oxygen gives 2 volumes of water vapor

[1] It is, of course, now recognized that compounds with analogous chemical formulas do not always have the same crystalline form. However, the fact that they often do was of considerable aid in the early history of chemistry.

correctly, as

2 molecules of hydrogen + 1 molecule of oxygen gives 2 molecules of water

he would have seen that, since 1 molecule of oxygen is used in forming 2 molecules of water, the oxygen molecule must be broken in two and hence it must contain an even number of atoms. The fact that in the study of many reactions the oxygen is never broken into four or more parts leads one to write the formula as O_2.

In the same way, the formulas of other elementary gases may be found. If oxygen then is assigned the atomic weight 16, it is possible by comparing the vapor densities to get the atomic weight of any other element whose formula in the vapor state is known. But the usefulness of the method does not stop here, for it is possible to find the molecular weights of compounds, and having done this it is possible to find, by chemical analysis, the weight of any given element combined in one mole[1] of a certain compound. If, then, it should be possible to find a number of compounds of a single element, which can be obtained in the gaseous state, the smallest weight of that element which can occur in one mole would become known. This smallest weight will naturally be supposed to be the weight of one gram atom of this substance, *i.e.*, the atomic weight.

Furthermore, if the atomic weights of a series of elements have been found, it is possible to determine the smallest weight of another element that will combine with one gram atom of any one of the series of elements whose atomic weights are known. This smallest weight will be presumed to be the atomic weight of the new element. This method can of course be used, no matter how the series of known atomic weights is obtained, but is effective only if these atomic weights are *really known.*

1.6. The Periodic System.—By means of all the methods outlined above, it was finally possible to arrive at a reasonably reliable system of atomic weights. The final confirmation of these atomic weights came with Mendeléeff's discovery of the periodic law. This uncovering of the regularity of nature left little doubt that the system of atomic weights which led to it

[1] It is now, of course, possible to define a mole as the amount of substance that occupies a definite volume in the gas phase under standard conditions.

was correct. The developments of modern physics provide a theoretical foundation for the periodic system, which will be discussed in later chapters, and which, of course, strengthens the conviction that the table of atomic weights is correct. Furthermore, modern methods, such as the use of X rays in the study of crystals, have fully confirmed the chemical formulas used by the pioneers in the determination of atomic weights.

1.7. Faraday's Law.—If two electrodes of the same metal dip into a solution of a salt of that metal, an electric current can be forced through the solution by means, say, of a battery applied between the two electrodes, as shown in Fig. 1. It is found that, in such an electrolysis, metal disappears from one electrode (the anode) and goes into solution, while it is deposited from the solution on the other electrode. These phenomena were studied by Faraday, and, translating his fundamental discovery into modern terminology, we may say that he found that for every gram atom of material deposited on, or dissolved from, an electrode in this manner a certain definite quantity of electricity, *i.e.*, 1 faraday ($= 9.650 \times 10^4$ coulombs) or a small integral multiple of a faraday, depending upon the material of the electrodes, passes through the circuit. This is explainable in terms of an atomic theory of electricity in much the same way that the laws of chemical combination are explainable in terms of an atomic theory of matter in general. It indicates that electricity can enter into combination with matter much as different kinds of matter enter into combination with each other.

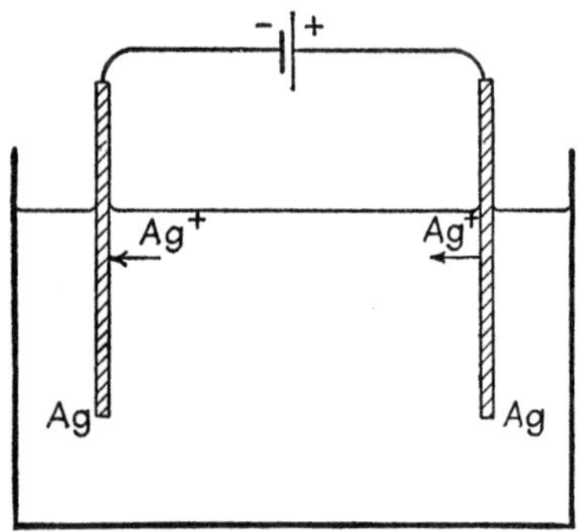

FIG. 1.—Electrolysis of a silver solution.

Since all substances are electrically neutral, they must contain equal amounts of positive and negative electricity. But it is now known, of course, that positively or negatively charged ions can exist in solution, although the solution as a whole is neutral. Deposition of a positively charged ion on an electrode, for example, requires that it be furnished with one or more electrons, depending upon its charge. Thus a silver ion, Ag^+, requires an electron before it can be deposited as neutral metallic silver. These electrons are produced by some chemical process at another electrode (*e.g.*, solution of some metal, in the battery which furnishes the current for the circuit) and travel through a wire to the place where they are needed, or, rather, the electrons produced by the chemical process exert a "push" down the wire which results in electrons already at the other end being released for the chemical process taking place there—there is no bodily transfer of the particular electrons, which take part in the reaction at the one electrode, to the other electrode, but only a displacement of electrons down the wire.

In the foregoing paragraph, we have outlined a rather highly developed theory. It did not, of course, spring into being all at once, but developed gradually, as the result of numerous experiments on the properties of electrolytic solutions and the properties of electrically charged particles in gaseous discharges. We cannot go into all the ramifications of this development, but some of them will be touched on briefly in the next chapter, which deals with the properties of individual atoms and electrons and describes some of the methods by which these properties have been discovered.

Exercises

1. Show in detail how the three laws of chemical combination follow from Muir's statement, page 3.

2. Deduce the law of reciprocal proportions from the atomic theory.

3. Discuss the volume relations in some reaction which enables one to conclude that the chlorine molecule contains an even number of atoms.

4. 1 liter of a gaseous compound of a certain element at 1 atm. and 25°C. contains 2.45 g. of the element. 1 liter of another gaseous compound under the same conditions contains 1.47 g. of the element. On the basis of these data, assuming the perfect gas laws, what is the largest possible value of the atomic weight of the element? State another value of the atomic weight consistent with these data.

CHAPTER II

THE CONSTITUTION OF MATTER

According to the modern theory of the constitution of matter, each atom is composed of a positively charged nucleus, which has most of the mass of the atom and is surrounded by a number of electrons, there ordinarily being just enough electrons so that the positive charge of the nucleus is exactly neutralized. The chemical properties are determined by the behavior of the electrons, and depend to a great extent upon their number. The number of electrons is determined in turn by the positive charge of the nucleus, but the latter has only an indirect influence on the chemical properties of the atom. One of the chief aims of this book will be the correlation of chemical properties and electronic structure. The first step in carrying out this purpose is a description of the properties of the electron itself. This description will be given in the present chapter, and the most important experimental evidence, which leads to the establishment of the most general features of the above-outlined theory of the constitution of matter, will be reviewed.

2.1. The nature of the electron was first learned from the experiments of J. J. Thomson and of Lenard on electric discharges through highly evacuated tubes[1] (pressure less than 10^{-2} or 10^{-3} mm.). Such a tube is illustrated in Fig. 2. C is the cathode, A is the anode, and B is a thick metal disk which is connected to ground. Both A and B have slits through them. D and E are plates across which an electrical potential can be applied. If, now, a moderate potential is applied between C and A, a phosphorescent spot will appear at p, which is apparently due to something that has come through the slits in A and B. If the tube is arranged so that some solid object can be placed between B and p, the phosphorescence at p ceases. If an electrical potential is applied across D and E, the spot p moves down if E

[1] See, *e.g.*, J. J. THOMSON and G. P. THOMSON, "Conduction of Electricity through Gases," 3d ed., vol. I, Cambridge University Press, 1928.

is positively charged, up if D is positively charged. It therefore seems reasonable to suppose that the phosphorescence at p is due to negative particles (cathode rays, or electrons) which have been released from the negative electrode C, accelerated by the positive electrode A, and have passed through the slits in A and B.

Negatively charged particles can also be obtained from other sources. Thus they are observed in the emissions from radioactive bodies, and they are ejected from glowing metals and metallic oxides. The latter sources may be used in electron tubes, of which there are a large variety. Sommerfeld in his book "Atomic Structure and Spectral Lines"[1] gives a description of a tube in which the cathode is furnished with a spot of calcium oxide (a substance that readily emits electrons), which is heated

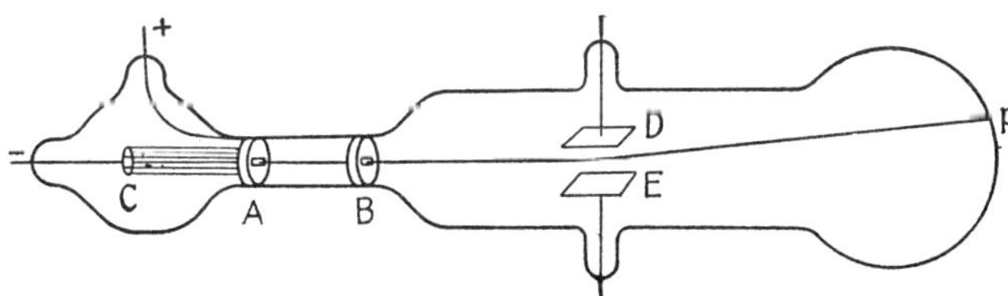

FIG. 2.—A discharge tube. *D* shown positively charged. (*Adapted, by permission, from Fig.* 55, *of J. J. Thomson* and *G. P. Thomson*, "*Conduction of Electricity through Gases*," 3d. *ed.*, *vol.* I, *Cambridge University Press*, 1928.)

by the discharge. This tube contains gas at 0.1 mm. pressure, which makes it possible to follow the path of the stream of electrons, since they ionize the gas and cause it to glow. The gas in the tube also causes practically the entire potential drop to take place in the immediate neighborhood of the cathode, so that the electrons reach their full velocity almost immediately, start out at right angles to the cathode surface (*i.e.*, the calcium oxide surface), and travel in a straight line unless acted upon by an external field. The effect of an applied electric field may be easily followed in a tube such as described by Sommerfeld, since the path of the electrons is outlined by the glow of the gas. It can be easily observed that they behave as negatively charged particles, but accurate measurements cannot be made in a tube containing 0.1 mm. of a gas, because the gas is rendered conducting by the electrons, and this interferes with the action of the electric field.

[1] English translation, 3d ed., vol. 1, Chap. I, §3, Methuen & Co., Ltd., 1934.

The tube shown in Fig. 2 may, however, if highly evacuated, be used for quantitative measurements, and by observing the effect of a magnetic field, and comparing it with the effect of an electric field, it is possible to determine the specific charge (ratio of the charge of the particles to their mass) and the velocity of the electrons, as will be shown immediately.

2.2. The Action of Electric and Magnetic Fields on Charged Particles.—The action of an electric field on a charged body is known from experiments on macroscopic bodies, and it is assumed that the same laws hold for microscopic bodies. In stating these laws, we assume that the definitions of all the fundamental electrical quantities are understood, it being beyond the scope of the present book to consider the relations between the fundamental experiments upon which these definitions are based and the laws we wish to discuss. A summary of the most important laws about electrical forces is given in Appendix III.

A magnetic field bears the same relation to magnetic poles as an electric field bears to electric charges. The magnetic field due to a pole of strength q_1 at a distance r is q_1/r^2 and the force exerted in vacuum on another pole of strength q_2 is q_1q_2/r^2, being attractive or repulsive according to whether q_1 and q_2 are different types or the same type of magnetic pole. q_1 and q_2 are in such units that two unit poles 1 cm. apart exert a force of 1 dyne on each other. The field strength is then given in gausses.

The action of a magnetic field on a charged particle may be inferred from its action on a wire carrying electric current. The latter is known from direct experimental measurements, which may be summarized as follows. If a wire carrying a current i is acted upon by a magnetic field H, then (if i is expressed in electrostatic units and H in gausses) the force f acting on unit length of wire in vacuum is given by

$$f = \frac{iH}{c}\sin\theta,$$

where θ is the angle between the direction of magnetic field and the direction of the current, and c the velocity of light.[1] If we represent the field and the current as vectors of length H and i,

[1] If both i and H were expressed in electromagnetic units, as is more usual in treatises on electromagnetism, the velocity of light would not appear in the expression.

respectively, f is equal to the area of the parallelogram thus defined (Fig. 3) divided by c. The direction of the force f is at right angles to the directions of both i and H. If i gives the direction of positive current, then in Fig. 3 the force is directed into the paper. However, the question of positive and negative signs and directions of force can for the most part be disregarded.

Now a stream of electrons moving close together in the same direction and with the same velocity will produce a current equal to nev, where n is the number of electrons per unit length in the stream, e the charge of one electron, and v the velocity. Assuming then that the action of a magnetic field on this stream of electrons is the same as on a current in a wire, we may write for the force per unit length of the stream

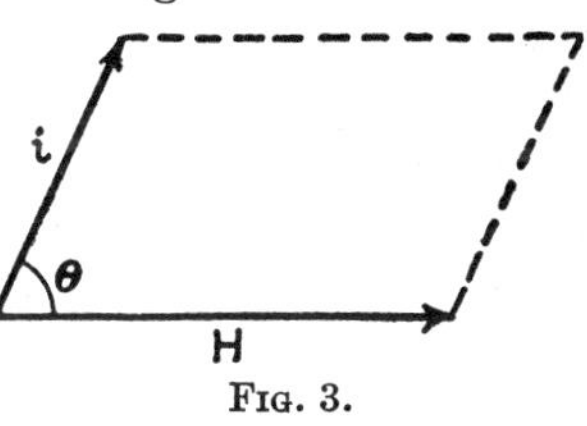

Fig. 3.

$$f = \frac{nevH}{c} \sin \theta$$

whence the force per electron f_0 is given by

$$f_0 = \frac{evH}{c} \sin \theta. \tag{1}$$

If the electron is moving in a uniform magnetic field in a path at right angles to the field, since the force on it will also be in the plane perpendicular to the magnetic field, the electron will continue to move indefinitely in this plane. Since the force is also at right angles to the velocity of the electron, the path will be a circle in this plane. The force exerted by the magnetic field will be balanced by the centrifugal force arising from the motion of the electron in the circle.[1] If the circle has a radius ρ, the centrifugal force is mv^2/ρ, where m is the mass of the electron. As $\sin \theta$ is 1 in this case we have

$$\frac{mv^2}{\rho} = \frac{evH}{c}$$

or

$$\frac{e}{mc} = \frac{v}{\rho H}. \tag{2}$$

[1] Our treatment neglects the relativity of mass of the electron, which may be important if its velocity is very great. It also neglects any effect of radiation, which is always entirely insignificant.

2.3. Determination of e/m of an electron was first carried out, using a direct method, by J. J. Thomson.[1] If the tube of Fig. 2 is placed in a magnetic field which, to the right of the slit B, is perpendicular to the long axis of the tube and has a uniform value, and which is zero to the left of B, the electron beam will be deflected in the plane perpendicular to the field. The radius of curvature of its path is readily found, through a simple geometrical calculation, by observing the deflection of the fluorescent spot at p. It is thus possible to obtain e/m from Eq. (2), provided a method for obtaining v can be found. This can be done by applying an electric field across the plates D and E. If a field E is thus applied, the force on each electron is eE. This force acts for the time t during which the electron is between the plates. If the length of the plates along the line of motion of the electrons is l, then $t = l/v$. During the passage between the plates, the electron will acquire a component of velocity $v_\perp$, given by

$$mv_\perp = \frac{eEl}{v}, \tag{3}$$

in a direction perpendicular to its original velocity v. [Equation (3) follows from equating the momentum $mv_\perp$ to the force eE times the time l/v during which it acts.] The ratio $r = v_\perp/v$ can be determined from the geometry of the tube by observing the deflection of the spot p. From (3), we may write

$$v^2 = \frac{eEl}{rm}. \tag{4}$$

Elimination of v from (2) and (4) makes possible the determination of e/m, since all other quantities are measured experimentally.

The experiment for the determination of e/m, as outlined, is of course an ideal experiment; in actual practice, various complications (nonuniform electric and magnetic fields and the effects of any residual gas in the tube, for example) have to be allowed for. The effect of the mutual repulsion of the electrons, which might be expected to cause a spreading of the stream, is negligible.

[1] See J. J. Thomson and G. P. Thomson, "Conduction of Electricity through Gases," 3d ed., vol. I.

There are various modifications that may be made in the experimental details. For example, in a highly evacuated tube, the second relation between v and e/m may be obtained by noting the total potential drop V (equal to the average field E times the distance), through which the electron falls. The energy thus imparted to the electron is eV (equals average force eE on the electron times the distance through which it travels), so that we have instead of (4) the relation

$$\tfrac{1}{2}mv^2 = eV \tag{5}$$

provided the electron starts out with zero kinetic energy. The best method for producing the stream of electrons is by the use of a hot cathode, usually a wire heated by an electric current, as then a very high vacuum may be used. The best recent value of e/m, obtained by deflection methods with all necessary refinements, is 5.27×10^{17} e.s.u. per g.[1]

It is important to note that cathode rays have been produced in tubes in which the electrodes have been constructed of a variety of different materials, and in which the residual gas has been of different kinds. Always, in all these cases, the value of e/m has been found to be the same within the limits of the experimental error, which indicates that the same kind of particles is always produced. Also, the β-rays produced in radioactive decomposition, and the particles produced by the action of light on metals, have been shown to be of the same nature. It thus appears that there is but one kind of electron which occurs as a constituent of ordinary material systems, and it is of very widespread occurrence in nature.

2.4. Determination of the Charge of an Electron.—The charge of an electron e as distinguished from the charge per unit mass e/m has been very accurately determined by Millikan by means of the famous oil-drop experiment.[2] Fine drops of oil are sprayed

[1] The values of this and other constants given in this chapter are taken from "An Outline of Atomic Physics," 2d ed., pp. 389–391, by the University of Pittsburgh Staff, John Wiley & Sons, Inc., 1937. However, on account of the large number of calculations in the literature using the values given by Birge, Physical Review Supplement (*Rev. Mod. Phys.*) **1,** 1 (1929), it has been found necessary, in order to avoid inconsistencies, to use these earlier values for calculations presented in other parts of the text. The errors thus caused are not important.

[2] See Millikan, "Electrons (+ and −), Protons, Photons, Neutrons, and Cosmic Rays," University of Chicago Press, 1935.

between two electrodes, which consist of plates placed horizontally so as to produce a vertical electric field, with the positive plate on the top. Electrons (or negative ions) and positive ions are produced in the space between the plates by the action of X rays on the air between the plates, and one or more of them may be caught by an oil drop. The electron or ion thus captured will impart a charge to the drop, which will move with a constant velocity under the combined action of the electric field, gravity, and the frictional resistance of the air. Its motion may be observed by means of a microscope. Under gravity alone (electric field turned off), the velocity v_1 will be given by Stokes' law of fall[1]

$$Mg = 6\pi r\eta v_1, \tag{6}$$

where M is the mass of the oil drop (corrected for the buoyancy of the air), Mg the force exerted by the gravitational field on the oil drop, r the radius of the oil drop, and η the viscosity of the air. If the oil drop has a negative charge ϵ and an electric field of strength E is turned on, the oil drop will be pulled upward toward the positive plate by a force ϵE. If this exceeds the force of gravity, the oil drop will move upward with a velocity v_2 given by Stokes' law, which will now have on the left-hand side the new expression for the force exerted by the combined electric and gravitational fields on the drop:

$$E\epsilon - Mg = 6\pi r\eta v_2. \tag{7}$$

From (6) and (7),

$$\epsilon = \frac{Mg(v_1 + v_2)}{Ev_1}. \tag{8}$$

But the mass of the drop may be found by use of Eq. (6) provided the density of the oil is known so as to make it possible to express r in terms of M, for g and η are known constants, while v_1 is found experimentally. Everything on the right-hand side of Eq. (8) is then known or experimentally determined so that ϵ may be evaluated.

It was found by experiments of this kind that the charge on the oil drop changed only by whole multiples of a certain definite

[1] For a discussion of this law, and certain modifications required, see Millikan, *op. cit.*, especially Chap. V.

quantity of electricity, which was therefore presumed to be the smallest possible charge that any body could have and assumed to be equal to e, the charge on the electron. The present accepted value of e is 4.80×10^{-10} e.s.u.[1]

From the values of e/m and e, it is possible to determine m, which turns out to be 9.12×10^{-28} g.

2.5. Positive Ions.—We shall now turn to the consideration of positive ions.[2] These are readily produced as "canal rays" in a discharge tube in which a hole is bored through the cathode, and which contains gas at a low pressure. A pencil of light will appear behind the hole in the cathode, with phosphorescence at the point where it strikes the glass. These effects are produced by positively charged ions, as may be shown by the direction in

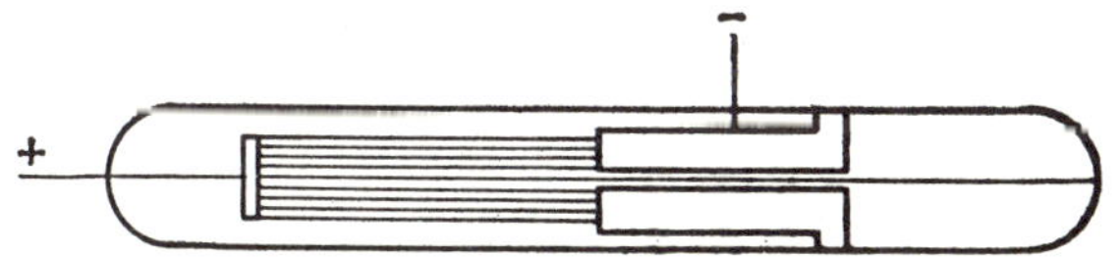

Fig. 4.—Schematic drawing of a tube for the production of positive ions.

which they are deflected by electric and magnetic fields. The amount of deflection, however, is extremely small, corresponding to a value of e/m for these particles, which is very much smaller than the value of e/m for the electrons.

In order to measure the e/m of these particles accurately, it is necessary to have the gas in the tube at a very low pressure, not more than a few thousandths of a millimeter of mercury. In order to measure the amount they are deflected, it is best to allow them to impinge on a photographic plate, which they affect at the point where they strike. It is found that the value of e/m depends upon the gas used in the tube, the largest value of e/m being obtained when hydrogen is used. This value is nevertheless about 1835 times smaller than the e/m for the electron;[1] if it is assumed that this is due to the mass of these ions being larger than the mass of the electron rather than the charge being smaller, and that the particle whose e/m is thus measured is H^+ (a positively ionized hydrogen atom) then the

[1] See footnote 1, p. 15.

[2] See, *e.g.*, J. J. Thomson and G. P. Thomson, "Conduction of Electricity through Gases," 3d ed., vol. I.

mass of H^+ is seen to be about 1835 times that of the electron. The value of e/m for H^+ found by deflection experiments agrees within the limit of error with that found in electrolysis by measuring the weight of hydrogen (or, better, a metal) deposited at an electrode when a given amount of electricity passes through the circuit. (It is the latter, more accurate value, that is actually quoted).

The technique of measuring e/m has been greatly developed by Aston[1] and, more recently, by others. e/m has now been determined for a large number of positive ions with great accuracy. On the assumption that the charge of a positive ion is always equal either to that of an electron or to a small integral multiple of it, it has been possible to determine the masses of many different positive ions. It is found that an element may really consist of a number of different substances called "isotopes." The isotopes of a given element all have, in general, practically identical chemical properties, but each respective isotope has its own characteristic atomic weight. When one takes the average atomic weight of the different isotopes, weighting according to the amount of each isotope present, it is found that the averages for the different elements are in the same ratio as the atomic weights of those elements obtained from chemical determinations. This comparison furnishes most important evidence as to the correctness of our view on the nature of positive ions and the atomic theory of matter.

2.6. The Constitution of the Atom.—All the experiments which have been considered in this chapter are consistent with the view that an atom is composed of a positively charged nucleus whose charge is exactly neutralized by electrons. A positive ion is produced when an atom loses one or more electrons. Since the electrons are very light compared with the positive nuclei, the mass of an ion will be practically equal to the mass of the atom from which it is formed.

We shall now consider the experiments of Rutherford,[2] which indicate that all the positive charge of an atom is in *one* extremely small nucleus and that the electrons form a rather diffuse cloud

[1] See Aston, "Mass Spectra and Isotopes," Longmans, Green & Company, 1933.

[2] Rutherford, Chadwick, and Ellis, "Radiations from Radioactive Substances," Cambridge University Press, 1930.

outside the nucleus. These experiments also make possible an estimate of the positive charge of the nucleus and hence the number of electrons in the atom. They were performed before the work of Aston, and it was Rutherford's work, in fact, that laid the foundations for the modern theory of atomic structure.

Rutherford used α-particles as tools in his investigation. α-particles are the positively charged ions emitted by radioactive bodies; before we can understand Rutherford's work, we must know the nature of these particles which he used as instruments of research.

The value of e/m can be measured for the α-particle as for any other positive ion, and it is found to be approximately half as great as for H^+. It is further found that all the α-particles from any given radioactive substance have a certain definite velocity (except that in some cases there are several groups of α-particles, all the particles in a given group having the same velocity). A bit of a radioactive substance therefore furnishes a very convenient source of charged projectiles with a constant and very high energy.

α-particles can be individually observed, due to the fact that a flash is seen every time one hits a zinc sulfide screen. This makes it possible to determine the number of α-particles emitted by a given amount of a radioactive substance in unit time. It is also possible to collect all the α-particles emitted in a given time through a certain solid angle and measure the total charge given to an electrometer by these particles. Since the rate of emission of particles has already been found by counting flashes, it is possible to determine how many α-particles are required to produce the given charge, and hence the charge on a single α-particle may be found. It is just twice the charge on an electron. The mass of the α-particles must therefore be approximately four times that of a hydrogen ion. All the information about the properties of α-particles needed in order to use them as tools in the study of atoms is thus at hand.

That α-particles differ in no way from doubly ionized helium atoms was proved by Rutherford and Royds, who collected them in a helium-free chamber, allowed them to be neutralized, and then showed that helium was present. It may be remarked parenthetically, that, since the weight of an α-particle is four times that of a hydrogen atom, and since the molecular weight of helium (known from the gas density) is twice that of hydrogen, helium gas must be monatomic if hydrogen is diatomic.

If an α-particle is allowed to shoot into a gas, it will be slowed up and will stop after it has penetrated a certain distance, which will be of the order of several centimeters if the gas is at a pressure somewhat near normal. For an α-particle of definite energy, the range is inversely proportional to the concentration of the gas, if the same kind of gas is always used. The stopping power per atom of gas is thus a constant, characteristic of the gas. α-particles will penetrate sheets of metals or other solid substances, provided they are thin enough. The stopping power per atom of substance is roughly proportional to the square root of the atomic weight.

When α-particles go through a thin metallic screen, it is found that most of the α-particles emerge without much change (3° or less) in the direction of their motion. A few, however, are observed to be highly deflected. It is these latter which are of interest to us here, for we shall see that these deflections can be explained on the assumption that the α-particles pass near massive charged bodies; and in order to explain the results, the charge on such a body must be taken to be equal to the charge on the electron times the atomic number (*i.e.*, the number that gives the position of the element composing the screen in the periodic system). The experiments therefore give direct evidence as to the structure of atoms, and we shall now proceed to discuss them in some detail.

It is obvious that when the α-particle is deflected through a large angle it must be repelled by a relatively heavy body. A light body, like an electron, would merely be pushed out of the way, and this, in fact, is happening all the time as the α-particle goes along its path. The α-particle gradually loses energy, and is eventually stopped, which accounts for its finite range, but it suffers no sharp change in its motion unless it penetrates very close indeed to the massive and highly charged nucleus. In fact, if we assume that we are here dealing with the ordinary inverse-square-law repulsion of two positively charged bodies, it can be readily shown that, if the nucleus has a charge not larger than the charge of an electron multiplied by the atomic number of the heaviest nuclei, the α-particle must come within a distance of the order of 10^{-11} cm. or less.[1] This distance is very much smaller than the distance between centers of adjoining atoms, which is of the order of 10^{-8} cm. Therefore it is reasonable to suppose that the α-particle, if it penetrates close enough to be appreciably deflected, is so much closer to the deflecting nucleus than to any other body which might affect it that the effect of the latter can be neglected. Further, since the atomic weights of most substances, which have been studied in this way, are much higher than that of helium, we can as a first approximation assume that the α-particle is simply deflected from a fixed center of force.

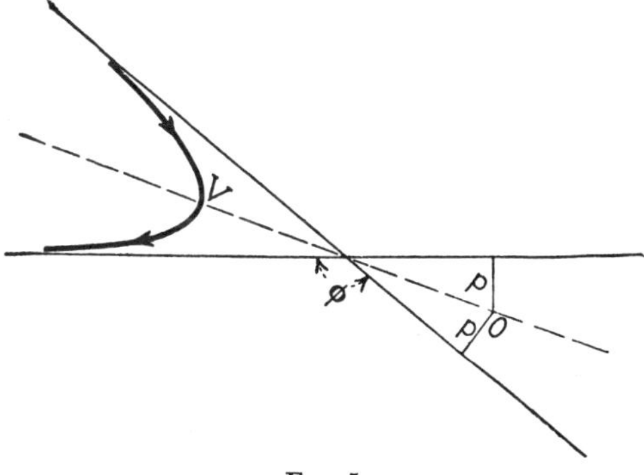

Fig. 5.

The orbit described by a particle that is repelled from a fixed point O by a force varying inversely as the square of the distance between them is a hyperbola with O as a focus. We shall discuss, without giving proofs, the characteristics of this hyperbola. The fixed point, being a focus, is, of course, symmetrically placed with respect to the two asymptotes. Let the distance from O to either of the asymptotes be p (see Fig. 5). p is the distance at which the particle would pass by O, if no force were acting. Let e be the charge on the electron. Then the charge on the α-particle is $2e$ (we disregard sign), and the charge on the nucleus will be designated as Ze, Z being the atomic number. We let M be the mass of the α-particle and v its velocity. ψ is, as shown in Fig. 5, the angle through which the α-particle is deflected.

[1] This is given approximately by p of Eq. (0). It is necessary for a typical α-particle (velocity, say, 1.5×10^9 cm. per sec.) to come at least this close to a nucleus of atomic number around 90 to be deflected 30° or more.

Then it may be shown[1] that

$$\tan\frac{\phi}{2} = \frac{2Ze^2}{Mv^2p}. \tag{9}$$

It is thus seen, as might be expected, that the α-particle is deflected more the less its kinetic energy, the smaller the distance it passes from the nucleus, and the greater the charge on the nucleus.

The quantity p characterizes the hyperbola and determines ϕ, Z and v being fixed in any given experiment. But it is easy to determine how many α-particles will pass a nucleus at a distance lying between p and $p + dp$, where dp is a small difference (or rather would so pass, if no deflection occurred). Since p determines ϕ, the number of α-particles which one would expect to be deflected at any given angle can be calculated and compared with experiment. The number thus obtained depends upon Z, which can thus be calculated, and when the proper value of Z is used the experimental results are reproduced by the theoretical equation. The values of Z thus obtained agree with the atomic number of the element, as determined independently by use of other considerations. This gives strong support to the theory that atoms are composed of positive nuclei with charge equal to the atomic number and neutralized by electrons which surround the nucleus within a distance of the order of 10^{-8} cm.

It may not be amiss to give some details of the above-indicated calculation. The chance that an α-particle would pass a distance between p and $p + dp$ from a given nucleus is proportional to the area included in the annular space between p and $p + dp$, in the plane perpendicular to the flight of the α-particle, namely, $2\pi p\,dp$. Now suppose that there are N atoms per unit volume in the screen which is used and that the thickness of the screen is t. Since the α-particle is moving normally to the screen, the effect is just the same as if it were striking a plane containing Nt atoms per unit area. This assumes that no atom is directly behind another; appreciable deflections occur for such small values of p, and the screens used are so thin that for our purposes this assumption is justified. The chance that the α-particle would pass, in its flight, the required distance from some atom is equal to the ratio of the total area of the annular spaces of the atoms in unit area to unit area, or $2\pi pNt\,dp$. It may be assumed that no α-particle is deflected twice through a reasonably large angle; it is a rare event for it to be so deflected once. The formula, which is thus derived on the assumption that a given α-particle is not deflected more than once,

[1] RUTHERFORD, CHADWICK, and ELLIS, *op. cit.*, p. 193.

naturally does not hold for very small values of ϕ, since for a small deflection the α-particle does not need to approach the heavy nucleus so closely, but t can be made small enough so it holds for $\phi \sim 3°$.

Now, as noted before, the angle of deflection ϕ is a function of p, and the fraction of the particles that will be deflected through an angle between ϕ and $\phi + d\phi$ will be equal to the fraction incident at a distance from a scattering center between p and $p + dp$, where p and $p + dp$ are the distances that correspond to ϕ and $\phi + d\phi$. This fraction is equal to $2\pi pNt\,dp = 2\pi pNt(dp/d\phi)\,d\phi$. The quantity $|2\pi pNt\,(dp/d\phi)|$, then (the lines indicating that the quantity is to be taken positive), is the fraction per unit angle scattered at the angle ϕ, which corresponds to the distance p. From Eq. (9),

$$p = \frac{2Ze^2 \cot(\phi/2)}{Mv^2}$$

and

$$-\frac{dp}{d\phi} = \frac{Ze^2}{Mv^2 \sin^2(\phi/2)}$$

whence

$$\left|2\pi pNt\frac{dp}{d\phi}\right| = \frac{4\pi NtZ^2e^4 \cos(\phi/2)}{M^2v^4 \sin^3(\phi/2)}. \tag{10}$$

It is seen that the number of particles per unit angle deflected at a given angle will depend upon quantities all of which are known except Z; the latter may consequently be determined. Since the shape of the curve is independent of Z, it is necessary to measure the actual number of α-particles falling on unit area of the screen, and the actual fraction of them scattered at the various angles.

Equation (10) has been thoroughly tested in Rutherford's laboratory and found to hold under all attainable conditions for elements heavier than copper. In these heavier elements, the closeness of approach of the α-particle is limited because of the strong repulsive force of the highly charged nucleus. In lighter elements, where under favorable circumstances (fast α-particles and large angles of deflection) closer approach is possible, deviations have been observed. This is due to breakdown of the inverse-square law of repulsion upon which Eqs. (9) and (10) are based. In the case of aluminum, this breakdown occurs when the distance of approach of the α-particle (OV of Fig. 5) is about 10^{-12} cm. It is, of course, not at all surprising that the inverse-square law should break down at such small distances. It simply means that at such distances the actual structure of the nucleus is of importance, and it is no longer permissible to regard it simply as a point charge, but a more elaborate description becomes necessary.

2.7. Summary.—The results, the experimental basis of which is discussed in Chaps. I and II, may be briefly summarized as follows. All substances are composed of atoms which are in most instances combined into molecules. An atom is composed of a positively charged nucleus, which has most of the mass of the atom, and a surrounding cloud of negatively charged electrons. An electron has a charge of 4.80×10^{-10} e.s.u. and a mass of 9.12×10^{-28} g. The mass of the nucleus of the hydrogen atom, the lightest of all nuclei, is about 1835 times as great as the mass of an electron; other nuclei are heavier, ranging up to more than two hundred times this amount. The charge on the nucleus is equal to a multiple of the charge on the electron, the whole atom, in general, being neutral. The atomic number gives the charge on the nucleus in terms of the electronic charge, and it is also equal to the number of electrons in the neutral atom. The radii of nuclei are of the order of 10^{-12} cm., whereas the radii of atoms, including the cloud of electrons, are of the order of 10^{-8} (generally, 2 to 3×10^{-8}) cm. The nuclei are themselves of complex structure, but for the purpose of the present work it is not necessary to discuss their structure.

CHAPTER III

WAVE AND CORPUSCULAR PROPERTIES OF RADIATION AND MATTER

In the preceding chapter, we have considered experiments that may be described by treating the electron as a corpuscle. There are other experiments which show that electrons also have wave properties. Such experiments involve changes in the motion of the electron which take place in a distance of the order of atomic dimensions, 10^{-8} cm or less. When distances of this order of magnitude are of importance, it is necessary to deal with the wave properties of electrons; so it is obvious that these properties will play an important role in the theory of atomic structure. These distances are actually of the order of (and, in fact, determined by) the wave length of the electron waves. Just as we must use physical optics rather than geometrical optics when we consider distances of the order of the wave length of light, so we must take into account the wave properties of electrons when distances of the order of the wave length of the electron become important. But the wave-corpuscle dualism contains something more profound, in the case of both optics and electricity, than the difference between physical and geometrical optics. It will be convenient to begin the discussion of this subject with the case of light. This offers a natural introduction to the study of the wave properties of the electron, but it is also of importance in itself, since the light emitted by atoms provides one of the best instruments we have for the study of their properties.

3.1. The Wave Properties of Light.—The phenomena of diffraction and interference have long been explained on the assumption that light consists of waves which travel outward in all directions from the source. These waves may be thought of as displacements in an elastic medium, the ether, though the modern views do not admit the reality of this medium. The ideas involved may be illustrated by a simple interference experiment, shown in Fig. 6. At O, we have a source of light that falls on a screen, which is perpendicular to the plane of the paper, the setup being shown in section in the figure. In this screen, there are two openings, at A and B, that act as secondary sources from which the light spreads out in all directions, and the interference phenomena appear on the second screen. Suppose C is a point such that

$$AC - BC = n\lambda,$$

where λ is the wave length of the light and n is a whole number.[1] Then the light waves from A and B are in the same phase at C, the "displacement" represented by the wave from B is always in the same direction as that represented by the wave from A, and the result is that reinforcement occurs and there is a bright spot at C. On the other hand, if

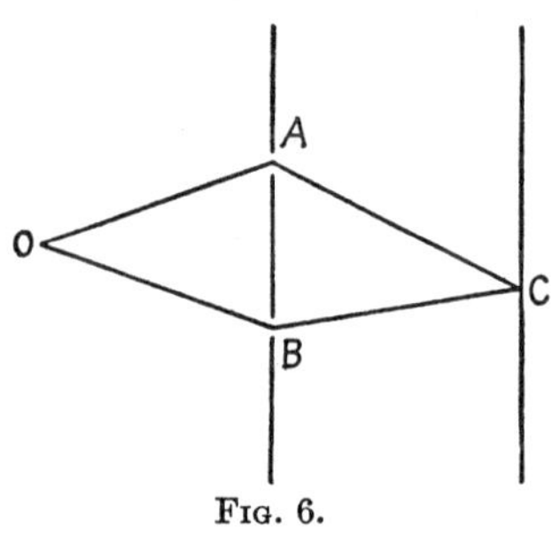

Fig. 6.

$$AC - BC = (n + \tfrac{1}{2})\lambda,$$

then the displacements at C are in opposite directions, and cancel each other, so that there is a dark spot at C. Points on the screen that do not correspond to either of the extremes described receive intermediate amounts of light.

A few other well-known phenomena that are explainable on the basis of the wave theory of light are the colors of thin films, the spreading of light when it goes through a very small opening (which explains why the small openings A and B of the experiment we have discussed do act as secondary sources), the reflection of light from a grating, etc.

3.2. Reflection of X Rays from a Crystal.—The question which we shall treat in this section is not only one that illustrates very well the type of calculation which may be made by aid of the wave theory of light but is one that we shall find of importance later on, as it furnishes a most powerful tool for the investigation of crystal structure. Ordinary visible light has a wave length ranging around 4000 to 7500 Å. (1Å. = 10^{-8} cm.). This is large compared with atomic distances. X rays, however, are well known to be light of extremely short wave length; in fact, their

[1] It is assumed that the source O is equidistant from A and B, so that the light waves start out in phase from these points.

wave lengths are of the order of atomic dimensions. The result is that crystals can be used as gratings in the study of X rays, and conversely, X rays may be used to study the structures of crystals.

The characteristic property of crystals is that the atoms are arranged in regular rows and planes. The nature of these planes of atoms (*i.e.*, planes in which lie the centers of gravity of a series of atoms) will be more fully appreciated from the figures of Appendix IV. A beam of X rays impinging upon a crystal is reflected from these planes, just as a beam of light is reflected from a plane mirror, with the angle of reflection equal to the angle of incidence. This comes about because of destructive

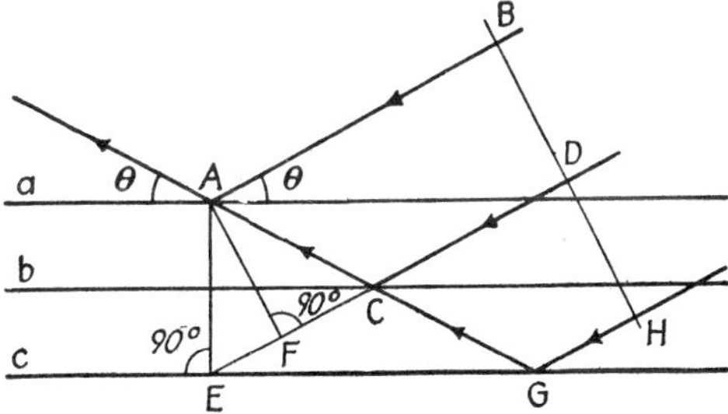

Fig. 7.—Reflection of X-ray beam from atomic planes in a crystal.

interference of light scattered at any other angle. The proof of this statement will not be given here, but it will be accepted as the basis of some further deductions.

Only a small proportion of an X-ray beam will be reflected from any given plane of atoms in the crystal. Most of the beam passes through such a plane, and a small proportion of that which passes through will be reflected from the next parallel plane, and so on. This series of reflections gives rise to interference effects, which may be quantitatively described, and furnishes the basis for a theory of the reflection of X rays from crystals.

In Fig. 7, we represent a crystal with a beam of X rays traveling in the direction indicated by the arrow. The figure shows a section formed by cutting along the plane which is perpendicular to the planes of atoms in the crystal and which passes through the line AB which gives the direction of motion of the X rays. The lines a, b, c represent in sectional view planes of atoms in the crystal. From each of these planes, a small fraction of the beam of X rays will be specularly reflected in the direction CA, as indicated in the figure. Now let us consider the wave front BD. The portion at B travels to A, and a certain fraction is reflected there. The portion at D travels to C, and the same fraction is reflected there. (So little is reflected at any one plane that we may consider that the original beam passes from plane to plane practically undiminished in intensity, and so the same amount is reflected from each plane.) The wave front moves from C to A and arrives at A a certain distance behind the part that came from B. If this distance (ACD − AB) is equal to $n\lambda$, where n is an integer and λ is the wave length of the X rays, then the part coming from D will arrive at A just in phase with a later wave from B, and we shall have reinforcement. But it may be readily seen that the distance

$$\mathrm{ACGH} - \mathrm{AB} = 2(\mathrm{ACD} - \mathrm{AB}) = 2n\lambda,$$

so that the portion starting at H will also be in phase at A; and so for X rays reflected from all planes below c. But if ACD − AB differs from $n\lambda$ by only a very small amount, there will be waves in all possible phases[1] meeting at A, and the result will be that cancellation will occur. So the reflections from individual planes of atoms destroy each other, and the result is that under circumstances like this no reflected beam of X rays can be observed coming from the crystal. Reflection can be observed only if the angle θ is such that $\mathrm{ACD} - \mathrm{AB} = n\lambda$. But $\mathrm{ACD} - \mathrm{AB} = 2d \sin \theta$, where d is the distance between planes. For if we drop AF perpendicular to EC, then it is seen that EF, which is equal to EA $\sin \theta$ or $2d \sin \theta$, is equal to

$$\mathrm{ECD} - \mathrm{FD} = \mathrm{ECD} - \mathrm{AB},$$

and we see from the figure that ECD = ACD. Therefore, we may write, as the condition for reflection from the crystal,

$$2d \sin \theta = n\lambda, \tag{1}$$

[1] This will be the case because only a very small amount of reflection takes place from any one plane and the light in the reflected beam is contributed by reflections from many planes, even deep in the crystal. The phase of the portion reflected from each plane may differ but slightly from that reflected from immediately neighboring planes, but when the whole reflected beam is considered it is seen that light in any given phase can be matched (and hence canceled) by light in just the opposite phase.

which is the well-known Bragg law of reflection. n may take any integral value that does not make $\sin \theta$ greater than 1, so several angles of reflection are possible. n is called "the order of the reflection."

When X rays pass *through* crystalline slabs, diffraction phenomena also occur. This is the basis of the von Laue method of investigation of crystal structure. In this method, instead of using X rays of a single wave length, a beam containing a wide continuous range of wave lengths is used. It is found that if a pencil of such X rays is passed through a crystal to a photographic plate the plate will show a heavy central spot, which is due to the X rays that have passed through without alteration in their direction, surrounded by a series of lighter spots due to X rays whose direction of propagation has been changed by passage through the crystal. These deflected beams make perfectly definite angles with the original beam and the crystal axes, and each of the deflected beams is homogeneous, *i.e.*, it consists of a single wave length only. So this method selects out not only particular directions for the deflected beams, but particular wave lengths. The theory of this phenomenon rests on the same general principles as the theory of reflection from a crystal face, but we shall not consider the details.

In a modification of the von Laue method, a monochromatic X-ray beam is used and the crystal is rotated. Refraction then occurs only for definite orientations of the crystal, and the refracted light goes off at certain definite angles.

3.3. The Photoelectric Effect and the Corpuscular Theory of Light.—When ultraviolet light or X rays are allowed to fall on a metal, it is observed that electrons are emitted from the metal. These electrons emerge with various velocities, but with monochromatic light there is a certain maximum, depending upon the frequency of the light, and most of the electrons move with a speed not very much lower than the maximum velocity. It is found that the kinetic energy corresponding to the maximum velocity is given by the relation.

$$\tfrac{1}{2}mv^2_{\max} = h\nu - W_0, \qquad (2)$$

where h is a universal constant (Planck's constant, equal to 6.628×10^{-27} erg sec.[1]), which has an exceedingly far-reaching

[1] See footnote 1, p. 15.

importance in modern physics, and W_0 is a constant that is characteristic of the particular metal used in the experiment. The *number* of electrons emitted from the metal per unit time is dependent on the intensity of the light, but the maximum energy (and also the distribution of velocities for any given metal) depends solely upon the frequency of the light and not at all on its intensity. If the frequency of the light is such that $h\nu$ is less than W_0, practically no electrons are emitted. If $h\nu = W_0$, the frequency is called the threshold frequency. The threshold frequency is in the visible or infrared for the alkali metals and in the ultraviolet for other substances.

Elster and Geitel[1] showed that the proportionality between the number of electrons emitted and the intensity of the light persisted even when the intensity of the light was cut down to 3×10^{-9} erg cm^{-2} sec^{-1}. This is a very remarkable result. Since there are about 10^{15} atoms per sq. cm. at the surface of the metal, an electron emitted with an energy of the order of 1 electron volt[2] (1 electron volt being equal[3] to 1.602×10^{-12} erg) would somehow have to collect, presumably from the incident light, energy equivalent to that falling in 1 sec. on the space occupied by 10^{11} or 10^{12} atoms. It is very difficult, on the basis of any wave theory, to see how this light could be collected, concentrated, and transmitted to a single electron, especially as this does not occur when $h\nu$ is only slightly less than W_0. Furthermore, there is practically no time lag during which this collection of energy could occur. Marx and Lichtenecker,[4] illuminating a sensitive potassium cell by light from a revolving mirror, showed that, when the time of illumination was reduced to 10^{-7} sec., the proportionality between the number of electrons emitted and the intensity of light held even when the latter was reduced to 0.56 erg per sq. cm. per sec.

These phenomena, however, as was pointed out by Einstein, all become very clear if we assume that light consists of corpuscles moving perpendicularly to the wave front with energy

[1] Elster and Geitel, *Phys. Zeit.*, **17**, 268 (1916); see Hughes and DuBridge, "Photoelectric Phenomena," p. 31, McGraw-Hill Book Company, Inc., 1932.

[2] An electron volt (or, less properly, a volt) is the energy acquired by an electron on dropping through a potential difference of 1 volt.

[3] See footnote 1, p. 15.

[4] Marx and Lichtenecker, *Ann. Physik*, **41**, 124 (1913).

depending on the frequency and equal, in fact, to $h\nu$, and that the number of these corpuscles falling on unit area per unit time is proportional to the intensity of the light. An electron, then, may be ejected if it is struck by and receives the energy of one of these corpuscles, or light quanta, as they are called. The electron requires an amount of energy equal to at least W_0 to get out of the metal (the energy required may be greater than W_0 if the particular electron that gets the energy is not right at the surface). The electron, therefore, emerges with kinetic energy equal to $h\nu - W_0$, or less.[1] The number of electrons ejected will obviously be proportional to the intensity of the light.

3.4. Reconciliation of the Wave and Corpuscular Theories of Light.—We have seen in the last few pages that there are two theories, which at first sight seem contradictory. How can light consist of waves and particles flying through space at the same time? Yet the one theory seems to be necessary to explain some phenomena, the other appears necessary to explain others. Some indication of the meaning of this apparent paradox is afforded by noting which experiments require the wave theory and which the corpuscular theory. As has been illustrated by the limited number of examples presented in this chapter, those experiments which require the wave picture involve the *propagation* of light, while the corpuscular theory is necessary to treat those which involve *transfers of energy to individual electrons or atoms.* A possible way to resolve the paradox lies in the hypothesis that light actually does consist of quanta or corpuscles, but that the motion of those corpuscles is guided by the waves.

Let us return again to the interference experiment considered at the beginning of this chapter. The wave theory says that certain spots on the screen[2] at C will be light and others dark. The light corpuscles, therefore, move in such a way that they fall on the light spaces and avoid the dark ones. In other words, the density of light quanta at any point in space is proportional

[1] It is not strictly true that there is a definite maximum velocity, for there are some electrons in the metal that already have high velocities, and this velocity may be added to the velocity imparted by the light quantum. However, only a few electrons will be emitted with energies exceeding the "maximum."

[2] See Fig. 6.

to the energy per unit volume which would be given by the wave theory. If the energy per unit volume at any point calculated by the wave theory is E then the number of quanta per unit volume at the same point is $\mathsf{E}/h\nu$, since the energy per quantum is $h\nu$.

A rather curious situation arises if the intensity of light is very small. Suppose, for example, that in the interference experiment, the intensity is so small that only rarely is there even a single quantum between O and C. To speak of the density of quanta then has no meaning. But the energy density calculated from the wave theory still is proportional to the *probability* per unit volume that we shall find a light quantum in any given element of volume. If we wait a very long time, the fraction of the time that the given element of volume contains a quantum will be proportional to the energy density calculated from the wave theory.

3.5. Wave Properties of Electrons.[1]—In Chap. II, we consistently treated electrons as particles. If, however, a beam of electrons is reflected from the surface of a crystal, a diffraction phenomenon very similar to that found in the case of X rays may be observed. Various complications arise in the case of electrons which are not important in the case of X rays; thus the electron does not penetrate into the crystal as readily as X rays, and so reflection takes place from only a few crystal layers; furthermore, the energy of some of the electrons is changed owing to interaction with the atoms of the crystal, and other effects also occur owing to the interaction of the electrons with the individual atoms of the crystal. It is beyond the scope of this book to deal in detail with these complicating factors, and we need emphasize only the important qualitative fact that the appearance of diffraction indicates that electrons have a dual wave-corpuscular nature similar to light.

Diffraction phenomena also occur when electrons of several thousand volts energy pass through very thin films of various crystals, particularly metals. The phenomena which occur in

[1] The wave quantum theories, to be considered in this and following chapters, are based on the work of de Broglie, Schrödinger, and Heisenberg. The early experimental work was due to Davisson and Germer and G. P. Thomson and Reid. See Thomson, "Wave Mechanics of Free Electrons," McGraw-Hill Book Company, Inc., 1930.

this case are very similar to those found in the case of X rays, and the wave length λ of the electrons may be determined with rather good accuracy from such experiments, provided the distances between planes of atoms in the crystal are known. These distances have already been determined by the help of X rays and density measurements. It has thus been found (as was already anticipated theoretically) that there is a connection between the wave length and the velocity v of the electrons, as follows;

$$\lambda = \frac{h}{mv}. \tag{3}$$

This formula is not limited to electrons, but can be applied to any body if the appropriate mass is used.

Although this is apparently of quite different form from the relation between the energy and the frequency of light, it is really closely related to it. This may be seen if we accept the equivalence of energy and mass as deduced from relativity theory. According to this theory, mass and energy are interconvertible. For example, if an electron could be imagined to be annihilated, a corresponding amount of energy would have to appear in some other form. On the other hand, if the energy of a body is increased in any manner, its mass increases correspondingly. If all quantities are expressed in c.g.s. units, the mass multiplied by the square of the velocity of light gives the corresponding energy. Thus for a light quantum,

$$h\nu = mc^2, \tag{4}$$

where m is the mass of the light quantum and c the velocity of light. But $\nu = c/\lambda$, where λ is the wave length of the light, whence it is readily seen that

$$\lambda = \frac{h}{mc}, \tag{5}$$

exactly the same relation that holds for the electron, as c is the velocity of the light quantum. Therefore,

$$\lambda = \frac{h}{M}, \tag{6}$$

where M is momentum, for either light quantum or electron, the momentum being equal to the mass of the body times its velocity in relativity theory as in classical theory.

It is thus seen that there is considerable similarity between electrons and light quanta. Both of them may be considered to be corpuscles whose motion is guided by waves. In spite of this, however, there are probably more points of difference than of similarity. One of the most important of these differences is due to the fact that an electron has a mass, and hence possesses considerable energy, even when it is at rest. When it is set in motion, its mass increases, corresponding to the extra kinetic energy. If the mass of the electron when at rest is m_0, then, according to the theory of relativity, its mass when it is moving with a velocity v is given by

$$m = m_0\left(1 - \frac{v^2}{c^2}\right)^{-1/2}, \tag{7}$$

where c is the velocity of light. We may expand the square root by means of the binomial theorem, and if v is very small compared with c, we have the approximation

$$m = m_0 + \frac{\frac{1}{2}m_0v^2}{c^2}.$$

Multiplying through by c^2 to get the energy gives

$$mc^2 = m_0c^2 + \tfrac{1}{2}m_0v^2;$$

that is to say, the energy is equal to the rest energy plus the kinetic energy, which has the usual form $\frac{1}{2}m_0v^2$, provided v is small enough compared with c. As v increases, this expression breaks down, and it is seen from (7) that m approaches infinity as v approaches c. c is thus an upper limit for v. However, a light quantum has the velocity of light, and yet its mass is finite. This can mean only that the rest mass of a light quantum is zero. The energy of a light quantum is uniquely determined by its true mass. The energy of an electron is also determined by its mass, but since it has a fixed rest mass and its velocity can approach but never reach c, its energy may just as well be considered to be determined by its velocity.

3.6. The properties of the electron waves will be treated in bare outline only, giving just what is necessary for our purposes. The wave, considered as a function of the coordinates of the space in which the electron moves, will be designated by the symbol ψ. We have already discussed one of its important properties, namely, the connection between its wave length and the velocity of the electron. The term "wave" and the mention of "wave length" already indicate to some extent the nature of the function ψ. It will be a sinusoidal function of the space coordinates, or a similar type of function which is periodic or nearly periodic. It may also be a function of time, and in this case it will be a periodic function of the time. However, in the cases in which we shall be interested, ψ may be treated as independent of time—a function of the space coordinates only. It may be real or complex.[1] We designate its conjugate complex

[1] It may be well to summarize the mathematical conceptions used here. The unit of imaginaries i is defined as the square root of minus one, $\sqrt{-1}$. Any multiple of this unit by a real number, as bi, where b is real, is a pure imaginary. A complex number, such as $a + bi$, where a and b are real, is the sum of a real number and a pure imaginary. The number $a - bi$ is the

as ψ^*. In practically all the cases we shall have to consider, however, ψ will be real, in which case $\psi^* = \psi$.

The physical significance of the function ψ is that it determines the probability that an electron should be found at a given point in space. Now ψ itself can be partly imaginary or negative, while the probability that an electron will be found at a given place can obviously be only real and positive. This probability, therefore, cannot be determined by ψ itself. The function $\psi\psi^*$, however, has the right properties, and according to the theory, it is this product $\psi\psi^*$ which determines the probability that an electron be at a given point. This statement may be made more precise. Consider a region in space defined so that the x-coordinate lies between x and $x + dx$, the y-coordinate between y and $y + dy$, and the z-coordinate between z and $z + dz$. dx, dy, and dz are so small that the function ψ has essentially a constant value $\psi(x,y,z)$ throughout the element of volume thus defined. Then the probability that an electron be found in the volume considered (or, if the density of electrons is high, the number of electrons per unit volume) will be proportional to $\psi(x,y,z)\psi^*(x,y,z)\,dx\,dy\,dz$.

The function $\psi(x,y,z)$ is determined as the solution of a certain differential equation, the wave equation, which may be found if the equation of motion of the electrons under the conditions of the given experiment is known. As will appear in the next chapter, it is not necessary to use this differential equation to find the wave function when the velocity of the electron under consideration is constant. It is necessary, however, when the velocity is not constant, and especially if it changes within a distance short compared with a wave length of the electron, under which conditions Eq. (3) naturally loses some of its meaning.

In the following chapters, we shall consider the type of wave functions that occur in several special cases, though we shall endeavor to give a qualitative rather than a quantitative picture, and use a kind of hybrid of classical and wave theories. It must be noted that in a stricter theory even dynamical quantities, like energy and momentum, must be defined in terms of the wave equation.

---

conjugate complex of $a + bi$. The product of a number and its conjugate complex is real. Thus

$$(a + bi)(a - bi) = a^2 - b^2i^2 = a^2 - b^2(-1) = a^2 + b^2.$$

CHAPTER IV

ELEMENTARY QUANTUM THEORY

4.1. Nature of Atomic Spectra.—One of the most powerful means of investigating atoms and molecules is the study of the light they emit or absorb, especially when they are in the gaseous state. Such a study has shown that not all frequencies are emitted or absorbed by a given species of atom or molecule, but that in general there appear spectral lines of definite frequency. Each substance has its own characteristic set of spectral lines.

It was early found that the frequencies of the spectral lines could be advantageously expressed as differences of certain quantities, called "term values"; this is the so-called "Ritz combination principle." Thus suppose that for some given atom there was a series of term values $\tau_1, \tau_2, \tau_3, \tau_4,$ such that $\tau_1 > \tau_2 > \tau_3 > \tau_4 \ldots$; then the following frequencies might be observed:

$$\nu_{12} = \tau_1 - \tau_2, \; \nu_{13} = \tau_1 - \tau_3, \; \nu_{14} = \tau_1 - \tau_4, \; \nu_{23} = \tau_2 - \tau_3, \text{ etc.}$$

The utility of this scheme lies in the fact that it expresses a certain relationship which may be observed among the frequencies (thus in the example given, $\nu_{12} + \nu_{23} = \nu_{13}$) and that the number of term values necessary to describe a given spectrum is much smaller than the number of spectral lines. This latter statement holds true, despite the fact that, owing to certain "selection rules," not all the spectral lines which might be expected from a given set of term values actually occur.

On the basis of the quantum theory of light, outlined in the preceding chapter, these term values receive a simple interpretation if it is supposed that when an atom emits light it emits just one quantum. It is thus assumed that the term values represent energy levels in the atom; this is described by saying that the atom is "quantized." An atom, then, we infer, can have only certain definite energies; it can occasionally lose energy, dropping from a higher to a lower energy level, and emitting thereby a light quantum whose $h\nu$ is equal to the difference between the energies of the atomic states.

We may be a little more specific. Our picture of an atom supposes that it consists of a positively charged nucleus surrounded by electrons. These electrons must be moving in various paths or orbits about the positive nucleus; and we may assume that a stationary state[1] of an atom is defined by the conditions of motion of the electrons in that atom. Thus if an atom is in one of its higher energy states, this may be owing to one or more of its electrons having more energy than ordinarily. Such an atom is said to be in an excited state.

It is one of the triumphs of the wave theory of electrons that it has been able to give an explanation of the existence of stationary energy states in atoms, and in many cases it is possible to make an actual calculation of the allowed energy values. In order to do this, it is necessary to impose certain restrictions on the wave function ψ of any electron. In general, it is assumed that, in addition to being a solution of a certain type of differential equation, it must be a continuous and single-valued function of the coordinates x, y, and z of the electron it represents, and it must be everywhere finite. These restrictions, reasonable though they are, must be taken as additional hypotheses, whose claim to correctness, as the claim of the wave theory as a whole, rests on the fact that by means of them it is possible to explain the experimental facts. Such a theory, it should be emphasized, is of value, not because it is to be considered as more important or more fundamental than the experimental facts, but because it is able to explain a great many experimental facts and bring out the relationship between phenomena which at first sight appear to be unrelated.

4.2. Theory of an Electron in a "Box."—The way in which the assumptions just outlined make it possible to explain the quantization of atoms will be illustrated by considering a simple case, which does not correspond to any known atom, but which brings out the general principles involved very nicely. Suppose an electron[2] is moving in the potential energy field shown in Fig. 8.

[1] The term "stationary state" is often used instead of "quantum state," as descriptive of a condition in which an atom can exist for some length of time.

[2] The same considerations will apply to any other particle if the proper mass is used.

In the region between $x = 0$ and $x = a$, there is no force acting on the electron, and the potential is constant. At $x = 0$ and $x = a$, the potential suddenly rises to infinity. The electron is thus enclosed between two potential walls, and so may be said to be in a box. In a rectangular three-dimensional box, the range of motion of an electron will also be confined in the y- and z-directions. It will be supposed that at $y = 0$ and $y = b$ and at $z = 0$ and $z = c$ the potential also suddenly rises to infinity. The electron will then be confined to that portion of space for which $0 < x < a$, $0 < y < b$, and $0 < z < c$. According to the classical theory, the electron will move with a constant velocity till it strikes one of the walls, at which point one component of the velocity, though remaining the same in magnitude, will reverse its direction, and this process will recur indefinitely.

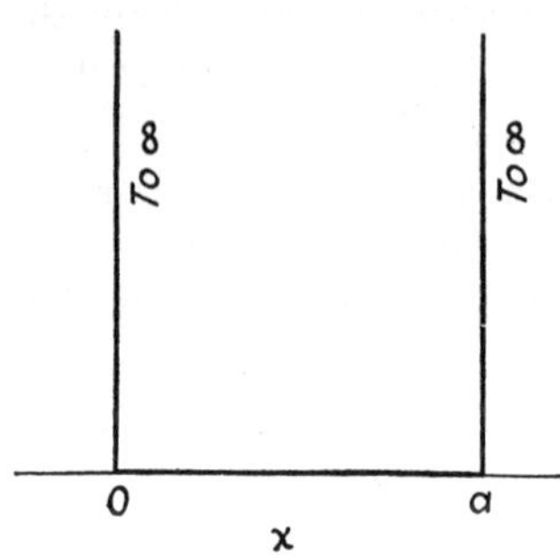

Fig. 8.—Potential energy curve for an electron in a box.

Let us now consider the corresponding wave picture. We shall deal first with the x-component of the velocity of the electron. To indicate the possibility of doing this, let us consider a simple illustration. Suppose an electron is moving in some arbitrary direction in space, making an angle α with the x-axis, and suppose that the distance between the wave fronts, *i.e.*, the wave length, is equal to λ. The electron, of course, moves in a direction perpendicular to the wave front, and the distance between wave fronts is measured along this perpendicular. It is readily seen (Fig. 9) that the distance between wave fronts along the x-axis is given by $\lambda_x = \lambda/\cos\alpha$. Since the x-component of velocity is given by $v_x = v\cos\alpha$, it is seen that the formula (3), Chap. III, holds for the *components* of the velocity and the wave length as well as for these quantities themselves. Thus the x-component of the velocity can be treated as separate from and independent of the other components.

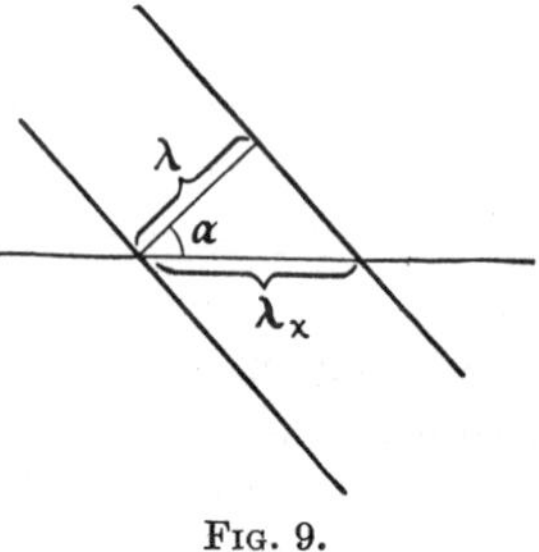

Fig. 9.

However, a set of wave fronts such as indicated in Fig. 9 cannot apply directly to an electron enclosed in a box.[1] For Fig. 9 implies a definite direction of motion for an electron at any particular point in space, whereas in a box, an electron changes its direction of motion every time it strikes one of the walls. But the point that the discussion was intended to illustrate, namely, the possibility of considering the x-, y-, and z-components of the motion as independent, still holds good. In a rectangular box a certain simplicity is introduced because, although a component of velocity changes *direction* on a collision with a wall, it does not change in *magnitude*, and hence the wave length associated with it does not change. Furthermore, if the potential energy within the box is constant, as has been assumed, the magnitude of the x-component of the velocity and hence the corresponding wave length are independent of x. Since this is true, the wave function of the electron will be a simple sinusoidal function of x; it may be written in the form

$$\psi = A \sin \frac{2\pi x}{\lambda_x} + B \cos \frac{2\pi x}{\lambda_x}, \tag{1}$$

where A and B are constant amplitudes (assumed real, so that $\psi^* = \psi$ and $\psi\psi^* = \psi^2$). But the electron cannot be in the region outside $x = 0$ and $x = a$. Therefore,

$$\psi = 0, \qquad \text{if} \qquad x < 0 \qquad \text{or} \qquad x > a. \tag{2}$$

The only way the wave function can be continuous (one of the conditions that is imposed upon it) is for the right-hand side of (1) to be 0 for $x = 0$ and $x = a$. If it is to be 0 at $x = 0$, we must have $B = 0$. If it is to be zero at $x = a$, we must have

$$\sin \frac{2\pi a}{\lambda_x} = 0 \tag{3a}$$

or

$$\frac{2\pi a}{\lambda_x} = n\pi, \tag{3b}$$

where n is a whole number giving the number of half waves between $x = 0$ and $x = a$; n is called the "quantum number"

[1] At least, as the next sentence indicates, they cannot describe a stationary state of the electron.

of the particular state that it defines. Thus the following condition is imposed on λ_x:

$$\lambda_x = \frac{2a}{n}. \tag{3c}$$

From Eq. (3), Chap. III, this is equivalent to the following condition on the velocity:

$$v_x = \frac{nh}{2am}. \tag{3d}$$

For the kinetic energy, we have

$$\tfrac{1}{2}mv_x{}^2 = \frac{n^2h^2}{8a^2m}. \tag{3e}$$

It is thus seen that the part of the kinetic energy due to the x-component of the velocity can have only certain definite values; all others are excluded by the conditions imposed upon the wave functions. It is usual to designate the allowed wave functions as ψ_n, giving n the appropriate value to indicate to which one of the allowed functions reference is had, though if no confusion is possible the subscript may be omitted. The allowed functions are generally called "eigenfunctions," or "characteristic functions," though also referred to simply as "wave functions."

The function ψ_n is of course a function of x only. It is, in fact, really only a factor of the wave function, that part dealing with the x-coordinate only, and $\psi_n{}^2\,dx$ may be taken as representing the probability that, if an electron is in the state n, it will be found between x and $x + dx$. This interpretation makes it possible to evaluate A_n, the amplitude of the eigenfunction ψ_n. The probability that the electron will be between $x = 0$ and $x = a$ is obviously equal to 1, since the electron must be somewhere in this range. Therefore,

$$\int_0^a \psi_n{}^2\,dx = A_n{}^2 \int_0^a \sin^2\frac{2\pi x}{\lambda_x}\,dx = 1. \tag{4}$$

Now $\int_0^a \sin^2(2\pi x/\lambda_x)\,dx$ is, by direct integration, equal to $a/2$, provided a is a whole or half integral multiple of the wave length λ_x; if λ_x has one of the values given by (3c), this relation, together with (4), gives

$$A_n = \sqrt{\frac{2}{a}}. \tag{5}$$

We may, therefore, write

$$\psi_n = \sqrt{\frac{2}{a}} \sin \frac{\pi n x}{a}. \tag{6}$$

When A_n is given the value indicated in Eq. (5), ψ_n is said to be normalized.

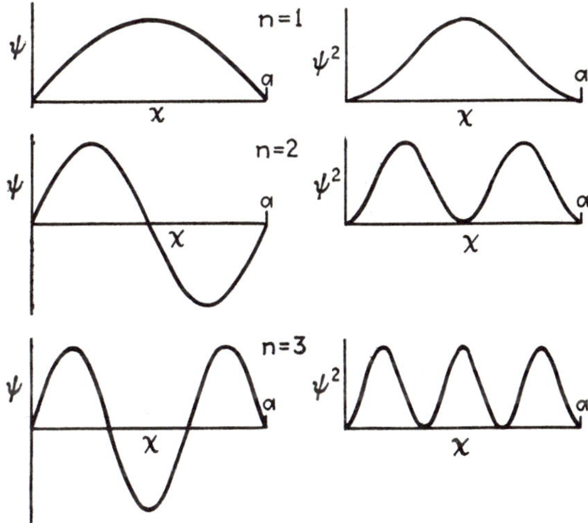

Fig. 10.—Wave functions for the first three energy levels of a particle (one dimensional) in a box, and the squares of the wave functions.

As has already been noted for $y < 0$ and $y > b$ and for $z < 0$ and $z > c$, the potential energy also becomes infinite. Restrictions similar to those found for the x-part of the kinetic energy exist for the y- and z-parts, and only very definite values of that part of the energy contributed by the motion in these directions are possible. In other words, its entire energy is quantized.

Corresponding to the three directions of space, there are three wave functions, similar to that defined by Eq. (6), say

$$\left.\begin{aligned} X_n &= \sqrt{\frac{2}{a}} \sin \frac{\pi n x}{a} \\ Y_k &= \sqrt{\frac{2}{b}} \sin \frac{\pi k y}{b} \\ Z_l &= \sqrt{\frac{2}{c}} \sin \frac{\pi l z}{c}. \end{aligned}\right\} \tag{7}$$

Here the notation has been changed slightly: X_n is the same function previously designated as ψ_n. In order to specify exactly the state of the electron, it is necessary to give three quantum numbers n, k, and l. Once these numbers are given, $X_n^2\,dx$ gives the probability that the electron be between x and $x + dx$, $Y_k^2\,dy$ the probability that it be between y and $y + dy$, and $Z_l^2\,dz$ the probability that it be between z and $z + dz$. The probability that it be simultaneously between x and $x + dx$, y and $y + dy$, z and $z + dz$ is given, according to a well-known theorem of the theory of probability, by the product of the separate probabilities, namely, $X_n^2Y_k^2Z_l^2\,dx\,dy\,dz$. We may therefore write for the complete eigenfunction, which is a function of x, y, and z,

$$\psi_{nkl} = X_nY_kZ_l, \tag{8}$$

and it is seen that $\psi_{nkl}^2\,dx\,dy\,dz$ then gives the probability that the electron be in the specified volume element.

It may seem strange that, when the electron is in a certain definite energy level, there are certain places where there is no probability of finding the electron and other places where there is a large probability of finding it. It seems at first sight contrary to the usual conception of the nature of mechanical motion. There is, however, strong evidence from a variety of sources that the wave picture is correct, and that the commonly held ideas are merely approximations which are, however, very good if the bodies that are considered are large. For such a small body as an electron, these approximations break down.

It is also evident that the picture of an electron in a stationary state (*i.e.*, a quantized energy level) is quite different from our ordinary picture of a moving body, in that there is a probability of finding the electron at various places which extend all the way from one side of the box to the other, and the motion of the electron is not brought at all into evidence by regarding the waves themselves. It is possible, however, to reproduce the appearance of a moving electron, by means of the waves, and the way in which this is done may be briefly outlined. As has been stated above, for our purposes any dependence of the wave functions on the time can in general be neglected, but in order to reproduce this appearance of a moving electron, it is necessary to consider the time dependence of the wave functions. Without going into too great detail, we may note that this is accomplished[1] by multiplying ψ_{nkl} by $e^{2\pi i E_{nkl} t/h}$, where E_{nkl} is the energy of the state nkl, t is the time, and[2] $i = \sqrt{-1}$. Since, by a well-known theorem in function theory,[3]

$$e^{2\pi i E_{nkl} t/h} = \cos (2\pi E_{nkl} t/h) + i \sin (2\pi E_{nkl} t/h),$$

it is seen that this exponential is a periodic function of the time with frequency $\nu_{nkl} = E_{nkl}/h$, the customary relation between the energy of the particle and the frequency of the waves. Now it is possible to set up an expression for the wave function of the system, of the form

$$\psi = \Sigma_n \Sigma_k \Sigma_l C_{nkl} \psi_{nkl} e^{2\pi i E_{nkl} t/h}, \tag{9}$$

by choosing the proper values of the constants C_{nkl}, such that at time $t = 0$ the electron is confined to some limited region of space, and such that the center of gravity of this wave packet will move with a certain velocity. Such a wave packet thus behaves very much like a moving electron, except that it has a tendency to spread, *i.e.*, the region in which the electron is likely to be found gets larger and larger. But this effect is easily accounted for, since the ψ_{nkl} for various values of n, k, and l, and hence corresponding to *various energies*, are included in the sum (9). Thus it is not certain that the electron represented by the wave packet has a definite velocity, but there is only a certain probability that its velocity lies within any given limits, just as one can only say that there is a certain probability that the electron is located within certain limits of space. It is found that the smaller the space region in which there is an appreciable probability of finding the electron, the greater is the range of

[1] This is discussed in some detail in Condon and Morse, "Quantum Mechanics," p. 25, McGraw-Hill Book Company, Inc., 1929.

[2] See footnote 1, p. 34.

[3] See, *e.g.*, Slater and Frank, "Introduction to Theoretical Physics," pp. 22–23, McGraw-Hill Book Company, Inc., 1933.

possible velocities. If the electron is confined to a region of space whose linear dimensions are of the order of 10^{-8} cm., the range of probable velocities is quite appreciable. If, however, the electron is confined to a region whose linear dimensions are of the order of 10^{-4}, then the range of probable velocities is, or at least may be, extremely small. Ordinarily, distances less than 10^{-4} cm. cannot be measured, and in general, experiments are designed in such a way that the probable region for an electron is at least this large. If this is so, it is quite possible for the range of probable velocities to be within the limits of experimental error. In such a case, the wave mechanics gives essentially the same answer to any problem as ordinary mechanics. Although we can say only that there is a certain probability that the electron is in such and such a position, and has such and such a velocity, there is a very small probability of finding the electron outside a certain region the size of which is so tiny that it would be beyond the power of our instruments to measure it. And the position, at any time, of the region in which the electron must very probably be may be calculated to within a very small error by assuming that it is moving with a certain velocity, which is therefore identified with the velocity of the electron. However, when dealing with atoms and atomic distances, the situation is very different, and when an electron is in a certain stationary state, nothing more can be said about where that electron is than may be learned from the eigenfunction belonging to that stationary state. In other words, nature forbids us to know too much about any electron. This principle, due to Heisenberg and known as the "uncertainty principle," has been discussed quantitatively in numerous treatises on wave mechanics,[1] but it is outside the scope of this book to go into it beyond the cursory and qualitative discussion that has just been given.

4.3. The Number of Quantum Conditions and the Separation of Variables.—The example of the particle in the box illustrates a general principle of considerable importance, which will be applied later in this chapter. It will have been noted that three quantum numbers, corresponding to three independent quantum conditions, are necessary to define the state of the system. This corresponds to the fact that three coordinates are necessary to

[1] See, *e.g.*, Heisenberg, "Physical Principles of the Quantum Theory," University of Chicago Press, 1930.

determine the position of the particle in space. In general, the number of independent conditions necessary to define the state of a system (which may consist of more than one particle or be subject to various constraints) is equal to the number of coordinates necessary to completely specify the positions of all parts of the system in space. This number is described as the number of degrees of freedom of the system. The coordinates used need not necessarily be rectangular coordinates but may be polar or other types of coordinates; however, the *number* of coordinates necessary to specify the state of the system is independent of the kind used.

The application of the quantum conditions is readily effected only if the equations of motion of the system are *separable* in one set of coordinates or another. The equations of motion of the particle in the box, for example, are separable in the ordinary Cartesian coordinates x, y, and z. This evidences itself in the fact that the kinetic energy (the only part that can vary) can be written in the form $\frac{1}{2}mv_x^2 + \frac{1}{2}mv_y^2 + \frac{1}{2}mv_z^2$, and that each part has a value which is constant in the course of the motion, and which is entirely independent of the other parts. There are thus three independent constants of the motion. There are other constants of the motion, *e.g.*, the absolute values of the components of the momentum in the three directions, but these are not independent of the kinetic energy. It is seen that the three independent constants of the motion are just the quantities that are quantized and, in general, when desirous of finding the quantities to be quantized, one will look for the independent constants of the motion. These may be energies, momenta (in the general sense of Appendix I), etc., but the number of independent constants will be equal to the number of degrees of freedom of the system.

4.4. The Quantization of Rotational Motion.—We shall now turn our attention to the wave functions and quantization of another highly idealized mechanical system. Suppose that a body of mass m is attached to a rod of negligible mass and constrained by it to move in a circle of radius r about a fixed point. The position of the body is described by the angle χ made by the rod with a fixed direction in the plane of the circle in which the mass is constrained to move (see Fig. 11). Since this variable completely describes the position of the system, it is a system with one degree of freedom, requiring one quantum condition to fix its state of motion.

Once the body is set in motion, it will continue to move with constant angular velocity. This suggests setting up an expression for the wave function similar to Eq. (1), but involving the angle χ instead of the distance x. Thus we write

$$\psi = A \sin \frac{2\pi\chi}{\lambda_\chi}, \tag{10}$$

where λ_χ is what may be called a wave angle. (We refrain from using the more general form $A \sin (2\pi\chi/\lambda_\chi) + B \cos (2\pi\chi/\lambda_\chi)$, for this is quite equivalent to writing $A \sin (2\pi\chi/\lambda_\chi + \delta)$ where δ is another constant, which can be eliminated simply by changing the starting point for the measurement of χ.)

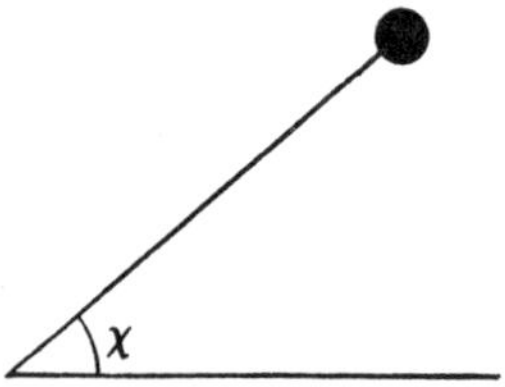

Fig. 11.

Proceeding on the basis of an analogy which will become evident immediately, we set

$$\lambda_\chi = \frac{h}{mvr} = \frac{h}{m\dot{\chi}r^2} \tag{11}$$

(where $\dot{\chi}$ is the angular velocity[1]). Multiplying through by r gives $\lambda_\chi r = h/mv$. Now $\lambda_\chi r$ is a length—it is the length of arc, on the circle described by the moving body, which is included between wave fronts, *i.e.*, between the successive angles that correspond to maxima, let us say, of ψ. It is seen that this wave length has the usual relation to the velocity of the particle.

It may appear that the introduction of Eq. (10) and the use of a wave angle which obeys Eq. (11) are arbitrary procedures. This is quite true. It must be remembered that the object of theoretical research is not to give an ultimate explanation of experimental facts, but to find relations between them. The usual procedure is to make arbitrary assumptions and find out where they lead. They are satisfactory if they give results consistent with experiment, and they are useful if they make it easier to correlate and understand the relations between different experimental facts. It is often desirable to make as few assumptions as possible and from them to derive as many experimental

[1] In general the dot represents differentiation with respect to time, *e. g.*, $\dot{\chi} = d\chi/dt$.

facts as possible. Thus it could be shown that Eqs. (1) and (10) are both solutions of special cases of a general wave equation, the assumption of the validity of which could be taken as our starting point. For our purposes, this would introduce unnecessary mathematical abstractions, and we have contented ourselves with a less ambitious program in which special assumptions are introduced for special cases. These special assumptions are still useful in correlating the facts, and an attempt is made to show that they are not entirely distinct, but that there are definite relations between the assumptions made in various cases—that they at least have a certain similarity. The results achieved in the special cases will also be seen to have certain characteristics in common. In other more complicated cases, then, we shall be content to go directly, or almost directly, to the results, merely showing that they still have the common characteristics, and referring the student to more detailed treatises for a more rigorous treatment.

Returning from this digression, let us proceed to the quantization of ψ. It has previously been noted that a wave function must be single valued and continuous. In order for ψ to have these properties, it must repeat itself on going around the circle; that is to say, increasing the angle χ by 2π should bring back the original value of ψ, which means that

$$\sin \frac{2\pi(\chi + 2\pi)}{\lambda_\chi} = \sin \frac{2\pi\chi}{\lambda_\chi}. \tag{12a}$$

This gives

$$\frac{2\pi(\chi + 2\pi)}{\lambda_\chi} = \frac{2\pi\chi}{\lambda_\chi} + 2\pi n, \tag{12b}$$

where n is a whole number, giving the number of waves in the circle.

From this, it is seen that

$$\lambda_\chi = \frac{2\pi}{n} \tag{12c}$$

and the energy, which is equal to $\frac{1}{2}mv^2 = \frac{1}{2}m\dot{\chi}^2r^2$, is from Eq. (11) given by

$$E = \frac{h^2n^2}{8\pi^2mr^2}. \tag{12d}$$

In this case, the value $n = 0$ is a possibility. This makes λ_χ infinite, and ψ is simply a constant[1] (and so, obviously, single valued). E and the momentum are zero, which is consistent with the infinite wave length. Furthermore, each of the other energy levels is double, *i.e.*, really consists of two coincident levels, since the body can rotate either clockwise or counterclockwise.

The results just obtained may be expressed in a useful form in terms of the angular momentum. Inasmuch as the angular momentum is defined (see Appendix I) as the product of (1) the mass, (2) the distance r from the fixed point, and (3) the component of velocity perpendicular to the direction of the line joining the fixed point to the moving body (in this case, the velocity itself), it is given by

$$p_\chi = mrv = mr^2\dot{\chi}. \tag{13}$$

Using (11) and (12c)

$$p_\chi = \frac{nh}{2\pi} \tag{14}$$

for the nth quantum state.

It is thus seen that the quantization may be expressed in terms of p_χ as an independent constant of the motion, as well as in terms of E; these expressions are, of course, not independent, one being derived from the other.

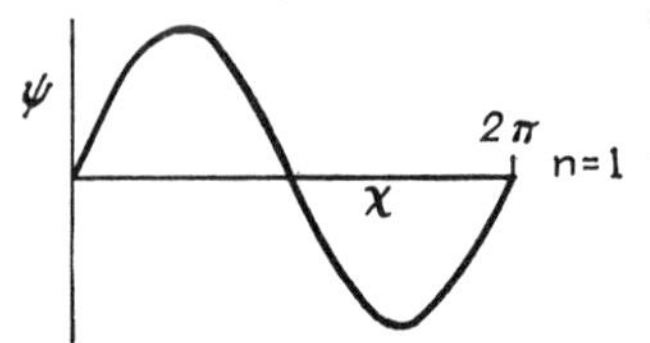

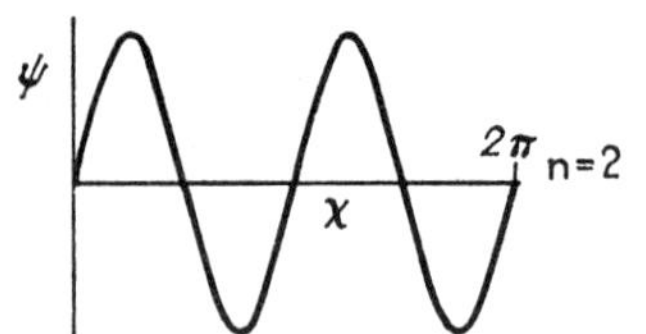

FIG. 12.—Wave functions for the first three quantum states of the rotator.

It may appear strange that in the preceding discussion the starting point for the measurement of the angle χ should be arbitrarily chosen, for that means that the maxima and minima of the probability function ψ^2 occur at arbitrary positions in space, which, unlike the case of the particle in the box, are not determined by the actual physical situation. The answer to this objection is to be found in the fact that the expression for ψ given in Eq. (10) is really not quite complete. Actually, this is a problem in which the physical reality can be better represented by assuming ψ to be complex, writing instead of (10)

$$\psi = A \sin \frac{2\pi\chi}{\lambda_\chi} + Ai \cos \frac{2\pi\chi}{\lambda_\chi},$$

[1] If ψ is constant λ_χ is obviously infinite, and Eq. (10) does not hold.

which may be shown to be a solution of the wave equation as well as expression (10). It must be remembered that the general form of the probability function is $\psi\psi^*$ rather than ψ^2. In this case,

$$\psi^* = A \sin \frac{2\pi\chi}{\lambda_\chi} - Ai \cos \frac{2\pi\chi}{\lambda_\chi}$$

and

$$\psi\psi^* = A^2 \sin^2 \frac{2\pi\chi}{\lambda_\chi} + A^2 \cos^2 \frac{2\pi\chi}{\lambda_\chi} = A^2,$$

which is independent of χ and states that the probability of finding the particle anywhere is independent of the angle. Obviously the new expression for ψ has the same periodicity as before, and Eqs. (12b), (12c), and all equations based on them will be unchanged. For the quantization of the motion, the important fact is the existence of a wave angle and the conditions on the wave function, not the exact form of the wave function. The latter may be of importance in other connections, but need not be considered unless the particular experiment under consideration requires it.

4.5. Quantization of the Space Rotator.—In Sec. 4.4, we dealt with a two-dimensional rotator, *i.e.*, one that was constrained to

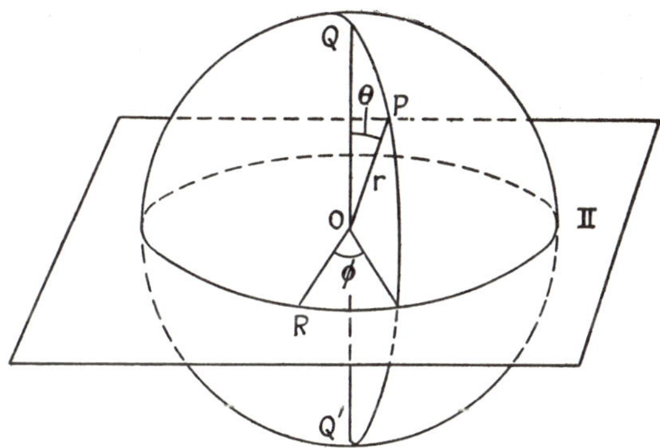

Fig. 13.—Polar coordinates.

move in a definite, fixed plane. The results may be extended to the case of a rotator which is similar in every respect, except that it is free to move in space. The position of the revolving mass may be described by means of a system of polar coordinates, with origin at the fixed center of rotation, such as is shown in Fig. 13. The distance of the rotating body from the center of rotation is, of course, determined by the length of the weightless rigid rod which connects it to this center. Its position, designated as P in Fig. 13, is then completely determined by the angles used in polar coordinates, namely, the angle θ, made by the line

OP with the axis QOQ′, and the angle ϕ, made by the intersection of the plane QPQ′ and the plane Π (which is perpendicular to QOQ′ at the origin) with an arbitrary line OR in Π.

Since two coordinates ϕ and θ are necessary to determine the position of the rotator, it has two degrees of freedom. It is therefore necessary to find two independent constants of the motion in order to effect the quantization.

One of these constants of the motion may be taken as the total angular momentum. This is constant, as in the example of the plane rotator. In fact, once the body is set into motion, it will continue to move in the same plane until it is disturbed, for there will be no forces at right angles to its direction of motion tending to pull it out of the plane determined by its direction of motion and the rod joining it to the origin.

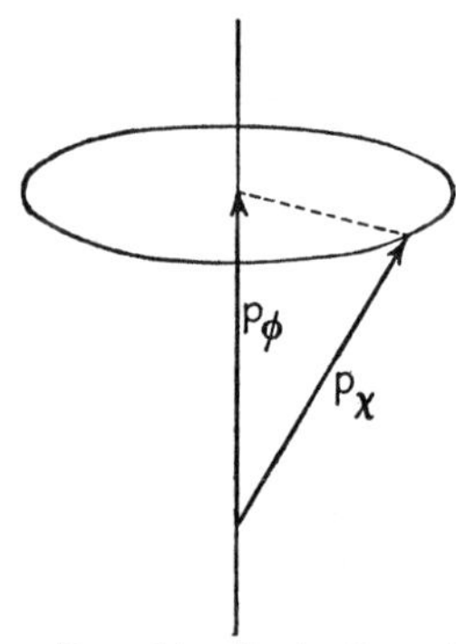

FIG. 14.—Projection of angular momentum on polar axis.

The angular momentum may be taken as a vector quantity perpendicular to the plane of motion, and of magnitude p_χ determined, as in Sec. 4.4, as the product of the distance r of the body from the center of rotation, its mass m, and its velocity $r\dot{\chi}$ where $\dot{\chi}$ is the angular velocity (χ is the angle in the plane of revolution). The direction of the angular-momentum vector is such that, if the direction of rotation of the rotator appears counterclockwise to the observer, the angular-momentum vector points toward the observer (see Fig. 78, Appendix I). Giving the angular momentum thus determines the plane of the rotation and its tilt in space, as well as the speed of the rotation.

It should be noted that when we say that the angular momentum is quantized, we refer to its magnitude. Assuming the *vector* to be quantized would be equivalent to quantizing the three components, and there are only two quantities to be quantized. The second constant of the motion to be quantized may, to be sure, be (and generally is) taken as the projection or component of the angular momentum along the polar axis. This projection is called p_ϕ.

It is seen, as indicated in Fig. 14, that p_ϕ will not uniquely determine the direction of the angular-momentum vector, since the latter may describe a cone and still give the same value of p_ϕ. This simply means that these various directions are allowed by the quantization, and the wave description of the phenomenon can in fact be used to give the *probability* of any such direction once p_χ and p_ϕ are determined.

More disturbing, perhaps, is the fact that the quantization depends on the polar axis which is, of course, arbitrarily chosen. There is thus something arbitrary about the process of quantization. However, this arbitrariness is simply an expression of the fact that it is not possible to distinguish between the directions in space unless they can be correlated with some physical reality. Under some circumstances, there may be a physical reason for selecting some particular polar axis. Suppose, for example, the rotating body were electrically charged. It would then interact with a magnetic field. Such a magnetic field would, according to the classical theory, cause the angular-momentum vector to precess about an axis in the direction of the field, *i.e.*, to revolve about the axis at a constant inclination, forming a cone and leaving the component of the angular momentum in the direction of the field constant. In a case like this, it is clear that the polar axis must be taken so as to coincide with the direction of the field, since p_ϕ should be the component to remain constant. Even an extremely small magnetic field, whose effect on the energy of the system is practically vanishing, would, according to classical theory, cause a precession which would eventually cause the vector to take all positions in the cone. Then p_χ and p_ϕ would be the only two independent dynamical quantities to remain constant, and, therefore, only these two could conceivably be quantized (at least other quantities would not be *independently* quantized). If the magnetic field is zero, then it is, to be sure, possible to choose the polar axis arbitrarily. But the wave functions for one set of axes can be expressed in terms of the wave functions for another set; if there is a certain quantum state for which the projection of the angular momentum on the polar axis has a given value in one set of axes, this means that there is a certain *probability* that it have any given value if taken along some other polar axis.

This statement may seem paradoxical, but the fact is that it is not possible to define the axis of quantization without either expressed or implied reference to some definite experiment which measures p_ϕ along that axis, and it so happens that it is not possible to perform an experiment that will give information about the projection of angular momentum on more than one axis at a time. This is another aspect of the uncertainty principle, and will be discussed further in Chap. VI, after the consideration of some actual atomic systems.

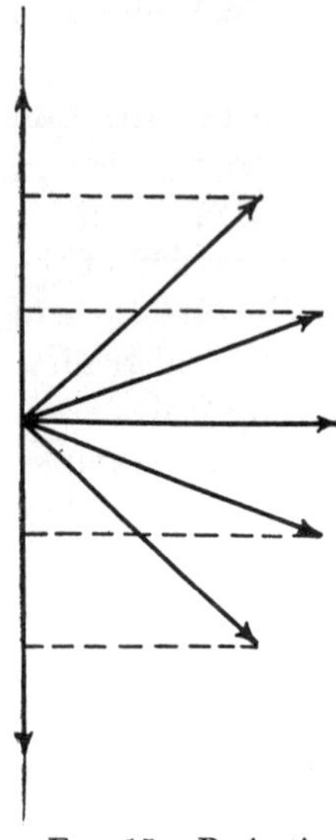

FIG. 15.—Projection of l-vector for $l = 3$. In this diagram l is taken as a vector with length = 3 and all projections on the axis are shown, which are whole multiples of the unit length. The lengths of the projections are the values of m_l.

The polar axis having then been determined or arbitrarily selected, it is desired to know the actual values of p_χ and p_ϕ which are allowed by the quantization. In this case, the wave function depends upon the two variables ϕ and θ, and it may be shown, by considering the differential equation of which it is a solution, that it is, indeed, a product of the type $\Phi(\phi)\Theta(\theta)$, where $\Phi(\phi)$ depends only on ϕ and $\Theta(\theta)$ depends only on θ. However, the momenta and velocities involved, when expressed in terms of ϕ, θ, $\dot{\phi}$, and $\dot{\theta}$, are sufficiently complicated so that it would be difficult to make any predictions as to the form of $\Phi(\phi)$ or $\Theta(\theta)$ without solving the wave equation (though it does turn out that $\Phi(\phi)$ is a simple sinusoidal function). It will be necessary to merely give the results in this case, and to show later that they are at least consistent with certain general rules which will be developed. It is found that p_χ and p_ϕ may have the following values:

$$p_\chi = \sqrt{l(l+1)}\frac{h}{2\pi}, \tag{15}$$

where $l = 0, 1, 2, \ldots$
and

$$p_\phi = \frac{m_l h}{2\pi}, \tag{16}$$

where[1] $m_l = -l, -l+1, \ldots, 0, \ldots l-1, l$.
The limit on the possible values of m_l follows from the fact that

[1] Do not confuse the quantum number m_l with the mass m.

p_ϕ cannot, being a projection of p_χ, be greater than p_χ in absolute value. It is seen that when the value of l is fixed, p_χ is determined and there are $2l + 1$ possible values of p_ϕ. The situation for $l = 3$ is illustrated in Fig. 15 by means of a vector diagram.

The energy of the rotator that we have been considering is determined by p_χ. The kinetic energy is $\frac{1}{2}mv^2$, and by Eq. (13), which applies equally well here, this is equal to $p_\chi^2/2mr^2$. As there is no potential energy the expression for the energy becomes

$$E = \frac{l(l+1)h^2}{8\pi^2 mr^2}. \tag{17}$$

The value of m_l has no influence on the energy; m_l simply determines the tilt of the plane of revolution with respect to the polar axis.

4.6. Quantum States and Phase Space.—In the preceding pages, the motion of a particle under various conditions of constraint has been considered. In Sec. 4.2, the case of a free particle (*i.e.*, free aside from the fact that it was confined in a box) was treated, and the particle had three degrees of freedom; in Sec. 4.4, the case of a particle constrained to rotate in a plane at a definite distance from a fixed center was considered—this particle had one degree of freedom; finally in Sec. 4.5, a particle, constrained to rotate a definite distance from a fixed point, but free as to its plane, and having, hence, two degrees of freedom, was treated. In the case of the free particle, the position of the particle was given by the three coordinates x, y, z with corresponding momenta p_x, p_y, and p_z, which could be written in terms of the velocities as

$$p_x = m\dot{x},$$
$$p_y = m\dot{y},$$

and

$$p_z = m\dot{z}.$$

In the case of the plane rotator, the position of the particle is described by the single coordinate χ, and the corresponding momentum (angular momentum in this case) is given by

$$p_\chi - mr^2\dot{\chi}.$$

In the case of the space rotator, the position of the particle is given by the polar angles ϕ and θ, and again there are corresponding momenta. The momentum corresponding to ϕ is just the angular-momentum projection p_ϕ already considered. In terms of the velocities and coordinates, it is given by

$$p_\phi = mr^2\dot{\phi}\sin^2\theta, \tag{18}$$

as is shown in Appendix I. Corresponding to θ is a special momentum p_θ which is given by

$$p_\theta = mr^2\dot{\theta} \tag{19}$$

(see Eqs. (7c) and (7b) of Appendix I).

If there is but a single particle, there cannot be more than three degrees of freedom. However, in a general case, involving more than one particle, there may be any number of degrees of freedom. If there are N degrees of freedom, which means that it requires N independent coordinates to describe the position of all parts of the system completely, there will be, correspondingly, N momenta. The N coordinates do not need to be simple rectangular or polar coordinates, and the corresponding momenta may be very complicated functions of the coordinates and velocities. The *number* of coordinates and momenta does not depend, however, on the way the coordinates are set up.

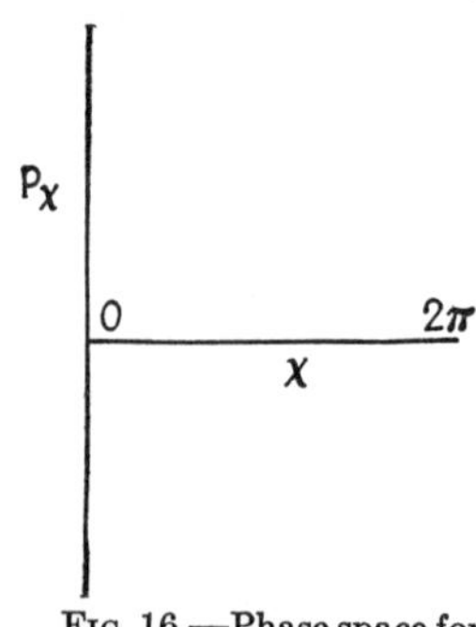

FIG. 16.—Phase space for plane rotator. Positive and negative p_χ's correspond to the two directions of rotation.

Suppose, having a system with N degrees of freedom, we construct a new space with $2N$ dimensions, the *rectangular* coordinates of which are the coordinates *and* the momenta of the system. Such a space is called the "phase space" of the system. It is important to note that in the phase space for the space rotator, for example, the values of such quantities as ϕ, θ, p_ϕ, p_θ are marked off on axes which are set up at right angles to each other. In a $2N$-dimensional space, it is, of course, possible to set up $2N$ mutually perpendicular axes, though it is not possible to draw a figure that will adequately illustrate the situation. In Fig. 16, however, an attempt is made to bring out the essential particulars by showing the axes in the phase space for the plane rotator. It will be noted that the axis for the angle χ is limited to a length of 2π, because the angle ranges only from 0 to 2π.

The usefulness of the phase space will become evident from the following considerations. According to classical mechanics, the state of a dynamical system at any time may be represented by a single point in an appropriate phase space, as that determines the q's and p's of the system, and from the q's and p's the position and velocity of every particle in the system may be calculated. The motion of the system would then, according to classical theory, be determined for all time,[1] and this motion of the system could be represented by a corresponding motion of the phase point. According to the wave mechanics, however, which deals in probabilities, one cannot trace the behavior of the system in such detail, but if a system is in a given stationary quantum state, we may say in a general sort of a way that it is confined to a certain region of the phase space. In general, it may be stated that in a system with N degrees of freedom a quantum state is equivalent to a hypervolume of $2N$ dimensions in the phase space equal in magnitude to h^N. (It may be readily verified that the dimensions of h are the same as the dimensions of a coordinate times the corresponding momentum, namely, energy times time, so the dimensions of h^N are correct for the phase space.)

This may be illustrated by a consideration of the example discussed earlier in this chapter, the motion of a particle in a "box."[2] In this case, as noted above, there are three coordinates—the ordinary Cartesian coordinates—and the three corresponding momenta, so the phase space is six dimensional. It is possible to deal, however, with the two-dimensional sub-phase spaces, each of which is defined by the pair of axes representing a coordinate and its corresponding momentum. Consider first the subspace defined by x and p_x. p_x will be positive if the particle is moving toward larger values of x, negative if it moves in the opposite direction. As the particle moves along, in the nth quantum state let us say, its p_x remains constant, according to the classical picture, till the particle comes to one end of the box, when it

[1] It is a well-known principle in the classical mechanics, arising from the fact that the equations of motion are second-order differential equations, that if the mutual positions and velocities of all parts of a system are known, the state of the system may (at any rate in principle) be foretold for any future time.

[2] Another example is discussed by Sommerfeld, "Atomic Structure and Spectral Lines," 3d English ed., vol. 1, p. 78, Methuen & Co., Ltd., 1934.

reverses sign. Thus the representative point in the two-dimensional sub-phase space will have a path such as is shown in Fig. 17 (solid lines ABDEA). Equation (3d) page 40, gives the x-component of the velocity of the particle, but it does not, of course, take into account the sign, giving only the absolute value. From (3d), the absolute value of p_x when the particle is in the nth state will be given by

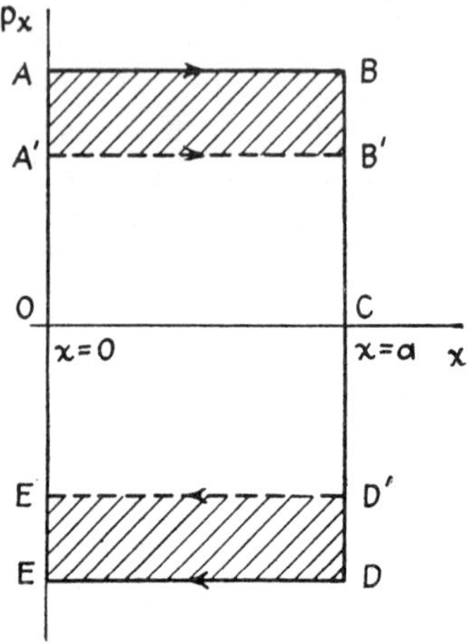

FIG. 17.—Phase diagram for particle in a box (one dimensional). In the nth quantum state the particle follows the path ABDEA, the area enclosed being the value of $\int p_x dx$ (special case of $\int pdq$—see p. 57) for the motion of the particle in the nth quantum state.

$$|p_x|_n = \frac{nh}{2a}.$$

When the particle is in the $(n-1)$th state, the path of the representative point will be given by the dotted lines A′B′D′E′A′ of Fig. 17. The space between the two lines, shown shaded in Fig. 17, may be taken as the measure of the region in the sub-phase space occupied by one quantum state. It is equal to

$$2 \times \left[\frac{nh}{2a} - \frac{(n-1)h}{2a}\right] \times a = h.$$

At the same time, the y and z motions of the particle will be quantized and occupy a region equal to h in their respective subspaces. The areas in the x, y, and z sub-phase spaces will all be perpendicular to each other and may thus be considered as the projections of a six-dimensional space whose hypervolume is h^3.

This result may be used to find the approximate number of quantum states in any designated region of the six-dimensional phase space. We shall illustrate this by obtaining a result that will be found useful in later chapters. We shall calculate the approximate number of quantum states for a region of the phase space for which the total kinetic energy

$$T = \frac{1}{2m}(p_x^2 + p_y^2 + p_z^2) \tag{20}$$

(which is, in this case, the total variable energy) is less than some given value T_0. The volume in the phase space for which the energy T is less than T_0 may be taken as the product VP_0 of two factors, a space factor V, and a momentum factor P_0. V is naturally equal to abc, where a, b, and c are the lengths of the sides of the box, *i.e.*, V is the volume of the box. It is seen from Eq. (20) that T is constant on the surface of any sphere in the momentum space, the radius of the sphere being $\sqrt{2Tm}$, which is the magnitude p of the momentum vector (see Fig. 18). The volume of the sphere bounded by the T_0 surface is

$$\tfrac{4}{3}\pi p_0{}^3 = \tfrac{4}{3}\pi(2T_0m)^{3/2} = P_0.$$

The number n_0 of quantum states in this region for which the energy is less than T_0 is thus given by

$$n_0 = P_0Vh^{-3} = \tfrac{4}{3}\pi(2T_0m)^{3/2}Vh^{-3}. \tag{21}$$

It is convenient, however, whenever possible, to consider a two-dimensional projection, as was done with the example just considered. In general, the hypervolume of a quantum state in a $2N$-dimensional phase space can be thought of as having a cross section of area h on each of N mutually perpendicular planes, determined by the axes of a coordinate q and its conjugate momentum p_q. This relationship may be expressed in terms of the phase integral, which is defined as $\int p_q\,dq$, where the variable of integration is allowed to range over the values taken during one period of the motion, *i.e.*, until it returns to the starting value. This integral represents an area on the p_q-q plane. Thus in the case of the particle in the box, it is clear that the area OABC (see Fig. 17) is the contribution to the integral for the motion of the particle from one side of the box to the other, whereas the area CDEO is the contribution to the integral for the reversal of the motion and completion of the period of motion. The latter part of the integral is positive, for although p_q is nega-

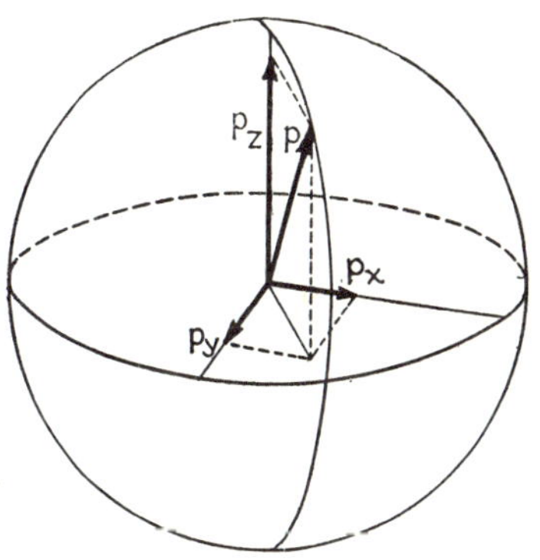

FIG. 18.—Momentum vector and its projections in momentum space.

tive, the increments dq are likewise negative while the particle is reversing its motion.

In order that the area between neighboring quantum states should be equal, or approximately equal, to h, it is clear that, for the nth quantum state, $\int p_q\,dq$ should be equal, or approximately equal, to nh, and that, in any event, there should be but one quantum state for which $\int p_q\,dq$ lies between nh and $(n+1)h$. Sometimes $\int p_q\,dq$ may depend on several quantum numbers; in this case, a slight alteration of this rule is necessary, as will be

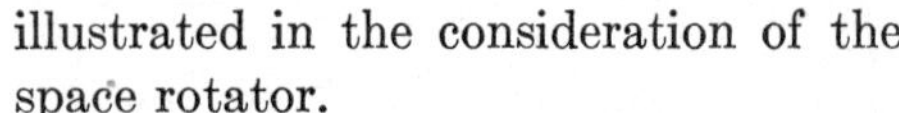

illustrated in the consideration of the space rotator.

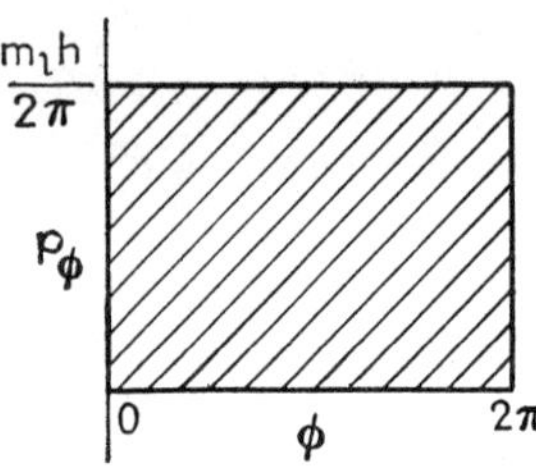

FIG. 19.—Phase diagram for ϕ and p_ϕ. Shaded area = $\int p_\phi\,d\phi$.

In the case of the space rotator, there are, as already noted, four dimensions in the phase space. The quantities that are quantized are the constants of the motion p_ϕ and p_χ, whereas the phase integrals are $\int p_\phi\,d\phi$ and $\int p_\theta\,d\theta$. Since p_ϕ is equal to $m_l h/2\pi$ (where m_l is the quantum number) and ϕ goes from 0 to 2π while the particle executes one period of its motion, it is seen that $\int p_\phi\,d\phi = \int_0^{2\pi} \frac{m_l h}{2\pi}\,d\phi = m_l h$. The integral, therefore, has the expected values.

With the other integral, the situation is slightly more complicated. It is necessary, in order to evaluate the phase integral, to express p_θ in terms of p_ϕ and p_χ. In Eq. (4) of Appendix I, it is shown that

$$r^2\dot{\chi}^2 = r^2\dot{\theta}^2 + r^2\dot{\phi}^2 \sin^2\theta.$$

Therefore, by Eqs. (13), (18), and (19),

$$p_\chi\dot{\chi} = p_\phi\dot{\phi} + p_\theta\dot{\theta}.$$

Since the dot means time differentiation, multiplying this through by dt gives

$$p_\chi\,d\chi = p_\phi\,d\phi + p_\theta\,d\theta,$$

whence

$$\int p_\theta\,d\theta = \int p_\chi\,d\chi - \int p_\phi\,d\phi$$
$$= \int_0^{2\pi} \frac{\sqrt{l(l+1)}\,h}{2\pi}\,d\chi - \int_0^{2\pi} \frac{m_l h}{2\pi}\,d\phi$$

$$= \sqrt{l(l+1)}h - m_l h$$

if l and m_l are the quantum numbers. In order to understand this result, the quantum number m_l should be held constant and l should be varied. It can then readily be seen, since l must be equal to or greater than m_l, that the area enclosed between adjacent l values is approximately h, and these quantum levels are correctly spaced. The hypervolume in the four-dimensional space, occupied by any given quantum state, will be at least approximately equal to the area on the p_ϕ-ϕ plane between adjacent m_l values times the area on the p_θ-θ plane between adjacent l values, since these are mutually perpendicular cross sections of the cell in the phase space assigned to this particular quantum level. This product is always close to h^2, as expected. This deduction may be considered as a rough verification of the results cited in Sec. 4.5.

Exercises

1. Calculate the wave length of an electron with 1 electron volt of energy; of a proton with the same energy.

2. Calculate the wave length of a body with a mass of 1 g., moving with a velocity of 1 cm. per sec. What is its energy in electron volts?

3. Calculate the energy in ergs and in electron volts of an electron and of a proton in the first quantum state of a one-dimensional box of length 1 cm.; of length 2×10^{-8} cm. Calculate the velocity and energy in ergs and electron volts of a body of mass 1 g. in the first quantum state of a box of length 1 cm. Repeat all calculations for the fifth quantum state.

4. From Eq. (3e), find an approximate formula for the number of energy levels in a box for which the energy lies between E_1 and E_2, where E_1, E_2, and $E_2 - E_1$ are very large compared with $h^2/8a^2m$ but $E_2 - E_1$ is very small compared with E_1 and E_2.

In these problems the potential energy may be set equal to zero.

CHAPTER V

THE HYDROGEN ATOM

The hydrogen atom is the simplest of all atoms. It consists of a single electron revolving about a positively charged nucleus which has a mass about 1835 times that of the electron. The electron is attracted to the nucleus by a force equal to e^2/r^2, where e is the charge on the electron and r the distance between electron and nucleus. The mass of the nucleus is so great compared with that of the electron that to a very good approximation it may be considered to be a fixed point to which the electron is attracted by the force noted.

As was first shown by Bohr, whose work lies at the basis of all modern atomic theory, the motion of the electron about the nucleus is quantized. An exact description requires the use of the wave mechanics. However, the general procedure in case of atomic systems is to consider first the classical motion of the electron and then see how it is affected by the quantum theory. This method has been used in several instances in Chap. IV, and will be applied to the present case. The classical treatment not only provides a basis for further discussion and a convenient nomenclature, but is a sufficiently good approximation for many purposes. It will, therefore, be given in some detail, after which the quantization of the motion will be considered, and finally some aspects of the wave picture will be presented.

5.1. Classical Motion of the Electron.—Since the electron is attracted to the positive nucleus by a force that varies inversely as the square of the distance, its motion, on the classical picture, is exactly similar to the motion of the planets about the sun. Thus in atomic mechanics, many of the results of celestial mechanics may be taken over bodily. Most important for our purpose are the first two of Kepler's laws, which may be transcribed to fit the case at hand, as follows:

1. The electron moves in an ellipse with the positive nucleus at one of the foci.

2. The line joining the electron and the positive nucleus sweeps out equal areas in equal times.

The first of these laws was shown by Newton to be a result of the inverse-square law of attraction and the general laws of motion. As the proof of it is as much a matter of geometry as dynamics, and since it is available in many treatises on dynamics and celestial mechanics, it will not be given here. The second of these laws, as shown in Appendix I, is an expression of the fact that the angular momentum of the electron in its orbit remains constant. It was shown by Newton to follow merely from the fact that the force is directed to a central point; his proof is given in Appendix I.

5.2. Circular Orbits.—A circle is a special case of an ellipse, in which the two foci coincide at the center.[1] It is a specially simple kind of orbit, and it will be helpful to consider the properties of such an orbit first. As before, let r represent the distance from the moving electron to the attracting center; in this case r is the radius of the orbit. The electron rotating in its orbit has a centrifugal force equal to mv^2/r, where m is its mass and v the magnitude of its velocity. This is balanced by the force of electrical attraction e^2/r^2, where e is the magnitude of the charge on the electron (taken positive) so that

$$\frac{mv^2}{r} = \frac{e^2}{r^2}, \tag{1}$$

from which it is seen that with a circular orbit (r constant) v must be constant and the electron has a kinetic energy given by

$$\tfrac{1}{2}mv^2 = \frac{e^2}{2r}. \tag{2}$$

The total energy is the sum of the kinetic and potential energies. The potential energy depends only on the distance between the electron and the positive nucleus, *i.e.*, it is a function of r alone, and it does not depend at all on whether the orbit is circular or not. Since the electron and proton have equal and opposite charges of magnitude e, the potential energy is equal

[1] Bohr treated only circular orbits; the extension of the old quantum theory to elliptical orbits in space was due to Sommerfeld and to Wilson. See Sommerfeld, "Atomic Structure and Spectral Lines," 3d English ed., vol. 1, pp. 109*ff*., Methuen & Co., Ltd., 1934.

to $-\frac{e^2}{r}$ (see Appendix III). Since it is necessary to do work on the electron and thus increase its potential energy in order to move it to a position of zero potential energy (taken according to the usual convention to be $r = \infty$), its potential energy at any finite distance r must be negative. Therefore, the total energy E of an electron moving in a circular orbit is, from Eq. (2), given by

$$E = \tfrac{1}{2}mv^2 - \frac{e^2}{r} = -\frac{e^2}{2r}. \tag{3}$$

The angular momentum is given by

$$mvr = e\sqrt{mr}. \tag{4}$$

5.3. Energy and Angular Momentum of Elliptical Orbits.—It will be shown in this section that the energy of an elliptical orbit

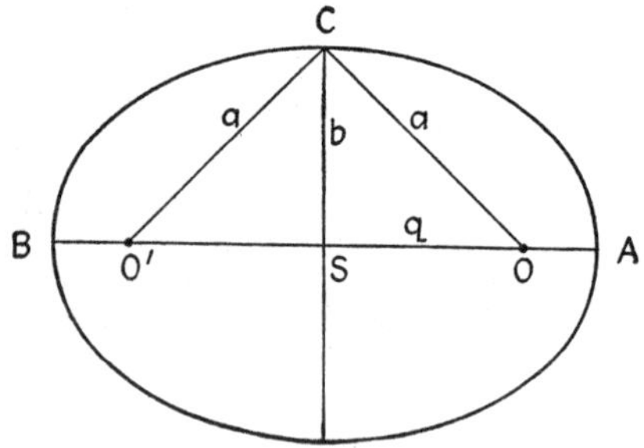

Fig. 20.—Elliptical orbit, attracting center at O.

is dependent only upon the major axis of the ellipse and is, therefore, equal to the energy of a circular orbit if the diameter of the circle is equal to the major axis of the ellipse. However, with a given major axis, the angular momentum of the orbit is smaller, the smaller the minor axis of the ellipse. The energy of an orbit may, therefore, be said to depend only upon its size, in so far as that is measured by the major axis, whereas the angular momentum depends upon the shape of the ellipse as well.

Consider the ellipse shown in Fig. 20 with semimajor axis $a(=\text{SA})$, semiminor axis b, and the attracting center at the focus O. By a well-known geometrical property of the ellipse, the sum of the distances to the foci from any point on the ellipse is equal to the major axis. Thus, $\text{CO}' = \text{CO} = a$. We designate the distance SO as q and let the velocity of the electron at A be v_A and the velocity at B be v_B. When it is at A, the distance r of the electron from O is $a - q$, and when it is at B, its distance from O is $a + q$. Also at A and B, the electron is moving perpendicularly to the line joining it with O. Therefore, on account of the conservation of angular momentum, which says that the angular momentum is the same at all points in the orbit,

$$m(a - q)v_A = m(a + q)v_B \tag{5}$$

or

$$v_B = v_A \frac{a - q}{a + q}. \tag{6}$$

Since the total energy of the electron must be the same at any point in its orbit, we have, adding kinetic and potential energies at A and equating to the similar sum for B,

$$\frac{1}{2}mv_A^2 - \frac{e^2}{a - q} = \frac{1}{2}mv_B^2 - \frac{e^2}{a + q}. \tag{7}$$

Substituting the value of v_B from (6) into (7) and rearranging the latter slightly gives

$$\frac{1}{2}mv_A^2\left(1 - \frac{(a - q)^2}{(a + q)^2}\right) = \frac{e^2}{a - q} - \frac{e^2}{a + q}.$$

But $1 - \dfrac{(a - q)^2}{(a + q)^2} = \dfrac{4aq}{(a + q)^2}$. Hence

$$\frac{1}{2}mv_A^2 = \frac{e^2}{4aq}\left[\frac{(a + q)^2}{(a - q)} - (a + q)\right] = \frac{e^2}{2a}\,\frac{a + q}{a - q}. \tag{8}$$

Substituting (8) into the left-hand side of Eq. (7) gives for the energy

$$E = -\frac{e^2}{2a}, \tag{9}$$

which shows that the total energy is determined solely by the major axis of the ellipse.

The value of the angular momentum, which is equal to $m(a - q)v_A$, may be found by substituting into this expression the value of v_A obtained from Eq. (8), giving

$$m(a - q)v_A = e\sqrt{\frac{m(a^2 - q^2)}{a}}.$$

By Fig. 20, $a^2 - q^2 = b^2$. Therefore

$$m(a - q)v_A = eb\sqrt{\frac{m}{a}}. \tag{10}$$

5.4. The Quantization of the Hydrogen Atom.—In the case of the hydrogen atom, there are three constants of the motion, the energy E, the total angular momentum p_χ, and the projection of the angular momentum along some axis p_ϕ. The quantization of the latter two quantities proceeds in the same way as for the space rotator (see Sec. 4.5). p_χ can take on the values p_l, where

$$p_l = \sqrt{l(l+1)}\frac{h}{2\pi}, \tag{11}$$

where $l = 0, 1, 2, \ldots$, and p_ϕ can take on the values p_{m_l}, where

$$p_{m_l} = \frac{m_l h}{2\pi} \tag{12}$$

and $m_l = -l, -l+1, \ldots, 0, \ldots, l-1, l$. The energy E can take on the values E_n given by

$$E_n = -\frac{2\pi^2 me^4}{n^2h^2} \tag{13}$$

where n (the "total" quantum number) $= l + 1, l + 2, \ldots$.

The statement that n is equal to $l + 1$, or greater, is a natural consequence of considering the quantization of angular momentum before the quantization of energy. It is often desired to find out how many quantum states there are with a given energy, and in this case it is more convenient to state that n may take on the values 1, 2, 3, . . . , and place the restriction on l, *i.e.*, l can take on the series of values $0, 1, 2, \ldots, n - 1$.

These results are very easily visualized from the classical picture of the motion in an ellipse. It was shown in Sec. 5.3 that the energy of such an elliptical orbit depends only on the major axis, the larger the major axis, the greater being the energy ($E = -e^2/2a$, therefore, the larger the a, the smaller the $-E$, or the greater is the actual value of the energy). The size of the orbit is determined by n; only a discrete set of values for the major axis is permitted by the quantum conditions. The angular momentum depends on the minor axis of the ellipse, as well as the major axis. The greater the minor axis, the greater the angular momentum. But the minor axis cannot possibly become greater than the major axis. It is not surprising, then, that the greater the major axis, *i.e.*, the greater the energy, the greater the number of values of the angular momentum which will be allowed by the quantum conditions. There are n different values of l for each value of the energy and $2l + 1$ different values of m_l for each value of the angular momentum. It is, therefore, clear that the number of quantum levels having the same energy increases rapidly with the energy. This is illustrated in the following table, which gives the various combinations of quantum numbers for the different quantum states for which n is equal to 3 or less.

TABLE OF QUANTUM LEVELS

n	l	m_l	Type
1	0	0	1*s*
2	1	1	2*p*
	1	0	2*p*
	1	−1	2*p*
	0	0	2*s*
3	2	2	3*d*
	2	1	3*d*
	2	0	3*d*
	2	−1	3*d*
	2	−2	3*d*
	1	1	3*p*
	1	0	3*p*
	1	−1	3*p*
	0	0	3*s*

The *type* of quantum level, given in the last column of the table, designates the quantum state in terms of a notation that is often used to classify the energy levels into states that have the same values of n and l. The numeral stands for the value of n, whereas, if $l = 0$, the atom is said to be in an *s*-state; if $l = 1$, the state is called a *p*-state; $l = 2$ is a *d*-, $l = 3$ an *f*-state; from this point on, the notation proceeds alphabetically: *g*-state, *h*-state, etc. If the quantum number n of the electron is 2, for example, and the quantum number l has the value 1, then this information may be expressed by saying that the atom is in a 2*p*-state. This statement, which does not specify m_l, is sufficient for many purposes. This peculiar nomenclature is to be traced back to the early study of atomic spectra, when the relationship between spectral lines and energy levels was not understood; unfortunately, perhaps, it became so widely used that it has persisted to the present time.

5.5. The Quantum States and the Phase Integrals.—The proof that Eqs. (11) and (12) are consistent with the condition on $\int p_q\,dq$, discussed in Sec. 4.6, follows exactly as for the space rotator. The expression (13), for the energy, may be shown to conform to the quantum condition on $\int p_r\,dr$, where p_r is the momentum conjugate to r. According to Eq. (5) of Appendix I, the kinetic energy of the electron will be given by

$$T = \tfrac{1}{2}mv^2 = \tfrac{1}{2}m\dot{r}^2 + \tfrac{1}{2}mr^2\dot{\chi}^2.$$

The total energy is the sum of the kinetic and potential energy. The latter is equal to $-\dfrac{e^2}{r}\cdot$ Therefore,

$$E = \tfrac{1}{2}m\dot{r}^2 + \tfrac{1}{2}mr^2\dot{\chi}^2 - \frac{e^2}{r}.$$

This can be expressed in terms of the momenta p_r and $p_\chi = p_l$ [see Eqs. (7a) and (3) of Appendix I] as follows:

$$E = \frac{p_r^2}{2m} + \frac{p_l^2}{2mr^2} - \frac{e^2}{r}. \tag{14}$$

By Eq. (11), then,

$$E = \frac{p_r^2}{2m} + \frac{l(l+1)h^2}{8\pi^2mr^2} - \frac{e^2}{r} \tag{15}$$

or

$$p_r = \pm\sqrt{2mE - \frac{l(l+1)h^2}{4\pi^2r^2} + \frac{2me^2}{r}}. \tag{16}$$

When the notation of Fig. 20 is used, it is seen that r fluctuates between the values $a + q$ and $a - q$. While r is increasing, *i.e.*, while dr is positive, p_r is also positive since $\dot{r}$ is positive; for decreasing values of r, on the other hand, p_r is negative and the negative value of the square root [Eq. (16)] must be used; p_r goes through its zero values when r is neither increasing nor decreasing, at $a + q$ and $a - q$. The part of the integral for which dr is positive contributes just the same amount to the total value as the part for which dr is negative. We can, therefore, integrate from $a - q$ to $a + q$ only and double the result. This gives

$$\int p_r\,dr = 2\int_{a-q}^{a+q}\sqrt{2mE - \frac{l(l+1)h^2}{4\pi^2r^2} + \frac{2me^2}{r}}\,dr. \tag{17}$$

This integral may be evaluated and yields

$$\int p_r\, dr = -\sqrt{l(l+1)}\, h + \frac{2\pi\sqrt{m}\, e^2}{\sqrt{-2E}}. \tag{18}$$

Substituting from Eq. (13) into Eq. (18), we get

$$\int p_r\, dr = h[n - \sqrt{l(l+1)}], \tag{19}$$

it being quite obvious then that the difference of successive values of $\int p_r\, dr$, if l is fixed, is exactly equal to h, so that the expected condition on $\int p_r\, dr$ is fulfilled. Equation (19) also shows why n must always be greater than l. For otherwise $\int p_r\, dr$ would have a negative value, whereas it has been shown that it is always positive.

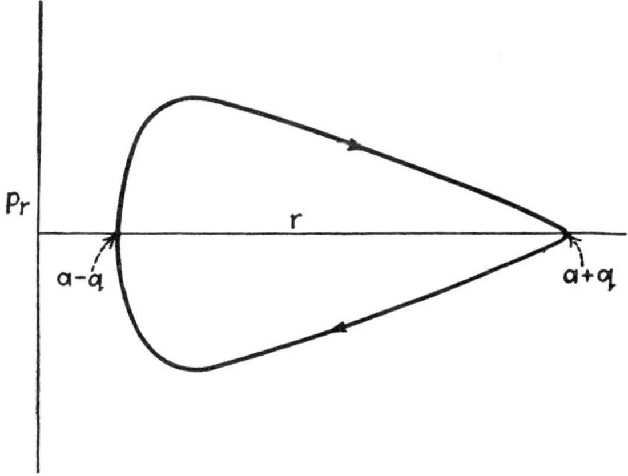

FIG. 21.—Schematic phase diagram for the radial motion of the electron in the hydrogen atom. The enclosed area is $\int p_r\, dr$.

5.6. Positive Energies of the Hydrogen Atom.—So far we have considered only the case in which the total energy is negative, *i.e.*, the kinetic energy is smaller in absolute value than the potential energy which is always negative. It is of course possible for the total energy to be positive. If this is so, the kinetic energy will have a finite positive value even if r is taken to be infinite. In other words, the electron could go an infinite distance from the positive nucleus (this would require an infinite time, of course) and still be moving. An electron that starts at $r = \infty$ with a definite velocity will be attracted toward the nucleus, swing in toward it with increasing velocity, fly past it, and proceed with decreasing velocity to infinity in some other direction. Its path, according to the classical theory, will be a hyperbola. Such an electron is not attached to the positive nucleus in the same sense as an electron that is moving in an elliptical orbit in which, of course, it never gets farther than a certain finite distance from the nucleus. The electron moving in a hyperbolic orbit is said to be a free electron, and its motion is not quantized. Only negative energies are quantized according to Eq. (13)—all positive energies are allowed. (Of course, if the system is put in a box, then a certain quantization will occur, as in the last chapter, but if the box is considerably larger than atomic dimensions, the energy levels will be so close together that they may be considered to form a continuum.)

5.7. Wave Picture of the Radial Motion of the Electron.—The picture of the electron orbit given above is, of course, not strictly correct. It is sufficiently near to being correct to suffice for many purposes, but, according to the wave mechanics, it is *too* detailed. It is not actually possible to follow an electron in its orbit, either theoretically or by any conceivable experiment, however idealized, even though it be imagined that absolutely perfect instruments are available. It is possible to predict only the probability of the electron being in any particular position, this probability being given by the wave function. Further, it is the wave function that determines the allowed quantum levels. The results of the quantization have, however, as a matter of fact, been taken from the wave mechanics so that no error enters into them from this cause.

This particular case of the hydrogen atom offers an especially good example of the way the properties of the wave functions determine the quantum levels, and it seems worth while to consider this matter a little more closely, though still not attempting to give quantitative details. We shall confine ourselves to a consideration of the wave picture for the radial motion of the electron, *i.e.*, the motion having to do with the coordinate r. It will be observed from Eq. (14) that, for a fixed value of l, the energy depends on a term $p_r^2/2m$, which has the form of a kinetic energy, and a term $\dfrac{p_l^2}{2mr^2} - \dfrac{e^2}{r}$ or $\dfrac{l(l+1)h^2}{8\pi^2mr^2} - \dfrac{e^2}{r}$ which depends, on r. Though the term $p_l^2/2mr^2$ really originated as a kinetic term, it acts exactly like an addition to the potential energy in a virtual one-dimensional motion involving only the coordinate r and its derivatives with respect to time. If the effective potential energy $U_r = \dfrac{l(l+1)h^2}{8\pi^2 mr^2} - \dfrac{e^2}{r}$ is plotted as a function of r, the curves shown in Fig. 22 are obtained for the cases $l = 0$ and $l = 1$, respectively. In this figure, the positions along the energy axis, determined by the quantum number n, are marked by the horizontal lines. Parenthetically, it may be remarked that the reason n cannot be less than 2 if $l = 1$ is graphically brought out by Fig. 22. The energy determined by $n = 1$ is seen to lie below the potential-energy curve; this would necessitate that

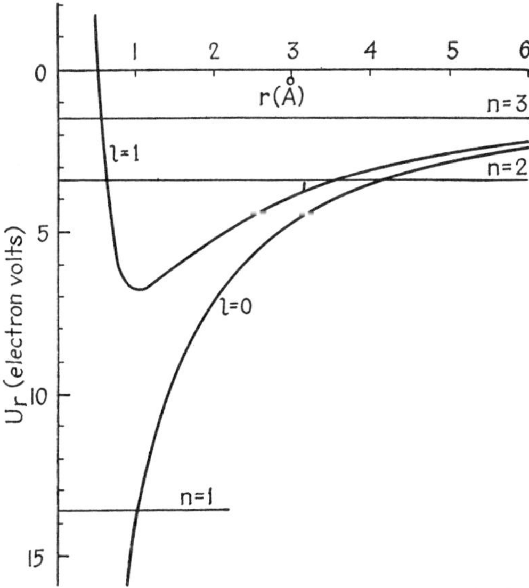

Fig. 22.—Effective potential-energy curves and energy levels for hydrogen atom.

the term $p_r^2/2m$ should always be negative, which would mean that the velocity would be imaginary.

The wave functions ψ of the hydrogen atom are functions of r, θ, and ϕ. Just as in the case of an electron moving in a box the wave functions could be expressed as products of three factors, each one of which was a function of one only of the coordinates x, y, and z, so in this case we can write $\psi = R\Theta\Phi$, where R is a function of r alone, Θ a function of θ alone, and Φ a function of ϕ alone. In Fig. 23 are exhibited the functions R for the various energy levels that are shown in Fig. 22.

In polar coordinates, an element of volume defined by θ and $\theta + d\theta$, ϕ and $\phi + d\phi$, and r and $r + dr$ has the magnitude

$r^2 \sin\theta\, d\theta\, d\phi\, dr$. The probability of finding an electron in this volume element is proportional to the volume and to ψ^2 and so is given by

$$\psi^2 r^2 \sin\theta \; d\theta\, d\phi\, dr = R^2\Theta^2\Phi^2 r^2 \sin\theta \; d\theta\, d\phi\, dr. \tag{20}$$

In order to normalize this expression (see pp. 40*f*.), we can set

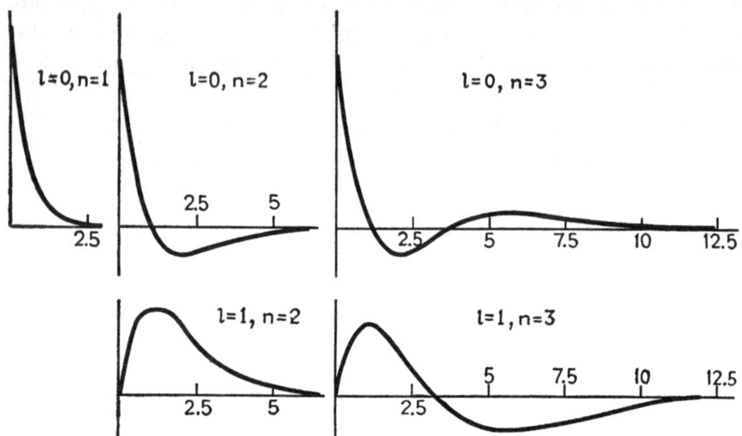

FIG. 23.—Radial wave functions, R, for various energy levels of the hydrogen atom. R (ordinates) in arbitrary units, different for each diagram. r (abscissas) in Å.

the integrals of its three independent products separately equal to 1; thus

$$\int_0^\infty r^2 R^2\, dr = 1$$

$$\int_0^\pi \Theta^2 \sin\theta \; d\theta = 1$$

and

$$\int_0^{2\pi} \Phi^2\, d\phi = 1.$$

The probability of finding the electron between r and $r + dr$, and anywhere as far as θ and ϕ are concerned, is obviously given by integrating over the whole range of the latter two variables. Since integration over ϕ and θ gives factors of 1, the probability in question is $R^2 r^2\, dr$. Similarly, the probability that it lie between θ and $\theta + d\theta$ is $\Theta^2 \sin\theta\, d\theta$, and, likewise, the probability that it lie between ϕ and $\phi + d\phi$ is $\Phi^2\, d\phi$. The probability that it lie within all three limits simultaneously, namely, that it lie within the volume element defined by them is, of course, the product of the three independent probabilities, as is indicated in the preceding expression, (20).

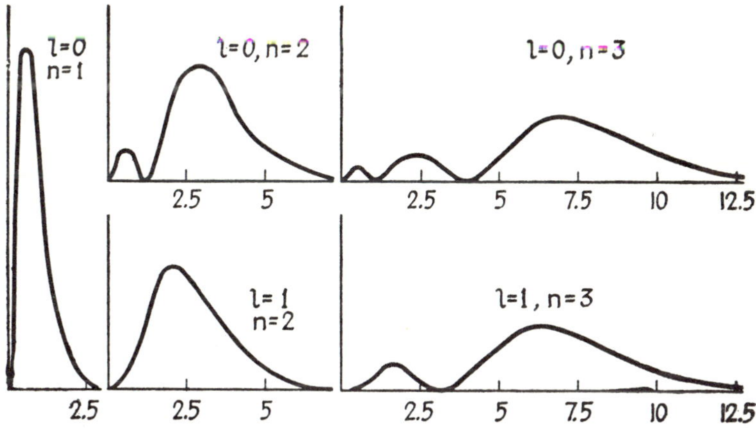

FIG. 24.—r^2R^2 for various energy levels of the hydrogen atom. r^2R^2 (ordinates) in arbitrary units, but same for each diagram. r (abscissas) in Å.

The function r^2R^2 which gives the probability per unit distance that the electron be found in any range of r is obviously an important function, and it is shown for various energy levels in Fig. 24.

It will be observed from Figs. 23 and 24 that the number of waves increases as n increases. In a general way, the most

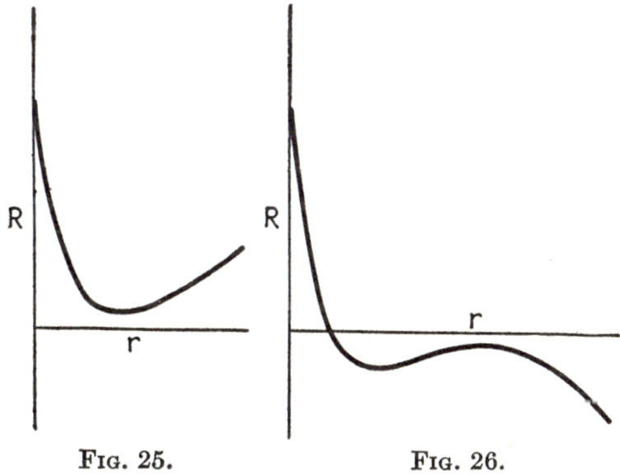

FIG. 25. FIG. 26.

probable positions for the electron lie roughly within the range through which the electron would move according to the classical theory, but the details of the picture are very different.

An interesting situation arises if we try to construct a wave function to fit the differential equation which determines R, for some energy that does not coincide with one of the allowed energies given by Eq. (13). For example, an attempt to construct such a wave function for an energy just below the energy given by $n = 1$ for $l = 0$ gives a wave function like that shown in Fig. 25. Instead of asymptotically going to zero as r becomes infinite, R approaches the axis, does not reach it, and eventually becomes infinite. If the energy lies between that which corresponds to $n = 1$ and that for $n = 2$, a curve something like that shown in Fig. 26 is obtained. In general, R will become infinite, which means that the electron would practically always be found at infinity. In excluding such cases, the possibility of having any energy excepting one of those given by Eq. (13) is ruled out. Only in these cases are the wave functions everywhere finite. This qualitative discussion will illustrate in a rough way the manner in which the more exact analysis of the wave mechanics yields Eq. (13).

5.8. Experimental Confirmation of the Hydrogen Energy Levels.—In the preceding pages, we have given a theoretical account of the quantum states and energy levels of the hydrogen atom. The experimental verification of these results comes from observation of the spectrum of hydrogen, and, when certain fine points of the theory are taken into account,[1] the agreement between experiment and theory is as complete and beautiful as any in the whole range of science. What is observed, of course, as will be clear from Sec. 4.1, is the differences of energy which exist in the hydrogen atom, rather than the actual energy levels themselves. The experimental method is, briefly, as follows. If hydrogen is brought into a situation in which it can receive

[1] The results presented fail to take account of two corrections, which produce slight changes. In the first place, the nucleus was assumed to be so heavy that it could be regarded as a fixed center of force. Secondly, it was assumed that the electron always moves so slowly compared with the velocity of light that the relativity correction is negligible. These effects produce very slight shifts in the energy levels which, for accurate spectroscopic work, need to be taken into account, but which, for most purposes, may be entirely neglected. The relativity correction is, however, as shown by Dirac, closely related to the spin of the electron, a property that will be taken up in the next chapter. The details of this relationship are outside the scope of this book.

energy, as for example, in an electric discharge, it may become dissociated, *i.e.*, some of the molecules of hydrogen, H_2, may become split up into their constituent atoms. These atoms then may receive further energy, so that instead of all of them existing in the lowest possible energy level some of them may be excited up to higher energy levels. The latter can then emit radiation, dropping at the same time to a lower energy level, the energy going out in the form of a light quantum. There exists between the energy of the light quantum (which is equal to an energy difference in the hydrogen atom) and the frequency of the light the relation discussed in Chap. IV. It is, therefore, possible to get a verification of the theory just discussed by measuring the frequencies or the wave lengths of the light emitted by the hydrogen, *i.e.*, by a study of its spectrum. This has been done, and as has been stated, the theory has been extremely successful.

There is one point in the theory, however, that at first sight would seem to be extremely difficult of verification. It will be remembered that the theory predicts, in general, that there will be a number of different quantum states, all of which have the same energy, the number of levels with a given energy rapidly increasing as n increases. Now we may ask the question: How in practice can these various levels be separated? If we effectively measure only the energies of the atom, it would appear that they could not be separated. But suppose the atom is observed in a strong electric field. Such an electric field will slightly change the energy of the levels of the atom. But the effect of the electric field will depend not only upon the size of the ellipse, but also upon its shape and its orientation in space. Thus some of the components of the multiple energy state will be shifted by different amounts. It is true that in fields of strengths readily obtained there are still some states whose energies coincide, but the multiple nature of the energy levels is readily demonstrated. The amount each quantum state should shift is calculable, and once more theory and experiment are in excellent agreement.

In the course of such experiments, it becomes evident, however, that transitions with emission of light do not occur between all possible quantum states; and, in fact, such transitions will not occur unless $l' = l'' \pm 1$ and $m_l' = m_l'' \pm 1$ or $m_l' = m_l''$, where l'' and m_l'' are the rotational quantum numbers of the initial state, *i.e.*, the state of higher energy, and l' and m_l' are the quantum numbers of the final state. This so-called "selection rule" may also be explained theoretically; for our purposes, however, it is not necessary to consider this matter further.

5.9. Summary.—The results on the energy levels and quantization of the hydrogen atom will be frequently referred to, so it seems desirable to summarize them at this point.

The *energy* of the electron of the hydrogen atom in its elliptical orbit depends upon the *major axis*, hence the *size*, of the ellipse. It is given by the quantum number n, according to the equation

$$E = -\frac{2\pi^2 me^4}{n^2h^2},$$

where n can take the values 1, 2, 3 . . .

The *angular momentum* of the electron depends upon both the *major* and the *minor axis* of the ellipse, hence upon the *shape* of the ellipse. It is given by the quantum number l, according to the equation

$$p_l = \sqrt{l(l+1)}\frac{h}{2\pi}.$$

If the value of n is given, then l can take the values $0, 1, 2, \ldots, n-1$, there being thus n values of l possible. If $l = 0, 1, 2, 3, 4, \ldots$, the atom is said to be in an *s*-, *p*-, *d*-, *f*-, *g*-, . . . state.

The *projection of the angular momentum* on an arbitrary axis depends upon the *tilt in space* of the plane in which the electron is moving. It is given by the quantum number m_l by the equation

$$p_{m_l} = m_l\frac{h}{2\pi}.$$

If the value of l is given, then m_l can take the value $-l, -l+1, \ldots, -1, 0, 1, \ldots, l-1, l$, there being thus $2l+1$ values of m_l possible.

Exercises

1. Calculate in ergs and electron volts the energy necessary to just remove an electron from the lowest orbit of the hydrogen atom.

2. Calculate the value of a (semimajor axis of the ellipse) for the lowest state of the hydrogen atom, by use of Eqs. (9) and (13).

CHAPTER VI

ELECTRON SPIN, ANGULAR MOMENTUM, AND MAGNETIC MOMENT

6.1. Electron Spin.—In the preceding chapter, it has been shown that many of the characteristics of the hydrogen spectrum can be explained on the basis of the assumption that the state of each electron is determined by three quantum numbers n, l, and m_l. This holds also for the spectrum of many other atoms, though the complications naturally are much greater. But many of the features of the spectra of the more complex atoms, and the effect of a magnetic field on hydrogen itself, have made it necessary to assume that the electron has a fourth quantum number. This fourth quantum number is the spin quantum number.[1]

It is assumed, in short, that the electron is not simply a point charge, but that (to speak very loosely) it has a certain extension in space and is spinning with a constant angular momentum. It is not the actual angular momentum, which is of greatest importance for the applications to be considered, but its projection along some particular but arbitrary line in space (for our present purpose, we may take this line in space as being the same axis with respect to which the orbital angular momentum p_l is oriented); this projection is loosely referred to as the "spin angular momentum." The spin quantum number s takes the values $\frac{1}{2}$ or $-\frac{1}{2}$ so that the angular momentum is $\pm\frac{1}{2}h/2\pi$. The same direction is taken as positive as in the case of the projection p_{m_l} of the orbital angular momentum. It is of interest to note that the two possible values of the projection of the angular momentum differ by $h/2\pi$, just as do the angular momenta corresponding to two neighboring values of m_l (say m_l and $m_l + 1$) in the orbital case, and that these are the projections

[1] UHLENBECK and GOUDSMIT, *Naturwiss.*, **13**, 953 (1925); *Nature*, **117**, 264 (1926); BICHOWSKY and UREY, *Proc. Nat. Acad. Sci.*, **12**, 80 (1926); PAULI, *Zeits. Physik*, **43**, 601 (1927).

which, by analogy with the orbital case, would be expected to result from a total angular momentum of $\sqrt{\frac{1}{2}\left(\frac{1}{2}+1\right)}\frac{h}{2\pi} = \sqrt{\frac{3}{4}}\frac{h}{2\pi}$.

From our point of view, the chief justification of these rather elaborate assumptions is that they make it possible to explain a multitude of facts in spectroscopy, though they have been given a theoretical foundation.[1] We cannot go into these spectroscopic and theoretical details, but shall see later in the development of the periodic system of the elements how useful this concept of the spinning electron is.

6.2. The Magnetic Moment of Spinning and Rotating Electrons.—As remarked in the preceding paragraphs, the spinning of the electron manifests itself experimentally in large part through its magnetic effects. Although it is beyond the scope of this book to go into the details of the effect of a magnetic field on the spectra of atoms, it will be of interest to consider some of the elementary magnetic properties of both the orbital motion of the electrons and their spin. For this purpose, a discussion based on classical theory is sufficient; quantum theory yields the same results.

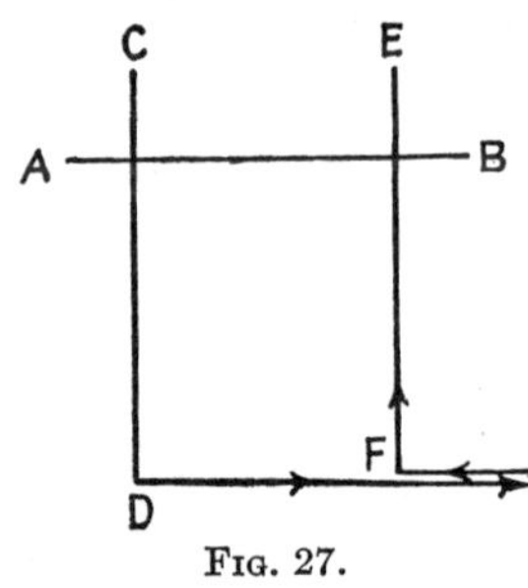

FIG. 27.

As seen in Chap. II, if a wire carrying an electric current is placed in a magnetic field, the field exerts on the wire a force f per unit length of wire, which is perpendicular to the wire and the magnetic field, and which is given by the equation

$$f = \frac{iH}{c}\sin\theta, \tag{1}$$

where i is the magnitude of the current, H the strength of the magnetic field, θ the angle between the direction of H and that of the wire, and c the velocity of light.

Consider a circuit such as shown in Fig. 27, with the movable wire AB placed across the fixed wires CD and EF and a current i flowing through the circuit, and suppose there is a magnetic field of strength H directed perpendicular to the plane of the paper.

[1] DIRAC, *Proc. Roy. Soc.*, **A117**, 610 (1927); **A118**, 351 (1928).

Under these circumstances, there results a force on AB equal to iHl/c, where l is the length of AB from the wire CD to the wire EF. This force will be in the plane of the paper and perpendicular to AB, and we shall suppose that i and H are so directed that the force opposes motion of AB from CE to DF. Then the amount of work done in moving AB from CE to DF is equal to the force times the distance moved, which is $(iHl/c)w$, where w is the distance from CE to DF. But lw is the area A of the circuit which is thus opened out, and the work done is equal to iHA/c.

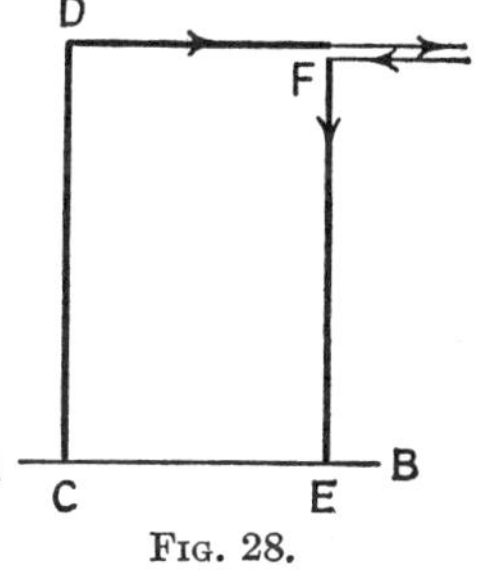

Fig. 28.

Suppose now, with the wire AB in the position CE, the whole circuit to be turned over bodily to the position shown in Fig. 28. Work is done on the circuit in this process on account of the effect of the magnetic field. The same amount of work is done, however (since the final and initial conditions of the circuit are the same) if, starting with the circuit as in Fig. 27, (1) the wire is brought from CE to DF, (2) the circuit turned over, and (3) the wire brought from DF to CE, so that the situation is as shown in Fig. 28. In process (1), the work done is equal to iHA/c; in (2), no work is done, since the electrical circuit now consists of two wires between D and F, practically coincident in position and carrying current in opposite directions; in process (3), since the wire is actually moving in the same direction as in process (1), work is again done on the wire, in the same amount as in (1). The total amount of work done is therefore $2iHA/c$.

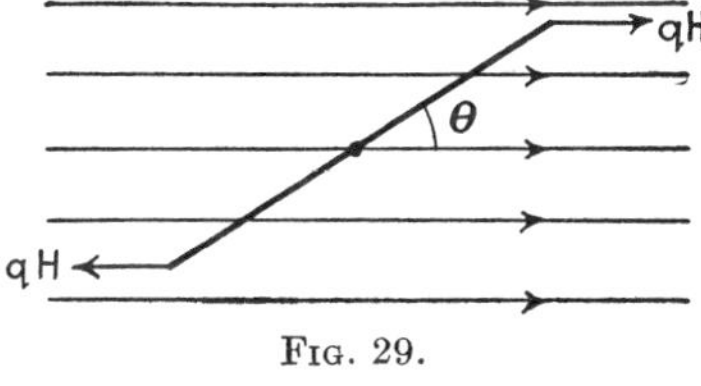

Fig. 29.

Let us compare this with the amount of work necessary to turn over a bar magnet in a magnetic field. Let the magnet have poles of strength q, separated by a distance δ. Suppose for simplicity and definiteness that the magnet is pivoted at its center, though this is not really necessary and is not to be taken as indicating any lack of generality in our considerations. If the magnet is oriented at an angle θ with respect to the magnetic field, as shown in Fig. 29, then there is a force equal to qH exerted on each pole, as indicated in the figure. The force on one of the poles is in the direction of the field, that on the other is in the opposite direction. These two forces may be balanced, assuming the magnet remains rigidly pivoted at the center, by a force g, exerted at one of the poles, directed along the circumference of the circle described by the poles of the magnet and equal to $2qH \sin \theta$. If θ is equal to zero when the magnet is in its normal position of equilibrium (as is the case with the forces in the directions shown in the figure), then work must be done on the magnet in order to increase θ. The work done by the force g on the magnet in moving it through an angle $d\theta$ is equal to the force times the distance moved, the latter being equal to $\frac{1}{2}\delta\, d\theta$. The total amount of work done on the magnet when it is turned through an angle of 180°, to the position in which it is parallel, but opposed, to the field, from the position in which it is normally at rest, is thus

$$\int_0^\pi qH\delta \sin \theta \; d\theta = 2qH\delta = 2\mu H,$$

where $\mu = q\delta$ is the so-called magnetic moment of the magnet. If this is compared with the amount of work necessary to turn over the electric circuit in a magnetic field, it is seen that the latter behaves in a magnetic field as though it has rigidly attached to it a bar magnet with its long axis perpendicular to the plane of the circuit and with magnetic moment equal to iA/c.

For the purpose of investigating its magnetic behavior, an electron revolving about in an orbit may be considered as equivalent to a little closed circuit. If the frequency of revolution of the electron in its orbit is ν_0, the current is given (disregarding sign) by $e\nu_0$, since this is the amount of electricity that passes a given point in the orbit per unit time. The electron moving about in its orbit thus has associated with it a magnetic moment μ equal to $e\nu_0 A/c$, where A is the area included in the orbit. But $\nu_0 A$ is just the area described by the radius vector of the electron per unit time, and is, thus, as shown in Appendix I, equal to $\frac{1}{2}p_l/m$ where p_l is the total angular momentum and m the mass of the electron. Therefore, still disregarding sign,

$$\mu = p_l \frac{e}{2mc} = \sqrt{l(l+1)} \frac{h}{2\pi} \frac{e}{2mc}, \tag{2}$$

p_l being given by Eq. (11) of Chapter V.

The quantity μ is a vector quantity directed perpendicularly to the plane of the orbit of the electron. Its projection on some axis will bear the same relation to the total magnetic moment as the corresponding projection of the angular momentum bears to the total angular momentum. Thus we may write

$$\mu_{m_l} = p_{m_l}\frac{e}{2mc} = \frac{m_l h}{2\pi}\frac{e}{2mc}. \tag{3}$$

The fundamental magnetic moment $he/4\pi mc$ is known as the "Bohr magneton."

This equation gives the magnetic moment due to the orbital motion only. In the case of spin, the spectroscopic facts show that the ratio of the magnetic moment to the angular momentum is twice as great as this, and if we let p_s be the spin angular momentum about the same axis, we may write (disregarding sign)

$$\mu_s = p_s\frac{e}{mc} = \frac{1}{2}\frac{h}{2\pi}\frac{e}{mc}. \tag{4}$$

In the preceding discussion, we have adhered to the assumption that the orbital motion and the spin are oriented independently of each other with respect to the direction of the magnetic field. This is what happens when the magnetic field is sufficiently strong. In a weak magnetic field, the spin is generally oriented by the magnetic field created by the orbital angular momentum in the direction defined by this field, and then the resulting spin-orbit combination is oriented by the external field. In an atom with several electrons, the orbits are usually oriented first with respect to each other, giving a resultant orbital magnetic moment; similarly, there is a resultant spin moment which is oriented with respect to the resultant magnetic field due to the orbital motion, the whole finally being oriented with respect to the external field. This is known as Russell-Saunders coupling, and is an approximate description of the state of affairs in many atoms, giving a fairly good account of their spectra. For our purposes, we are most interested in the number of states with similar energies, and this number may be obtained without the details of the interactions between the various orbital motions and spins in a complicated atom.

6.3. The Stern-Gerlach Experiment.—The fact that the different orientations of the orbit are in general associated with different values of the projection of the magnetic moment along a specified direction in space provides a means by which the atoms having various values of this projection may be separated. A method for effecting such a separation was first invented by Stern and Gerlach. It consists essentially in allowing a beam of atoms to pass down a nonhomogeneous magnetic field, *i.e.*, a magnetic field that varies from point to point in space. The nature of the action of such a field may readily be understood by considering the force it exerts on a small bar magnet. Suppose, for simplicity, that the direction of the field is along the z-axis and that the field itself is a function of z. Suppose further that the bar magnet has its long axis at an angle θ to the z-axis. Let the field at one of the poles of the magnet, say the one with the smaller value of z, be H; the field will exert a force of qH, where q is the pole strength, on this pole. The field at the other end of the magnet will be given very closely by $H + \frac{dH}{dz}\delta \cos \theta$, where δ is the length of the magnet, $\delta \cos \theta$ being the distance *along* the field from one pole to the other, and the force on this pole will be in the opposite direction and equal to $q\left(H + \frac{dH}{dz}\delta \cos \theta\right)$. The net force on the magnet will be given by the difference of these forces, which is equal to $q\delta\frac{dH}{dz} \cos \theta$ or $\mu \frac{dH}{dz} \cos \theta$. It is thus seen that the magnet will be propelled in one direction or the other by a force varying with its magnetic moment μ, the degree of nonuniformity of the field, and the relative orientation of the bar magnet and the field. An atom with a magnetic moment will behave just as the equivalent bar magnet in such a field.

Stern and Gerlach have designed a magnet, which gives a nonhomogeneous field, whose pole pieces are as shown in Fig. 30. Figure 30a gives a cross-sectional view, and Fig. 30b gives a perspective view. The magnetic field is very strong just at the tip of the triangular pole piece and grows weaker away from this neighborhood, the magnetic lines of force being somewhat as indicated in Fig. 30a. A beam of atoms is allowed to pass along the triangular pole piece (in vacuum to avoid molecular collisions), as indicated in Fig. 30b. If z measures the distance in the vertical direction, it is seen that in the region where the beam goes the field is a function of z only, and it will separate out the atoms whose z-components of magnetic moment are different, *i.e.*, it will separate out the atoms whose orbits are oriented in different ways with respect to this particular z-direction. Of course, the magnetic moment due to spin also has its effect and tends to complicate the situation, especially when the interactions between spin and orbit are taken into account. In Fig. 30b, the paths indicated are illustrative of what would happen if the atoms passing down the field were hydrogen atoms in their lowest state. Here there is no orbital angular momentum, and therefore no corresponding magnetic momentum. There are, however, the two possible orientations of the spin of the electron, in which the equivalent bar magnets are oppositely placed. One of them is thus displaced in one direction and the other in the other direction, resulting in two beams of hydrogen atoms, in one of which the electron spins are in one direction and in the other

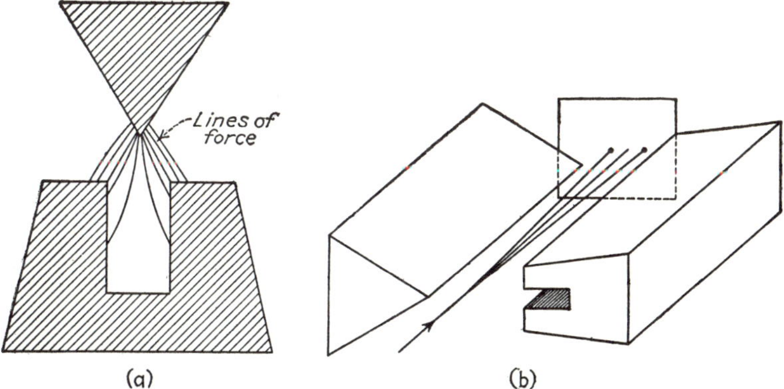

Fig. 30.—The Stern-Gerlach experimental arrangement. The term *vertical direction*, used in the text, refers to Fig. 30a. It is the direction of inhomogeneity of the magnetic field and actually appears horizontal in Fig. 30b. (*From Ruark and Urey, "Atoms, Molecules, and Quanta."*)

of which the spins are opposite. The central line is the undeflected path which would be taken if there were no magnetic field.

It is seen that in the preceding experiment the direction along which the projection of the angular momentum is quantized is automatically determined; the z-direction may be said to be experimentally defined. Once the experiment is set up, this direction is no longer arbitrary, but the way in which we choose to set it up is of course as arbitrary as ever. It is instructive to consider what happens if we "change our minds" as to what the z-direction should be. Suppose that after passing through the magnet shown in Fig. 30 the beam of hydrogen atoms is allowed to pass through another magnet, placed at an angle, as shown in Fig. 31, but with its long axis still in the same direction. Between the two magnets, let a metal plate with a small hole in it be so placed that only one of the separated atomic beams emerging from the first magnet can enter the second magnet. Suppose further that the change from the magnetic field of the first magnet is very sudden—that there is no gradual transition from one magnet to the other. If this condition is fulfilled, then the beam will again be analyzed into different components in the second magnet. That is to say, we requantize with respect to the new z'-axis. The first magnet and the metal plate selected out a definite beam, *i.e.*, one which contains atoms with a certain definite angular-momentum-projection quantum number with

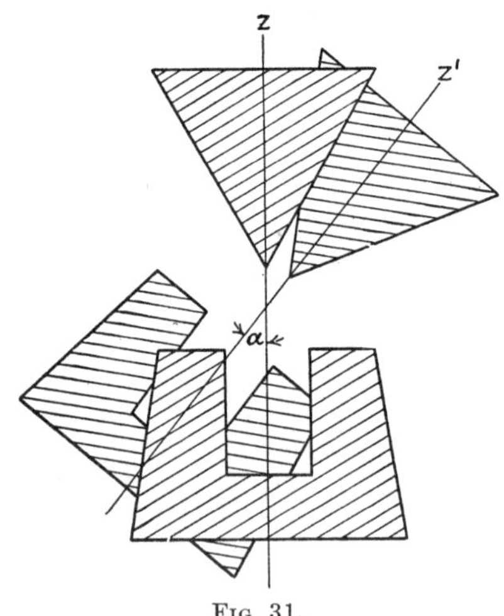

Fig. 31.

respect to the z-axis. But on examining the quantization of this beam with respect to the z'-axis, it is found that there is a certain probability that it have any given quantum number. The probability of any given atom having any particular quantum number on the z'-axis depends upon the angle α which defines the inclination of the one axis with respect to the other. In particular, if $\alpha = 0$, quantization along the z'-axis is the same thing as quantization along the z-axis, and the beam goes through unchanged. Likewise, if $\alpha = \pi$, the beam goes through unchanged but now is deflected in the opposite direction. It may be remarked that it is really not at all surprising that defining the projection of the angular momentum for the z-axis will not in general be sufficient to define it exactly for a z'-axis, if it is remembered that, as shown in the last chapter, the actual direction of the orbit in space is by no means *completely* defined when the quantum number giving the projection of the angular momentum is given.

The hypothetical experiment just described should add something to the definiteness of the ideas involved in selecting out a certain z-axis. In making such a selection, we always have, in the back of our minds, some such experiment which actually selects out the atoms with the various quantum numbers. For many purposes, it is not necessary to state the nature of the experiment in detail, and when it is only desired to count up the *number* of quantum states possible, as is often the case, the z-axis can be quite arbitrarily chosen. In other cases, however, the z-axis to be chosen is definitely specified by the circumstances of the problem.

The reanalysis by a second magnet of one component of a beam that has already been analyzed by one magnet has an analogy in the case of polarized light. As is well known, when a pencil of light is allowed to fall normally on a calcite crystal, it is broken up into two beams, one of which, the ordinary beam, goes straight through the crystal and the other of which, the extraordinary beam, is bent to one side. These beams are polarized at right angles to each other. If one of these beams is selected and allowed to pass through a second calcite whose crystal axes are parallel to the axes of the first crystal, the light passes through unchanged. If the axes of the two crystals are at an angle to each other, however, the beam is analyzed by the second crystal into two beams, in just the way the original unpolarized light would be analyzed by a crystal placed in this position, except that the intensity of the two beams is different depending upon the relative orientation of the two crystals. Although the analogy is by no means exact, the separation of the light polarized in definite directions with respect to the crystal axes is in some respects similar to the separation of the different quantum states of the atom by a magnetic field parallel to some particular line in space. The fact that the light has already been analyzed into its polarized components by one crystal does not prevent its reanalysis *de novo* by a second crystal.

CHAPTER VII

MANY ELECTRON ATOMS AND THE PERIODIC SYSTEM

7.1. Atoms with Many Electrons.—The hydrogen atom was considered in some detail in an earlier chapter, and later on we shall make a fairly extended study of the next simplest atom, the helium atom. In the present chapter, the general features of the many-electron atom and the rules that govern the electronic structure of these atoms will be discussed.

It is convenient in treating atoms having many electrons to consider the electrons to be added to the nucleus one by one. The first electron is thus attracted by a single center of force, and the problem is the same as the problem of the hydrogen atom, except that the potential energy between the electron and the nucleus is given by the expression $-\frac{Ze^2}{r}$, where Ze is the charge on the nucleus, e being the charge on the electron, with the result that wherever e^2 occurs in the formulas for the hydrogen atom it is replaced by Ze^2.

When the second electron is added, the situation is somewhat more complex, as it is necessary to take into account the forces between the two electrons. However, as a first approximation, if the first electron is in its lowest energy state and hence is moving very rapidly, it may be treated as a spherically symmetrical smear of negative electricity, surrounding the positive nucleus. Then the motion of the added electron is handled by assuming that it moves in the centrally symmetric but non-Coulomb potential field[1] produced by the nucleus together with the distribution of electricity from the first electron. As electrons are added, one by one, we treat all those which have been added previously as part of the spherically symmetrical smear. Now it is well known that a spherically symmetrical distribution of electricity exerts the same force on an electrically charged

[1] By the term "non-Coulomb potential field" we mean one in which the potential does not vary inversely as the distance from a fixed point.

body completely outside of it as if all the electricity in the distribution were concentrated at the center of symmetry. Thus, if the added electron is at a great distance from the nucleus and completely beyond the orbits of the electrons already added, the latter simply act as a partial screen for the charge of the nucleus, so that the whole distribution exerts the same force as a point charge $(Z - N)e$, where N is the number of electrons already in their orbits about the nucleus. If the last electron penetrates within the smear of other electrons, then these exert a smaller shielding effect. That part of a spherically symmetrical distribution which is at a greater distance from the center of symmetry than a given point charge exerts no force on the latter; so if the electron under consideration penetrates completely inside the others, it will experience the full force of the nuclear attraction.

For any electron that moves under the influence of a centrally symmetrical field, no matter what the law of force may be, the angular momentum and its projection are determined by the quantum numbers l and m_l. The proof, given in Appendix I, of the constancy of the angular momentum of a body attracted to a center of force does not make any assumptions as to the law of force except that it be centrally directed. The angular momentum, being constant, can be quantized in exactly the same way as before, with the same results. The only change comes in calculating the energy levels. If the electron did not penetrate into the cloud of electrons surrounding the nucleus at all, it would be always moving in the field of an effective charge of $(Z - N)e$ and its energy would be given by[1]

$$-E = \frac{2\pi^2 m(Z - N)^2 e^4}{n^2 h^2}. \tag{1}$$

If the electron does penetrate, however (as in general it will, at least to a small extent), the energy corresponding to any given value of the quantum number n may be expected to be lowered below the value given by Eq. (1) (*i.e.*, E will have a greater nega-

[1] In Eq. (13) of Chap. V, it is possible to show that the factor e^4 is really to be considered as a product, $e^2 \times e^2$, one factor of which comes from the charge on the electron, the other from the charge on the nucleus. In the hydrogen atom, both of these happen to be equal. When the charge on the nucleus has a value different from e, one of the factors is changed accordingly.

tive value) because the electron is part of the time under the influence of a stronger attractive force. The greater the amount of penetration, the more the energy will be lowered. With a given value of n, the penetration will be the more marked the smaller the value of l. This is because the small values of l correspond to the orbits of low angular momentum. In the case of the hydrogen atom, where the orbits are elliptical, it will be remembered that although the length of the major axis of the ellipse is fixed by the quantum number n, the minor axis depends

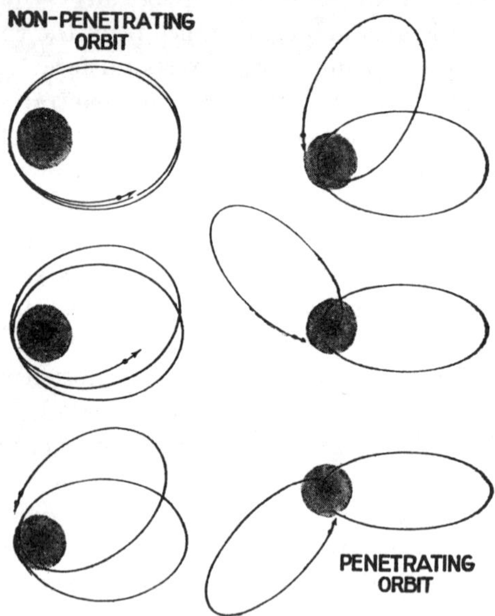

FIG. 32.—Showing the change in electron orbits with increasing penetration. (*From White, "Introduction to Atomic Spectra."*)

on l and is small if l is small. In an ellipse of large major and small minor axis, the foci of the ellipse are of course near the ends of the major axis. Since the nucleus is at one of the foci, this means that with the length of the major axis (determined by n) given, the electron comes nearer the nucleus, the smaller the minor axis, *i.e.*, the smaller l. This means greater penetration for an *s*-orbit than for a *p*-orbit, for example. This remains true in spite of the fact that, in the case of an electron moving in a centrally symmetric but non-Coulomb field, the orbit is not an ellipse, but instead becomes a rosette, the path of the electron being bent in sharply by the large force exerted when the electron is close to the nucleus. Portions of some orbits of this sort are illustrated in Fig. 32, which shows the transition between a deeply penetrating orbit and one that penetrates only slightly. The orbits may be approximately represented as ellipses with rotating major axes.

It will, of course, be understood that in speaking so definitely of the path of the electron we are here, as always, making only an approximate statement of the true state of affairs. Actually, of course, to be strictly correct, the phenomena should be described in terms of the wave theory. The wave picture, however, will show that an electron in a penetrating state will have a relatively large probability of being near the nucleus, and so "penetrates" in the wave theory also. This should be clear qualitatively from the discussion of the wave functions of hydrogen for various values of l, given in Sec. 5.7.

A more exact understanding of the effect of penetration of the electron's orbit on the total energy may be obtained from a consideration of Eq. (17) of Chap. V. Suppose we set the potential energy of the electron in the centrally symmetrical but non-Coulomb field equal to V, a function of r which is equal to $-\frac{(Z - N)e^2}{r}$ at great values of r and to $-\frac{Ze^2}{r}$ at small values of r. This quantity V, then, takes the place of the factor $-e^2/r$ in the last term under the radical in Eq. (17) [originally introduced from Eq. (15)], and Eq. (17) becomes

$$\int p_r\,dr = 2\int_{a-q}^{a+q} \sqrt{2mE - \frac{l(l+1)h^2}{4\pi^2 r^2} - 2mV}\,dr. \tag{2}$$

In this integral, $a + q$ is the farthest distance of the electron from the nucleus in its rosette orbit and $a - q$ the smallest distance, just as was the case with the elliptic orbits of the hydrogen atom. It will be understood that substituting V for $-e^2/r$ does not alter any of the relations which do not involve the potential energy, nor does it change those which do involve it very profoundly. Thus p_r, the integrand of Eq. (2), goes through its zero values when $r = a + q$ and $r = a - q$ as in Chap. V.

If V were equal to $-\frac{(Z - N)e^2}{r}$ for all values of r, then Eq. (2) would give just the value of E given by Eq. (1). Were V equal to $-\frac{Ze^2}{r}$ throughout the range of r, we should get the value of E given by setting $N = 0$ in Eq. (1). If V changes, E must have such a value that $\int p_r\,dr$ retains its value, independently of the form of V (it will continue to be at least approximately true that $\int p_r\,dr$ is always given by Eq. (19) of Chap. V and so depends only on n and l). If V varies between $-\frac{(Z-N)e^2}{r}$ and $-\frac{Ze^2}{r}$, then it is obvious that, in order for $\int p_r\,dr$ to have the proper value for the given l and n, the value of E must lie between the corresponding values $-\frac{2\pi^2 m(Z-N)^2 e^4}{n^2h^2}$ and $-\frac{2\pi^2 mZ^2e^4}{n^2h^2}$, and so will be smaller (have a greater negative value) than the first of these. Furthermore, the larger l is, the greater is the lower limit $a - q$ at which the integrand becomes zero. This is obvious from inspection of Eq. (2), since increasing l increases the second term in the square root, and this term is negative. The quantity under the square root is, of course, positive as a whole. Of course, the region for which r is less than $a - q$ contributes nothing to the integral; since making l greater increases $a - q$, it is clear that it will cut out a contribution to the integral from a region in which V deviates from $-\frac{(Z-N)e^2}{r}$. Hence the greater l, the less the energy will deviate from the value given by Eq. (1), and the less the effect of penetration.

7.2. Pauli's Exclusion Principle.—Since the amount of energy available per atom at ordinary temperatures is small[1] compared with the difference in energy between a quantum state for which $n = 1$ and one for which $n = 2$, it would be natural to expect that all the electrons in any given atom would go to a quantum state for which $n = 1$. This, however, turns out not to be the case. On the basis of certain spectroscopic evidence, Pauli enunciated the following rule, known as the "Pauli exclusion principle," which seems to be of universal validity: *In any atomic or molecular system, no two electrons can have all four quantum numbers the same.* The basis of this rule is purely empirical. Pauli, as stated, originally derived it from spectroscopic considerations, but it has a profound effect on all properties of atoms, and its verification may be said in a sense to rest on the entire experience of mankind, for it would be a profoundly different world if Pauli's principle did not hold. It will appear, in the following account, how the properties of the chem-

[1] More precisely expressed, we may say that the energy difference ΔE between quantum states for which $n = 2$ and those for which $n = 1$ is large compared with kT at ordinary temperatures, where k is the Boltzmann constant (gas constant divided by Avogadro's number); therefore, the relative probability $\left(= e^{-\frac{\Delta E}{kT}}\right)$ that the state for which $n = 2$ will be occupied is very small. See Appendix II, pp. 457*f*. For purposes of orientation as to orders of magnitude, it may be said that at room temperature kT is about 0.025 electron volts.

ical elements depend upon it and, following Bohr, Stoner, and Main Smith, how the periodic law may be understood.

When we attempt to build up an atomic structure by adding electrons to the nucleus one by one, we see, in the light of Pauli's principle, that the energy levels must be filled up in the order of their energies. There is not, in general, sufficient energy available at ordinary temperatures to excite an electron from one energy level to another, but if any given energy level is occupied, it is not possible to put another electron into it. The state of the last electron added to the nucleus can be described by the same four quantum numbers n, l, m_l, and s as if the other electrons were not there, though the energy of this state may be changed, as we have seen, by the presence of the other electrons.

7.3. The helium atom is, with the exception of hydrogen, the simplest of all atoms, as it contains only two electrons, neutralizing the charge of $2e$ on the nucleus. The first electron will go into a $1s$-state, as in hydrogen, such a state having the lowest energy. There are, however, two $1s$-states, for there are the two possible directions of the spin, which will always, of course, simply double the number of quantum states shown in the table of Sec. 5.4. The structure of the helium atom may therefore be written as $1s^2$, where the superscript indicates the number of electrons in the $1s$-state.

It will be of interest to consider the energy of the helium atom in its lowest state. The energy necessary to remove the second electron (*i.e.*, the last one added, therefore the first one to be removed) is 24.5 volts, which lies between the energy 13.5 volts necessary to remove an electron from hydrogen and the energy[1] 54.1 volts necessary to remove the first (remaining) electron. That the value is thus intermediate is to be expected. If the first electron completely shielded the second one from half of the charge of the nucleus, the energy to remove the second one should be just the same as that necessary to remove an electron from hydrogen, whereas if there were no shielding, it should require as much energy to remove the second electron as the first one. There is one thing which should be noted, however; namely, the two electrons, once they are both in $1s$-states in

[1] These are the older values. The value given in "An Outline of Atomic Physics" (see footnote 1, p. 15), for the removal of an electron from hydrogen, based on the more recent values of the physical constants, is 13.60 volts.

helium, are exactly alike and may really be considered to shield each other, which does not appear very clearly from our approximate picture of one electron rotating in the field of the nucleus and all the other electrons. Whichever electron is taken away first will require 24.5 volts for its removal, and the one that is left will require 54.1. As a matter of fact, this is the worst case for our approximate method of treatment, for this is the only atom in which all the electrons are in the same state. The helium atom will be considered again in some detail in Chap. X.

7.4. The First Row of the Periodic System.—There cannot be more than two electrons in the $1s$-level, so that lithium, which has three electrons, must have one in a higher energy level. This third electron will naturally seek the lowest energy available to it; this will be a level with $n = 2$. On account of the penetration effect, an s-electron has a lower energy than a p-electron. The third electron in lithium will therefore go into a $2s$-state, and we may designate the structure of the lithium atom by the symbol $1s^2\ 2s^1$.

Beryllium has four electrons. As there can be two electrons in $2s$-states, as in $1s$-states, its structure will be $1s^2\ 2s^2$.

In the case of boron, with five electrons, one electron will again have to be placed in a higher energy level, as the $1s$- and $2s$-levels are filled by two electrons in each. The next lowest energy level is a $2p$-energy level. Boron therefore has the structure $1s^2\ 2s^2\ 2p^1$. From the number of possible orientations (the number of possible values for the projection of the angular momentum) of an orbit for which $l = 1$, and from the fact that there are always two possible values of the spin quantum number, it is seen that there may be six $2p$-electrons (compare the table in Sec. 5.4.). The electronic structures of the next five elements are therefore as follows:

C	$1s^2\ 2s^2\ 2p^2$
N	$1s^2\ 2s^2\ 2p^3$
O	$1s^2\ 2s^2\ 2p^4$
F	$1s^2\ 2s^2\ 2p^5$
Ne	$1s^2\ 2s^2\ 2p^6$

With the rare gas neon, the shell for which $n = 2$ is filled.

7.5. A Remark on Notation.—In the last sentence, we have used the word "shell." The group of quantum states for which n has a definite value is called a shell. The shells are often designated by giving the value of n, or by means of letters, beginning with K and proceeding alphabetically. Thus the $n = 1$ shell is also called the K-shell, the $n = 2$ shell is called the L-shell, etc. A group of quantum states for which the values of both n and l are fixed is called a "subshell." However, when no confusion can arise, we shall often call a subshell a shell, and also speak, for example, of the 2*s*-shell.

For reasons that will presently become obvious, the electrons in the outermost shell and sometimes some electrons in the next outermost shell are called "valence electrons," provided these shells are not completely filled. For example, nitrogen has five valence electrons, while copper is sometimes said to have two valence electrons. (For electron structure of copper see below.)

7.6. The Remaining Rows of the Periodic Table.—The structures of all the elements are shown in Table 1, which gives the number of electrons of each kind, when the element is in its normal state. In the second row of the periodic table, the 3*s*- and 3*p*-levels are filling up, as shown.

TABLE 1.—PERIODIC TABLE AND OUTER ELECTRONS OF THE ELEMENTS

	H																	**He**
1*s*	1																	2
	Li	**Be**											**B**	**C**	**N**	**O**	**F**	**Ne**
2*s*	1	2											2	2	2	2	2	2
2*p*													1	2	3	4	5	6
	Na	**Mg**											**Al**	**Si**	**P**	**S**	**Cl**	**A**
3*s*	1	2											2	2	2	2	2	2
3*p*													1	2	3	4	5	6
	K	**Ca**	**Sc**	**Ti**	**V**	**Cr**	**Mn**	**Fe**	**Co**	**Ni**	**Cu**	**Zn**	**Ga**	**Ge**	**As**	**Se**	**Br**	**Kr**
3*d*			1	2	3	5	5	6	7	8	10	10	10	10	10	10	10	10
4*s*	1	2	2	2	2	1	2	2	2	2	1	2	2	2	2	2	2	2
4*p*													1	2	3	4	5	6
	Rb	**Sr**	**Y**	**Zr**	**Cb**	**Mo**	**Ma**	**Ru**	**Rh**	**Pd**	**Ag**	**Cd**	**In**	**Sn**	**Sb**	**Te**	**I**	**Xe**
4*d*			1	2	4	5	(5)	7	8	10	10	10	10	10	10	10	10	10
5*s*	1	2	2	2	1	1	(2)	1	1	0	1	2	2	2	2	2	2	2
5*p*													1	2	3	4	5	6
	Cs	**Ba**	**La**[a]	**Hf**	**Ta**	**W**	**Re**	**Os**	**Ir**	**Pt**	**Au**	**Hg**	**Tl**	**Pb**	**Bi**	**Po**	—	**Rn**
4*f*				14	14	14	14	14	14	14	14	14	14	14	14	14	14	14
5*d*			1	2	3	4	5	6	7	9	10	10	10	10	10	10	10	10
6*s*	1	2	2	2	2	2	2	2	2	1	1	2	2	2	2	2	2	2
6*p*													1	2	3	4	5	6
	—	**Ra**	**Ac**	**Th**	**Pa**	**U**												
6*d*			(1)	(2)	(3)	(4)												
7*s*	1	2	(2)	(2)	(2)	(2)												

Data from Horsberg, "Atomspektren und Atomstruktur," Theodor Steinkopff Verlag, Dresden and Leipzig, 1936.

Numbers in parentheses are doubtful.

[a] The rare earths come between La and Hf. They all have 0 or 1 5*d*-electrons, 2 6*s*-electrons, no 6*p*-electrons, and 1 to 14 4*f*-electrons.

When $n = 3$, it is possible for the first time to have $l = 2$, *i.e.*, d-states. It might now be expected that after the $3p$-states those of next lowest energy would be the $3d$-states. However, an s-electron penetrates into the inner shells so much more than a d-electron that the lowering of energy due to this cause more than counterbalances the increase of energy caused by going from $n = 3$ to $n = 4$, the result being that the $4s$-levels are lower in energy than the $3d$. Potassium therefore has the structure $1s^2\,2s^2\,2p^6\,3s^2\,3p^6\,4s^1$, and calcium has the structure $1s^2\,2s^2\,2p^6\,3s^2\,3p^6\,4s^2$. The energy of a $3d$-level, however, is lower than that of a $4p$-level, so the $3d$-levels are filled up before the $4p$-levels. The electronic structures of the rest of the elements in this row of the periodic table may be seen by reference to Table 1. The $3d$-level fills up regularly from scandium to nickel. In chromium and copper, there are slight irregularities. This must mean that under some circumstances the $3d$-level to be filled has a lower energy than the $4s$-level. It is not surprising that irregularities like this should occur, as the relative energies of various levels must be affected in some measure by the electrons already in the atom, and in any case the difference in energy between $3d$- and $4s$-levels is not great. Besides, there is undoubtedly a certain special stability connected with a completed subshell, and there appears also to be a special stability about a half subshell, as in chromium. The fact that copper has but one $4s$-electron is deduced from its spectrum, as is, indeed, the structure of any atom. The details of the method by which these conclusions are reached are part of the theory of complex spectra and are beyond the scope of this book. We shall see later, however, how the electronic structure is connected with other properties of the element.

The fourth row of the periodic table is built up much like the third row, as will be seen from Table 1. It will be seen that the same type of irregularity occurs as in the third row; in fact, the irregularities are a little more frequent.

With the completion of the fourth row, the $5p$-shell is filled, but the $4f$-orbits are still vacant. The preceding rows of the periodic table would lead one to expect that the $6s$-levels would be filled before the $5d$-levels, and it turns out that the $6s$-levels also have a lower energy than the $4f$. Then in lanthanum, an electron goes into the $5d$-level. In the case of cerium, however, the added electron goes into the $4f$-level, and in the immediately succeeding elements the $4f$-shell fills up. These elements form the group of rare earths, all of which have almost identical chemical properties. After the $4f$-shell is complete, the $5d$-levels fill up, with some irregularity, as seen in Table 1. In this section of the periodic system, the available energy levels are very close together, and small perturbations are apt to change their order; we have seen, for example, that the addition of one electron to the $5d$-shell raises the energy of the rest of the levels of that shell sufficiently so that the $4f$-levels become lower, and then the $4f$-shell fills up before the $5d$.

The periodic system as we know it ends with the filling up of the $7s$-level and part of the $6d$-level. In none of these elements is there a $5f$-electron in the normal state.

The order in which the electrons enter in the periodic table may be summarized as follows: $1s$, $2s$, $2p$, $3s$, $3p$ ($4s$, $3d$), $4p$ ($5s$, $4d$), $5p$ ($6s$, $5d$, $4f$), $6p$, $7s$, $6d$. The energy levels in parentheses have about the same energy and change their order in different elements. The states within the parentheses get their first electrons in the order in which they are written, but they are finally filled up in just the reverse order.

7.7. Recapitulation.—In the first chapter, we briefly reviewed the work of the early chemists, their discovery of the laws of combination of the chemical elements, and the determination of atomic weights. It was seen how the study of the properties of the elements and the measurement of their atomic weights led to an arrangement of the elements, practically in the order of atomic weight, which gave also a description of their properties.

The study of matter was then taken up from another point of view, namely, we inquired how it is built up from its fundamental constituents. The properties of electrons, nuclei, and atoms were considered, and it was noted that the properties of matter may be deduced from atomic spectra. The simplest of atoms, the hydrogen atom, which contains one electron bound to a relatively heavy nucleus with equal and opposite charge, was then considered in some detail. Passing to more complicated structures which contain many electrons, we saw how, by relatively simple assumptions, a series of electronic structures capable of explaining the spectra of the known elements can be built up. This leads to an arrangement of the elements that corresponds exactly to the arrangement based on their chemical properties. Even without a detailed consideration of the way the chemical properties may be explained on the basis of the electronic structure, this happy correlation greatly strengthens our belief both in the correctness of the chemical findings, in particular the atomic weights found by the chemists, and in the essential correctness of the physical theory which has led to this result. This theory may therefore be used with confidence as a foundation for the discussions to follow.

CHAPTER VIII

SOME PROPERTIES OF THE ELEMENTS AND THEIR CONNECTION WITH ELECTRON STRUCTURE

8.1. Effects of Penetration of Electron Orbits into Underlying Shells.—In the preceding chapter, we have discussed the effect of the electrons already in the atom on the motion of a new electron and have seen how the fact that an electron can penetrate into the inside shells affects its energy. It may not be amiss, for the sake of orientation as to the magnitudes of the quantities involved and their relation to the periodic arrangement of the elements, to consider this matter in a little more detail in the case of those elements which have a single valence electron, namely, the alkali metals and the copper group. Such elements resemble the hydrogen atom, and if there were no penetration of the valence electron, its energy would be given by the hydrogen formula

$$-E = \frac{2\pi^2 me^4}{n^2h^2} \tag{1}$$

Of course, it is understood that n in this formula cannot take on a value already claimed by electrons in one of the closed subshells. Thus in cesium, if the outer electron is in an s-state, the value of n could not be less than 6, but if it should be in a d-state, for example, it could have a value 5 since only the $3d$- and $4d$-states are occupied by inner electrons.

If there is penetration of the outer electron, then as seen in the last chapter, Eq. (1) does not hold; it has been customary to write

$$-E = \frac{2\pi^2 me^4}{n^{*2}h^2} \tag{2}$$

where n^*, the so-called effective quantum number, is the value it is necessary to insert into the equation in place of the true total quantum number n in order to get the actual value of the energy. A comparison of n^* and n will then give a measure of the effect of penetration. In Table 2 are given the values of n^* for the lowest *s*-, *p*-, *d*-, and *f*-states of the alkali metals and copper, silver, and

TABLE 2.—VALUES OF n^*

	s	*p*	*d*	*f*
Li	1.59 (2)	1.96 (2)	3.00 (3)	4.00 (4)
Na	1.63 (3)	2.12 (3)	2.99 (3)	4.00 (4)
K	1.77 (4)	2.23 (4)	2.85 (3)	3.99 (4)
Rb	1.80 (5)	2.28 (5)	2.77 (4)	3.99 (4)
Cs	1.87 (6)	2.33 (6)	2.55 (5)	3.98 (4)
Cu	1.33 (4)	1.86 (4)	2.98 (4)	4.00 (4)
Ag	1.34 (5)	1.87 (5)	2.98 (5)	3.99 (4)
Au	1.21 (6)	1.72 (6)	2.98 (6)	

gold.[1] These may be compared with the corresponding values of n (given in parentheses). It will be seen, for example, how much greater the effect is for a deeply penetrating *s*-electron than for a *p*-electron which does not penetrate so deeply, how much greater it is for a heavy atom than for a light one which has not so many electrons in the inner shells for the outer electron to penetrate, and how much greater it is for an atom in which the first shell the outer electron has to penetrate is a shell of 18 rather than a shell of 8.

The difference between n^* and n is almost independent of the latter as the values for sodium, given in Table 3, will indicate.

TABLE 3.—VALUES OF n^* FOR SODIUM

n	*s*	*p*	*d*	*f*
3	1.63	2.12	2.99	
4	2.64	3.13	3.99	4.00
5	3.65	4.14	4.99	5.00
6	4.65	5.14	5.99	6.01

8.2. The ionization potential, *i.e.*, the energy necessary to remove the most loosely bound electron from an atom in its

[1] See HUND, "Linienspektren," pp. 30, 39, Julius Springer, Berlin, 1927; WHITE, "Introduction to Atomic Spectra," pp. 89, 90, McGraw-Hill Book Company, Inc., 1934.

lowest energy level,[1] is one of the most important properties of the atom from the point of view of chemistry. This energy is obtained directly from Eq. (2) if n^* is given the value appropriate to the most loosely bound electron when the atom is in its lowest state. The considerations involving penetration are of utmost importance in determining the ionization potential of an element. The energy necessary to remove the most loosely bound electron in its lowest energy state from the ion that is left after one electron has already been removed is called the ionization potential of the ion or the second ionization potential of the corresponding atom; in like manner, the ionization potential of the

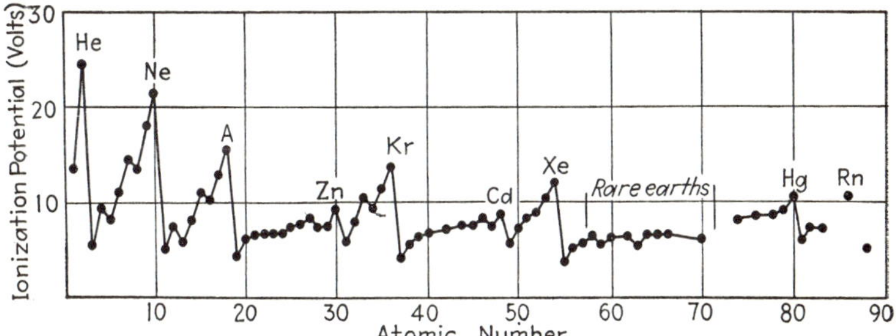

FIG. 33.—First ionization potentials of the elements.

ion formed by removing two electrons is the third ionization potential of the atom, etc. The first, second, and third ionization potentials of the elements as far as they are known are given in Table 4. The first ionization potentials are also shown graphically in Fig. 33.

In looking over the trends of the ionization potentials,[2] we note that along any row of the periodic table the ionization potentials tend to increase from left to right. This general tendency is due to the increasing nuclear charge as the atomic number increases, and to the fact that all the electrons involved in the building up of any row of the table are at roughly similar distances from the nucleus. The result is that the outermost

[1] It will, of course, be understood that this refers to an isolated atom, *i.e.*, an atom in a dilute gas.

[2] When the term "ionization potential" is used without qualification, it means the first ionization potential, or the energy necessary to remove the outermost electron from the atom.

TABLE 4.—IONIZATION POTENTIALS
(In electron volts)

H	13.53		
He	24.47	54.14	
Li	5.36	75.27	121.84
Be	9.28	18.12	153.11
B	8.25	25.00	37.74
C	11.21	24.26	47.64
N	14.47	29.46	47.3
O	13.55	34.94	54.7
F	(18)	34.81	62.35
Ne	21.46	40.91	63.3
Na	5.11	47.07	71.0
Mg	7.61	14.96	79.73
Al	5.96	18.73	28.31
Si	8.10	16.26	33.34
P	11.0	19.7	30.02
S	10.31	23.29	34.88
Cl	12.96	23.4	39.70
A	15.69	27.74	40.73
K	4.32	31.66	47.
Ca	6.09	11.82	50.96
Sc	6.7	12.8	24.62
Ti	6.81	13.6	27.6
V	6.76	14.1	26.4
Cr	6.74	16.6	(27)
Mn	7.40	15.70	(32)
Fe	7.83	16.16	
Co	8.5	17.3	
Ni	7.61	18.2	
Cu	7.68	20.2	29.5
Zn	9.36	17.89	39.5
Ga	5.97	20.43	30.6
Ge	8.09	15.86	34.1
As	10.5	20.1	28.1
Se	9.70	21.3	33.9
Br	(11.5)	19.1	25.7
Kr	13.94	24.47	36.8
Rb	4.16	27.36	(47)
Sr	5.67	10.98	(42.8)
Y	6.5	12.3	20.4
Zr	6.92	13.97	24.00
Cb	.....	.....	24.2
Mo	7.2		
Ru	(7.7)		
Rh	7.7		

TABLE 4.—IONIZATION POTENTIALS.—(*Continued*)

Pd	8.3	19.8	
Ag	7.54	21.7	(32)
Cd	8.96	16.84	38.0
In	5.76	18.80	27.91
Sn	7.30	14.52	30.49
Sb	8.35	(18)	24.7
Te	8.96	.....	30.
I	10.5	19.4	
Xe	12.08	(23)	32.0
Cs	3.87	23.4	(35)
Ba	5.19	9.95	(35.5)
La	5.59	11.38	(20)
Ce	6.54	14.8	
Pr	(5.76)		
Nd	(6.31)		
Sm	(6.55)	11.4	
Eu	5.64	11.4	
Gd	(6.65)		
Tb	(6.74)		
Dy	(6.82)		
Yb	(6.23)	12.05	
Hf	.....	(14.8)	
W	8.1		
Os	8.7		
Pt	8.88		
Au	9.19	19.95	
Hg	10.38	18.66	34.3
Tl	6.07	20.31	29.7
Pb	7.38	14.96	31.91
Bi	7.25	16.60	25.43
Rn	10.70		
Ra	5.25	10.10	

From Landolt-Börnstein, "Tabellen," Julius Springer, Berlin, and Herzberg, "Atomspektren und Atomstruktur," Verlag von Theodor Steinkopff, Dresden and Leipzig, 1936. Values from these sources were averaged, except where a choice appeared indicated.

Doubtful values in parentheses.

See footnote, p. 89.

electron is not very effectively screened or shielded by the other outer electrons and so is affected by the increase in nuclear charge.

On passing from a rare gas structure to the next alkali element, a great decrease in ionization potential occurs. The added *s*-electron, being in a completely new shell, is relatively well shielded from the nuclear charge, in spite of a considerable tendency to penetrate the inner shell. Going from an alkali element to the adjacent alkaline earth element brings a noticeable increase in the ionization potential, for one of a pair of *s*-electrons does not shield the other one very well against the increased nuclear charge of the alkaline earth element as compared with the alkali element.

A *p*-electron outside the 2*s*-subshell is much better shielded, as it does not penetrate so strongly as an *s*-electron, and boron and aluminum have lower ionization potentials than beryllium and magnesium, respectively. However, a *d*-electron outside a completed *s*-subshell is not so well shielded (compare, *e.g.*, scandium with calcium). At first sight this may seem strange as a *d*-electron should penetrate still less than a *p*-electron, but the *d*-electron added in this case has a lower quantum number n than the outer *s*-electrons. It is for this reason that the *d*-level has a low energy and, in fact, it is because of this that the electron enters the *d*-level in the first place.

In addition to those mentioned above there occur other exceptions to the general rule that the ionization potentials increase across the rows of the periodic table. There are rather marked and perhaps rather unexpected irregularities at oxygen and sulfur. Other irregularities, the causes of which are rather obvious from the electron structures of the elements, occur at gallium, indium and thallium.

Turning now to a comparison of the first, second, and third ionization potentials of the same element, we note that the second ionization potential is always much larger than the first and the third always much larger than the second. It is, of course, to be expected that it will be harder to remove an electron from a positively charged ion than from a neutral atom, and the difficulty increases with the charge on the ion. It is to be noticed, however, that there are certain striking differences in this respect between sodium, magnesium, and aluminum, for example. In the case of sodium, after one electron is removed, a rare-gas-like electron configuration is left and it is very difficult to remove the second electron. In the case of magnesium, two electrons must be removed before getting to the rare-gas structure and these two electrons are relatively easy to remove, whereas in the case of aluminum even the third electron is more easily removed than the second electron in sodium. These facts are related to the fact that sodium is univalent in its compounds, calcium bivalent, and aluminum trivalent.

8.3. The electron affinity is defined as the energy that is given out when an electron is added to an atom to form a negative ion. The second electron affinity is the energy liberated when an electron is added to a negative ion to give a doubly charged negative ion. The electron affinities of only a few elements are known, some of these are given in Table 5. They are determined in a way to be discussed later, in Sec. 14.7.

TABLE 5.—ELECTRON AFFINITIES
(In electron volts)

F	4.12
Cl	3.78
Br	3.55
I	3.22
O	−7.2[a]
S	−4.0[a]
Se	−4.6[a]

[a] For the addition of *two* electrons.

The halogens, if they add one electron, give ions with a rare-gas structure. The latter are, therefore, very stable, and the electron affinities of the halogens are correspondingly high. The electron affinities of the elements of the oxygen group are presumably also positive, but all that can be measured is the sum of the first two electron affinities, which is negative in spite of the rare-gas structure of the doubly charged ion thus formed, because the last electron must be added to the singly charged ion against an electrostatic repulsion.

8.4. Electropositivity and Electronegativity of the Elements.—When he uses the term "electropositive" or "electronegative" in describing the characteristics of an element, the chemist refers to a complex array of properties, not all of which parallel each other completely. Nevertheless, in a general way, an element is electropositive if it has a relatively great tendency to lose an electron (or, perhaps better, a relatively small attraction for an electron), and it is electronegative if it has a relatively great electron affinity. It is, in fact, possible to formulate a reasonably good definition of electropositivity or electronegativity in terms of these two properties of the elements; this will be done in Sec. 12.3. In a general way, it can be seen from Tables 4 and 5 that the general trends of electropositivity and electronegativity defined in this way will be that which the chemist has always found. Elements become more electropositive as we go to the left of the periodic table and down; they become more electronegative as we go to the right and up.

8.5. X-ray spectra of the elements are of great historical interest because it was by means of them that Moseley first found a completely objective and unequivocal method of ordering the elements in the periodic table. The X-ray lines he studied are lines of very small wave length emitted by the elements. The wave lengths of these lines do not, on progressing from element to element, show any changes that correspond to the periodic properties of the elements but instead change gradually and uniformly. It is possible to assign to each element an ordinal number, representing the position, in this series, of its X-ray line of a given type, a number that turns out to be equal to the atomic number of the element.[1] Further, the frequency of the X-ray line, if it belongs to the so-called K-series (explained below), is approximately proportional to the square of the atomic number.

These relationships are easily understood in the light of present-day theories. X-ray lines arise from transitions in the inner shells of the atoms. Suppose, for example, a metal is bombarded by high-speed electrons, as is the case in an X-ray tube. It is possible for such a high-speed electron to knock an electron out of the K-shell of the atom. The K-electrons of an atom are those near the nucleus, and as most of the time they are much closer to the nucleus than to most of the other electrons in the atom, the latter have only a relatively small shielding effect. We can get an approximate value for the energy of a K-electron by neglecting the shielding entirely; this yields the expression $-\frac{2\pi^2 m Z^2 e^4}{h^2}$, the quantum number n being 1. Now suppose an L-electron drops into the K-shell, filling up the hole which is there. In this process, an X-ray line is emitted; it belongs to the K-series because the electron drops back into the K-shell. If as a very rough approximation we neglect shielding for the L-electron too, its energy is $-\frac{2\pi^2 m Z^2 e^4}{4h^2}$, since $n = 2$, and the fre-

[1] The term "atomic number" might be defined in this way. We consider it better, however, to define it in terms of the charge on the nucleus.

quency of the X-ray line is $(2\pi^2 m Z^2 e^4/h^3)(1 - \frac{1}{4})$; since the energy of the L-electron contributes only a relatively unimportant term to this expression, the approximation made in neglecting shielding will be sufficiently good. The corrections that should be made to this expression will depend on which one of the L-electrons actually makes the jump, as they are not all alike, but we may neglect these differences and compare always the same X-ray line for the different elements. The frequency will thus be roughly proportional to the square of the atomic number, as actually found. Of course, in the case of hydrogen and helium which have no L-electrons, lines of this sort cannot occur, but certain possible electron transitions correspond pretty closely. Elements that have M-, N-, etc., electrons will also present the possibility of jumps from these higher levels to a hole in the K-shell, and there can also be transitions between the higher shells. Similar relations between frequency and atomic number will hold for all these types of transition provided the comparison is made between corresponding lines of the different elements.

Exercises

1. Explain the breaks in the value of the ionization potential that occur at gallium, indium, and thallium.

2. Rubidium is more electropositive than silver. Explain. Discuss the difference in chemical properties from the point of view of the electronic structure.

3. It is not so easy to remove the valence electron from normal cesium as it is to remove the valence electron of lithium after the latter has been excited to the 6s-level. Explain.

4. Copper has a smaller atomic volume than potassium, the alkali metal in the same row of the periodic system. Explain.

5. Calculate the fourth ionization potential of beryllium.

6. From measurements on crystals, such as described in Chap. XIV, it is found that the size of the rare-earth ions (of the type Ce^{+++}) progressively decreases as the 4*f*-level fills up (the "lanthanum contraction"). Explain.

CHAPTER IX

MOLECULAR POTENTIAL-ENERGY CURVES AND MOLECULAR MOTION

We turn now to the consideration of a problem that is central to the whole of chemistry. It will be the purpose of this chapter to discuss qualitatively, and with the aid of simple examples, the general ideas that are involved in the study of chemical combination, as well as the types of motion that occur in molecules.

9.1. The Formation of Compounds.—Perhaps the simplest type of chemical-compound formation is the formation of polar compounds, and for purposes of illustration, we shall discuss the forces involved in cases of this kind. Let us consider, for example, the formation of a gaseous sodium chloride molecule from a gaseous sodium ion, Na^+, and a gaseous chlorine ion, Cl^-, *i.e.*, the reaction

$$Na^+ + Cl^- \rightarrow NaCl \tag{1}$$

It would, of course, be possible to consider the formation of a molecule of sodium chloride from a neutral sodium atom and a neutral chlorine atom (all being supposed to be in the gaseous state), writing the reaction

$$Na + Cl \rightarrow NaCl \tag{2}$$

However, it will appear later, in Chap. XIV, that this latter reaction is best treated as the sum of a series of steps

$$\begin{aligned} Na &\rightarrow Na^+ + E^- \\ Cl + E^- &\rightarrow Cl^- \\ Na^+ + Cl^- &\rightarrow NaCl \end{aligned}$$

where E^- is used as the chemical symbol for a free electron. At present, however, it is merely desired to illustrate the process of compound formation by means of an example in which the attractive force is of a particularly simple type, and for this purpose we confine our attention to reaction (1).

When the sodium and chlorine ions are sufficiently far away from each other, the force that exists between them is of a purely electrostatic nature and is the same as that existing between an electron and a positively charged nucleus, namely, e^2/r^2 where e is the charge on the electron and r the distance between the centers of charge (which are the same as the centers of gravity) of the sodium and chlorine ions. If it is assumed that the potential energy is zero when r is infinite, then the potential energy at any distance r is given by the usual expression $-e^2/r$. This is true at any rate as long as r is great enough. However, it will break down at smaller distances. For it must be remembered that the sodium and chlorine ions are not point charges, but that they have shells of electrons that extend out from the nuclei to distances of the order of 10^{-8} cm. When the two ions come so close that interpenetration of the electron shells begins to take place, other forces come into play. These result in a resistance against bringing the nuclei closer together; *i.e.*, a repulsive force appears which works against the attraction due to the gross electric charges. The repulsive forces probably arise principally from the following circumstance.[1] It was shown in Sec. 4.6 that a quantum state occupies a region of a definite size in the phase space, and, by Sec. 7.2, there can be only one electron in a single quantum state. If the electron shells are pushed together, this means that more electrons are forced into a given volume (in ordinary space). Since each electron is in its own quantum state, the result is that there are more quantum states per unit volume of ordinary space. This means that each quantum state must occupy a greater region in the *momentum* part of the phase space. Thus the momentum, and therefore the energy, corresponding to each quantum state is raised; just as in the case of the electron moving in a box, considered in Sec. 4.2, the energy of all the quantum states goes up if the box is made smaller. It might be argued that instead of more electrons on the average occupying the given volume some of the electrons would simply be pushed out of the volume, and this is partly true; but this pushing-aside process itself requires energy. The net result is that as soon as the two ions are close enough

[1] See, *e.g.*, Jensen, *Zeits. Physik*, **101**, 164 (1936); Gombás, *ibid.*, **93**, 378 (1935). The type of repulsive force considered here is always operative, but may be said to be typical of cases where ions with closed shells come into contact. That, in some cases, repulsive forces can also arise in a somewhat different way will be seen in Chap. X.

together so that the electron shells affect each other, a repulsive force sets in, and this repulsion then increases very rapidly when the distance is decreased only a very little, and it quickly overcomes the attractive force. The effective potential energy between the two ions is shown schematically as a function of r in Fig. 34. At great distances, the curve starts off as an inverse first-power curve; at smaller distances, the effect of the repulsive forces begins to become evident. At the distance r_0, the repulsive and attractive forces just balance and the potential-energy curve has a minimum at that point; at smaller distances, the repulsive force predominates, and since it now requires work to push the

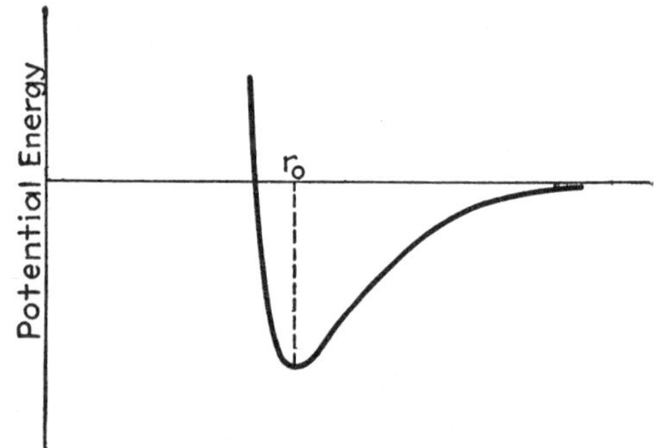

Fig. 34.—Molecular potential-energy curve.

particles together the potential rapidly increases as the distance decreases.

It is, of course, not necessary for the attractive forces to be of the same nature as those in the example just considered. It is quite possible to have strong attractive forces between neutral atoms, due to the interaction of the electrons in the atoms, as we shall see in the next chapter; other types of repulsive force than those considered above may be operative.[1] In such cases, the attractive force will not manifest itself at such great distances as in the case where the attraction is between two ions; but in general, attractive forces, if they occur at all, do manifest themselves at greater distances than the repulsive forces. The latter, whatever their origin, always predominate at very small distances, and one gets potential curves which resemble Fig. 34 qualitatively in that, as r decreases from infinity, the potential first decreases to a

[1] See Chap. X.

minimum and then increases very rapidly to an extremely high value.

In looking at the matter from the point of view of experiment, the mere fact that molecules exist with a fairly definite distance between atoms is indication that the potential-energy curve has somewhat the shape shown in Fig. 34. If there were no repulsive forces at small distances, the atoms in the molecule would not remain a definite distance apart; instead, their positions would tend to coincide. The existence of a position of equilibrium implies a minimum in the potential-energy curve. At ordinary temperatures, a molecule tends to have a low energy; this means that there will be but little kinetic energy and the atoms will have to be at positions of low potential energy; *i.e.*, the value of r will not, in the course of the motion of the atoms, become very different from r_0, where the potential-energy curve has its minimum. The shape of the potential-energy curve will be further discussed, in connection with the experimental material, in Secs. 9.4 and 9.5. The nature of the attractive and repulsive forces operative in solids will be discussed in later chapters; in particular, the forces in the ionic type of crystals, such as NaCl, will be discussed in some detail in Sec. 14.5, and the forces in gaseous molecules of this type treated more or less quantitatively in Sec. 14.9. It may be remarked that every available line of evidence on the repulsive forces verifies the relatively great steepness of the potential curve in the region where the interatomic distance is less than r_0.

Leaving then, for the present, the more detailed discussion of the experimental facts, a somewhat more exact formulation of the theoretical side of the problem may be attempted. When we consider a single atom, we must deal with the motion and the quantization of a number of electrons attracted to a nucleus which may be treated approximately as a fixed center of force. In a case of a molecule consisting of two atoms, it is necessary to consider the effects of *two* centers of force. If these two centers of force are far removed from each other, each electron can be thought of as attached to one or the other of them and the problem reduces to two problems of a single center of force. On the other hand, if the two centers of force are close together, then it is not possible to say definitely whether a given electron, especially an outer electron, belongs to one center or the other, and the two-center problem must be considered in its entirety. The approximation made if it is assumed that each electron is definitely attached to one or the other of the centers of force is much worse in some cases than in others. It is not necessary to go into this matter at the present time; what we wish to bring out now is that each electron occupies a certain quantum state with a certain energy in the two-center problem just as in the one-center problem. But in the case of the two-center problem, the energy of any quantum state depends upon the distance between centers. Whether the atoms will appear to attract or repel each other depends upon whether the sum total of all these electron energies is increased or decreased when the distance between the atoms is increased. From this point of view, then, the effective potential energy of the two atoms is merely the sum of the energies of the quantum states of all the electrons.[1,2] Subtracting from this sum the value of the total energy of the quantum states of the two atoms at infinite separation is equivalent to changing the zero of potential energy so that it occurs when the two atoms are infinitely separated; this was done in Fig. 34. This method of setting the zero of potential energy is sometimes inconvenient, however, because there is more than one possible potential-energy curve for a pair of atoms; we often wish to compare the energies of the different curves and, therefore, desire that there be a common energy zero for them all, instead of having a special zero for each one.

If one of the electrons is excited, it will be in a quantum state that has a different energy than the lowest or normal state, and the total energy of all the electrons will be a different function of the distance between the atoms. Thus, in Fig. 35 are pictured schematically some potential-energy curves belonging to a single pair of atoms. The zero of potential energy is taken arbitrarily as convenient, but it is the same for all curves. The difference between the curves at any value of r shows how much energy is necessary to excite an electron to the particular excited

[1] A rather similar case of an effective potential energy occurs in Sec. 5.7.

[2] This statement involves the assumption that the nuclei move so slowly compared with the electrons that the latter always have time to adjust their motion to that expected for fixed nuclei at the distance apart at which the nuclei happen to be at any given instant. Actually, a very good approximation to this assumption is always true because of the great mass of the nuclei as compared with the electrons.

state considered, when the atoms are that distance apart. When r is very great, this simply means exciting the electron of one or the other of the atoms; the asymptotic difference between the curves when r is great therefore represents the energy necessary to excite an electron of a particular one of the atoms to some definite state of higher energy. And, of course, states can also occur in which more than one electron is excited.

As long as there is no transition from one potential-energy curve to another, we can, if we are interested in the motion of the nuclei, consider them as point masses moving in a potential-

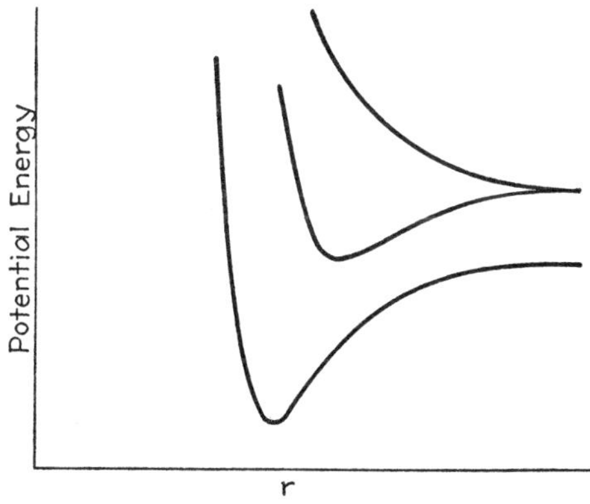

FIG. 35.—Normal and excited potential-energy curves for a pair of atoms. It will be noted that one of these curves is depicted without a minimum. It is quite possible for electronic states to exist in which no attraction between the atoms occurs. Such a state, obviously, will not result in molecule formation.

energy field which is a function of the distance between them, the forces between them being directed along the line joining the particles. It is a well-known mechanical principle that the center of gravity of the system may move with a constant velocity without affecting the relative motion of the two masses (see Appendix I). The relative position of the two masses may be given by specifying three coordinates r, the distance between them, and the polar angles θ and ϕ giving the direction in space of the line joining them. The relative motion is given if r, θ, and ϕ are known as functions of the time. It may be shown (Appendix I, pp. 451*f*.) that, as far as their relative motion is concerned, the system of two particles (of mass m_1 and m_2) can be replaced by a single particle of mass $\mu = m_1m_2/(m_1 + m_2)$ attracted to (or repelled from) a fixed center with a force that is the same function of the distance as the force between the two particles of the system is of the distance between them. The coordinates r, θ, and ϕ, of the representative particle, taking the fixed center as the origin, are then the same functions of the time as the coordinates r, θ, and ϕ defined above for the two-particle system. The mass μ is called the "reduced mass" of the original pair of particles.

9.2. Molecular Energy Levels.—Since the relative motion of the atoms in a molecule can be replaced by the motion of a representative particle of mass μ, acted upon by a central field of force, the quantization of the molecule is very similar to that of the hydrogen atom,[1] the only essential difference being that the potential $-\frac{e^2}{r}$ is replaced by a potential V which is a more complicated function of r. No account need be taken of the motion of the center of gravity of the molecule, since the relative motion of the atoms is quite independent of it. The quantization of the angular momentum and its projection along an arbitrary axis proceeds just as in Sec. 4.5, provided the vector sum of the orbital and spin angular momenta of all the electrons vanishes when the motion of the nuclei is not taken into account. If there is a resultant electronic angular momentum, it has a gyroscopic effect on the motion of the molecule, but for our purposes consideration of this may be omitted. We shall use the quantum numbers j and m_j to designate the nuclear momentum and its projection, respectively, writing for the total angular momentum of the jth state (j, corresponding to l of Sec. 4.5, can have any integral value from 0 to ∞)

$$p_j = \sqrt{j(j+1)}\frac{h}{2\pi} \tag{3}$$

and for its projection

$$p_{m_j} = \frac{m_j h}{2\pi}, \tag{4}$$

[1] For detailed accounts of molecular motions and molecular spectra, the reader may be referred to the following books: Jevons, "Report on Band Spectra," The Physical Soc., London, 1932; Sponer, "Molekülspektren," vol. II, Julius Springer, Berlin, 1936; and Weizel, "Bandenspektren," Akademische Verlagsgesellschaft, Leipzig, 1931, etc.; as well as portions of many general works on quantum mechanics.

where m_j, corresponding to m_l of Sec. 4.5, can take integral values from $-j$ to j.

In quantizing the energy, account must be taken of the radial motion, just as in the case of the hydrogen atom, though the actual nature of the radial motion of the atoms in a molecule is very different from the radial motion of the electron in the hydrogen atom. It has been remarked in the preceding section that for a molecule in a low-energy level the value of the coordinate r never becomes greatly different from r_0. This is illustrated

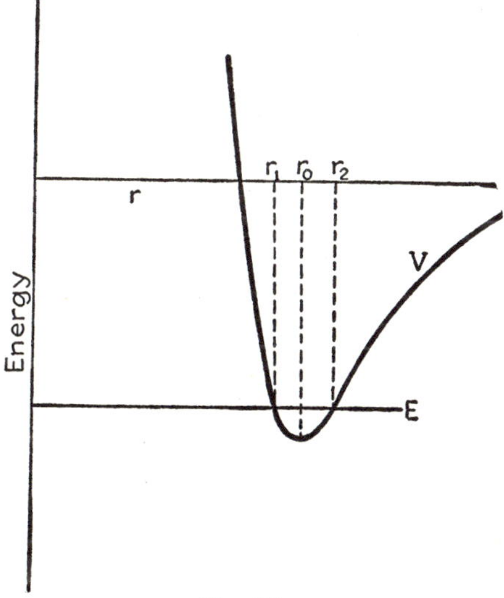

FIG. 36.

by Fig. 36. For simplicity, we shall first consider the case in which $j = 0$, so that there is no rotational energy. The direction of the line joining the two atoms remains fixed in space, *i.e.*, θ and ϕ are constant, but the atoms oscillate back and forth about positions of equilibrium. Since the pair of atoms can be replaced by a single representative particle of mass μ, the kinetic energy associated with this oscillation or vibrational motion is $\frac{1}{2}\mu\dot{r}^2$. If E represents the total energy of the pair of atoms (excluding, of course, any energy of translation of the center of gravity) and V, the potential energy, the difference between them $E - V$ is the kinetic energy. In the case sketched in Fig. 36, this is negative to the left of r_1 and to the right of r_2. According to the classical picture, then (and the description heretofore given has adhered to the classical picture) values of r less than r_1 or greater than r_2 cannot occur. According to wave mechanics, there is a certain probability that r might be less than r_1 or greater than r_2, but these classically forbidden cases are not very probable, and it is approximately correct even on the wave mechanical basis to say that r is confined between r_1 and r_2. In general, for low molecular energies, $r_2 - r_1$ is not more than about 10 per cent as great as r_0, so the oscillation may be treated as a small oscillation.

Let the value of the potential energy at its minimum be designated as V_0. The potential energy V at any point r may be expressed in terms of V_0 as a power series in $r - r_0$, a process very commonly employed in mathematical considerations, in the following manner:

$$V = V_0 + a(r - r_0) + b(r - r_0)^2 + c(r - r_0)^3 + \dots \tag{5}$$

where a, b, c, . . . are constants. Differentiating with respect to r gives

$$\frac{dV}{dr} = a + 2b(r - r_0) + 3c(r - r_0)^2 + \cdots \tag{6}$$

but since V has its minimum at $r = r_0$, dV/dr must be equal to zero there. Hence it is seen that $a = 0$. Further, if $r - r_0$ is not too large (and as noted above it does not become large if E is not too high), it is possible to neglect the higher terms in $r - r_0$. Then Eq. (5) becomes

$$V = V_0 + b(r - r_0)^2. \tag{7}$$

Motion in a potential-energy field given by an equation of this type is called "simple harmonic motion."

The increase dV in the potential energy, when the representative particle is displaced a small distance dr, is equal to $-fdr$, where f is the force on the particle. (The minus sign arises because f is taken as positive if it acts in the direction of increasing r, and the potential energy increases if the force is *opposed* to the displacement.) Since, from these remarks, $f = -\frac{dV}{dr}$, it is seen from Eq. (6) that, approximately,

$$f = -2b(r - r_0). \tag{8}$$

This can be set equal to the reduced mass times the acceleration, giving as the classical equation of motion

$$-\mu\ddot{r} = 2b(r - r_0). \tag{9}$$

The solution of this differential equation is

$$r - r_0 = A \sin\left(\sqrt{\frac{2b}{\mu}}t + \delta\right), \tag{10}$$

where A and δ are constants of integration. This shows that, according to the classical picture and making the approximations noted (simple harmonic motion), the value of r oscillates back and forth past the position of equilibrium r_0 as a sinusoidal function of the time. r returns to its original value after a time τ such that $\sqrt{2b/\mu}\tau = 2\pi$. The reciprocal of τ is the frequency of vibration ν_0 which is thus given by

$$\nu_0 = \pi^{-1}\sqrt{\frac{b}{2\mu}}. \tag{11}$$

It should be noted that b depends only upon the potential-energy curve, the broader the minimum being, the smaller b. μ, of course, depends only on the masses of the atoms. From Eq. (11), it is seen that ν_0 depends, then, at least to the approximation to which Eq. (7) is valid (*i.e.*, for sufficiently low energy levels) only on the potential energy and the masses of the atoms and is independent of the total energy, which depends upon the amplitude of the motion (*i.e.*, upon the largest value $r - r_0$ can take in the motion). From Fig. 36, it is clear that the higher the energy level, the greater this amplitude will be, and the total energy is closely related to the magnitude of A in Eq. (10).

The preceding account gives a fairly complete description of the vibration from the classical point of view. Quantum theory, we know, will restrict the possible energies of the system. In the case of the simple harmonic oscillator described above, the energy levels may be shown to be given by the following formula:

$$E_v - V_0 = (v + \tfrac{1}{2})h\nu_0, \tag{12}$$

where $v = 0, 1, 2. \ . \ . \ .$

The treatment of the vibrational motion outlined above was carried out for the case where there is no rotational motion.

If there is rotational motion, it may be shown that, to a first approximation, it is independent of the vibrational motion, and the energy due to the rotational motion may simply be added to the energy given by Eq. (12). Since the amplitude of the vibrational motion is small, the rotational motion may be handled as though the two atoms were at a fixed distance r_0. This means that Eq. (17), of Sec. 4.5, will hold without alteration other than insertion of the appropriate quantities for the case at hand, which gives for the rotational energy

$$E_j = \frac{j(j+1)h^2}{8\pi^2\mu r_0^2} = \frac{j(j+1)h^2}{8\pi^2 I}, \tag{13}$$

where $I = \mu r_0^2$ is the moment of inertia of the two nuclei about their center of gravity, assuming them held rigidly fixed at the distance r_0. Since this equation involves a number of approximations, it is not generally applicable when the energy of vibration or rotation is not low, in contrast to Eqs. (3) and (4), which are not subject to these limitations. m_j, of course, has no effect on the energy.

A better understanding of the relationship (12) above, may be had by a consideration of the integral $\int p_r\,dr$. As stated above, the problem of the diatomic molecule is formally the same as that of the hydrogen atom, except that the potential energy does not have the form $-\frac{e^2}{r}$. If, therefore, $-\frac{e^2}{r}$ is replaced by $V = V_0 + b(r - r_0)^2$ and l is replaced by j, and m is replaced by μ, Eq. (15), page 66, will hold in the present instance, and the expression for the total energy may be written

$$E = \frac{p_r^2}{2\mu} + \frac{j(j+1)h^2}{8\pi^2\mu r^2} + V_0 + b(r - r_0)^2. \tag{14}$$

This may be solved for p_r and the result placed in $\int p_r\,dr$. The integral must be taken over a complete oscillation of the particle; *i.e.*, it must be taken over the motion of the particle while it moves from the smallest classically possible value of r, *i.e.*, r_1, to the largest value r_2, and back again. p_r is positive when dr is positive, and negative when dr is negative. This means that $p_r\,dr$ is always positive, and we write for the integral in the particular case that $j = 0$,

$$\int p_r\,dr = 2\int_{r_1}^{r_2} p_r\,dr = 2\int_{r_1}^{r_2} \sqrt{2\mu[E - V_0 - b(r - r_0)^2]}\,dr, \tag{15}$$

r_1 and r_2 are the values of r for which $\dot{r}$ and hence p_r, the integrand of (15), becomes zero, so that $r_1 = r_0 - \sqrt{(E - V_0)/b}$ and $r_2 = r_0 + \sqrt{(E - V_0)/b}$. The integral in Eq. (15) may be readily evaluated, giving

$$\int p_r\, dr = \pi(E - V_0)\sqrt{\frac{2\mu}{b}} = \frac{(E - V_0)}{\nu_0}, \tag{16}$$

the latter relation following from Eq. (11). It is now readily seen that Eq. (12) represents the type of condition on $\int p_r\, dr$ demanded by Sec. 4.6.

The rotational energy levels may be readily taken care of in this scheme. if it is noted that the term $\frac{j(j+1)h^2}{8\pi^2\mu r^2}$ in Eq. (14) acts just like an addition to the potential energy. If the range of values from r_1 to r_2 is small enough, it may be considered as practically constant and written $\frac{j(j+1)h^2}{8\pi\mu r_0^2}$. This term, identical with the right-hand side of Eq. (13), then finally appears as an addition to the total energy.

9.3. Molecular spectra offer, as might be expected, the best source of information on molecular energy levels. The energy, of course, is not observed directly, but as usual the spectral lines represent differences between energy levels. The spectra are much simplified by the fact that the quantum number j obeys certain selection rules; namely, it can change only by ± 1 or, in some cases, 0. The quantum number m_j can change only by ± 1 or 0, but since it does not affect the energy it is not important for our present purposes.

Three kinds of molecular spectra are observable.[1] In the very far infrared, transitions involving only changes in j and no other quantum numbers are observed. On account of the selection rules mentioned above, these spectra are very simple, consisting of a series of lines equally spaced on the frequency scale, but because they lie so far in the infrared, their observation requires a rather difficult technique. Since far infrared means waves of very low frequency, and hence quanta of low energy, the fact that these spectra appear in this region indicates that the energies

[1] Another important method of investigation of molecular energy levels is the study of Raman spectra. These spectra result from the absorpion of a light quantum and its reemission in the same step with altered frequency, the difference in energy between the absorbed and reemitted light quantum being equal to an energy difference in the molecule. They are especially useful for the study of vibrations of complex molecules (see Sec. 9.6). See, *e.g.*, Kohlrausch, "Der Smekal-Raman Effekt," Julius Springer, Berlin, 1931, and Ergänzungsbd., 1938.

of different rotational states with the same value of v lie very close together.

In the near infrared, it is possible to observe spectra that arise from simultaneous changes in v and j. Since higher frequencies are involved, it is evident that vibrational energy levels are spaced farther apart than rotational levels.[1]

The most important spectra occur at still higher frequencies, in the visible and ultraviolet. They involve electronic transitions, being due to transitions from vibrational and rotational levels belonging to one potential-energy curve to levels belonging to another curve.

From analyses of these spectra, it is possible to learn much about the details of the energy levels of molecules. Since the energy may be described approximately as a sum of a term such as given by Eq. (12) and a term such as given by Eq. (13), the statement that the energy is separated into vibrational and rotational parts is experimentally verified, at least approximately. Since, as just seen, the difference in energy between adjacent rotational levels with given v is in general small compared with the difference between adjacent vibrational levels, the energy states appear to occur in groups or bands, all those with given v and different j's being grouped together.

From the spacing between vibration levels, it is possible to find ν_0 from Eq. (12), and ultimately b from Eq. (11), since μ is known. This gives information about the shape of the potential-energy curve near its minimum. From the spacing of the individual rotational levels in a given band, it is possible to find I from Eq. (13), and from I, the interatomic distance r_0 may be found. The study of molecular spectra (band spectra) thus yields very important information about the molecule.

9.4. The Dissociation Energy.—A still more important molecular constant, the energy necessary to pull the two atoms

[1] This is directly connected with the fact that the atoms of the molecule vibrate through a space which is small compared with the distance between the atoms, while in making a rotation the atoms move through a distance comparable with the distances between them. Therefore, a vibrational state has a relatively small extension in ordinary space, so it must occupy a relatively large region of momentum space, in order that it may have its quota of phase space. In order for the vibrational state to have a relatively large region in momentum space, the energy levels must be relatively far apart. See Sec. 4.6.

apart, can also often be obtained by the study of band spectra. As already noted, it is only an approximation to consider a molecule as a harmonic oscillator. The energy levels of a

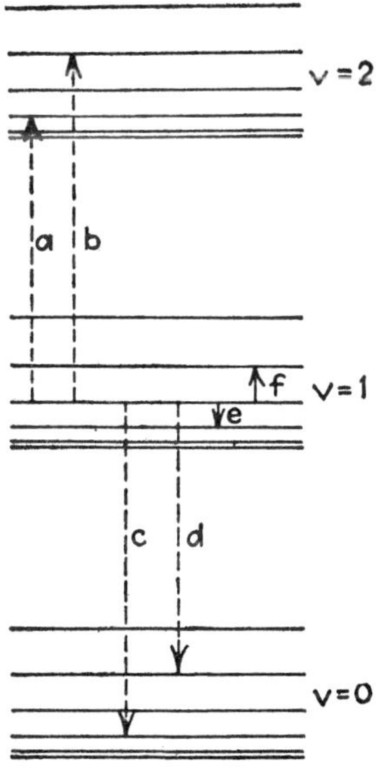

FIG. 37.—Illustrating spacing of molecular vibration-rotation levels, showing positions of bands for which $v = 0$, 1, and 2, above the minimum of potential energy V_0. Levels for j with values from 0 to 5 shown in each band. To illustrate the transitions which occur in molecular spectra, the possible quantum jumps starting from the level $v = 1$, $j = 3$ are shown. (We assume j can change by ± 1.) An increase in energy means a quantum must be absorbed, a decrease of energy means a quantum is given out. The transitions shown are classified as follows:

a, b—near infrared absorption.
c, d—near infrared emission.
e —far infrared emission.
f —far infrared absorption

Visible and ultraviolet transitions cannot be shown because the bands belonging to only one electronic state appear in the figure.

harmonic oscillator are spaced at equal intervals, but in actual molecules it is observed that the higher the energy, the closer together the levels become. When the allowed energies become closer together, the difference in momenta between adjacent levels also becomes smaller, which means that an energy level occupies a smaller volume in the momentum space than it would were the molecule a harmonic oscillator with a frequency corresponding to the difference in energy for the low-energy levels. Therefore, in order that each energy level should have its regular quota h in the complete phase space, the distance over which the oscillator can move in coordinate space must be greater. Thus the actual potential-energy curve must be more spread out than a harmonic-oscillator curve would be. The situation is illustrated in Fig. 38. At the energy indicated, a classical harmonic oscillator would have a range of motion from A to B, whereas with the other curve, which resembles an actual molecular-potential curve, the classical range of motion at the same

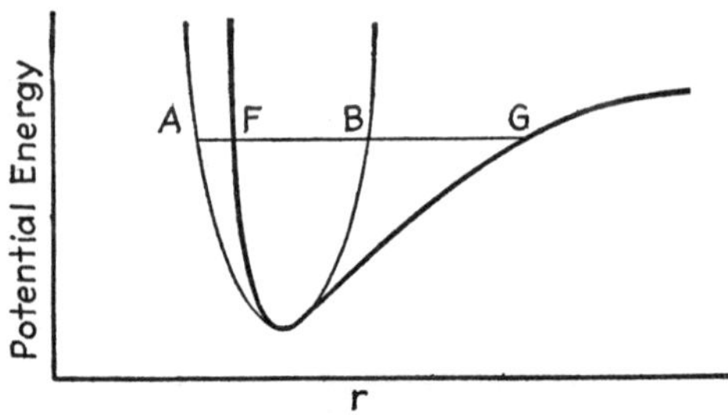

FIG. 38.—Actually observed type of molecular energy curve, contrasted with curve for harmonic oscillator.

energy would be from F to G, a much greater distance. This spreading out of the potential-energy curve is in harmony with the tacit assumption, which has been made throughout, that a potential-energy curve will approach a definite limiting value as r is indefinitely increased; in other words, it will require only a finite amount of energy to pull the atoms in a molecule completely apart. It will be observed that the real potential-energy curve in Fig. 38, as well as the potential-energy curves illustrated previously, has been so drawn as to be approaching an asymptotic value for large distances of separation. If the total vibrational energy of the pair of atoms exceeds this value, the classical range of the relative motion of the two atoms becomes infinite; *i.e.*, the atoms are free. Above this energy, the energy levels are infinitely close together; in other words, the spectrum becomes continuous, just as in the case of the hydrogen atom when the electron has energy greater than the ionization energy and is therefore a free electron. If the energy at which the continuous spectrum sets in can be observed, it is possible to determine the dissociation energy of the molecule. In general, rather than to observe this directly, it is more accurate to observe the discrete vibration levels as near as possible to the dissociation potential, and extrapolate to the energy at which the difference in energy between two successive levels becomes zero.

The actual energy necessary to dissociate a molecule in its lowest vibrational state will not be exactly equal to the difference between the lowest point on the curve and the asymptote, because, as will be seen from Eq. (12), even when v is zero, $E - V_0$, the energy of the oscillator measured from the lowest point of the curve, is *not* zero, but is equal to $\frac{1}{2}h\nu_0$, which is known as the "zero-point energy."

It should be borne in mind that in speaking of the dissociation energy of a molecule reference is had to a molecule in some definite electronic state. A molecule has many electronic states, to each of which corresponds a potential-energy curve, and each potential-energy curve has its own dissociation energy. In other words, the dissociation energy of a molecule depends upon the electronic state it is in. When no statement to the contrary is made, however, it is usual to use the term "dissociation energy" to mean the dissociation energy of the molecule in its lowest electronic state.[1]

9.5. The shape of the potential-energy curves in the case of actual molecules is not a problem that needs to be considered in great detail. An account of a few of the qualitative features will be sufficient for our purposes. The fact that the vibrational levels become closer together at higher energies means that the actual potential-energy curve must spread out at these higher energies more than the corresponding harmonic-oscillator curve, but it does not say just how it will spread out. In fact, it is possible to find an infinite number of potential curves to fit any given set of energy levels. It will be noted, however, that in Fig. 38 the curve, which has been claimed to have the qualitative features of actual potential curves, has been drawn so that it is steeper for small values of r than the harmonic-oscillator curve,

[1] Should some of the potential-energy curves cross, the dissociation energy is the energy necessary to go from the lowest undissociated state to the lowest dissociated state, even though it belongs to a different curve.

the increasing width at high energies of the "actual" curve being due to its behavior at large values of r. The steepness for small values of r has already been mentioned in Sec. 9.1 as being in accord with various lines of experimental evidence. Further evidence regarding the shape of the potential-energy curves may be obtained from a study of the rotational spectrum of molecules. The treatment of the rotational energy levels given above was, of course, a very crude one. It was recognized that it cannot give an entirely correct result to simply add to V the constant quantity $\frac{j(j+1)h^2}{8\pi^2\mu r_0^2}$, and especially will this be in error if the potential-energy curve is not symmetrical. In the case of the nonharmonic curve shown in Fig. 38, it is quite obvious that it would be a better approximation for the higher energy levels if, while still assuming that the effective addition to the potential energy did not vary with r, we should replace r_0 by a greater value. Thus the rotational levels may be expected to be closer together, the greater the value of the *vibrational* quantum number. This is actually observed to be the case, and provides further evidence that the type of curve shown in Fig. 38 is correct in its qualitative features.

9.6. The rotation and vibration of polyatomic molecules present a much more complicated problem than the case of diatomic molecules.[1] The translational motion of the whole molecule, however, is handled just as easily as before; the coordinates of the separate atoms are referred to the center of gravity of the molecule, and the translation of the molecule is just the same as that of a mass equal to the sum of the masses of all the atoms in the molecule concentrated at the center of gravity.

Consider first the case of a "rigid" molecule. In such a molecule, each of the atoms vibrates about a position of equilibrium and all of the equilibrium positions are definitely fixed with respect to each other, forming a "rigid framework" for the molecule. In general, such a molecule has three degrees of freedom of rotation, *i.e.*, it is necessary to specify three angles, as well as the coordinates of the center of gravity, in order to completely determine the position of the rigid framework of equilibrium positions in space. The state of rotation of the

[1] For a more complete account see Pauling and Wilson, "Introduction to Quantum Mechanics," pp. 282*ff.*, McGraw-Hill Book Company, Inc., 1935.

molecule is given by specifying three rotational quantum numbers which determine, among other things, the rotational energy. In order to completely specify the positions of the atoms, we can then fix their relative positions by giving the Cartesian coordinates of each one, taking its own equilibrium position as the origin for the coordinates of any particular atom. If there are N atoms in the molecule, there are, in all, $3N$ such coordinates. However, only $3N - 6$ are *independent.* For example, if all the atoms were displaced in the same direction and by the same distance, this would amount simply to a translation of the whole molecule, *i.e.*, the rigid framework to which the coordinates of the separate atoms are referred would be shifted along. There being six independent ways in which the position and orientation of the rigid framework can be changed, this implies six conditions on the $3N$ coordinates if the motion of the atoms is to be such as not to include a translation or rotation of the framework. There are thus $3N - 6$ independent degrees of freedom of relative nonrotational motion of the atoms, *i.e.*, $3N - 6$ vibrations, each with its own frequency, and each having a quantum number.

If all the atoms are arranged in a straight line, then two angles are sufficient to specify the direction in space of this line, and there are only two degrees of freedom of rotation. In this case there are, then, $3N - 5$ degrees of freedom of vibration.

A rigid polyatomic molecule with three rotational degrees of freedom has, as is shown in treatises on mechanics, three principal axes which are at right angles to each other. If the body is set rotating about one of these axes, it will continue to rotate about this axis forever, provided it is not acted upon by external forces, whereas if it is set to rotating about any other axis, the instantaneous axis of rotation continually changes. There are three principal moments of inertia, one for each of the principal axes. In terms of these, the motion of the system is readily described, and the moment of inertia about any arbitrary axis can be readily calculated. The principal moments of inertia enter into the quantization of the rotational motion, and an analysis of the rotational spectrum of the molecule will sometimes give values for the principal moments of inertia. These moments of inertia depend upon the arrangement of the atoms in the molecule, and they in turn can, if the molecule is simple enough, give information about the arrangement of the atoms, and thus furnish a valuable aid in the determination of the structure of the molecule.

We shall not discuss the rotational spectrum or energy levels of polyatomic molecules further but shall give a brief account of the vibrational motion. It will be assumed that the vibrations are small, that the center of gravity of the molecule is at rest, and that the molecule is not rotating. This procedure is justified, for it may be shown that to a first approximation the motion of the molecule may be considered to be separated into translational, rotational, and vibrational motions which are independent of each other.

In discussing the vibration of a polyatomic molecule, the mathematical details will not be given, but an attempt will be made to describe the motion qualitatively according to the classical picture. In a polyatomic molecule, the force acting upon any given atom depends not only on the position of that particular atom, but also upon the position of all other atoms. When all the atoms are simultaneously in a position of equilibrium, the force on any one of them is, of course, zero. If any atom is displaced, not only are forces exerted on it, but it must necessarily exert forces on its neighbors. In general, this set of complex interactions results in a very complicated motion. However, it is possible to set a polyatomic molecule into vibration in such a way that all the atoms vibrate about their equilibrium positions with the same frequency and in phase with each other, *i.e.*, they all pass through the equilibrium position at the same time. Such a state of motion is known as a normal mode of vibration. There are as many different normal modes of vibration as there are vibrational degrees of freedom. Any actual vibrational motion, however complex, can be analyzed into (*i.e.*, be shown to be built up by superposition of) some or all of the normal modes of vibration.

The nature of the normal modes of vibration may be illustrated by means of a simple example, the carbon dioxide molecule,[1] in which the atoms are known to be arranged in a straight line, thus, O—C—O. This molecule has $3N-5=4$ vibrational degrees of freedom. In two of the normal modes of vibration, the motion of the atoms will be in the line of centers, as illustrated in Fig. 39

[1] See SPONER, "Molekülspektren," vol. I, p. 75, for a synopsis and references.

(a) and (b). One of them (a) consists of motions of the oxygens only about their positions of equilibrium, the two oxygens moving simultaneously toward or away from the carbon atom. In the other (b), the oxygens move simultaneously in one direction while the carbon moves in the opposite direction. These are stretching vibrations. There are also two bending vibrations, one of the type shown in Fig. 39 (c), the other just like it but at right angles to it. A bending at some other angle is simply a

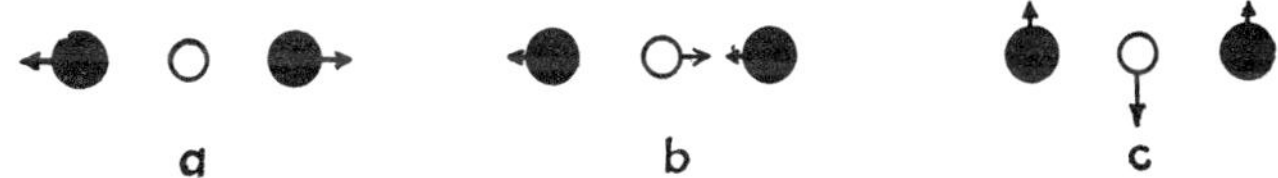

FIG. 39.—Normal vibrations for carbon dioxide. Carbon atoms white, oxygen atoms black.

superposition of the other two, and so is not an independent mode of vibration. In general, bending frequencies are much lower than stretching-type frequencies connected with the same type of bond.

In Fig. 40, the normal modes of vibration of the water molecule are illustrated. Here there are $3N - 6 = 3$ vibrational degrees of freedom, since it is not a straight-line molecule. The oxygen atom, being so much heavier, will move very little compared with the hydrogen atoms. This is indicated roughly by the length of

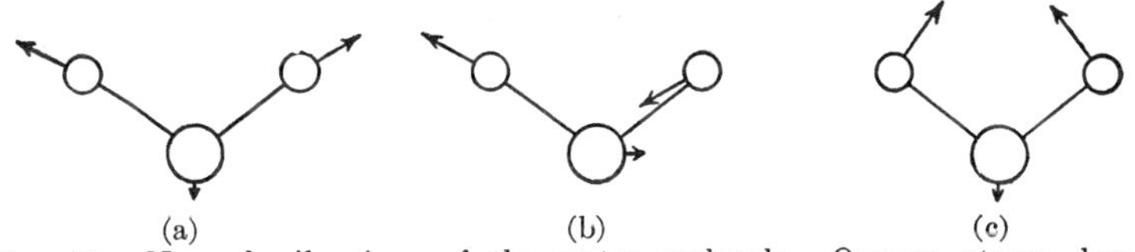

FIG. 40.—Normal vibrations of the water molecule. Oxygen atoms, large circles; hydrogen atoms, small circles.

the arrow. It will be seen that vibrations (a) and (b) involve stretching of the bonds, whereas (c) is the bending type.

In any molecule, each normal mode of vibration has its own characteristic frequency and each one is quantized independently of the others. The allowed energies of any normal mode of vibration are given in terms of the frequency in the same way as for a simple vibration. The total vibrational energy of the molecule is the sum of the separate energies of the various modes of vibration.

Though each normal mode of vibration has its own frequency, there are often relations between these frequencies, since the various types of vibration involve motions of the same bond. Sometimes these relations can give valuable information concerning the structure of the molecule. Thus, it can be readily shown that in a straight-line triatomic molecule the ratio of the two stretching frequencies depends only on the masses of the atoms, provided the forces between the two outside atoms can be neglected, and the ratio can be calculated theoretically. If approximately this ratio is found experimentally, it lends evidence that the molecule is a straight-line molecule. Otherwise it has a kinked structure.

The method can be applied, however, only to relatively simple polyatomic molecules, since the actual analysis of the spectrum is generally very difficult.[1] In making such an analysis, use is made of all known properties of the spectrum. In particular, selection rules, which depend very largely on the symmetry properties of the molecule, have proved very helpful.

Exercises

1. The maximum distance between the atoms (*i.e.*, the distance between them at the end of the vibration) of the molecule C_2 would be 0.058Å. greater than the equilibrium distance, if they vibrated classically with an energy equal to the energy of the lowest vibrational state. Calculate b, the frequency, and the energy between energy levels, assuming simple harmonic oscillation.

2. The equilibrium distance between atoms in the molecule HI in the normal state is 1.62Å.; in I_2, it is 2.66Å.; in C_2, it is 1.31Å. Calculate the energies of the first five rotational states of each of these molecules.

3. A body of mass m moves (one dimensionally) in a potential-energy field such that if $x > 0$, the potential energy $U = ax$, whereas if $x < 0$, then $U = -ax$. How many energy levels are there with energy less than E? Answer this question numerically if m is the mass of a hydrogen atom, E is 0.5 electron volt, and a is 1 volt per Angstrom (see Chap. IV).

4. Suppose that in far infrared absorption the quantum number j changes from a value j_0 to $j_0 + 1$. From Eq. (13), find a formula for the frequency of the light absorbed as a function of j_0. Show that the spectrum consists of a series of lines of equally spaced frequencies.

[1] For a fairly detailed discussion of a number of molecules, see Penney and Sutherland, *Proc. Roy. Soc. (London)*, **A156**, 654 (1936); Stuart, "Molekülstruktur," pp. 295 *ff.*, Julius Springer, Berlin, 1934; Schaefer and Matossi, "Das ultrarote Spektrum," pp. 225*ff.*, Julius Springer, 1930.

CHAPTER X

THE HYDROGEN MOLECULE

The hydrogen molecule is the simplest example of a nonpolar or nonionic molecule. The combination in this instance is between two neutral atoms, which are exactly alike, so that one has no more tendency than the other to form either positive or negative ions. Of course, ultimately the forces involved are electrostatic forces, for the hydrogen nuclei are bound together by the electrons; but it is obvious that this statement implies that the electrons which effect the binding are between the protons, and the reason that they stay in this advantageous position is understood only by a study of the wave functions of the electrons. Ordinary mechanics is completely incapable of solving this problem, and the solution of it by the use of wave mechanics has been one of the great triumphs of that theory. We may, however, start our discussion by using the corpuscular picture of the electron.

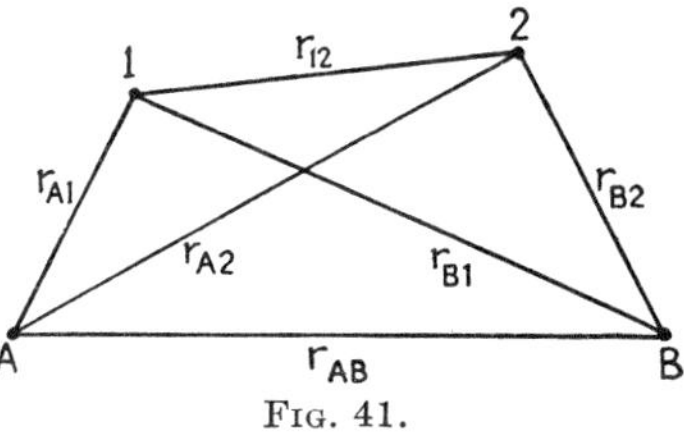

FIG. 41.

Suppose that we have two positive hydrogen nuclei or protons A and B and two electrons 1 and 2. The distances between the various particles are designated as shown in Fig. 41. The potential energy of the whole system is given in this case by the expression

$$\frac{e^2}{r_{12}} - \frac{e^2}{r_{\mathrm{A1}}} - \frac{e^2}{r_{\mathrm{B2}}} - \frac{e^2}{r_{\mathrm{A2}}} - \frac{e^2}{r_{\mathrm{B1}}} + \frac{e^2}{r_{\mathrm{AB}}} \tag{1}$$

If, however, we consider the problem of the electrons moving under the influence of the two centers of force at A and B, we see that only the first five of the terms in the foregoing potential expression enter into the problem of determining the electronic energy. The last term, which arises from the repulsion of the nuclei, is a term that is simply added to the electronic energy to get the total energy of the system. Starting now with the two nuclei an infinite distance apart, let us suppose them to be slowly pushed together. At the beginning, the problem is that of two separated atoms, but as the nuclei approach each other it is necessary to consider the effect of both nuclei on both electrons. Finally, the two nuclei may be thought of as merged together. In this case, the *electronic* energy is just the same as for the helium atom, since there are two electrons moving in the field of a center of force of twice the strength of a proton. This suggests that a study of the energy levels of the helium atom may be of some assistance in understanding the nature of the binding forces in hydrogen, and this is indeed the case.

10.1. The helium atom was treated by means of the quantum mechanics by Heisenberg. We shall consider a helium atom in which one of the electrons is in the lowest energy level, a $1s$-state, but in which the other electron is in some excited energy level. At first, we shall neglect the Coulomb repulsion between the electrons—*i.e.*, the motion of each electron under the influence of the positive nucleus is treated as though the other electron were not there. The wave functions of the electrons are thus hydrogenlike wave functions, modified, of course, on account of the double charge on the nucleus. We shall denote the wave function for the unexcited state as ψ and the wave function for the excited state as ϕ. The two electrons will be designated as 1 and 2, respectively. If electron 1 is in the unexcited state and electron 2 in the excited state, this will be indicated by writing $\psi(1)$ and $\phi(2)$, and if the electrons are exchanged we shall write $\psi(2)$ and $\phi(1)$. Wave functions for the whole system are found by taking products of ψ and ϕ. Thus if $(\psi(1))^2\,dx_1\,dy_1\,dz_1$ is the probability of finding electron 1 in a certain element of volume of size $dx_1\,dy_1\,dz_1$ and $(\phi(2))^2\,dx_2\,dy_2\,dz_2$ is the probability of finding electron 2 in an element of volume of size $dx_2\,dy_2\,dz_2$, then the probability that they will both simultaneously be in the indicated volume elements is $(\psi(1))^2(\phi(2))^2\,dx_1\,dy_1\,dz_1\,dx_2\,dy_2\,dz_2$ $= (\psi(1)\phi(2))^2\,dx_1\,dy_1\,dz_1\,dx_2\,dy_2\,dz_2$. Thus $\psi(1)\phi(2)$ can be taken as the combined wave function for both electrons. It should be noted that $dx_1\,dy_1\,dz_1$ and $dx_2\,dy_2\,dz_2$ refer to two different volumes in ordinary space, but they are both referred to the same set of coordinate axes, the subscript 1 referring to the coordinates of electron 1 and the subscript 2 referring to the coordinates of electron 2. In the *phase space* (Sec. 4.6) for the two electrons, however, x_1 and x_2, y_1 and y_2, and z_1 and z_2 would all be marked off along different axes.

But the preceding wave functions make no mention of the spin. The state of the spin can be indicated by means of one of two *spin wave functions*,[1] α which says that the spin of the electron is in one direction and β which says that the spin is in the other direction. Thus the wave function $\psi(1)\phi(2)\alpha(1)\beta(2)$ tells us that electron 1 is in the unexcited state and has its spin in the direction indicated by α, and that electron 2 is in the excited state and has its spin in the opposite direction. Now, of course, either electron can be the one in the unexcited state, and either electron can have its spin in either direction. There are eight possible combinations, as follows:

$$\begin{array}{l} \psi(1)\phi(2)\alpha(1)\alpha(2) \\ \psi(1)\phi(2)\beta(1)\beta(2) \\ \psi(1)\phi(2)\alpha(1)\beta(2) \\ \psi(1)\phi(2)\beta(1)\alpha(2) \\ \phi(1)\psi(2)\alpha(1)\alpha(2) \\ \phi(1)\psi(2)\beta(1)\beta(2) \\ \phi(1)\psi(2)\alpha(1)\beta(2) \\ \phi(1)\psi(2)\beta(1)\alpha(2) \end{array} \tag{2}$$

But on closer thought it will be realized that these functions do not really furnish satisfactory wave functions for the problem. For they assume that the two electrons are distinguishable, whereas in reality it is quite impossible to distinguish any given state from the state in which the electrons are interchanged; it is to be expected that two such states, which are entirely indistinguishable and which have exactly the same energy, would pass freely from one into the other. Such a situation should be exhibited by the wave functions themselves, and it is necessary to find wave functions that do take it into account when attempting to calculate the effect of the action of the two electrons on each other, even in rough approximation. Now the wave func-

[1] The spin wave function is not a function of the ordinary coordinates of the electron, but may be thought of as a function of a special spin coordinate, which describes the internal state of the electron. Its exact nature does not need to concern us here.

tions, as written, are solutions of the wave equation, but it may be shown that the sum of two solutions, or, more generally, any linear combination of the various solutions which represent states having the same energy, is also a solution.[1] This suggests that we should look for linear combinations of the solutions, which have the property that they involve the two electrons in the same way. Such a wave function will be so constituted that it either remains unchanged (the wave function is *symmetric*) or changes in sign (the wave function is *antisymmetric*) when the two electrons are interchanged. It is allowable for the wave function to change in sign, because any calculation of actual probabilities involves a square of the wave function and is unchanged when the electrons are interchanged even though the wave function itself changes in sign. There are eight independent wave functions, as has been seen, and there are eight independent linear combinations which have the foregoing necessary properties. These combinations are given herewith, and the reader may verify, by actually interchanging the symbols 1 and 2 for the two electrons, that they have the properties claimed for them.

Symmetric combinations:[2]

$$\psi(1)\phi(2)\alpha(1)\alpha(2) + \phi(1)\psi(2)\alpha(1)\alpha(2) = \alpha(1)\alpha(2)\{\psi(1)\phi(2) + \phi(1)\psi(2)\}$$

$$\psi(1)\phi(2)\beta(1)\beta(2) + \phi(1)\psi(2)\beta(1)\beta(2) = \beta(1)\beta(2)\{\psi(1)\phi(2) + \phi(1)\psi(2)\}$$

$$\psi(1)\phi(2)\alpha(1)\beta(2) + \phi(1)\psi(2)\beta(1)\alpha(2)$$

$$\psi(1)\phi(2)\beta(1)\alpha(2) + \phi(1)\psi(2)\alpha(1)\beta(2) \tag{3}$$

Antisymmetric combinations:[2]

$$\psi(1)\phi(2)\alpha(1)\alpha(2) - \phi(1)\psi(2)\alpha(1)\alpha(2) = \alpha(1)\alpha(2)\{\psi(1)\phi(2) - \phi(1)\psi(2)\}$$

[1] A more detailed discussion of this statement is beyond the scope of this book, but it may be remarked that it depends on the fact that the wave equation is a linear differential equation.

[2] These combinations are not normalized [see statement just following Eq. (6) of Chap. IV], that is to say, each one should properly still be multiplied by a constant factor. Inasmuch as only relative probabilities for the location of an electron are important to us in this chapter, we shall, in general, not trouble to normalize the wave functions considered.

$$\psi(1)\phi(2)\beta(1)\beta(2) - \phi(1)\psi(2)\beta(1)\beta(2) = \beta(1)\beta(2)\{\psi(1)\phi(2) - \phi(1)\psi(2)\}$$

$$\psi(1)\phi(2)\alpha(1)\beta(2) + \psi(1)\phi(2)\beta(1)\alpha(2) - \phi(1)\psi(2)\alpha(1)\beta(2) - \phi(1)\psi(2)\beta(1)\alpha(2) = \{\alpha(1)\beta(2) + \beta(1)\alpha(2)\}\{\psi(1)\phi(2) - \phi(1)\psi(2)\}$$

$$\psi(1)\phi(2)\alpha(1)\beta(2) - \psi(1)\phi(2)\beta(1)\alpha(2) + \phi(1)\psi(2)\alpha(1)\beta(2) - \phi(1)\psi(2)\beta(1)\alpha(2) = \{\alpha(1)\beta(2) - \beta(1)\alpha(2)\}\{\psi(1)\phi(2) + \phi(1)\psi(2)\} \quad (4)$$

Now it is found in nature that only antisymmetrical functions occur; for example, in our study of the spectrum of helium we shall see that it may be understood by taking into account only the antisymmetric wave functions. This statement is a somewhat more general statement of the Pauli principle (see Sec. 7.2). Consider, for example, the special case in which ϕ is the same as ψ. Then, according to the Pauli principle, since both electrons are in the same state aside from spin, their spin must be different. It is seen that the first two symmetrical wave functions do not have this property. And if ψ is the same as ϕ, the last two symmetrical wave functions both reduce to $\psi(1)\psi(2)\{\alpha(1)\beta(2) + \beta(1)\alpha(2)\}$. Though the spin state $\alpha(1)\beta(2) + \beta(1)\alpha(2)$ is originally built up of states in which one electron has one spin and another the other spin, this *combination* is generally classified, for reasons into which we cannot go here, with those spin states in which both electrons have the same spin. Thus all the symmetrical states must be discarded in this case. But if ψ is the same as ϕ, all the antisymmetrical wave functions automatically vanish, except the last one, in which the spin function $\alpha(1)\beta(2) - \beta(1)\alpha(2)$ is the one that is recognized as having the electrons in opposite spin states.

In our discussion up to this point, the action of the electrons on each other has been neglected. We now have to see how this interaction affects the results. Only the electrical interaction needs to be considered. The magnetic forces between the magnetic moments associated with their spin and orbital rotations (which have hitherto not even been mentioned in this chapter, but which were briefly discussed at the end of Sec. 6.2) are much smaller, and may still be neglected for the present. The electrical force which we wish to take into account is the electrostatic repulsion between the two electrons. Since it is a repulsion, it is to be expected that the energy of any state will be higher than that obtained by neglecting it, but the exact amount the energy will be raised for any given ϕ and ψ depends upon which one of the possible antisymmetrical combinations is considered. We note that the antisymmetrical wave functions are divided into two factors, a spin factor and a coordinate (or, as it is often called, orbital) factor. If the wave function is antisymmetric in the spin part, it is symmetric in the orbital part, and vice versa. The first three of the wave functions as written are symmetric in the spin and antisymmetric in the orbital part, while only the last one is antisymmetric in the spin and symmetric in the orbital part. The coordinate part is most important in determining the energy to be associated with any given wave function, for it is this part of the wave function which determines the probability that the electrons shall be any given distance r_{12} apart, and the mutual potential energy of the electrons is given by e^2/r_{12}. Now the antisymmetrical orbital factors vanish if the coordinates of the two electrons are the same. For the values of $\psi(1)$ and $\phi(2)$ depend only upon the coordinates of the respective electrons, and if the coordinates of the two electrons are the same, $\psi(1)\phi(2)$ must have the same value as $\psi(2)\phi(1)$. Thus a wave function with an antisymmetrical orbital factor gives a probability of zero for the two electrons to be in the same place and a small probability for them to be near each other. On the other hand, a wave function with a symmetric orbital factor allows the electrons to come close together. When the electrons are close together, their mutual energy is high, so the energy of a state whose wave function is symmetrical in its orbital part is higher than the energy of the corresponding state that is antisymmetrical in the orbital part. As has been seen, there are three states with antisymmetrical orbital factors. On account of the small interaction of the spin magnetic moments and the magnetic moments due to the orbital rotations, these three states may have slightly different energies, but they will all be considerably lower than the energy of the state with a symmetric orbital factor. Corresponding to any given excited state ϕ of one of the electrons there is, then, a singlet state, and with a somewhat lower energy, the members of a triplet.

Of course, the considerations of the preceding paragraph concern chiefly the *difference* between states symmetrical and antisymmetrical in the orbital parts, but built up out of the same original wave functions ψ and ϕ; the actual total energy, however, is determined principally by the particular states that ψ and ϕ represent.[1] Consider, for example, all those term values or energy levels of helium for which ψ is the lowest s-state, namely, the 1s-state, and ϕ represents some other s-state. To distinguish the various ϕ's, they may be written as ϕ_n, using the total quantum number n, which goes with the particular ϕ, as a subscript. The energy levels, which go with some of the lower values of n, are shown schematically in Fig. 42, the vertical height of the horizontal line representing the energy of the particular state.

FIG. 42.—Energy levels of He, with one electron in the 1s-state and the other in some s-state (total quantum number given by n). The energy of the lowest state of He^+ is taken as the zero of energy (top horizontal line). The indicated separation of the members of triplet states is purely schematic. When both electrons are s-electrons, the members of a triplet actually coincide in energy, if the atom is not in an external field; in any case, the true separation would scarcely be visible on the scale of the diagram.

A somewhat more detailed consideration of the energies of the different states will be in order. $n = 1$ represents a very special case, as $\phi_1 = \psi$, and as we have seen, there is but one state whose wave function is different from zero, namely, the one for which the coordinate part is symmetrical. Now the ionization potential of He^+, *i.e.*, the energy necessary to just remove the remaining

[1] It will be recognized that this is the basis of the considerations of Chap. VII. There we assumed that the energy of a given electronic configuration depended upon the quantum numbers of the various electrons, and we did not allow for exchange of electrons at all. This is a sufficiently good approximation for many purposes, as will be evident from the comparison of ionization potentials for comparable singlet and triplet states, given in the next paragraph; at the same time, it will be seen that the exchange effect is by no means vanishingly small.

electron, is 54.1 volts. If there were no interaction between the two electrons of He, the energy necessary to remove both electrons would be twice this, or 108.2 volts. Actually, the energy necessary to remove one electron from He is only 24.5 volts, the sum of the two ionization potentials being 78.6 volts. It is thus seen that the mutual repulsion of the electrons has a large effect on the energy. The energy necessary to remove one electron is considerably greater than the ionization potential of the hydrogen atom which is 13.5 volts; it is obvious, however, that one electron to some considerable extent shields the other electron from the charge on the nucleus,[1] as was already seen in Sec. 7.3. If one of the electrons is in an excited state, then we should expect the electron in the lower state to have a still greater shielding effect. Thus in the state represented by ϕ_2, the energy necessary to pull away the electron in the $2s$-state is 4.74 volts if the atom is in the triplet state and 3.95 if it is in the singlet state, as compared with 3.38 volts to pull off a $2s$-electron from hydrogen. Parenthetically, it may be remarked that in the particular case under consideration, where ϕ_2 represents an s-state, the energies of all three states of the triplet coincide. In other cases, the three triplet levels differ from each other by very small amounts, but for a few states the differences have been measured.

The preceding discussion will have given some idea of the nature of the helium spectrum. We shall now proceed to consider Heitler and London's approximate theory of the hydrogen molecule,[2] making use of many of the ideas developed in connection with the helium atom. In particular, the device of setting up certain approximate wave functions, in which certain potential-energy terms are neglected, will again be used. As in the case of the helium atom, these approximate wave functions will be sufficiently good to make possible a rough estimate of how the neglected potential-energy terms affect the energy levels. Again we shall not attempt to carry out quantitative calculations, but

[1] To say that there is mutual repulsion between the electrons, and to say that one electron shields the other from the nuclear charge, are clearly two ways of expressing the same thing.

[2] A useful review of the early work on the hydrogen molecule and the hydrogen molecule ion was written by Pauling, *Chem. Rev.*, **5**, 173 (1928), which may be consulted for original references. Other accounts are to be found in some of the books listed on pp. 476*f*.

shall be content with the qualitative picture this procedure gives us.

10.2. The Hydrogen Molecule.—Reference is once again made to Fig. 41 for the notation. We shall start our discussion of the hydrogen molecule by supposing that it is a simple combination of two atoms with one of the electrons definitely attached to nucleus A and the other to nucleus B. It will be supposed, as a first rough approximation, that these two atoms do not affect each other at all, an approximation which is good if they are far apart, but becomes increasingly poorer as they come closer together. If it is electron 1 that is attached to nucleus A and electron 2 that is attached to nucleus B, this approximation may be thought of as the result of neglecting the terms

$$-\frac{e^2}{r_{A2}} - \frac{e^2}{r_{B1}} + \frac{e^2}{r_{12}}$$

in Eq. (1); whereas, if electron 2 is on nucleus A and electron 1 on nucleus B, the terms $-\frac{e^2}{r_{A1}} - \frac{e^2}{r_{B2}} + \frac{e^2}{r_{12}}$ are neglected.

The wave function corresponding to the lowest energy of the atom formed from nucleus A, if all interaction with the other atom is neglected, will be designated as ψ_A; the corresponding wave function for the other atom will be ψ_B. These are simply wave functions for the lowest state of the hydrogen atom, the only difference between them being that they are centered on different points of space. If electron 1 is on atom A, the wave function ψ_A is a function of the coordinates of this electron, x_1, y_1, and z_1; to indicate this, we shall write $\psi_A(1)$; similarly, $\psi_B(2)$ is a function of x_2, y_2, and z_2, the coordinates of the other electron, and represents this electron in the lowest energy state of the atom formed by it with nucleus B. The wave function for the system is $\psi_A(1)\psi_B(2)$. This is very similar to the wave function for the helium atom and means, of course, that the probability that the coordinates of electron 1 lie between x_1 and $x_1 + dx_1$, y_1 and $y_1 + dy_1$, z_1 and $z_1 + dz_1$ and that, simultaneously, the coordinates of electron 2 lie between x_2 and $x_2 + dx_2$, y_2 and $y_2 + dy_2$, z_2 and $z_2 + dz_2$ is given by $(\psi_A(1))^2(\psi_B(2))^2\, dx_1\, dy_1\, dz_1\, dx_2\, dy_2\, dz_2$. This is anticipated because the two electrons are independent of each other in the approximation we are considering. If electron

2 is on A and electron 1 on B, then the wave function of the system is $\psi_A(2)\psi_B(1)$. It is necessary, in order to get a complete description of the system, to multiply by the proper spin functions, just as in the case of the helium atom.

The wave functions thus set up, which will be like those of expression (2) if ψ_A replaces ψ and ψ_B replaces ϕ, have a fault very similar to that of (2), namely, they do not allow for exchange of the two electrons. This can again be corrected for by taking combinations, and once more it will be only the antisymmetrical combinations that are allowed. In this case, these are the following:

$$\alpha(1)\alpha(2)\{\psi_A(1)\psi_B(2) - \psi_B(1)\psi_A(2)\}$$
$$\beta(1)\beta(2)\{\psi_A(1)\psi_B(2) - \psi_B(1)\psi_A(2)\}$$
$$\{\alpha(1)\beta(2) + \beta(1)\alpha(2)\}\{\psi_A(1)\psi_B(2) - \psi_B(1)\psi_A(2)\}$$
$$\{\alpha(1)\beta(2) - \beta(1)\alpha(2)\}\{\psi_A(1)\psi_B(2) + \psi_B(1)\psi_A(2)\} \qquad (5)$$

An exchange of the two electrons, each one changing its nucleus and, possibly, changing its spin, will cause a change of sign of each of these combinations.

Just as in the case of the helium atom, the wave functions with the antisymmetrical orbital part give a zero probability for the two electrons to be in the same position and a small probability for them to be close together. The wave function whose orbital part is symmetrical, on the other hand, gives a considerable probability for the electrons to be close together. This means, particularly if the distance between the nuclei is of the order of magnitude of the major axis of the ellipse for the lowest state of the hydrogen atom, that there is a considerable probability for both electrons to lie in the region between the nuclei; for, under these circumstances, both ψ_A and ψ_B have fairly large values in this region. On the other hand, the probability of a configuration in which both the electrons are, let us say, on the left of nucleus A in Fig. 41 will be relatively small, for ψ_B falls off rapidly as the distance from B increases (see Sec. 5.7, especially Fig. 23, case $n = 1$). The net result is that there is a marked tendency for the two electrons to lie between the nuclei for the wave function that is symmetric in the orbital part. This produces a lowering of the energy. For, although there is a repulsion between the two close electrons which causes a raising of the energy of the system, this is more than counterbalanced by the attractive forces between the electrons and nuclei (more specifically the attraction of each electron for the nucleus to which it was not originally attached) so that the electrons act as a sort of cementing bond between the nuclei. The forces thus considered are just those due to the terms $-\frac{e^2}{r_{A2}} - \frac{e^2}{r_{B1}} + \frac{e^2}{r_{12}}$ $\left(\text{or } -\frac{e^2}{r_{A1}} - \frac{e^2}{r_{B2}} + \frac{e^2}{r_{12}}\right)$ which were neglected in setting up the potential-energy expressions to determine the original wave functions, and it is seen that there are two negative terms and only one positive. As we have noted before, if the two centers of force are brought continuously closer together, the wave function will continuously change from that for the separated hydrogen atoms to a function representing a state of the helium atom, and the energy will go over from a value equal to the sum of the energies of two hydrogen atoms to that of the particular state of the helium atom, provided the repulsion between the nuclei is neglected. It is fairly obvious, and may be shown by quantum mechanical considerations, that the state just considered (two hydrogen atoms in their lowest states, orbital part of the wave function symmetrical) goes over into the lowest state of the helium atom, which, it will be remembered, was also symmetrical. If we take as our zero of energy that state in which all the particles, the two nuclei and the two electrons, are infinitely separated from each other, we see that the total energy of two widely separated hydrogen atoms is minus twice the ionization potential of hydrogen, or -27.1 volts. The energy of the helium atom in its lowest state is the negative of the sum of the ionization potentials of helium, or -78.6 volts. As the two nuclei, thought of as merely centers of force, are pushed together, the energy of the system continually decreases, as the immediately preceding considerations have indicated, changing from -27.1 to -78.6 volts.

This neglects, however, the nuclear repulsion. At larger distances, the attractive forces just considered prevail over the nuclear repulsion, but at smaller distances the electrons are, so to speak, "squeezed out" of the space between the two nuclei; so they no longer form such an effective bond, and the repulsive forces predominate over the attractive. This results in an effective potential-energy curve between the two nuclei of the general character of that shown in Fig. 34. (Of course, in

Fig. 34 the zero of potential energy has been made to coincide with the asymptotic part of the curve, whereas if the zero of energy is taken as described in the foregoing paragraph the asymptotic part of the curve in this case would come at −27.1 volts. However, this does not affect the shape of the curve.)

The minimum of the curve for the hydrogen molecule occurs when the nuclei are separated by 0.74Å., and the energy difference between the minimum and the asymptote (representing the energy of two separated hydrogen atoms) is 4.718 volts. The hydrogen in its lowest vibrational state has an energy (zero-point energy $= \frac{1}{2}h\nu_0$, see Sec. 9.4) 0.264 volt higher than the energy of the minimum, so the dissociation energy is 4.454 volts. It should be emphasized that we have described the situation resulting from the interaction of two hydrogen atoms in their lowest energy states, and with oppositely directed electron spins, *i.e.*, with antisymmetrical spin function, thus forcing the coordinate part of the wave function to be symmetrical, by the Pauli principle. It should also be stated that the values of the constants involved which have been given are experimental values. The approximate considerations that have been outlined, when carried through, do not yield exactly these values, but do show qualitatively how the attraction between two hydrogen atoms arises.[1]

10.3. The Lowest Repulsive State of the Hydrogen Molecule. If we start with two hydrogen atoms in their lowest energy levels, but with symmetrical spin functions and antisymmetrical coordinate functions, we find that the two electrons have but little tendency to be close to each other, and this results in there being but a small density of electricity in the region between the two nuclei, where the wave function for the center of force A overlaps that for the center of force B. If the repulsion between the two nuclei is neglected, it appears that the action of the two electrons is still to lower the energy of the system, because the two attrac-

[1] That there should be even qualitative agreement seems the more remarkable when it is noted that at the equilibrium position the resultant attractive potential is only a small fraction of the attractive potential when the nuclear repulsion is neglected. In other words, the true potential is a small difference between large quantities, as in the case of the hydrogen molecule ion considered below (Sec. 10.4, and see Fig. 44). That this is true will be obvious from Secs. 10.4 and 10.5.

tive terms $-\frac{e^2}{r_{A2}}$ and $-\frac{e^2}{r_{B1}}\left(\text{or } -\frac{e^2}{r_{A1}} \text{ and } -\frac{e^2}{r_{B2}}\right)$ overbalance the term e^2/r_{12} arising from the repulsion of the two electrons. The net attraction is not so great, however, as in the case considered in the last section, and the state connects with an *excited* level of the helium atom. It is, of course, a triplet state, as there are three symmetrical spin functions, and it will naturally connect with a triplet state of the helium atom. The connection is actually with the $1s\,2p$ triplet state of the helium atom, which has three energy levels in the neighborhood of -57.7 volts. We may describe this situation by saying that the electron that goes into the $2p$-state has been promoted. The attractive force has in this case been able to reduce the energy from -27.1 volts (that of the separated hydrogen atoms, which is of course the same as in Sec. 10.2) only to -57.7 volts instead of to -78.6 volts. On taking into account the repulsion of the nuclei, it is found that it more than counterbalances the attractive force at all distances of the nuclei. There is thus no tendency whatsoever for two hydrogen atoms in a triplet state to form a molecule.

Every state of a pair of hydrogen atoms is connected with some definite state of the helium atom, in a manner similar to that which we have discussed in the cases where both hydrogen atoms are in their lowest energy states. The excited states, which result in both attractive and repulsive curves of various types, are not of great interest for our particular purposes, and we shall therefore not go further into this subject. It will be found discussed in various papers and books on band spectra.

10.4. The Hydrogen Molecule Ion.—Experiments in which the positive ions resulting from an electrical discharge in hydrogen gas have been studied have demonstrated the existence of the hydrogen molecule ion, H_2^+. In this case, there is but one electron moving under the influence of two attracting centers of force. It is thus, in many respects, a simpler system than the hydrogen molecule.

Suppose that in Fig. 43 there are two nuclei at A and B and an electron at C, the distances being designated as indicated. The potential energy of the system, leaving out the repulsion between the nuclei, is equal to $-\frac{e^2}{r_A} - \frac{e^2}{r_B}$. It can be supposed, as a first approximation, that the electron is on nucleus A, uninfluenced by nucleus B, in which case the term $-\frac{e^2}{r_B}$ is neglected and we have a hydrogenlike wave function centered on A, namely, ψ_A; or we can suppose it is on nucleus B, neglecting the term $-\frac{e^2}{r_A}$, with wave function ψ_B. It, of course, makes no difference in the energy whether the electron is on nucleus A or nucleus B; we may therefore expect that the electron will pass freely from one nucleus to the other. To get approximate wave functions which indicate an equal probability of the electron being on either of the two nuclei, we may try the symmetrical and antisymmetrical[1] combinations $\psi_A + \psi_B$ and $\psi_A - \psi_B$. If ψ_A and ψ_B represent the lowest levels of the respective hydrogen atoms, then ψ_A depends only on the distance from A, and ψ_B only on the distance from B. On the plane equidistant from A and B, therefore, ψ_A and ψ_B will be equal, so that $\psi_A - \psi_B$ will be zero, whereas $\psi_A + \psi_B$ will be large. The symmetrical combination therefore gives a large probability that the electron will be between the nuclei, and the antisymmetrical combination gives but a small probability for this configuration. As might be expected, when the nuclear repulsion is allowed for, this results in a net attraction for the symmetrical combination and a net repulsion for the antisymmetrical. For if the electron has a large probability of being in the region between the nuclei where the neglected attractive potential $-e^2/r_A$ or $-e^2/r_B$ is larger in magnitude than the term e^2/r_{AB} arising from the nuclear repulsion, attraction results. Otherwise the nuclear repulsion predominates. $\psi_A + \psi_B$, therefore, represents the lowest state or ground level for H_2^+ and $\psi_A - \psi_B$ represents an excited state giving no molecule formation. The corresponding energies coincide, of course, when A and B are an infinite distance apart, but only then.

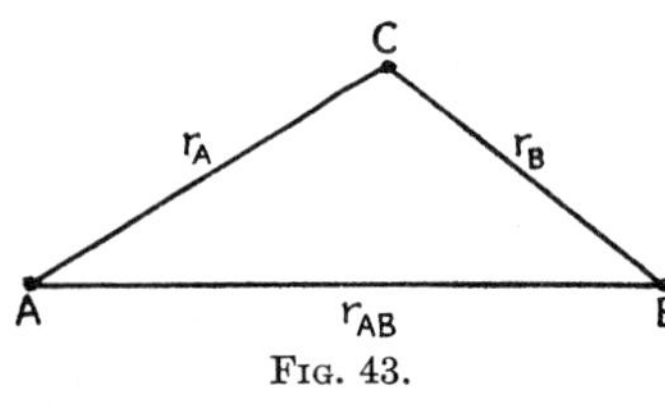

FIG. 43.

[1] Here the terms "symmetrical" and "antisymmetrical" are used in a slightly different sense than previously. As there is only one electron, no question of an exchange of electrons enters, but $\psi_A + \psi_B$ is symmetrical and $\psi_A - \psi_B$ is antisymmetrical with respect to a shift of the single electron from one nucleus to the other.

This shows, in a general way, how attraction results from the action of a single electron in the case of the hydrogen molecule ion. It is possible, however, in this simple case, to work out the complete wave mechanical problem with any degree of exactitude desired, by numerical methods. This has been done by Burrau for the lowest state of the hydrogen molecule ion.[1] Its potential-energy curve is similar in shape to that of the hydrogen molecule, with a minimum when the nuclei are 1.06Å. apart, the minimum being 2.78 volts below the asymptotic value of the curve. The dissociation energy, allowing for the zero-point energy, is 2.64.

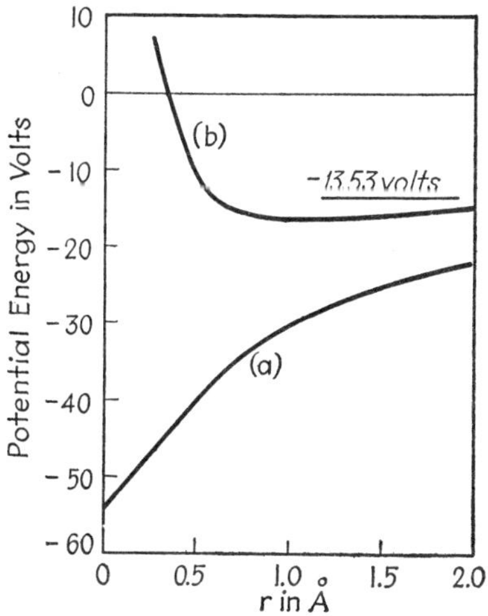

FIG. 44.—Potential-energy curves for the hydrogen molecule ion, (a) neglecting nuclear repulsion, (b) including nuclear repulsion.

This potential-energy curve, the term due to the nuclear repulsion being omitted, is shown schematically in Fig. 44(a). In this figure, the zero of energy is taken as the potential energy when the protons are an infinite distance apart and the electron is an infinite distance from both of them. The dissociation of the hydrogen molecule ion which has been considered, however, is not dissociation into two ions and an electron but dissociation into one ion and an atom. The asymptotic value for the potential energy at great distances of separation is therefore the negative of the ionization potential of hydrogen, −13.5 volts, as shown in the figure.

[1] See footnote 2, p. 132.

It is fairly obvious that when the distance between the nuclei is reduced to zero the state reached will be the lowest state of the helium ion, He^+. Thus when $r = 0$, the potential energy has reached the negative of the ionization potential of He^+, -54.1 volts. When we combine the attractive curve shown in Fig. 44(a) with the repulsion of the nuclei, we get the final curve shown in Fig. 44(b), in which the same zero of energy is used. This curve appears very flat on the scale used, and it is of interest to note that it represents a rather small difference between two large quantities, the attractive and repulsive potentials.

10.5. Alternative Treatment of the Hydrogen Molecule.—The treatment of the hydrogen molecule ion suggests another method (the Hund-Mulliken method—see Sec. 11.3) of considering the hydrogen molecule in which the problem is approached through a different set of approximations.[1] According to this point of view, we consider the two-center problem, and feed electrons one by one into the molecule, just as in our discussion of the building up of the periodic system we fed electrons in one by one in the single-center problem. Putting the first electron into the lowest energy level gives the hydrogen molecule ion, with the potential-energy curve (nuclear repulsion being neglected) shown in Fig. 44(a). (It will be remembered that the zero of energy is the energy when electrons and nuclei are separated.) Putting another electron into the lowest energy level would, provided the mutual interaction of the two electrons could be neglected, give a curve just twice as far below the axis as the curve of Fig. 44(a). The repulsive potential of the nuclei is the same when two electrons are in the molecule as when one is there, but the repulsive potential of the two electrons must be added. This can be taken into account approximately by assuming that the electrons are at the positions of the respective nuclei (which no doubt gives an underestimate since the electrons tend to be close together). This means, then, that the repulsive as well as the attractive potential is approximately doubled, hence the total potential is doubled, and the estimated potential-energy curve for H_2 lies just twice as far below the axis as the curve of Fig. 44(b). The calculated energy of dissociation is about

$$2 \times 2.64 = 5.28 \text{ volts}$$

[1] CONDON, *Proc. Nat. Acad. Sci.*, **13**, 466 (1927). The treatment is considerably more detailed than that given here.

which is about half a volt too large. This is reasonably good agreement, considering the approximations involved, and especially since the repulsions between the electrons were underestimated.

It will be recognized that this is a very suggestive and interesting method of treating the hydrogen molecule. In the general case of more complicated molecules, it is possible to start with two attracting centers of force of greater strength than protons and build up shells for the two-center problem, as has already been done for the one-center problem. Excited states of the molecule are states in which one, or more, of the electrons is not in its lowest possible orbit. For example, in the hydrogen molecule, if one of the electrons is in the state discussed in connection with the hydrogen molecule ion and is represented by $\psi_A + \psi_B$ and the other is in the state represented by $\psi_A - \psi_B$, which, as we saw in Sec. 10.4, corresponds to a higher energy, the molecule will be in an excited state. An exact correspondence cannot be found, however, between the excited states formed in this way and those discussed in Sec. 10.3. This is because both ways of representing the molecule are approximations, and since they are different approximations the results cannot be brought into exact coincidence. As a matter of fact, the method of the present section apparently yields *more* repulsive states starting with the wave functions ψ_A and ψ_B than the method of Sec. 10.3. For we can suppose that the function $\psi_A + \psi_B$ goes over to the $1s$-state of helium when the nuclei are made to coincide, and the function $\psi_A - \psi_B$ goes over to one of the $2p$-states of helium. If there are two electrons with opposed spins in the $\psi_A + \psi_B$ state, this gives the $1s^2$- or ground state of helium. However, if there is one electron in the $\psi_A + \psi_B$ state and one in the $\psi_A - \psi_B$ state, then as $\psi_A + \psi_B \rightarrow \psi$ and $\psi_A - \psi_B \rightarrow \phi$, where ψ and ϕ have the significance attached to them in Sec. 10.1 (ψ here representing the $1s$-state and ϕ a $2p$-state), all four combinations of Eq. (4) would be possibilities. Thus not only the triplet $1s2p$-state, as in Sec. 10.3, but also the singlet state is included. Furthermore, by putting both electrons in the $\psi_A - \psi_B$ state, we shall connect with a $2p^2$-state of helium. Thus, altogether, if we include the ground state, we get six helium levels out of ψ_A and ψ_B, instead of four. The reason for this apparent discrepancy will appear in the next section.

We may at this point insert a few words about the "electron pair." As we shall see very shortly, the paired-electron valence bond plays an important role in the theory of valence. The considerations of the present chapter show that a pair of electrons has a bonding action in the case of the hydrogen molecule. Though a single electron can give a bond, as in the hydrogen molecule ion, a pair of electrons produces a stronger bond. From the discussion of the preceding paragraph, it can be seen why three electrons do not usually give a strong bond. A third electron would have to go into a higher energy state, making the arrangement unstable.[1] Similar statements will hold also for polyatomic as well as diatomic molecules. A bonding electron between any pair of atoms in a polyatomic molecule may be said, with a certain approximation, to be in a definite quantum state. Only two electrons, with opposed spins, can go into the lowest level (in the term "level" we include *both* states with all quantum numbers, other than that of spin, the same). This does not mean that under certain circumstances there will not exist double and triple bonds in which more than two electrons are involved, or even three-electron bonds (see Sec. 16.13), but it does furnish a reason for the outstanding importance of the electron pair in chemistry.

10.6. Comparison of the Approximations Involved in the Two Methods of Treating the Hydrogen Molecule.—The wave functions for the electrons in the hydrogen molecule, obtained in the treatments outlined above, naturally differ from each other. The orbital part of the wave function for the lowest energy state, according to the Heitler-London method of treatment, is, as has been seen,

$$\Psi_{HL} = \psi_A(1)\psi_B(2) + \psi_B(1)\psi_A(2) \quad (6)$$

An approximation for the corresponding wave function in Mulliken's scheme is obtained by noting that when the hydrogen molecule is in its lowest energy level the state of both electrons is described by the lowest hydrogen molecule ion wave function $\psi_A + \psi_B$. Thus the combined wave function for the two electrons can be written as

[1] See Sec. 7.2. The Pauli exclusion principle holds for molecules as well as atoms. For further discussion see Sec. 11.3.

$$\Psi_M = \{\psi_A(1) + \psi_B(1)\}\{\psi_A(2) + \psi_B(2)\}$$
$$= \psi_A(1)\psi_A(2) + \psi_B(1)\psi_B(2) + \psi_A(1)\psi_B(2) + \psi_B(1)\psi_A(2) \quad (7)$$

Of course, neither of these is the exact wave function for the hydrogen molecule, for they both involve approximations. They differ from each other in that Ψ_M contains the terms $\psi_A(1)\psi_A(2)$ and $\psi_B(1)\psi_B(2)$ which represent a condition in which both electrons are on the same nucleus;[1] these terms are, in fact, of equal importance with the other terms. On the other hand, in the Heitler-London theory, such terms are deliberately excluded. Now on account of the strong repulsion between electrons, it is obvious that the true wave function will not give so great a possibility of the two electrons being on the same nucleus as the Mulliken method of approximation. On the other hand, there is *some* probability, even though it is small, of their being simultaneously on the same nucleus. This is entirely neglected in the Heitler-London method. This suggests that the true wave function would be better represented by an intermediate form,[2] a linear combination of Ψ_{HL} and Ψ_M,

$$\psi_A(1)\psi_B(2) + \psi_B(1)\psi_A(2) + a\{\psi_A(1)\psi_A(2) + \psi_B(1)\psi_B(2)\} \quad (8)$$

where a has some value between 0 and 1. This is indeed true, but the form (8) is still not the correct wave function, even though it is nearer to it than either (6) or (7). It still contains certain approximations, which may not all have been explicitly stated, but are, nevertheless, implied.[3]

[1] It is the presence of these two terms, which represent electronic states which are not considered in the Heitler-London approximation, that are responsible for the extra potential-energy curves that arise from the approximation of Sec. 10.5.

[2] Mulliken, *Phys. Rev.*, **41**, 65 *ff.* (1932); Slater, *Phys. Rev.*, **35**, 514 (1930).

[3] It is obvious that when the distance between the nuclei is very large the energy of the system with both electrons on one atom will be very large compared to that with the electrons on different atoms. The electron affinity of hydrogen is less than 1 volt, and the ionization potential is over 13; so the energy gained by putting an electron on one hydrogen by no means compensates the energy required to take it off the other. But for small internuclear distances, this is no longer the case, at least, to so great an extent, so if a is adjusted so that Eq. (8) gives a good approximation for small distances, it must give a very poor approximation when the distance between nuclei

It is seen that in (8), as written, the quantity a may be regarded as a parameter which may be varied until its best value is found. By extension of this method, using more parameters, and somewhat different forms of the wave function, better approximations can be secured, and the method may be developed into a series of successive approximations by means of which the true wave function may probably be approached with any degree of accuracy which time and patience will allow. Once the correct wave function has been found, the correct energy may be obtained also. A treatment of this sort has been carried out for the hydrogen molecule by James and Coolidge,[1] and these authors have found a value of the energy of dissociation which is very close to that observed experimentally. There is thus no doubt that quantum mechanics gives the correct result, though in many respects the exact calculation is less suggestive and the concepts less readily visualized than is the case with the approximate methods.

The more exact methods have also been applied to the calculation of the energy levels of helium, and to a few simple molecules other than hydrogen. Even in the case of hydrogen and helium, however, the calculations are exceedingly involved.

10.7. Properties of Hydrogen Atoms.—It may be well to insert at this point a brief account of the properties of hydrogen atoms as contrasted to hydrogen molecules.[2] As the two atoms in a molecule are bound together with considerable energy, it might well be expected that hydrogen atoms would be much more reactive than hydrogen molecules, and this is, indeed, the case. The large energy of binding is further reflected in the methods by which hydrogen atoms may be prepared, for it is necessary that large quantities of energy should be available.

Hydrogen atoms were first prepared by Langmuir, using a glowing tungsten wire. If a tungsten wire is heated by an electric current in the presence of hydrogen, more energy is required to keep it at a certain temperature than in vacuum, because heat

---

is large. This does not mean, however, that the wave functions $\psi_A \pm \psi_B$ are not good approximations, even when the distance is great, in the case of the hydrogen molecule ion which has only one electron.

[1] James and Coolidge, *J. Chem. Phys.*, **1**, 825 (1933).

[2] For a review, see Bonhoeffer, *Ergebnisse der exakten Naturwiss.*, **6**, 201 (1927).

is conducted away by the hydrogen. If the tungsten wire is heated above 2000°C., it is found that the cooling effect of the hydrogen becomes abnormally large. This is explained on the assumption that the hydrogen molecules are beginning to dissociate and are abstracting the necessary energy from the wire. Further evidence is the fact that if the hydrogen is present at a very small pressure (10^{-2} to 10^{-3} mm.) the gas may be noticed to gradually disappear (clean-up effect). It was found that the hydrogen disappeared into the glass walls of the vessel and could be driven out again by heating them. This occurs only if the tungsten wire is heated to a sufficiently high temperature. It obviously indicates that the properties of the gas have changed, and is readily explained by the assumption that hydrogen atoms are produced and absorbed by the glass. The enhanced chemical activity of the gas can also be studied at low pressures, but this is not a good way of producing hydrogen atoms for this purpose.

From the amount of the cooling of the tungsten wire, Langmuir found it possible to make an estimate of the equilibrium constant for the dissociation of hydrogen at various temperatures and so obtain the heat of dissociation. This calculation was based on the assumption that equilibrium was established between atoms and molecules absorbed on the tungsten, and involves consideration of the rate of transfer of atoms and molecules between gas phase and absorbed layer, as well as the diffusion of atoms from the wire. By measuring the amount of cooling of the wire at various pressures and temperatures, the percentage of hydrogen molecules dissociated and the heat of dissociation could be calculated. The value obtained was in reasonably good agreement with later more exact results obtained by spectroscopic methods. The results obtained at different temperatures and pressures were consistent with the assumption that an equilibrium was being measured. For orientation, Table 6, giving the approximate degree of dissociation at various temperatures and pressures, is inserted.[1]

Wood and later Bonhoeffer and others have prepared hydrogen atoms by passing hydrogen at low pressures (0.1 to 1 mm.) along a tube through which an electric discharge was maintained by means of a potential of 5,000 to 20,000 volts. The gas could be

[1] Langmuir, *J. Am. Chem. Soc.*, **37**, 442 (1915); and *Gen. Elec. Rev.*, **29**, 155 (1926).

TABLE 6.—DEGREE OF DISSOCIATION OF H_2

T(°K)	2000	3000	4000
At 760 mm.	0.0012	0.090	0.62
At 1 mm.	0.034	0.93	0.999

led away from the discharge, and was found to be very reactive. This reactivity would be retained after the gas had passed through considerable lengths of glass tubing provided there was a small amount of oxygen present, which poisoned the wall; otherwise the hydrogen atoms were absorbed on the wall and recombined there. That the abnormal reactivity of the gas thus produced is in reality due to the presence of hydrogen atoms is indicated by the fact that hydrogen atoms are known to be present in the discharge, because their spectrum is there, and there is no evidence to indicate that the effects are due to any other active body. In particular, the possibility of effects due to the presence of ions has been considered by Bonhoeffer. The rate at which the active bodies disappear, as discussed below, is also consistent with the assumption that they are atoms which on disappearing recombine.

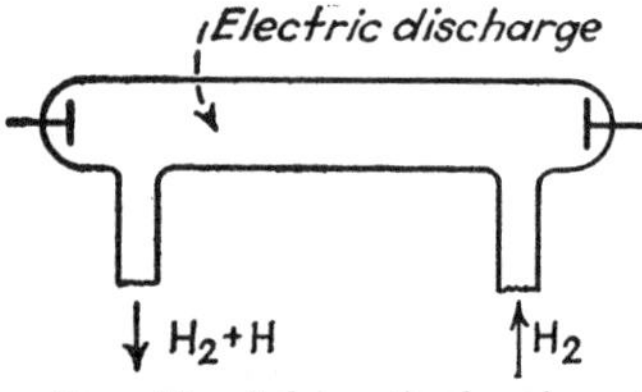

FIG. 45.—Schematic drawing of discharge tube for production of hydrogen atoms.

The gases from the discharge tube can be used to study the properties of hydrogen atoms. It is found that they are very reactive, and will undergo many reactions which hydrogen molecules will not. The atom will react with mercury liquid to form HgH gas; one sees over the surface of the mercury a bluish light in which the spectrum of this molecule is present, as well as the 2537Å. mercury line. The reactions that occur are supposed to be the following:

$$\text{Hg} + \text{H} \rightarrow \text{HgH (excited)}$$
$$\text{HgH (excited)} + \text{H} \rightarrow \text{Hg (excited)} + \text{H}_2$$

Hydrogen atoms react at room temperatures with oxygen to form hydrogen peroxide,[1] and they also react rapidly at room

[1] Water is also formed, but may be a secondary product. It should be remarked that even the hydrogen peroxide is undoubtedly formed as the product of a chain reaction involving unstable intermediates,

temperature and in the dark with chlorine, bromine, and iodine. Molecular hydrogen does not react with oxygen at room temperature, only slowly with chlorine in the dark, and extremely slowly, if at all, with bromine and iodine. Hydrogen atoms do not react with water, ammonia, or methane,[1] but they do react with various higher hydrocarbons, forming hydrogen molecules and free radicals. They react with hydrogen sulfide, arsine, hydrogen chloride, and hydrogen bromide. Hydrogen atoms reduce numerous solids, such as sulfur, phosphorus, arsenic, antimony, and a number of oxides and salts. The effect of hydrogen atoms on the oxides and salts was observed by passing hydrogen through the discharge and then over the substance to be investigated; reduction, when it occurred, could be seen to take place by means of the change in color.

In the case of sulfur, H_2S is formed, and as H_2S is volatile, the vessel containing the sulfur loses weight. This loss of weight can be used as an indicator of the rate of recombination of hydrogen atoms as the gas containing them passes down a tube. The vessel containing the sulfur is placed at varying distances along the tube as the gas runs down the tube at a constant streaming velocity (holding, of course, the conditions in the discharge constant), and the relative losses of weight are compared. In this way, knowing how fast the gas is moving down the tube, the rate of recombination can be estimated. Other experimental methods are also available.[2] It is found that only a small fraction of the collisions of hydrogen atoms (calculated assuming diameters of the order of 10^{-8} cm.) are effective in causing recombination. This low efficiency is, however, easily understood, for two hydrogen atoms obviously have enough energy to dissociate, else they would already be combined as a molecule. If the spins of the electrons are correct, they will attract each other when they come together in collision, but they will separate again unless another body is present to take away the excess energy. The recombination of hydrogen atoms

[1] This statement is based on the experiments of Boehm and Bonhoeffer, *Zeits. physik. Chem.*, **119**, 385 (1926), in which a stream of hydrogen containing hydrogen atoms from a discharge is mixed with the other gas. There seems to be no doubt that hydrogen atoms at least react much more slowly with water, ammonia and methane than with the other gases mentioned.

[2] See, *e.g.*, Amdur, *J. Am. Chem. Soc.*, **60**, 2347 (1938), which has references to other recent work.

thus requires a three-body collision. The third body need not be a hydrogen atom, or a gas molecule of any kind, however; it may, for example, be a solid. Hydrogen atoms combine very readily on the surfaces of many solids, in particular, metals, which are strongly heated by the energy of dissociation thus released.

10.8. Ortho- and Parahydrogen.—In preceding pages of this chapter, we have discussed the electronic wave functions of the hydrogen molecule. There are phenomena, of some interest, that are explained by reference to the *nuclear* wave functions.[1] It is natural to assume that the nuclear wave function, like the wave function of an electron, may be divided into two parts, a spin part and a coordinate part. Further, a proton, like an electron, also has a spin angular momentum, whose projection along an axis may take the values $\pm \frac{1}{2}h/2\pi$; so there will be three symmetrical spin wave functions and one antisymmetric spin wave function, with respect to interchange of the two nuclei.

The symmetry properties of the coordinate part of the nuclear wave functions must now be considered. If the orientation of the line joining the centers of the two nuclei is expressed by means of the usual polar coordinates ϕ and θ (see Fig. 13, page 49), it is seen that interchanging the two nuclei is equivalent to changing ϕ into $\phi + \pi$ and changing θ into $\pi - \theta$. In Chap. IX, we considered some of the properties of diatomic molecules, in general, and noted that since the force between the two atoms is directed along the line of centers the motion of these atoms has much in common with the motion of the electron in the hydrogen atom. In particular, the wave function may be written in the form $R\Phi\Theta$, where R is a function only of r, the distance between the atoms, and Φ and Θ are functions, respectively, of the two variables ϕ and θ. It will be clear from the statement made above that the symmetry properties of this wave function will depend upon the factor $\Phi\Theta$, and it is found that it is symmetrical if j, the rotational quantum number, is even, and antisymmetrical if it is odd. We shall not attempt a detailed explanation of this fact, but may call attention to an analogous case, which is somewhat simpler. If there were a molecule that could rotate only in

[1] For a more complete account of the subject treated in this section see Farkas, "Light and Heavy Hydrogen," Cambridge University Press, 1935.

a plane, then by introducing the reduced mass it could be treated as a plane rotator of the type considered in Chap. IV. Interchange of the nuclei would correspond to changing χ of Sec. 4.4 to $\chi + \pi$, and examination of Fig. 12 will show that wave functions are symmetrical if the rotational quantum number is even, antisymmetrical if it is odd.

If the complete nuclear wave function for the hydrogen molecule is to be antisymmetrical, then a symmetrical coordinate function must always be combined with an antisymmetrical spin function, and vice versa. The molecules of the type in which the spin function is antisymmetrical and the coordinate part symmetrical are called "parahydrogen"; the other combination gives "orthohydrogen." For parahydrogen the rotational quantum number j has the values 0, 2, 4, . . . ; for orthohydrogen $j = 1, 3, 5, \ldots$.

One would, therefore, expect a different proportion of ortho- and parahydrogen at high and low temperatures. If the temperature is very low, all the hydrogen will be in its lowest rotational energy level, if equilibrium is established, and hence will be parahydrogen. At high temperatures, the average molecule will have a large rotational quantum number. The number of orientations $2j + 1$ will not differ very much from one j to the next, if j is large enough, so that one might at first expect equal numbers of the two kinds of hydrogen at high temperatures. However, it must be remembered that there are three kinds of symmetrical *spin* functions and only one antisymmetrical. Therefore at high temperatures (room temperature is in this sense a high temperature), orthohydrogen should predominate in the ratio of 3:1. That this is actually true is indicated by the spectrum of hydrogen, which shows alternating intensities.

It is found, however, that if ordinary hydrogen is cooled to a low temperature the ratio of ortho- to parahydrogen actually remains unchanged over long periods of time; ortho- and parahydrogen thus behave much like separate compounds. The transition from one type to the other is very slow, except under conditions such that the atoms of an H_2 molecule are separated, or in the presence of a substance the molecules of which have a magnetic moment. Of course, if the atoms are separated they will recombine in such a way as to give an equilibrium mixture of ortho- and parahydrogen, even though the directions of the spins of the individual atoms are not changed; a magnetic substance is able actually to cause changes in the directions of the individual proton spins. Pure parahydrogen may be obtained at low temperatures by cooling hydrogen in the presence of activated charcoal, which acts as a catalyst for the conversion, probably due to magnetic interaction. If the charcoal is then removed, the parahydrogen can be heated up to room temperature, and will not go over into the ordinary mixture for a long time. Its properties differ from those of ordinary hydrogen, *e.g.*, it has a different specific heat and a different vapor pressure.

Other diatomic molecules, composed of two atoms of the same kind, are known from spectral observations to have similar properties. The phenomena are, in general, much more complicated, for the nuclei of other atoms than hydrogen consist not of a single proton, but of a number of particles closely cemented together. These matters are beyond the scope of this book, and will not be considered further here.

CHAPTER XI

THEORIES OF VALENCE

As has been indicated in previous chapters, there are at least two different kinds of chemical combination, (1) that which is due primarily to the attraction between a positive and a negative ion, and (2) that which is due to the formation of an electron-pair bond, as exemplified by the hydrogen molecule. The first type is known as "polar" or "ionic binding"; the second type is called "nonpolar" or "covalent binding." The distinction between the two types of bonds was probably first made by Abegg.

If a pair of atoms combines chemically, it may generally be expected that, if one of them has one or more electrons which are relatively easily removed, while the other has a considerable tendency to add one or more electrons, the bond will be polar. On the other hand, if the two have nearly equal tendencies to lose or add electrons, the bond will be covalent. If two atoms of the same kind combine, the situation is ideal for the formation of a nonpolar bond. In Chap. XII, these ideas will be given a more quantitative treatment, and the gradation between polar and nonpolar bonds will be considered in connection with the properties of simple gas molecules. For the present, it may be assumed that a bond can be classed as belonging to one type or the other.

The present chapter will be chiefly devoted to the formulation of electron structures associated with chemical bonds and, in particular, covalent bonds, though the place of polar bonds in the scheme will also be indicated. The electron-pair bond, similar to that formed in the hydrogen molecule, as discussed in Chap. X, will be the basis of this discussion.

In the present chapter, not much attention will be given to *properties* of the chemical bonds, such as bond energy. The bond energy and certain other properties will be considered in Chap. XII, and the discussion of other characteristics of the covalent bond, of a somewhat different nature, will be deferred to Chap. XV.

11.1. Lewis's Theory of Valence.—The importance of the electron pair was early recognized by Lewis,[1] long before the present concepts of the quantum theory which, as we saw in the last chapter, so clearly predict such a phenomenon, had crystallized. He called attention to the fact that practically all chemical compounds contain an even number of electrons. Almost the only common compounds existing at room temperature that have an odd number of electrons are NO, NO_2, and ClO_2. NO_2, as is well known, tends to polymerize, giving N_2O_4 which has an even number of electrons, and there is also some tendency for NO to double up. A few other compounds with an odd number of electrons are discussed by Lewis, but these are all unstable substances existing only under special circumstances.

Lewis also emphasized the tendency of atoms to fill up a shell of eight electrons. This idea, which was independently considered by Kossel and had been foreshadowed by Abegg and by Parson, has been greatly developed by Lewis and by Langmuir, and has been of much utility in the discussion of both polar and nonpolar compounds. Thus consider the formation of NaCl. The outer shell (M-shell) of electrons of a sodium atom contains but one electron, whereas the outer shell of a chlorine atom (also an M-shell) contains seven electrons. If the sodium loses an electron, the sodium ion that is left has an outer shell (an L-shell, the only M-electron having been removed) of eight electrons, whereas the chlorine ion also has an outer shell of eight. The electron configurations of these ions are exactly the same as those of the nearest rare gas atoms and, undoubtedly, represent very stable states. In the case of NaCl, we therefore expect sodium and chlorine ions to be formed as a preliminary step, and the combination to take place between them.

On the other hand, when a Cl_2 molecule is formed, the bond is an electron-pair bond, and Lewis assumed that each member of the pair was shared by both chlorine atoms and could complete the octet, or group of eight, of each one. Lewis adopted the scheme of writing formulas in which the electrons in the outermost shells were represented by dots. Thus the formula of Cl_2 would be $:\underset{\cdot\cdot}{\overset{\cdot\cdot}{Cl}}:\underset{\cdot\cdot}{\overset{\cdot\cdot}{Cl}}:$, the electrons being grouped together as pairs,

[1] See Lewis, "Valence and the Structure of Atoms and Molecules," Reinhold Publishing Corporation, 1923, for an extended account.

the middle pair belonging to both chlorine atoms equally. Much use will be made of this notation.

There are a great many nonpolar compounds in which one or more of the atoms involved have a shell of eight, when the shared electrons are counted for both atoms. We give a few examples:

```
      ..
     :Cl:                        H
  .. .. ..      .. .. ..         ..
 :Cl:C:Cl:     :O:S:O:         H:C:H
  .. .. ..      ..   ..          ..
     :Cl:                        H
      ..
```

In the case of CH_4, the hydrogen atoms have but two electrons in their shells, but this really corresponds to an octet in other atoms inasmuch as a K-shell can have but two electrons and the corresponding rare gas has but two electrons in its outer shell. There are many complex ions in which all the atoms have an outer shell of eight (or the equivalent). Thus we have

```
 ┌    ..    ┐--      ┌    ..     ┐-          ┌   H   ┐+
 │   :O:    │        │   :O:     │           │   ..  │
 │ .. .. .. │        │ .. ..  .. │           │ H:N:H │
 │ :O:S:O:  │   ,    │ :O:Cl:O:  │    and    │   ..  │
 │ .. .. .. │        │ .. ..  .. │           │   H   │
 │   :O:    │        │   :O:     │           └       ┘
 └    ..    ┘        └    ..     ┘
```

the net charge on the ion being indicated outside the brackets. In the case of a double or triple bond, Lewis assumes that two or three pairs of electrons, respectively, are shared between atoms in order to complete the shell of eight. Thus we have

```
 H.   .H
  :C::C:       and      H:C:::C:H
 H˙   ˙H
```

The formation of double and triple bonds is an indication of the tendency of the combining atoms to complete the shell of eight. Further examination of the SO_2 molecule shows that all the atoms could complete their shell of eight by the formation of a double bond, thus, O::S:O: (with dot pairs above and below the O atoms and above the S) As a matter of fact, this is probably nearer to giving a true description of this molecule than the formula previously given. It is rather unsymmetrical, but it is not to be expected that the double bond would always stay at a particular one of the S—O bonds. It is undoubtedly continually shifting from one side to the other. There are also other possibilities for the structure of SO_2. Thus either one of the oxygens could have but six electrons, instead of the sulfur. It may be that each one of the configurations has some grain of truth in it, and that the truth approximates some intermediate situation, or a tautomeric mixture of all possibilities,[1] with the two double-bonded structures (*i.e.*, the structures with the double bond to one or the other of the oxygens) predominating.

It should be noted that the character of the electron-pair bond is not dependent upon which one of the atoms involved originally furnished the electrons. Both electrons of a pair can come from one of the atoms (called by Sidgwick[2] the "donor," the other atom which shares these electrons then being called the "acceptor"), or one of the electrons may come from each atom, or in the case of ions, one of the electrons may have been furnished by an atom that is no longer present at all.

It is always possible, in purely formal fashion, to consider any electron pair as being formed by one electron from each atom. Thus SO_4^{--} may be considered to be formed from sulfur and oxygen ions, of structure $(\cdot\dot{\underset{\cdot}{S}}\cdot)^{++}$ and $(:\ddot{\underset{\cdot\cdot}{O}}\cdot)^{-}$, respectively. Whether this is a *good* way of viewing the situation, *i.e.*, whether the forces involved in the formation of the bond can really be considered to be the forces between the ions plus the ordinary force expected from an electron-pair bond, is another question—in this case, it may be a reasonably good approximation.

11.2. London's Theory of Valence.—As has been noted, Lewis stresses the stability of the octet of electrons. But there are compounds and ions in which it must be assumed that the central atom has more than an octet, as shown by PCl_5, SF_6, $(CoCl_6)^{---}$. London[3] has given a formulation of the theory couched in more modern language, which also allows for these cases, and a somewhat modified version of his theory will be given here.

London's theory deals primarily and directly with those cases in which the two electrons in the pair come from different atoms. A case like SO_2 could be handled by formulating it in terms of the ions, as was done above, so that each ion involved in the bond furnishes one of the electrons of the pair.

[1] This "resonance phenomenon" is certainly of importance in many instances, and is discussed in some detail later (see Sec. 12.5).

[2] Sidgwick, "The Electronic Theory of Valency," Oxford University Press, 1927. This book also contains a rather extended account of some of the aspects of the theory of valence.

[3] London, *Zeits. Physik.*, **46**, 455 (1928).

In order to have the possibility of forming an electron-pair bond, there must be one electron in each atom (or ion) in such a condition that it can pair with the other electron, so that their spins cancel each other. An electron that is already paired with another electron in the same atom is not available, while it is still so paired, for bond formation. Two electrons within a single atom are said to be paired if they have opposite spins and all their other quantum numbers are the same.

The question as to whether an electron that is already paired within an atom can become available for bond formation with an electron from another atom depends upon whether it can be unpaired without expenditure of too much energy. For purposes of discussion, London thus considers the unpairing process as an intermediate step in the formation of a compound. Now according to London's theory, an electron is readily unpaired, provided this can be done without change in the value of the total quantum number n of that electron. It is readily seen that this leads at once to the prediction that NCl_3 may be formed but not NCl_5, and that PCl_5 is possible. This is true, because the outer electrons of nitrogen are in the shell for which $n = 2$. Since two states form a pair provided all their quantum numbers except the spin quantum number are the same, we see that there are only four pairs of states for which $n = 2$, namely, one pair of s-states and three pairs of p-states. Since there are five electrons, it is obvious that only three can be unpaired if all of them are to remain in the $n = 2$ shell. Therefore there will be only three available to combine with the odd electron of a chlorine atom, so that NCl_3 is a possible compound, but not NCl_5, which is in accord with experimental facts. On the other hand, the value of n for the outer electrons of phosphorus is 3, and there are nine such states, one s-state, three p-states, and five d-states. It is therefore readily possible to have five electrons unpaired without going out of the $n = 3$ shell. In similar fashion, it is seen that it is possible to explain the existence of SF_6 and compounds in general in which one, or more, of the atoms has a shell of more than eight.

Some presumptive doubt, at least, as to the basis of this theory appears when an attempt is made to estimate the actual energy involved in the unpairing of the electrons. Thus the theory supposes that PCl_5 exists because it is possible to unpair the electrons in the phosphorus without leaving the $n = 3$ shell. But as a matter of fact, in the building up of the periodic system, the 4*s*-levels fill up *before* the 3*d*-levels. Now to be sure, we cannot say that the order of the levels in phosphorus will be the same as in K, the element in which a 4*s*-electron first appears, but it does seem likely that at least not much is gained by the fact that to unpair the electrons none have to be taken to the 4*s*-level. But another factor than the energy required to unpair the electrons enters into the problem, namely, the bond energy. It is altogether probable not only that the bonds which would be formed if some of the electrons were excited to the 4*s*-level would not be so strong as the bonds formed when this is not necessary, but also that this is a determining factor.

It is rather unfortunate, however, that it is often not really possible in the present stage of our knowledge to separate and evaluate the various factors that determine the stability of compounds. There seems to be no doubt that the stability of PCl_5 relative to PCl_3 is greater than the stability of NCl_5 relative to NCl_3, for PCl_5 vapor at 1 atm. dissociates only slightly at 100°C., whereas NCl_5 is unknown; but this difference may be due in large part to the mere fact that there is more room to pack atoms about the relatively large phosphorus atom, and what the relative importance of this factor as compared with the electronic structure may be is very difficult to decide. We shall have more to say, both about the coordination number (*i.e.*, the number of atoms that can surround a central atom) and about bond formation, in later chapters.

London summarizes the results of his theory by saying that, in agreement with experiment, fluorine should be only univalent, oxygen only bivalent, and nitrogen only trivalent. This is true, provided one is careful in his definition of valence, or if one attempts to apply it only in those cases in which one of the electrons of each pair comes from one of the atoms involved and the other from the other atom. The cases in which nitrogen is apparently pentavalent are always compounds in which one of the atoms is held primarily by ionic forces. Thus in NH_4Cl, for example, the Cl^- is held to the NH_4^+ by ionic forces, and the chlorine cannot be considered to be bound by a covalent link to nitrogen.

11.3. The Hund-Mulliken Theory of Valence.—We turn now to a consideration of the views advocated by Hund, Mulliken, Herzberg, Lennard-Jones, and others.[1] Their point of view is to consider the electronic orbits in the field of force produced by several centers of force. Each orbit in this complex field of force will be determined by definite quantum numbers, and will have a certain energy that depends upon the quantum numbers and also the relative positions of the various centers of force. We then feed electrons one by one into the quantum states, starting with those of lowest energy, just as we did with atoms in Chap. VII. The hydrogen molecule has already been considered from this point of view in Sec. 10.5.

It is very helpful to consider the gradual change in the energy of the quantum states on starting with a separate pair of atoms, and allowing them gradually to approach each other until they form a united atom with atomic number equal to the sum of the atomic numbers of the separate atoms. It was seen in Sec. 10.3 that an electron may be promoted in this process; that is to say, its value of the quantum number n may be greater in the united atom than in the separated atom. In the case of the hydrogen molecule, it was not necessary for promotion to occur. But in some cases, the united atom cannot be formed without the promotion of one or more electrons. For example, if two helium atoms in their normal states, each containing two $1s$-electrons, are united to form a beryllium atom, two of the electrons must be promoted to the L-shell. The process of promotion requires energy. In the intermediate situation, where there are neither separated atoms nor a united atom, promotion will have begun, and as it requires energy it contributes to the instability of the configuration. Thus we see why the He_2 molecule is not stable. Promotion is due to the operation of the Pauli exclusion principle, and is associated with repulsive forces. This repulsion

[1] See, *e.g.*, Hund, *Zeits. Elektrochem.*, **34**, 437 (1928); Mulliken, *Chem. Rev.*, **9**, 347 (1931); Van Vleck and A. Sherman, *Rev. Mod. Phys.*, **7**, 167 (1935). Hund, Mulliken, and others have also written many papers on molecular spectra, treating not only the lowest electronic state of molecules, but also excited states. As we are not primarily interested in excited states, this matter is not treated in the present volume. For a review, primarily of diatomic molecules, see Mulliken, *Rev. Mod. Phys.*, **2**, 60 (1930); **3**, 89 (1931); **4**, 1 (1932).

is essentially the same thing discussed in Sec. 9.1, though if the two atoms are pushed very close together, the direct electrostatic repulsion of the positive nuclei may play a role. It thus appears that the strength of a bond between two atoms will depend to a large extent on how many electrons have to be promoted (or at least are on the way to promotion) when the bond is formed. Electrons that are shared without promotion are bonding electrons, but promoted electrons have an antibonding action.

As we have remarked, the process of promotion is one that takes place gradually as the atoms are brought together. If we speak of an electron as promoted, it means that we consider that electron from the united atom point of view, for only in the united atom is it fully promoted. In many cases, this may be a good approximation, but in the case of the combination of many-electron atoms, it is evident that whether the separated atom or the united atom is the better approximation for the molecule at the actual molecular distance of separation will depend upon which of the electrons we are considering. Only the outer electrons are involved in valence phenomena, and it is obvious that the presence of another atom affects the inner electrons relatively little. The inner electrons may, therefore, best be considered from the separated atom viewpoint, and no promotion of these electrons is involved at the usual molecular distances. Of course, if the two atoms were pushed closer together, then in the case of many-electron atoms there would be many more electrons than could be accommodated in the inner shells, and wholesale promotion of electrons would be necessary. It is, of course, this threat of promotion that prevents the molecular distances from being smaller than they are.

The principles involved in this view of valence phenomena may be best brought out by considering a few examples. The hydrides of the light elements are particularly susceptible to this sort of treatment. Let us take the hydrides of boron, for example. The inner electrons of boron presumably do not come into the picture. It is necessary to consider only the outer electrons of boron and the electrons of hydrogen. All these electrons become part of one shell which may be considered to be the shell of the boron[1] atom for which $n = 2$. It is not necessary then to

[1] In the discussion we consider the possibility of promotion from the $n = 2$ shell of boron. An electron is shared with the $n = 1$ shell of a hydrogen

promote an electron until the $n = 2$ shell is filled, and so the following compounds might be expected to be possibilities: BH, BH_2, BH_3, BH_4, BH_5. Any compound with more than five hydrogens would be unstable as it would be necessary to promote one electron for each succeeding hydrogen atom that is brought in. It is, of course, not true that the molecules listed are actually known. BH has been found spectroscopically, but the others are actually unknown. Mulliken believes that all of them should be stable with respect to decomposition into atoms, but that they will undergo other reactions. Thus he believes that BH_5 will be stable with respect to

$$BH_5 \rightarrow B + 5H$$

but that the reaction

$$BH_5 \rightarrow BH_3 + H_2$$

will tend to go on account of the great stability of the hydrogen molecule. BH_3 itself does not exist as such, but only as a dimer B_2H_6. It will be noted that the electronic structure of BH_3, in so far as it resembles the united atom, is just the same as that of the oxygen atom. It has been suggested that the combination of BH_3's is a process very similar to the combination of oxygen atoms; however, there appears to be evidence against this point of view as we shall see in Sec. 16.13, and it seems likely that the bond between the borons resembles more closely a single bond.

Whether BH_5 is actually stable with respect to decomposition into the atoms is not surely known But whether it is stable or not, its prediction by a theory exhibits one of the characteristic features of that theory in a rather striking way. It is seen that in BH_5 there are four pairs of outer electrons and five hydrogen atoms. The hydrogen atoms are simply there, somewhere in the electron swarm, and they do not share particular electrons. The Hund-Mulliken theory in general lays stress upon the general electron configuration rather than the bonds between particular atoms.

According to the Hund-Mulliken theory, the series of hydrides of carbon should stop at CH_4. In this case, the existence of the various hydrides is pretty well established. CH is known spec-

---

atom, and promotion with respect to hydrogen does not occur. Of course, in the fully united atom, with all hydrogens merged with the boron, the situation would be different.

troscopically, and CH_3 has been demonstrated by chemical means to be an intermediate in many organic reactions taking place in the gas phase.[1]

We may now consider a number of other molecules from the Hund-Mulliken point of view. We turn first to Li_2. Each Li atom has two 1*s*-electrons and one 2*s*-electron. Now the average distance from the nucleus of the electron in hydrogen when in its 1*s*-state is about 0.5×10^{-8} cm. In lithium, on account of the larger nuclear charge, the distance of the 1*s*-electrons is even smaller, being about 0.2×10^{-8} cm. On the other hand, the distance between nuclei of Li_2 is known from spectroscopic data to be 2.67×10^{-8} cm. and is therefore seen to be very great compared with the region occupied by the 1*s*-electrons. In spite of this fact, James[2] has shown by quantum mechanical calculations that the 1*s*-electrons play a significant role in determining the bond energy. This raises some doubts as to the fundamental validity of our usual procedure of neglecting the inner electrons in discussing valence phenomena. It seems probable, however, that for qualitative discussions this assumption is justified. We therefore say that the 1*s*-electrons are not shared in Li_2. With such a small number of electrons, there is, of course, no question of promotion.

In the cases of N_2 and CO (which have identical electron structures), the 1*s*-electrons are not shared, and there are ten electrons to go into the $n = 2$ shell. But it will only hold eight, and two electrons must be promoted. In the case of CN (which is known spectroscopically and at high temperatures, though actually it is a molecule with an odd electron and not stable with respect to the reaction $2CN \rightarrow C_2N_2$), one electron must be promoted.

It may be reasonably expected that there will be more unshared electrons in molecules composed of atoms farther along in the first row of the periodic table. Mulliken believes that the 2*s*-electrons are not shared in NO, O_2, and F_2. This leaves six 2*p*-places held in common by the two atoms. Oxygen has six outer electrons of which two are 2*s*-electrons. There are thus in the O_2 molecule only $2 \times 4 = 8$ electrons to share the six places, and so two are promoted, whereas if the 2*s*-electrons were shared,

[1] See F. O. Rice and K. K. Rice, "The Aliphatic Free Radicals," Johns Hopkins University Press, 1935.

[2] James, *J. Chem. Phys.*, **2**, 794 (1934).

four would have to be promoted. In NO and F_2 with unshared 2*s*-electrons, one and four electrons, respectively, are promoted.

As far as the energy of the bond is concerned, it really makes little difference according to this theory whether the 2*s*-electrons are considered to be shared or unshared. For Mulliken considers that every electron that is shared but not promoted is a bonding electron, whereas every electron that is shared and promoted has an antibonding action, which, roughly speaking, counteracts the effect of a bonding electron. If the 2*s*-electrons are not shared, then the oxygen molecule, for example, has eight shared electrons, of which two are promoted. The net effect, therefore, is that of four bonding electrons. If the 2*s*-electrons are shared, then there are four more shared electrons, but two more must be promoted; the shared 2*s*-electrons, therefore, have no net effect. In any case, there are effectively four bonding electrons, equivalent to (but not actually forming—see Sec. 16.13) a double bond, in O_2.

It is readily seen why no molecule of the formula Ne_2 exists (at least it cannot be formed from neon atoms in their normal states). On the assumption that the 2*s*-electrons are not shared, there are twelve shared electrons, of which six are promoted, the net bonding effect being nil.

11.4. Comparison of the Theories of Valence.—The Hund-Mulliken theory of valence lays more stress than the other theories on the entire electron structure, and does not emphasize the idea of the electron-pair bond. According to the Hund-Mulliken theory, an electron if shared is either bonding or antibonding, and this does not depend on whether there is an electron pair or not. It must be stated that this view is certainly supported by the existence of the stable molecule H_2^+. On the other hand, it is true that in the great majority of compounds all the electrons are paired, and it certainly is convenient to speak of the electron-pair bond, even if in certain cases it occurs only incidentally. We saw in Sec. 10.5 why the electron pair is of frequent occurrence in valence phenomena, and in this connection it must be emphasized that the Hund-Mulliken theory is not opposed to the idea of the electron-pair bond, though it does not stress it; in fact, in Sec. 10.5 the hydrogen molecule was actually being considered from the Hund-Mulliken point of view. The Hund-Mulliken theory rather states that the electron-pair bond is a special case of a somewhat more general phenomenon. It is seen that the electrons in molecules form shells, which consist of quantum levels of somewhat similar energies, just as in atoms. Only when promotion occurs is the amount of energy required excessively high. This enables us to see very clearly why there are sometimes more than two electrons in a bond. However, we shall see in Chap. XV how double and triple bonds may also be understood from a point of view that may be considered an extension of the Heitler-London theory.

On the other hand, when we consider most polyatomic molecules, it is obvious that it is going to be very difficult to count up the shared and unshared electrons and decide which are promoted. Furthermore, there seems to be good evidence, from organic chemistry, for example, and from the chemistry of the complex ions which will be studied later, that there are definite valence bonds with definite directions. In all cases where there are such definite valence bonds, it would seem to be expedient, at least for our purposes, to use the electron-pair picture, though Mulliken has recently considered polyatomic molecules in some detail.

It is the belief of the author, also, that Lewis's rule of eight, in spite of the fact that there are many exceptions, is of definite value in deciding which compounds are likely to be most stable.

It seems very likely that the Hund-Mulliken theory is more appropriate for the discussion of the hydrides of the elements, because of the small size of the proton and the fact that its distance (obtained from spectroscopic data) from the element to which it is attached is so small that it may be considered to be within the shell of valence electrons, which is only slightly disturbed from its spherical shape. Every proton may thus be said to share the whole valence shell. This theory is also helpful, as we have seen, in considering the diatomic compounds of the light-elements.

Exercises

1. Give the Lewis formula for C_3H_8; for CH_3NNCH_3.

2. Discuss the molecules NO, F_2, CN, and the hypothetical molecule CF from the Hund-Mulliken point of view.

3. It has been claimed by Dennis and Rochow [*J. Am. Chem. Soc.*, **55**, 2431 (1933)] that they have evidence of a compound HFO_3, though Cady [*ibid.*, **56**, 1647 (1934)] does not believe their evidence is conclusive. If the compound does exist, is this necessarily inconsistent with London's ideas on valence? Write a possible Lewis formula for the compound.

CHAPTER XII

TRANSITION FROM COVALENT TO IONIC BINDING IN SIMPLE GASEOUS COMPOUNDS

Following the more or less formal description of chemical-valence phenomena in the last chapter, we are now ready to turn to a more detailed discussion of the physical properties of the chemical bond. This study will be initiated by a consideration of the properties of some of the simple molecules which commonly occur, at ordinary temperatures, in the gas phase. The binding within these molecules is generally of the covalent rather than the ionic type, but deviations from pure covalency occur and can be studied with the aid of a number of properties, the most important of which is the binding energy. The electric moment of the molecule also throws light on this question, as does the distance between atoms forming a bond. More detailed study of the latter property will be deferred to later chapters. Of importance, also, are the causes of deviation from covalent binding. These causes are bound up with the properties of the atoms forming the bond, such as their tendency to gain or lose electrons, and the ease with which they are distorted in an electric field. These various considerations will receive a systematic treatment in the present chapter.

12.1. Elementary Diatomic Gases.—The discussion may best be begun by a consideration of some of the properties of some typical molecules that are composed of two like atoms. In such a case, the valence forces are necessarily of a nonpolar character. Owing to the motion of the electrons, it is possible, even in such a case, for one of the atoms to be temporarily positively charged and the other, at the same time, negatively charged, but either of the atoms will be exactly as often positively as negatively charged. This temporary polarity, with displacement of electrons as often in one direction as in the other, is a part of the nonpolar bond, and any part of the forces that arises therefrom is included in the term "nonpolar force."

The magnitude of the valence forces is best measured by means of the energy necessary to break the molecule, when in its lowest electronic state, into its constituent atoms. Closely related, as it will appear, is the equilibrium distance between atomic nuclei for the lowest state of the molecule, and values of this quantity will also be given,[1] though, as stated above, a more detailed discussion is reserved for later chapters.

The equilibrium distance between the nuclei is generally obtained, as explained in Sec. 9.3, by a study of the rotational spectrum of the molecule. It also may be obtained by a study of the scattering of X rays or electrons, as outlined in Chap. XV. The energy of dissociation may be obtained by a study of the vibrational spectrum of the molecule (Sec. 9.4). It may also be obtained by application of the thermodynamic equation [Eq. (7) of Appendix II]

$$\frac{d \ln \kappa}{dT} = \frac{\Delta E}{kT^2},$$

in which κ is the equilibrium constant for the dissociation (the square of the concentration of dissociated atoms divided by the concentration of molecules), T the absolute temperature, k the gas constant per molecule, and ΔE the average energy of a pair of dissociated atoms minus the average energy of a molecule. Measurement of κ at different temperatures will yield ΔE. This latter quantity, the energy necessary to dissociate a molecule *on the average*, is not exactly the same as the energy of dissociation D obtained from band spectra, which is equal to the energy necessary to go from the lowest vibrational state to the asymptotic part of the potential-energy curve. For ΔE includes the energy due to thermal agitation; in other words, it takes into account the fact that at finite temperatures the molecules are not all in their lowest vibrational state and have, besides, rotational and translational energy, and the atoms also have some kinetic energy. The kinetic energy of the atoms does not exactly balance the average extra energy of the molecules, but both of these terms are small at ordinary temperatures, and we may for our purposes neglect the difference between ΔE and D.

[1] Actually, even when the molecule is in its lowest state, the atoms do not remain at a definite equilibrium position, owing to the zero-point energy (see p. 119). However, for all practical purposes, the interatomic distance may be taken as coinciding with the minimum of the potential-energy curve.

TABLE 7.—VALUES OF THE ENERGY OF DISSOCIATION AT ROOM TEMPERATURE FOR DIATOMIC MOLECULES
(Kilogram-calories per mole)

H_2				
103.2				
Li_2	C_2	N_2	O_2	F_2
26.4	83	169.6	116.7	62.9
Na_2		P_2	S_2	Cl_2
17.4		41.9	76.	57.2
K_2		As_2	Se_2	Br_2
11.6		34.3	72	45.5
Rb_2			Te_2	I_2
10.6			53	35.7
Cs_2		Bi_2		
10.1		18.8		

From Bichowsky and Rossini "Thermochemistry of Chemical Substances," except O_2 and the alkali molecules which are from data in the third supplement of Landolt-Börnstein, "Tabellen", and in Sponer, "Molekülspektren," vol. I, corrected to room temperature, and S_2, Se_2, and Te_2, which are from Goldfinger, Jeunehomme, and Rosen, *Nature*, **138**, 205 (1936), and C_2 which is from Herzberg, "Molecular Spectra and Molecular Structure."
Values in the table are ΔE of dissociation rather than ΔH (see Appendix II).

In Tables 7 and 8, we give values of ΔE (or D) and r_e, the equilibrium distance, for a number of diatomic molecules, some of which do not occur as ordinary substances in large quantity, but are known only through their spectra. These are arranged in such a way as to bring out the periodic relationships. The first feature which appears is that, in any column of the periodic table, the value of r_e increases and the value of D decreases on going from the lighter to the heavier elements. This is, of course, to be expected; the distance at which the repulsive forces between the

TABLE 8.—VALUES OF r_e FOR ELEMENTARY DIATOMIC MOLECULES
(In Angstroms)

H_2				
0.739				
Li_2	C_2	N_2	O_2	F_2
2.670	1.31	1.09	1.204	1.45[c]
Na_2		P_2	S_2	Cl_2
3.07		1.88	1.92[a]	1.983
K_2			Se_2	Br_2
3.91			2.19[b]	2.28
			Te_2	I_2
			2.59[b]	2.000

From SPONER, "Molekülspektren," vol. I, except as noted.
[a] MAXWELL, MOSLEY, and HENDRICKS, *Phys. Rev.*, **50**, 41 (1936).
[b] MAXWELL and MOSLEY, *ibid.*, **57**, 21 (1940).
[c] BROCKWAY, *J. Am. Chem. Soc.*, **60**, 1348 (1938).

nuclei set in increases as the number of electron shells surrounding the nuclei increases. This increase of the distances involved necessarily results in a decrease of the binding force; the electrons that effect the binding are far removed from the centers of force.

To get a qualitative understanding of the variations which occur across the rows of the periodic system, it is necessary to call to mind what has been learned about the electronic structure of the molecule. The Lewis formula for the fluorine molecule is $:\underset{\cdot\cdot}{\overset{\cdot\cdot}{\mathrm{F}}}:\underset{\cdot\cdot}{\overset{\cdot\cdot}{\mathrm{F}}}:$, involving a single bond. That this single bond will differ in many respects from the single bond in hydrogen is obvious, for the other electrons in the outer shells of the fluorine atoms must affect the electrons that form the bond. In the case of oxygen, however, there is a more profound difference, for in order to have a shell of eight around both oxygens we must write a double bond[1] $\dot{:}\underset{\cdot}{\dot{\mathrm{O}}}::\underset{\cdot}{\dot{\mathrm{O}}}\dot{:}$, and it is interesting that the dissociation energy of oxygen is roughly twice that of fluorine. Similarly, in the case of nitrogen we should write a triple bond $:\mathrm{N}:::\mathrm{N}:$, and here the dissociation energy is roughly three times that of fluorine. Looking at the matter from the point of view of the Hund-Mulliken theory of valence, we probably should not stress the qualitative aspects of the difference between those various molecules quite so much; but, as seen in the last chapter, when the antibinding effect of promoted electrons is taken into account, two effective bonding pairs are left in the case of oxygen and three in the case of nitrogen.

In view of the regularity of the values for the dissociation energy going from fluorine to nitrogen, there is a rather striking break at carbon. It is quite obvious that the dissociation energy of carbon is not what would be expected from a quadruple bond, and we shall see in Chap. XV reasons for doubting the possibility of four pairs of electrons being shared between two atoms and filling the outer shell of both of them to the desired number eight. The energy of C_2, as a matter of fact, is between that of O_2 and

[1] It is extremely doubtful that there is really a double bond in O_2; however, the binding energy is probably very close to what would be expected for a double bond. This is discussed further in Sec. 16.13. Attention, however, should be called to the fact that the strength of a single O—O bond, as found below in Sec. 12.8, is considerably smaller than that of the F—F bond.

that of F_2; so it appears that the bond is between a single and a double bond. The break in the value of the energy of dissociation that appears with C_2 in the first row of the periodic system appears in the elements of the nitrogen group in subsequent rows. It will be noted that P_2, As_2, and Bi_2 have relatively low energies of dissociation. However, the possibility of errors in these values is so great that too much stress should not be laid on them.

12.2. Elementary Polyatomic Gases.—A slight digression on the elementary polyatomic gases may be of interest at this point. A number of the substances at the right-hand side of the periodic table have polyatomic forms.[1] Among them are P_4, which is stable up to 1500°C. and then somewhat dissociated into P_2; O_4, which forms to some extent at liquid air temperature, and ozone, O_3; the various forms of sulfur, S_8, the form which predominates in the vapor at the boiling point of sulfur, and the molecules S_6 and S_2 occurring at higher temperatures (at 1000° sulfur vapor is largely S_2, and it is in large part dissociated into atoms at 2000°); Se_8 and Se_2; and perhaps other forms of sulfur and selenium. There is evidence of various polymers in liquid sulfur and selenium, which occur in a number of different forms. The fact that these elements exist in these various forms indicates a certain versatility in their electron structure as may also be indicated by the fact that their solids show allotropic modifications. (At least one solid form consists of polymer molecules, S_8.)

The electron structure of S_8 is

$$\begin{array}{c} :\ddot{\mathrm{S}}:\ddot{\mathrm{S}}:\ddot{\mathrm{S}}: \\ :\ddot{\mathrm{S}}:\quad :\ddot{\mathrm{S}}: \\ :\underset{\cdot\cdot}{\ddot{\mathrm{S}}}:\underset{\cdot\cdot}{\ddot{\mathrm{S}}}:\underset{\cdot\cdot}{\ddot{\mathrm{S}}}: \end{array}$$

It is a ring structure involving only single bonds, and P_4 is a tetrahedron involving only single bonds, each atom sharing an electron pair with each of the other three.[2] No such structure would be possible with the halogens, and only diatomic halogen molecules are known.

[1] These are discussed, *e.g.*, in Latimer and Hildebrand, "Reference Book of Inorganic Chemistry," pp. 25, 167*ff*., 188*ff*., The Macmillan Company, 1929; and in Ephraim, "Inorganic Chemistry," English ed., pp. 85*ff*., Gurney and Jackson, London, 1934.

[2] For S_8, see "Strukturbericht," vol. III, p. 4. The geometry of the ring structure is not intended to be indicated exactly by the formula shown. For P_4, see Maxwell, Hendricks, and Mosley, *J. Chem. Phys.*, **3**, 708 (1935).

The tetratomic molecule of oxygen would seem to be in a somewhat different category from the other polyatomic gases. Its energy of dissociation in two O_2 molecules is very low indeed, being only about 0.006 electron volt.[1] This suggests that the O_4 molecule may be thought of as two O_2 molecules rather loosely bound together by van der Waals forces, as discussed in Chap. XVII. Although this energy of binding is of the order of magnitude to be anticipated from this type of force, this evidence is not entirely conclusive, for we might readily suppose that no energy would be required to form four single bonds of O_4 from two double bonds[2] in O_2. However, certain peculiarities of the spectrum of O_4 are best explained on the assumption that it is composed of two O_2 molecules loosely held together,[3] and we may assume that this is actually the case. On the other hand, all the facts are against any similar assumption in the case of P_4, for example, as it is far too stable. (The reaction $P_4 \rightarrow 2P_2$ requires 1.2 electron volts.) The properties of the polymers of sulfur and selenium indicate that they also are stable. Of course, loosely bound molecules like O_4 undoubtedly occur with these substances also, but are not of importance, any more than is the case with oxygen, in affecting the gross physical and chemical properties.

Ozone is a different type of polymer from any of the others; it is very reactive and unstable as compared with O_2.

12.3. An Approximate Measure of Electronegativity.—Even in the case of a purely covalent bond, there is, as has already been noted, a certain chance for both the electrons of an electron-pair bond to be on one of the atoms, which causes a momentary polarity, but does not make the bond polar. In order for the bond to be wholly or partly polar, one of the bonded atoms must attract electrons more strongly than the other. If a measure for the attraction of an atom for electrons, *i.e.*, for its electronegativity, could be obtained, it would be possible to form some idea as to which compounds would tend to be polar. A method for getting a measure of the electronegativity of atoms has recently been suggested by Mulliken.[4]

[1] Lewis, *J. Am. Chem. Soc.*, **46**, 2027 (1924).

[2] See, however, footnote p. 166.

[3] Finkelnburg, *Zeits. Physik*, **90**, 1 (1934); **96**, 699 (1935); and other references cited by Finkelnburg.

[4] Mulliken, *J. Chem. Phys.*, **2**, 782 (1934).

Suppose that we had two atoms A and B a large distance from each other; in this case, it would be easy, at least theoretically, to tell whether it required more energy to take an electron from atom A and put it on atom B, or to take an electron from atom B and put it on atom A. The energy required in the first process is $I_A - F_B$, where I_A is the ionization potential of A and F_B is the electron affinity of B. The energy required in the second process is $I_B - F_A$, the meaning of the symbols being obvious. If these two quantities are equal, *i.e.*, if

$$I_A - F_B = I_B - F_A$$

or

$$I_A + F_A = I_B + F_B,$$

then it is just as easy to take an electron one way as the other. On the other hand, if

$$I_A + F_A > I_B + F_B,$$

then

$$I_A - F_B > I_B - F_A,$$

and it is easier to take the electron from B and put it on A than it is to take an electron from A and put it on B. Thus a comparison of $I + F$ for two different elements will show which element has the greater attraction for electrons. Mulliken proposes to take this quantity $I + F$ as a measure of the electronegativity of an element. It is, of course, a pure assumption to suppose that this remains a good measure of the electronegativity for an atom that is combined in a molecule. It does, however, give very reasonable results, as is seen from the second column of Table 9, which gives $I + F$ for the halogens and hydrogen.

TABLE 9.—ELECTRONEGATIVITY VALUES
(In volts)

Element	Original values	Revised values
F	22.	25.3
Cl	16.7	19.6
Br	15.1	18.1
I	13.7	16.5
H	14.25	14.25
Li	.....	5.7

The only feature of this table which may seem strange to chemists is the fact that hydrogen appears to be more electronegative than iodine, and this appearance is undoubtedly incorrect. The reason for it seems to reside in the fact that the state of a hypothetical I^+ ion in combination does not resemble sufficiently closely its state when free. In finding the ionization potential for the purpose of determining the electronegativity of iodine, we are not necessarily interested in the energy necessary to remove the electron and leave the resulting I^+ in its lowest possible state, but rather in the energy necessary to remove the electron and leave the I^+ in a state in which it would be capable of combining with H^-, say, to form HI. In I^+, there are four *p*-electrons to fill three *p*-states (six when spin is considered), and in the normal state of I^+ two of these electrons are in one of these levels and each of the other two is in a separate level; furthermore, these last two electrons have their spin in the same direction, and the H^- ion, in which the electrons have opposite spins, cannot combine directly with I^+ to form an electron-pair bond. Thus, in order to form HI from H^- and I^+, we have a situation which is the reverse of that considered in the last chapter in connection with London's valence theory; the electrons must be paired, or at least the resultant spin must be reduced to zero before the compound can be formed. It must be remembered, also, that there are a number of states of I^+ in which the resultant spin is zero, corresponding to different distributions of the electrons among the quantum levels. The actual state of combined iodine will not resemble any one of these as much as some average or composite of all of them. Thus in evaluating the ionization potential of I, we should really take the energy of the reaction

$$\mathrm{I} \rightarrow \text{electron} + \mathrm{I}^+(\text{composite excited state})$$

which will be greater than the energy for the reaction

$$\mathrm{I} \rightarrow \text{electron} + \mathrm{I}^+(\text{normal state})$$

On the basis of considerations such as these, Mulliken has made a revised estimate of electronegativities, as given in Table 9. To this, for comparison, the value of lithium has been added. The electron affinity of lithium has been roughly estimated, but since it is very small a very rough estimate will suffice. Thus the value of $I + F$ in the alkali metals is chiefly determined by I. It will be evident that the electronegativity of the alkali metals decreases, as expected, from lithium to cesium. Furthermore, the electronegativities of the copper group metals will be greater than those of the alkali metals. The electronegativities of both alkali and copper-group metals are very much less than that of hydrogen, and it is seen that from this point of view hydrogen should be classed with the halogens rather than with the alkalies.

Mulliken has considered the possibility of extending his electronegativity table to polyvalent elements. In such cases, however, it is necessary to consider the possibility of multiply charged ions, so that not only the first, but also the second and possibly higher ionization potentials and electron affinities need to be considered. The considerations thus become more involved and the conclusions less certain. Since there are other means of estimating electronegativity, which can also be applied to polyvalent elements, we shall not attempt to apply the present method to them.

12.4. Polarizability as a Criterion for Electronegativity.—Once we have a list of the electronegativities of the elements, we have a ready means of predicting which bonds will be more polar and which will tend rather toward the covalent type. The greater the difference between the electronegativities of two elements, the more polar the bond between them will be.

A succinct and useful set of rules for determining the relative polarity of valence bonds has been given by Fajans.[1] These rules may be conveniently formulated if we consider a polar bond as the norm, and inquire how far it departs from the pure polar type; this enables us to refer to the anion and the cation. The bond, then, will depart most greatly from the purely polar type if (1) the charge on the ions is large, (2) the cation is small, and (3) the anion is large.[2] The reasons for these rules are very easily seen. If the charge on the ions is large, the force between them is large, and the cation tends to draw over toward it some of the electrons of the anion, thus effectively decreasing the charge of

[1] FAJANS, *Zeits. Elektrochem.*, **34**, 507 (1928).

[2] We here consider only simple monatomic ions. Complex ions, such as SO_4^{--}, will be discussed later.

each. If the cation is small,[1] the center of force is able to get close to the anion, thus exerting a great force on it, and drawing electrons away from it; if the anion is large, the forces on the outer electrons due to its own nucleus are small, it is easily deformable, and the outer electrons can readily be displaced toward the cation. It will be seen that the last two rules are really simply qualitative expressions of the rule that a bond is less polar, the smaller the difference in the electronegativities of the atoms forming it. For the smaller the cation, the more force it will exert on electrons and the more electronegative it will be; similarly, the larger the anion, the less electronegative it will be, and thus (since, by definition, the anion is always the more electronegative of the two) the smaller will be the difference in electronegativities. As was noted in Sec. 12.3, it is necessary to consider the electronegativity of the atom when combined in the molecule; so it is not at all certain that the sum of the ionization potential and electron affinity of an atom will be a better measure of its actual electronegativity than would be given by the lack of deformability of its negative ion on the one hand, and by the ability of its positive ion to cause deformation on the other. However, a great ability of a positive ion to cause deformation will invariably be associated with a small deformability of the negative ion of the same element; so it might well be that in the case of the elements which tend to form negative ions the smallness of their deformability alone would be a reasonable measure of the electronegativity.

We are interested in the deformation of an ion in the presence of the electric field due to a neighboring ion. The deformability of any body in an electric field is commonly called its "polarizability,"[2] and this concept may be quantitatively formulated.

[1] We have remarked in the last section that the copper-group metals are less electropositive than the alkali metals. This is associated with the high ionization potentials of the copper-group metals, and it is also associated with a relatively great ability on their part to deform anions. It is not to be thought of as due to their ions having an especially small size (according to Table 16 in Chap. XIV, Cu^+ is about the same size as Na^+), but rather is connected with the fact that their electron shell has eighteen instead of eight electrons. This question will be discussed further in Sec. 16.11.

[2] The reader must take care to avoid being confused by this word. The larger the polarizability of an anion, the less polar is the bond formed by it with a cation.

We start with the consideration of a neutral atom, which is composed of electrical charges, but in which the total positive and negative charges are equal in amount and have, furthermore, the same center of gravity. If such an atom is placed in an electric field, there is a displacement of electric charge in the body, resulting in the separation of positive and negative charges, and producing an electrical dipole. The strength, or moment, of such a dipole is defined in a way that is similar to the definition of magnetic moment given in Sec. 6.2. If a negative charge of ϵ is placed a distance δ from an equal positive charge, the electric moment of the pair is $\epsilon\delta$. In the case of an atom, δ may be considered to be the displacement of the "center of gravity" of negative electricity with respect to the positive charge of the nucleus; the moment is then $Ze\delta$, where Z is the atomic number and e the elementary electronic charge. As far as dipole strength is concerned, this is equivalent to displacement of a single electronic charge through a distance $Z\delta$. If we are dealing with an ion rather than a neutral atom, the situation is not essentially changed; the displacement of charge is simply superposed upon the total charge.

Now, let the moment produced by an electric field applied to an atom or an ion be M. It is proportional to the applied field E, and the constant of proportionality is the polarizability, which is designated as α. We have

$$M = \alpha E. \tag{1}$$

In order to find the order of magnitude of α for an atom,[1] let us suppose, as a rough approximation, that the atom consists of a positive nucleus of charge Ze, surrounded by a sphere of negative electricity of uniform density and radius R. Suppose now that it is placed in a uniform electric field of strength E, and that this field causes a displacement of amount δ of the positive nucleus and the negative sphere with respect to each other, without distorting the latter, as shown in Fig. 46. If such a displacement occurs, then according to a familiar law of electrostatics, the positive nucleus is attracted back to the center by all the charge within a radius δ with the same force as if this charge were at the center of the sphere. Since the total charge is Ze, the amount of charge

[1] See SLATER and FRANK, "Introduction to Theoretical Physics," p. 275, McGraw-Hill Book Company, Inc., 1933.

in the small sphere of radius δ is $Ze(\delta^3/R^3)$. This attracts the positive charge with a force $\frac{Z^2e^2(\delta^3/R^3)}{\delta^2} = \frac{Z^2e^2\delta}{R^3}$. Or the situation may be described by saying that there are equal and opposite forces of this magnitude on the positive and negative charges which tend to pull them together. On the other hand, the field exerts equal and opposite forces equal to ZeE on the positive and negative charges, respectively, tending to pull them apart. Equilibrium occurs when these two tendencies balance, *i.e.*,

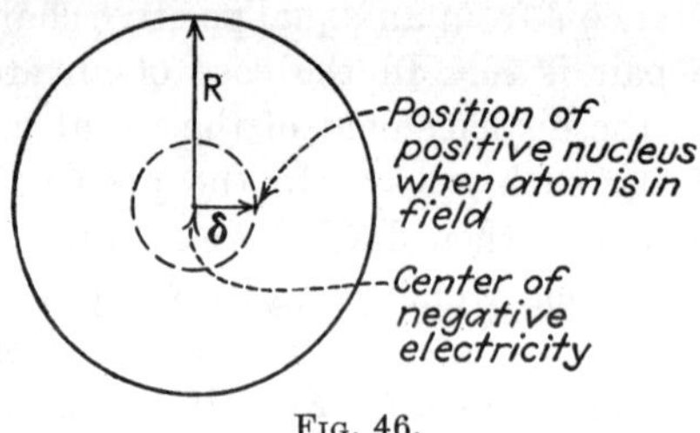

Fig. 46.

when $ZeE = Z^2e^2\delta/R^3$, and it is this condition which determines δ; from it, it is seen that

$$\delta = \frac{ER^3}{Ze}. \tag{2}$$

Now the electric moment produced by this displacement of charge is $Ze\delta$, which, by definition of α, must be equal to αE. Hence

$$\alpha = R^3. \tag{3}$$

This result is very rough, but should give the correct order of magnitude, and it is seen that there is a very close relation between polarizability and size.

α is generally measured indirectly by measuring the refractive index of light. A light beam produces an oscillating electric field, which polarizes the atoms of the substance through which it passes. The electric moment produced in the atoms naturally also oscillates and creates a field which interacts back with the field of the light beam. The index of refraction is closely related to this interaction. For details, the student must be referred to treatises on optics or electron theory; the relation that exists between polarizability and index of refraction is given by the formula[1]

$$\alpha = \frac{3}{4\pi N}\frac{n^2 - 1}{n^2 + 2}V, \tag{4}$$

where n is the index of refraction for light of infinite wave length and V is the volume containing N molecules. The theory indicates that this formula should be additive; *i.e.*, in a mixture, the quantity on the right-hand side of Eq. (4), which is directly measured experimentally, should give the average of the polarizabilities of the substances in the mixture. Thus if, of the N molecules, N_1 were of one kind and N_2 of another, and n were the refractive index of the mixture, then we should have

$$\alpha_{\text{average}} = \frac{N_1\alpha_1 + N_2\alpha_2}{N} = \frac{3}{4\pi N}\frac{n^2 - 1}{n^2 + 2}V. \tag{5}$$

However, the polarizability of an atom may be affected by the proximity of other atoms. In this case, (4) and (5) do not hold and corrections must be made. So in the case of ions dissolved

[1] The significance of this equation may perhaps be better understood if we write it

$$\alpha\frac{N}{V} = \frac{3}{4\pi}\frac{(n - 1)(n + 1)}{n^2 + 2}.$$

Now in vacuum n is equal to 1, and if the molecular concentration N/V is sufficiently small, n will be very close to 1, so that under these circumstances we may write

$$\alpha\frac{N}{V} \cong \frac{n - 1}{2\pi},$$

and the deviation of the refractive index from its value in vacuum is simply proportional to the density and to α. The more complicated expression is needed for high concentrations to take account of the effects of the atoms on each other. It must be emphasized, however, that this takes into account only the effects of fields, due to the polarization of the atoms, on neighboring atoms, and does not take into account any effect of one atom on the *polarizability* of its neighbors. Such an effect results in deviation from Eq. (4).

It should be noted that if N is equal to Avogadro's number so that V is the molar volume, the quantity $\frac{V(n^2 - 1)}{n^2 + 2}$ is what is commonly known as the "mole refraction."

See Lorentz, "Theory of Electrons," Chap. IV, G. E. Stechert & Company. Debye, "Polar Molecules," Chap. I, Reinhold Publishing Co., 1929.

in water, it might be expected on account of the large forces exerted by the ions on the water (see Chap. XIX) that the polarizability of the water would be affected. However, as a first approximation, in order to get a starting point, it is natural to assume Eq. (5) to be correct, which is the same thing as assuming that the polarizabilities of gaseous ions can be obtained from measurements on ions in solution. By measuring the refractive index of water, one can, from Eq. (4), obtain α for water. Having this value, it is then possible, assuming that (5) holds, to measure the index of refraction of dilute solutions and so obtain values for the polarizabilities of ions. In addition, the polarizabilities of the rare gases are known from measurements of their indices of refraction.

It must be noted, however, that the polarizability of a single ion cannot be measured, as an ion of opposite sign must always be present with it. Only the sum of the polarizabilities of a positive and negative ion, or the difference in polarizabilities of two positive ions or two negative ions (*e.g.*, the difference between Rb^+ and K^+ may be obtained from measurements on RbCl and KCl), can be found directly. However, certain relations between the polarizabilities may be inferred.[1] For example, consider the series Cl^-, A, K^+, all of which have the same electronic structure and differ only in the nuclear charge. The greater the nuclear charge, the more tightly the electrons will be bound. Tight binding of the electrons will be expected to result in a small size of the ion or atom and a small polarizability. Thus we may confidently write a series of inequalities for the polarizabilities of these substances, as follows: $\alpha_{Cl^-} > \alpha_A > \alpha_{K^+}$. Likewise, since the percentage change in nuclear charge on going from Cl^- to A is greater than that on going from A to K^+, it may be anticipated that the percentage change between the first pair will be greater than between the second pair. Thus we expect $\alpha_{Cl^-}/\alpha_A > \alpha_A/\alpha_{K^+}$. Similarly, we may write $\alpha_{Br^-} > \alpha_{Kr} > \alpha_{Rb^+}$ and $\alpha_{Br^-}/\alpha_{Kr} > \alpha_{Kr}/\alpha_{Rb^+}$. But since the percentage change from Cl^- to A is greater than from Br^- to Kr, we may also write $\alpha_{Cl^-}/\alpha_A > \alpha_{Br^-}/\alpha_{Kr}$. There is thus obtained a large set of

[1] FAJANS and JOOS, *Zeits. Physik*, **23,** 1 (1924). Some consideration of the relative magnitudes of the interactions of the various ions with the solvent also entered into these calculations.

inequalities, and it is found that these determine, within narrow limits, the way the average polarizability of a solution must be divided between the positive and negative ions. Furthermore, certain regularities among the various ratios appear, which hold well as long as we deal only with the fairly large ions, K^+, Rb^+, Cs^+, and Cl^-, Br^-, and I^-. When smaller ions are considered, the regularities break down. This is probably due to the fact that these smaller ions exert larger forces on the water, changing its polarizability sufficiently so that it cannot in these cases be assumed that Eq. (5) is valid, and it is, consequently, not possible to get the polarizabilities of the gaseous ions directly by measuring the refractive index of the solution. However, the regularities in the various ratios permit extrapolations, not only to the smaller ions, such as Na^+ and F^-, but also to ions of larger charge, where the interaction between ion and water is also so great that Eq. (5) breaks down. In this way, Fajans and Joos[1] have set up a table of ionic polarizabilities. This table has recently been slightly revised by Fajans,[2] and the results have been confirmed also by several theoretical investigations.[3] The values of most of the ions given in Table 10 are those of Fajans, except that the values for Li^+ and Be^{++} are the theoretically calculated values of Pauling,[3] as for these ions, Fajans's method does not give good results.

In the same table, we give in parentheses the cubes of the ionic radii as taken from Table 16 of Chap. XIV. It is seen that these values are of the same order of magnitude as the polarizabilities, and parallel them to a marked degree, but that the polarizabilities show a greater variation. Using the polarizability of the ions of the more negative elements as a measure of the electronegativity, we can arrange them in the following order of decreasing negativity:[4] F, O, Cl, Br, I, S, Se, Te. This order coincides with that obtained in Sec. 12.8 from energy considerations.

[1] Fajans and Joos, *Zeits. Physik*, **23**, 20 (1924).

[2] Fajans, *Zeits. physik. Chem.*, **B24**, 118 (1934).

[3] Born and Heisenberg, *Zeits. Physik*, **23**, 388 (1924); Pauling, *Proc. Roy. Soc.* (*London*), **A114**, 191 (1927); Mayer and Goeppert Mayer, *Phys. Rev.*, **43**, 605 (1933).

[4] In the case of the bivalent elements, such as sulfur, we do not attempt to make any correction for the higher charge on the ion, but we use the polarizability of the doubly charged ion taken directly from Table 10.

TABLE 10.—IONIC POLARIZABILITIES IN CUBIC ANGSTROMS
(In parentheses cube of radius from Table 16)

		He	Li^+	Be^{++}
		0.20	0.029	0.008
		(0.78)	(0.205)	(0.079)
O^{--}	F^-	Ne	Na^+	Mg^{++}
2.74	0.96	0.394	0.187	0.103
(5.43)	(2.51)	(1.40)	(0.86)	(0.55)
S^{--}	Cl^-	A	K^+	Ca^{++}
8.94	3.57	1.65	0.888	0.552
(10.5)	(5.92)	(3.65)	(2.35)	(1.64)
Se^{--}	Br^-	Kr	Rb^+	Sr^{++}
11.4	4.99	2.54	1.49	1.02
(12.4)	(7.4)	(4.83)	(3.25)	(2.30)
Te^{--}	I^-	Xe	Cs^+	Ba^{++}
16.1	7.57	4.11	2.57	1.86
(15.6)	(10.1)	(6.9)	(4.82)	(3.58)

12.5. The Transition between Covalent and Polar Bonds from the Point of View of Wave Mechanics.—Suppose there are two atoms A and B that we imagine to be held a fixed distance apart. Let us further suppose that these two atoms are held together by a single bond, consisting of a pair of electrons, and that the motion of these two electrons can be treated as in a two-center problem; *i.e.*, it is assumed that each electron moves under the influence of two centers of force, one at A and one at B, each centrally symmetric but neither one necessarily a simple inverse-square force. Since A and B are different atoms, they will have different laws of force, and one will attract the electrons more than the other. If now we should set up an approximate law of force, say a sort of average between the law for A and the law for B, and assume that both A and B had this same law of force, then we should get an approximate wave function, let us say ψ_c (a function of the coordinates of the two electrons and also depending on the distance between A and B), which represents a covalent bond between A and B On the other hand, if we assumed that the more electropositive of the atoms (let us suppose that this one is A) exerted no attraction on the electrons whatsoever, we should get a wave function ψ_p in which the electrons were always on the electronegative atom B, which could be said to represent a polar bond. The real wave function would be more closely approximated by some combination of the two wave functions, let us say $C_c\psi_c + C_p\psi_p$ where C_c and C_p are constants. These wave functions will have to fulfill the appropriate normalization conditions (see page 41)

$$\int \psi_c^2 d\tau = 1, \qquad \int \psi_p^2 d\tau = 1, \qquad \int (C_c\psi_c + C_p\psi_p)^2 d\tau = 1,$$

where $d\tau = dx_1\,dy_1\,dz_1\,dx_2\,dy_2\,dz_2$, where $x_1\ y_1$, z_1 and x_2, y_2, z_2 are the coordinates of the respective electrons, and the integration is carried over all space. These simply state that no matter what approximation is used, the probability[1] of finding the pair of electrons somewhere is 1. These equations are equivalent to one condition on C_c and C_p, but it is still possible to fix arbitrarily the ratio between C_p and C_c. This should naturally be done in such a way as to make the combination wave function resemble the true one as closely as possible, and the resulting value of C_p/C_c might then be taken as a measure of the polarity of the bond. The process of combining the two wave functions ψ_c and ψ_p is similar to the combining of the Heitler-London and Hund-Mulliken type of nonpolar wave functions as in Eq. (8) of Sec. 10.6. (It should be remarked that the wave function of Eq. (8) is *not* normalized and, therefore, really needs yet to be multiplied through by a constant.) The wave function ψ_c already consists of (or, better, is approximated by) a combination of the Heitler-London and Hund-Mulliken types of functions, and it is a *further* combination with a polar function ψ_p which is expected to give a still better approximation to the true molecular wave function.

But the idea of using linear combinations of approximate wave functions was considered even earlier in Chap. X, *e.g.*, when the symmetrical and antisymmetrical combinations of the functions $\psi_A(1)\psi_B(2)$ and $\psi_B(1)\psi_A(2)$ were taken. These two functions represented states with the same energy, and after taking into account the effects of some of the approximations, their combinations gave two states, one with much lower energy and one with much higher energy. In the case of the wave functions ψ_c and ψ_p now under consideration, the energies E_c and E_p which go with them are not necessarily equal. If they are not equal,

[1] The physical significance of the ψ-function is discussed in Sec. 3.6. In the present case, as in others we have discussed, the ψ-function may be assumed to be real and hence equal to its conjugate complex.

it may seem strange that ψ_c and ψ_p can be combined at all, as one would not expect the system to be able to switch back and forth between two states having different energies, apparently in violation of the law of conservation of energy. However, it will be remembered that in obtaining ψ_c and ψ_p and hence E_c and E_p special approximations were made about the forces exerted by the centers of force A and B on the bonding pair of electrons. This simply means that some of the energy of the electrons is neglected. If the error thus produced is of the order of magnitude of the difference of energy $E_c - E_p$, then it is quite possible for a linear combination of ψ_c and ψ_p to represent the true state of affairs better than either one alone.[1]

In the combination of $\psi_A(1)\psi_B(2)$ and $\psi_B(1)\psi_A(2)$ in Sec. 10.2, we had a symmetrical combination giving the lowest state of the system and there was also a second linearly independent, in a sense complementary, antisymmetrical combination representing an excited state of the system. So also, if a certain combination $C_c\psi_c + C_p\psi_p$ gives the lowest state of the system at present under consideration, there will be another possible linearly independent combination that gives an excited state of the system. If the low-energy state is predominantly polar, the complementary excited state will be predominantly nonpolar, and vice versa. Suppose $\mathfrak{E}_p$ is the energy of the more polar combination and $\mathfrak{E}_c$ that of the more covalent, then:

If $E_p > E_c$,
we have

[1] It will be noted that in the foregoing paragraphs we have described the system as being in a state represented by a linear combination $C_c\psi_c + C_p\psi_p$ of two wave functions, and as shifting back and forth between two states ψ_c and ψ_p. It is characteristic of the wave mechanics that it allows some ambiguity of this sort in the description of the system. It is equally correct to say that the system is in a state that is a sort of average of ψ_c and ψ_p, or to say that it is changing back and forth from ψ_c to ψ_p, the relative time it spends in either of the states being determined by the relative size of C_c and C_p. The system is said to "resonate" between these two states, and the phenomenon is known as the "resonance phenomenon." There have been many applications in the recent literature to problems in organic chemistry (see Sec. 12.11, and Pauling and Wilson, "Introduction to Quantum Mechanics," pp. 314*ff*., and pp. 377*ff*., McGraw-Hill Book Co., Inc., 1935, and Pauling, "The Nature of the Chemical Bond ," Cornell University Press, 1939).

and

$$\left.\begin{array}{l}\mathfrak{E}_p > E_p, \\ \\ \mathfrak{E}_c < E_c;\end{array}\right\} \quad (6)$$

but if $E_c > E_p$,
wo have

and

$$\left.\begin{array}{l}\mathfrak{E}_c > E_c, \\ \\ \mathfrak{E}_p < E_p.\end{array}\right\} \quad (7)$$

Thus, on improving the calculation, the energy levels are spread apart.

These results are taken from wave mechanics, and it is a little difficult to make them seem intuitively evident. They may be described in a more or less intuitive manner (and somewhat amplified) as follows. When a system is in its lowest energy level, the electrons tend to get into such a state as to make the energy as low as possible. Thus the combination of ψ_c and ψ_p which will make the energy as low as possible is the one that is chosen. If this combination is predominantly polar, then in its lowest state the molecule is polar. On the other hand, the other independent combination of the two wave functions is such as to make the energy as high as possible. Thus if the lower state is predominantly polar, we naturally expect the higher one to be predominantly nonpolar, as, if making the molecule polar decreases the energy, making it nonpolar should increase the energy.

The preceding account assumes that a good approximation for the true state of the molecule can be obtained by combining only *two* approximate functions. This is sometimes not truc, but if more than two are necessary the electrons nevertheless try to take up that position which makes the energy of the ground state as low as possible. This exceedingly important rule from the wave mechanics is entirely general[1] and will be used frequently in the subsequent pages.

The functions ψ_c and ψ_p, as remarked above, are functions of the distance between the atoms A and B, and so are the corresponding energies E_c and E_p. It may well be, for example, that

[1] See Pauling and Wilson, "Introduction to Quantum Mechanics," pp. 180*ff*.

E_c is greater than E_p at great distances of separation, and that at small distances the reverse is true. We then have the situation shown by the solid curves,[1] representing E_c and E_p, in Fig. 47. The dotted lines in Fig. 47 then represent $\mathfrak{E}_c$ and $\mathfrak{E}_p$. According to (6) and (7), the lower dotted line represents $\mathfrak{E}_p$ to the right of the point at which E_p and E_c cross, and to the left of the point it represents $\mathfrak{E}_c$, whereas the opposite is true of the upper dotted line. In any event, the state of the system which goes with the lower dotted line gradually changes in character from completely polar to practically completely nonpolar on going from greater to smaller distances of separation. The exact amount the dotted potential-energy curves depart from the solid curves depends

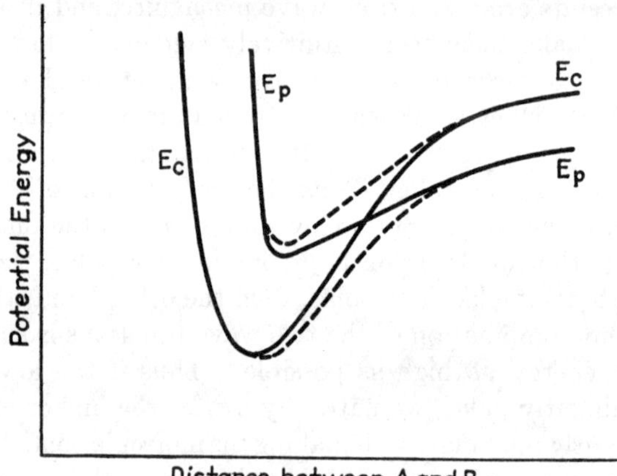

FIG. 47.—Crossing of potential-energy curves.

upon two things (1) the nature and the size of the approximations involved in setting up ψ_c and ψ_p and (2) the value of $[E_p - E_c]$. The smaller the latter quantity, other things being equal, the greater will be the deviation between solid and dotted curves.

12.6. Energies of Polar and Covalent Bonds.—The considerations of Sec. 12.5 suggest that an examination of the energies of various compounds might throw light on the polar or nonpolar character of the bonds involved in these compounds. If we knew the energy of formation to be expected if the bond involved were a strictly polar bond and if we knew that to be expected if it

[1] This phenomenon has been discussed in some detail for some special potential-energy curves by Pauling, *J. Am. Chem. Soc.*, **54**, 996 (1932). The general ideas were introduced early in the development of wave mechanics and discussed by many writers.

were a strictly nonpolar bond, then on the basis of Sec. 12.5 an actual bond which had some polar and some nonpolar character ought to have an energy lower than either.

These ideas may be illustrated by a consideration of the diatomic interhalogen compounds. Consider for example the compound ClF. Fluorine is more electronegative than chlorine; so if we regarded this as an ionic compound we would consider it to be formed from Cl^+ and F^-. Thus the reaction $Cl + F \rightarrow ClF$ is taken as the sum of the steps

$$Cl \rightarrow Cl^+ + \text{electron} \quad \text{(a)}$$
$$F + \text{electron} \rightarrow F^- \quad \text{(b)}$$
$$Cl^+ + F^- \rightarrow ClF \quad \text{(polar compound)} \quad \text{(c)}$$

The energy changes associated with the steps (a) and (b) are known (Tables 4 and 5, Chap. VIII), and the energy associated with (c) can be estimated; so an estimate for the over-all energy of the reaction can be made. From Table 33 of Chap. XVI, we can make a rough estimate of the Cl—F distance in the compound;[1] it turns out to be 1.63Å. We can make a rough guess of the energy of a polar compound of Cl^+ and F^- by assuming that the repulsive force starts in suddenly at this distance and at once becomes very great. The energy is then simply the electrostatic energy at that distance, and is 8.78 electron volts. When the energies of reactions (a) and (b) are added in, it is found that in the reaction $Cl + F \rightarrow ClF$ (polar compound) an energy of approximately 0.1 electron volt would be absorbed; thus considered as a polar compound, ClF would not be stable at all. It is therefore at once evident that the bond in ClF is preponderantly covalent. It may, however, have some slight polar character, which would make the energy of dissociation greater than it would be were the bond purely covalent. In order to decide whether this is actually the case or not, it is necessary to have some means of estimating what the strength of the bonds would be were it purely covalent. It seems very natural to suppose that if the bond were purely covalent the energy of dissociation should be some sort of a mean—as an approximation we may take

[1] The use of Table 33 is based on the assumption that the compound is covalent, which might seem a trifle inconsistent, but if we estimated from Table 16, even with a correction of the type indicated in Sec. 14.9, the distance would turn out to be greater than 1.63, and formation of the polar compound would appear to be even more endothermic.

the arithmetic mean—between the dissociation energies of Cl_2 and F_2. These latter are 2.48 and 2.72 volts, respectively (Table 7), the arithmetic mean being 2.60. The actual observed energy of dissociation is 3.74. It is thus seen that there is a difference Δ of 1.1 volts, and it is seen that this is too large to be much influenced by the way in which the estimate of the energy of a purely nonpolar bond was made. It seems most natural to suppose that this difference is due, at least in large part, to a certain amount of polarity in the bond. We shall hereafter use the arithmetic mean of the energy of the bonds between like atoms to get an estimate of the expected value of the purely nonpolar bond.[1] The chief justification of the procedure is the fact that over a wide range of compounds the actual bond energies are practically always, within the limit of accuracy, greater than or equal to that calculated for the nonpolar bond, thus indicating that the energy of the compound is less[2] than it would be if the bonds were purely nonpolar, as expected from Sec. 12.5. Another justification is the fact that it enables us to make a consistent arrangement of the elements in the order of their electronegativities, as we shall see in Sec. 12.8. The arrangement is consistent in that essentially the same order is obtained when different sets of compounds are used to estimate the relative electronegativities.

12.7. Definition of the Term "Bond Energy."—In the foregoing pages, the term "bond energy" has been used several times, without any definition having been formulated for it. In the case of diatomic molecules, its meaning will be clear; it is simply the dissociation energy, taken at room temperature, as a

[1] PAULING and YOST, *Proc. Nat. Acad. Sci.*, **18**, 414 (1932); PAULING, *J. Am. Chem. Soc.*, **54**, 3570 (1932). Recently Pauling and J. Sherman have suggested [*J. Am. Chem. Soc.*, **59**, 1450 (1937)] that a better estimate might be obtained from the geometric mean.

[2] There are some exceptions to this statement. For example, the gaseous alkali hydrides have dissociation energies ranging up to about 0.6 volt less than would be expected, according to the above method of calculation, for nonpolar bonds. However, the dissociation energies of the alkali double molecules, like Na_2, are very low, whereas the dissociation energy for H_2 is very high, and under such circumstances it is probable that the arithmetic mean is not a sufficiently good approximation to the expected nonpolar bond energy. In the case of certain nitrogen bonds and the C—I bond a similar but smaller discrepancy occurs.

matter of convenience.[1] It is only in the case of polyatomic molecules that any difficulty can occur. However, in the case of polyatomic atoms containing only one kind of bond, such as CH_4 or H_2O, there is still no difficulty. In this case, the bond energy is simply the energy required to dissociate the molecule completely into atoms, divided by the number of bonds. Thus the energy of the C—H bond in methane is one-fourth the energy of the reaction $CH_4 \rightarrow C + 4H$. It is to be noted that this is not the same as the energy of the reaction $CH_4 \rightarrow CH_3 + H$, as the bonds that are left in the unsaturated radical when one H is taken off are somewhat changed from their original condition. The latter quantity is of considerable importance in chemical kinetics, but is not essential to most of the problems treated in this book.

The significance of the term bond energy as applied to compounds in which more than one kind of bond occurs may perhaps best be brought out by considering some examples. The best set of examples to use for this purpose is that furnished by the series of paraffin chain saturated hydrocarbons, for here the experimental data are most extensive. The complete dissociation of a hydrocarbon C_nH_{2n+2} into $nC + (2n + 2)H$ obviously involves the breaking of $2n + 2$ C—H bonds and $n - 1$ C—C bonds. The usefulness of the concept of bond energy in this case lies in the fact that it is possible to assign a definite value to the energy of the C—H bond and a definite value to the energy of the C—C bond, and on multiplying the former by $2n + 2$ and the latter by $n - 1$ get the total energy required to break up the whole molecule into atoms to a reasonably good approximation, regardless of the value[2] of n.

A few words regarding the method of testing the hypothesis of constant bond energies in the case of hydrocarbons may not be out of place, since it is not possible actually to measure the energy of complete dissociation into atoms directly. The reaction which is actually studied is the combustion of the hydrocarbon[3]

[1] Theoretically, it would be better to use the dissociation energy at absolute zero, but the difference in the results obtained would be very slight.

[2] This *additivity* of bond energies was first observed by Thomsen. Thomsen and Berthelot were the pioneers in the study of heats of combustion.

[3] In this equation and throughout the discussion, all substances are in the gaseous state, unless otherwise indicated, except water, which is in the liquid state.

$$C_nH_{2n+2} + \frac{3n+1}{2}O_2 \rightarrow nCO_2 + (n+1)H_2O.$$

Following Thomsen, this can be thought of as broken up into the following steps:

$$C_nH_{2n+2} \rightarrow nC + (2n+2)H$$
$$nC + nO_2 \rightarrow nCO_2$$

and

$$(2n+2)H + \frac{n+1}{2}O_2 \rightarrow (n+1)H_2O.$$

Now let the bond energy of a C—H bond be E_{C-H} and that of a C—C bond be E_{C-C}. Further, let the energy given out when the reaction $C + O_2 \rightarrow CO_2$ takes place be E_{CO_2}, and the energy given out when the reaction $2H + \frac{1}{2}O_2 \rightarrow H_2O$ takes place be E_{H_2O}. Then the energy given out when the hydrocarbon C_nH_{2n+2} is oxidized is

$$E_n = nE_{CO_2} + (n+1)E_{H_2O} - (n-1)E_{C-C} - (2n+2)E_{C-H}. \tag{8}$$

Now E_{CO_2}, E_{C-C}, and E_{C-H} are not known independently, but E_{CO_2} is independent of n and E_{H_2O} is also. We cannot test the constancy of E_{C-C} and E_{C-H} separately; so the best way to proceed from this point seems to be to assume that E_{C-H} is independent of n and then check the constancy of E_{C-C} as obtained on the basis of this assumption. This can be done by comparing the value of E_n given by Eq. (8) with E_{n+1}, the energy of combustion of the next higher hydrocarbon. E_{n+1} is obtained by replacing n by $n+1$ in Eq. (8). It is readily seen that

$$\frac{n+2}{2}E_n - \frac{n+1}{2}E_{n+1} = E_{C-C} - \tfrac{1}{2}E_{CO_2}. \tag{9}$$

The left-hand side of this equation we shall call P_n. P_n may be obtained for different values of n, and since E_{CO_2} must be independent of n, P_n will be independent of n if E_{C-C} is, as required by the hypothesis of constant bond energies. If n is set equal to 1, then P_n gives the apparent energy of the carbon-carbon bond in ethane, minus $\frac{1}{2}E_{CO_2}$. Setting $n = 2$, the apparent additional energy of the new carbon-carbon bond in propane, minus $\frac{1}{2}E_{CO_2}$, is obtained, and so on. The values of P_n for different values of n are given in Table 11, which is computed from data of Rossini.[1] It is seen that the apparent value of E_{C-C} is not entirely independent of n, but that there is a definite trend. However, the trend is small so that for our purposes it may be neglected.

TABLE 11.—BOND ENERGIES IN THE NORMAL SATURATED HYDROCARBONS
(E_n in kilogram-calories per mole)

n	E_n	$-P_n$
1	211.61	53.9
2	371.33	50.5
3	528.79	49.8
4	685.87	49.6
5	842.90	

Another test of the hypothesis of constant bond energies may be made by using the energy of combustion of diamond.[2] The energy of combustion of diamond may be thought of as broken up into two steps, the evaporation of the solid to form carbon atoms and the combination of the carbon atoms with oxygen. Letting $-2P$ (for reasons that will presently become obvious) be the energy evolved in the reaction C (diam.) + $O_2 \rightarrow CO_2$ and S the energy absorbed on sublimation of diamond, we may write

$$-2P = -S + E_{CO_2}.$$

In diamond, each carbon atom has a bond to four other carbon atoms. Each bond, however, is common to two carbon atoms. There are, therefore, twice as many bonds as atoms, and we may set $S = 2E_{C-C}$, where E_{C-C} is the energy required to break a carbon-carbon bond in diamond. This gives

$$-2P = -2E_{C-C} + E_{CO_2}.$$

It is seen that, if E_{C-C} is to be the same for diamond as for hydrocarbons, not only should P_n be constant, but P and P_n should be

[1] ROSSINI, *Bur. Standards J. Res.*, **13**, 21 (1934).
[2] FAJANS, *Ber. deut. chem. Ges.*, **53**, 643 (1920).

equal. P is found experimentally[1] to be -47.2 kg.-cal. This agrees well with the value of P_n for the higher hydrocarbon and differs by 6.7 kg.-cal. (or 0.29 volt electron) from the value of P_1 (ethane). It thus appears that the constancy of bond energies becomes better for compounds containing a great many atoms (a diamond is a giant molecule containing practically an infinite number of atoms). It is probable, however, that reasonably good results will be obtained when we make our estimate of the nonpolar bond energy (as in Sec. 12.6, in order to calculate Δ) by consistent use of data for the compounds containing the smallest possible number of atoms. We are forced to do this in general, and so we shall also do it in the case of the carbon compounds.

Although we can make a test of the constancy of bond energies, it is not possible from available data to get the absolute values of the C—C and C—H bond energies with certainty. Of the energy quantities considered above, one, E_{H_2O}, is known, for the reaction $2H + \frac{1}{2}O_2 \rightarrow H_2O$ may be thought of as decomposed into the two steps, $2H \rightarrow H_2$ and $H_2 + \frac{1}{2}O_2 \rightarrow H_2O$, the energy of the first being known spectroscopically and that of the second calorimetrically. Assuming then, that E_{C-C} and E_{C-H} are constant, one of the three quantities E_{C-C}, E_{C-H}, and E_{CO_2} may be assigned arbitrarily and the two others evaluated. For, setting $n = 1$ and letting E_1 be the value of E_n for methane, we have

$$E_1 = E_{CO_2} + 2E_{H_2O} - 4E_{C-H},$$

and, similarly, for ethane

$$E_2 = 2E_{CO_2} + 3E_{H_2O} - E_{C-C} - 6E_{C-H}.$$

E_{CO_2} and E_{C-H} could be determined from these equations if E_{C-C} were known, but a third equation (for propane, for example) would not be independent. If the heat of vaporization of diamond were known accurately, we could obtain a value of E_{C-C}, as will be clear from the preceding discussion. Actually it seems about as satisfactory to assign it an arbitrary value of 3.00 electron volts (known to be of the right order of magnitude). It may be remarked that the value of Δ for the C—H bond is independent of the value assigned to E_{C-C}, for any change in

[1] Roth, *Zeits. Elektrochem.*, **21**, 1 (1915) (in agreement with early work of Berthelot).

E_{C-C} produces just half the change in E_{C-H} and

$$\Delta = E_{C-H} - \tfrac{1}{2}(E_{C-C} + E_{H-H}).$$

12.8. Polarity of Bonds and the Electronegativity Scale.—In Table 12 is given a comparison of actual *single*-bond energies (first value) and the values obtained theoretically on the assumption that the bonds are purely covalent (second value). Below these is given the difference, Δ, actual value minus the theoretical value for the hypothetical covalent bond. It is seen that Δ is positive as expected in almost every case, and in the exceptions the deviation is small and may be due to experimental error.

A few words regarding the method of determination of the single-bond energies in Table 12 will be in order. In many cases, it was possible to get single-bond energies from the dissociation energies of diatomic molecules, these being obtained spectroscopically; but sometimes thermochemical (calorimetric) measurements are also involved.

As an example, the determination of the ICl bond energy, which is the energy of the reaction

$$ICl \rightarrow I + Cl \tag{A}$$

may be considered. This may be obtained as half the sum of the energies of the following reactions:

$$2ICl \rightarrow I_2 + Cl_2 \tag{B}$$

(energy of reaction known from thermochemical measurements)

$$I_2 \rightarrow 2I \tag{C}$$

and

$$Cl_2 \rightarrow 2Cl \tag{D}$$

[energy of reactions (C) and (D) known from spectroscopic data].

Sometimes it is necessary to use already determined bond energies to get a new bond energy. Thus the energy for the O—O single bond is obtained in the following way. First the energy of the O—H bond is obtained. This is half the energy of the reaction

$$H_2O \rightarrow 2H + O \tag{E}$$

in which two O—H bonds are broken. The energy of reaction

TABLE 12.—BOND ENERGIES (FIRST ROW); CALCULATED NONPOLAR BOND ENERGIES (SECOND ROW); AND Δ VALUES (THIRD ROW)
(All in electron volts)

	H	C	N	P	O	S	F	Cl	Br	I
H	4.494	4.085 3.747 0.338	3.64 2.78 0.86	2.74 2.66 0.08	4.77 3.00 1.77	3.50 3.08 0.42	6.40 3.63 2.77	4.45 3.50 0.95	3.79 3.25 0.54	3.10 3.03 0.07
C		3.00	2.32 2.03 0.29		3.24 2.25 0.99	2.34 2.33 0.01	4.78 2.88 1.90	3.16 2.75 0.41	2.63 2.50 0.13	2.01 2.29 −0.28
N			1.06				2.99 1.91 1.08	1.60 1.78 −0.18	1.30 1.53 −0.23	
P				0.82				2.72 1.66 1.06	2.13 1.41 0.72	1.30 1.20 0.10
O					1.50		2.52 2.13 0.39	2.13 2.00 0.13		
S						1.66		2.83 2.08 0.75	2.44 1.83 0.61	
F							2.76	3.74 2.63 1.11		
Cl								2.505	2.29 2.25 0.04	2.22 2.04 0.18
Br									2.000	1.86 1.79 0.07
I										1.572

According to very recent data [Hughes, Corruccini, and Gilbert, *J. Am. Chem. Soc.* **61**, 2642 (1939)] value for N—N should be 0.86. This makes Δ for N—Cl and N—Br less negative, but does not have an important effect on Table 14. (Added in proof.)

(E) is obtained as the sum of

$$H_2O \rightarrow H_2 + \tfrac{1}{2}O_2 \quad \text{(thermochemical data)} \quad \text{(F)}$$
$$H_2 \rightarrow 2H \quad \text{(spectroscopic data)} \quad \text{(G)}$$
$$\tfrac{1}{2}O_2 \rightarrow O \quad \text{(spectroscopic data)} \quad \text{(H)}$$

Next, the energy of reaction

$$H_2O_2 \rightarrow 2H + 2O \quad \text{(I)}$$

is obtained. This is the sum of

$$H_2O_2 \rightarrow H_2 + O_2 \quad \text{(thermochemical data)} \quad \text{(J)}$$
$$H_2 \rightarrow 2H \quad \text{(spectroscopic data)} \quad \text{(K)}$$
$$O_2 \rightarrow 2O \quad \text{(spectroscopic data)} \quad \text{(L)}$$

The energy of reaction (I) is the O—O bond energy plus the energy of two O—H bonds, since H_2O_2 has the structure[1] H—O—O—H. The O—H energy being known, the energy of the O—O bond is found immediately. The ideas involved in the procedure are essentially similar to those used in obtaining the C—H bond energy.

In Table 13 is given a list of the reactions used in constructing Table 12. The first column gives the bond whose energy is sought, and in the last two columns are given the other bond energies and the dissociation energies used. In the last column, we also make note, by number, of any reaction energies which are used in calculating the bond energy in question, but which have been previously used in calculating some other bond energy, and so are not repeated in the tabulation. Sometimes a number of alternative methods for calculating a given bond energy are indicated; the results of these have been averaged, except in the case of the S—S bond where the results of the

[1] This is indicated chemically by the fact that the hydrogens are replaceable and the peroxide properties are always associated with a pair of oxygens. Further, X-ray analysis shows that in the crystals of BaO_2 and SrO_2 the oxygens lie close together in pairs ("Strukturbericht," Vol. III). These facts, of course, do not enable one to say whether the hydrogens are on different oxygen atoms, giving a structure with single bonds throughout, or if they are on the same oxygen, but the evidence seems to be in favor of the former. For a discussion of this point and references to the chemical considerations, see Mellor, "Comprehensive Treatise on Inorganic and Theoretical Chemistry," Vol. I, pp. 952*ff*. Longmans, Green and Co., 1927. See also the work on the structure of liquid H_2O_2 of Randall, *Proc. Roy. Soc.*, **159**, 83 (1937), and the theoretical work of Penney and Sutherland, *Trans. Faraday Soc.*, **30**, 898 (1934); *J. Chem. Phys.*, **2**, 492 (1934).

TABLE 13.—THERMOCHEMICAL DATA USED IN THE CALCULATION OF BOND ENERGIES

(All substances in the gaseous state unless otherwise indicated: l = liquid, s = solid, etc.)

Bond	Reaction	Energy absorbed, kg.-cal.	Other bond energies used	Other reaction energies used
N—H	(1) $NH_3 \rightarrow \frac{1}{2}N_2 + \frac{3}{2}H_2$	11.00	....	N_2, H_2
N—N	(2) $N_2H_4 \rightarrow N_2H_4$ (aq)	−14.1	N—H	N_2, H_2
	(3) N_2H_4 (aq) $\rightarrow N_2 + 2H_2$	−4.5		
P—P	(4) $P_4 \rightarrow 4P$ (I, yellow)	−13.2		
	(5) P (I, yellow) $\rightarrow$ P	31.6		
O—H	(6) $H_2O \rightarrow H_2 + \frac{1}{2}O_2$	57.801	....	H_2, O_2
O—O	(7) $H_2O_2 \rightarrow H_2 + O_2$	33.59	O—H	H_2, O_2
S—H	(8) S (rhomb.) $\rightarrow$ S	53.	....	H_2
	(9) $H_2S \rightarrow H_2$ + S (rhomb.)	5.3		
S—S	(10) $H_2S_2 \rightarrow H_2$ + 2S (rhomb.)	−8.94	S—H	H_2, (8)
S—S	(11) $\frac{1}{8}S_8 \rightarrow$ S (rhomb.)	−2.50	....	(8)
P—H	(12) $PH_3 \rightarrow$ P (I, yellow) + $\frac{3}{2}H_2$	2.3	....	(5), H_2
F—H	(13) $HF \rightarrow \frac{1}{2}H_2 + \frac{1}{2}F_2$	64.0	....	H_2
	(14) $\frac{1}{2}F_2 \rightarrow F$	31.75		
Cl—H	(15) $HCl \rightarrow \frac{1}{2}H_2 + \frac{1}{2}Cl_2$	22.06	....	H_2
	(16) $\frac{1}{2}Cl_2 \rightarrow Cl$	28.90		
Br—H	(17) $HBr \rightarrow \frac{1}{2}H_2 + \frac{1}{2}Br_2$ (l)	8.65	....	H_2
	(18) $\frac{1}{2}Br_2$ (l) $\rightarrow$ Br	26.88		
I—H	(19) $HI \rightarrow \frac{1}{2}H_2 + \frac{1}{2}I_2(s)$	−5.91	....	H_2
	(20) $\frac{1}{2}I_2(s) \rightarrow I$	25.59		
C—N	(21) $CH_3NH_2 \rightarrow$ C (diam.) + $\frac{1}{2}N_2$ + $\frac{5}{2}H_2$	7.3	N—H, C—H	H_2, N_2
	(22) $CH_4 \rightarrow$ C (diam.) + $2H_2$	18.24		
C—N	(23) $(CH_3)_2NH \rightarrow$ 2C (diam.) + $\frac{1}{2}N_2 + \frac{7}{2}H_2$	8.2	N—H, C—H	H_2, N_2, (22)
C—O	(24) $CH_3OH \rightarrow$ C (diam.) + $2H_2$ + $\frac{1}{2}O_2$	48.44	C—H, O—H	H_2, O_2, (22)
C—O	(25) $(CH_3)_2O \rightarrow$ 2C (diam.) + $3H_2$ + $\frac{1}{2}O_2$	46.4	C—H	H_2, O_2, (22)
C—S	(26) $CH_3SH \rightarrow$ C (diam.) + $2H_2$ + S (rhomb.)	3.6	O—H, C—O, S—H	O_2, (8), (24),
C—S	(27) $(CH_3)_2S \rightarrow$ 2C (diam.) + $3H_2$ + S (rhomb.)	8.0	C—C	(8)
	(28) $C_2H_6 \rightarrow$ 2C (diam.) + $3H_2$	20.96		
C—F	(29) $CF_4 \rightarrow$ C (diam.) + $2F_2$	163.	C—H	F_2, (22), H_2
C—Cl	(30) $CCl_4 \rightarrow$ C (diam.) + $2Cl_2$	25.9	C—H	Cl_2, (22), H_2
C—Cl	(31) $CHCl_3 \rightarrow$ C (diam.) + $\frac{3}{2}Cl_2$ + $\frac{1}{2}H_2$	23.6	C—H	Cl_2, H_2, (22)
C—Cl	(32) $CH_2Cl_2 \rightarrow$ C (diam.) + Cl_2 + H_2	21.7	C—H	Cl_2, H_2, (22)
C—Cl	(33) $CH_3Cl \rightarrow$ C (diam.) + $\frac{1}{2}Cl_2$ + $\frac{3}{2}H_2$	20.1	C—H	Cl_2, H_2, (22)
C—Br	(34) $CBr_4 \rightarrow$ C (diam.) + $2Br_2$ (l)	−12.	C—H	(18, (22), H_2
C—Br	(35) $CHBr_3 \rightarrow$ C (diam.) + $\frac{3}{2}Br_2$ (l) + $\frac{1}{2}H_2$	−6.	C—H	(18), H_2, (22)
C—Br	(36) $CH_2Br_2 \rightarrow$ C (diam.) + $Br_2(l)$ + H_2	1.	C—H	(18), H_2, (22)

TABLE 13.—THERMOCHEMICAL DATA USED IN THE CALCULATION OF BOND ENERGIES.—(*Continued*)

Bond	Reaction	Energy absorbed, kg.-cal.	Other bond energies used	Other reaction energies used
C—Br	(37) $CH_3Br \rightarrow C$ (diam.) $+ \frac{1}{2}Br_2(l) + \frac{3}{2}H_2$	8.5	C—H	(18), H_2, (22)
C—I	(38) $CHI_3 \rightarrow C$ (diam.) $+ \frac{3}{2}I_2(s) + \frac{1}{2}H_2$	−44.	C—H	(20), H_2, (22)
C—I	(39) $CH_2I_2 \rightarrow C$ (diam.) $+ I_2(s) + \frac{1}{2}H_2$	−25.	C—H	(20), H_2, (22)
C—I	(40) $CH_3I \rightarrow C$ (diam.) $+ \frac{1}{2}I_2(s) + \frac{3}{2}H_2$	−4.5	C—H	(20), H_2, (22)
N—F	(41) $NF_3 \rightarrow \frac{1}{2}N_2 + \frac{3}{2}F_2$	26.6	....	N_2, F_2
N—Cl	(42) NCl_3 (in CCl_4) $\rightarrow \frac{1}{2}N_2 + \frac{3}{2}Cl_2$	−55.0	....	N_2, Cl_2
	(43) NCl_3 (in CCl_4) $\rightarrow NCl_3$	6 (est.)		
N—Br	(44) $NOCl \rightarrow \frac{1}{2}N_2 + \frac{1}{2}O_2 + \frac{1}{2}Cl_2$	−12.8	N—Cl	Cl_2, Br_2
	(45) $NOBr \rightarrow \frac{1}{2}N_2 + \frac{1}{2}O_2 + \frac{1}{2}Br_2(l)$	−17.7		
P—Cl	(46) $PCl_3 \rightarrow P$ (I, yellow) $+ \frac{3}{2}Cl_2$	70.0	....	Cl_2, (5)
P—Br	(47) $PBr_3(l) \rightarrow P$ (I, yel.) $+ \frac{3}{2}Br_2(l)$	45.0	....	(18), (5)
	(48) $PBr_3(l) \rightarrow PBr_3$	10.0 (est.)		
P—I	(49) $PI_3(s) \rightarrow P$ (I, yellow) $+ \frac{3}{2}I_2(s)$	10.9	....	(20), (5)
	(50) $PI_3(s) \rightarrow PI_3$	30 (est.)		
O—F	(51) $F_2O \rightarrow F_2 + \frac{1}{2}O_2$	−5.5	....	F_2, O_2
O—Cl	(52) $Cl_2O \rightarrow Cl_2 + \frac{1}{2}O_2$	−18.25	....	Cl_2, O_2
S—Cl	(53) $S_2Cl_2 \rightarrow 2S$ (rhomb.) $+ Cl_2$	5.65	S—S	(8), Cl_2
S—Br	(54) $S_2Br_2(l) \rightarrow 2S$ (rhomb.) $+ Br_2(l)$	4.0	S—S	(8), (18)
	(55) $S_2Br_2(l) \rightarrow S_2Br_2$	11.9 (est.)		
F—Cl	(56) $ClF \rightarrow \frac{1}{2}Cl_2 + \frac{1}{2}F_2$	25.7	....	Cl_2, F_2
Cl—Br	(57) $BrCl \rightarrow \frac{1}{2}Cl_2 + \frac{1}{2}Br_2(l)$	−3.07	....	Cl_2, (18)
Cl—I	(58) $ICl \rightarrow \frac{1}{2}Cl_2 + \frac{1}{2}I_2(s)$	−3.46	....	Cl_2, (20)
I—Br	(59) $IBr \rightarrow \frac{1}{2}Br_2(l) + \frac{1}{2}I_2(s)$	−9.6	....	(18), (20)

All data from Bichowsky and Rossini, "Thermochemistry of Chemical Substances," Reinhold Publishing Corporation, 1936, except a few which were estimated and reaction (41) which is from the second supplement (1931) of Landolt-Börnstein, "Tabellen." Reaction (8) is corrected to conform to the value of the dissociation energy of S_2 given in Table 7. Reaction (2) is based on heat of vaporization from Landolt-Börnstein, "Tabellen," third supplement (1936), and heat of solution from Bushnell, Hughes, and Gilbert, *J. Am. Chem. Soc.*, **59**, 2142 (1937).

second method, involving S_8, have not been used. All energies in Table 13 are given in kilogram-calories taken directly from the various tabulations indicated, but in Table 12, in conformity with the usage of Pauling[1] in setting up his original table of electronegativities, electron volts are used. Results given in kilogram calories may be converted to electron volts by dividing

[1] PAULING, *J. Am. Chem. Soc.*, **54**, 3570 (1932).

by 23.06. (Concerning units, see note, page 461.) All energies and bond energies in Tables 12 and 13 are for room temperature.[1]

It will be noted that only those compounds have been used which involve elements in their lowest valence state. The table of electronegativities, which will be subsequently obtained from Table 12, therefore, refers specifically to elements in their lowest valence states, and will not be expected to be applicable in other cases. In finding the P—P bond energy, it is assumed that the phosphorus in P_4 is in its lowest valence state, as P_4 is known to have a tetrahedral structure, each phosphorus atom thus being connected with each of the others by single bonds, a total of six bonds being involved in the molecule. It is assumed that S_8 has a ring structure, each S atom being attached to two neighbors by a single bond. H_2S_2 presumably has a structure like H_2O_2, and S_2Cl_2 and S_2Br_2 are assumed to be[2] Cl—S—S—Cl and Br—S—S—Br. The structures of all other compounds used are obvious or well known and need not be discussed at this point. In the case of carbon bonds, only the simplest compounds involving the desired linkages have been used.

One of the most remarkable features brought out in Table 12 is the extremely low value of the bond energy for the N—N single bond and for the O—O single bond. The O—O bond has a strength less than that of S—S, which is an exception to the almost universal rule that under similar circumstances the binding between small atoms is tighter than that between larger atoms in the same column of the periodic table.[3] This will be discussed further in Sec. 14.11.

[1] All the bond energies listed in Table 12 are really too large by kT (or about 0.025 electron volt at room temperature), because the bond energy *plus* kT is what is obtained directly from tabulated values of heats of reaction (such as given in Table 13) which include the heat absorbed in order to supply energy for the work done against the atmosphere when the reaction is carried out at constant pressure [*i.e.*, tabulations give the ΔH instead of ΔE of the reaction, where the symbols are those used in Eq. (6) of Appendix II]. The inclusion of the extra kT in the tabulated values will obviously make no difference. It is to be noted that it is *not* included in the values of Tables 7 and 11, which may cause some apparent discrepancies between these tables and Table 13.

[2] See Ackermann and Mayer, *J. Chem. Phys.*, 4, 379 (1936).

[3] It is indeed possible that the value for the S—S bond is in some error, but in any event the O—O single-bond energy is, comparatively speaking, quite low. In the case of carbon, the situation is very different from that

We may now consider the setting up of a table of relative electronegativities. It is desired to place the elements in linear order, assigning to each one a number, which will express its electronegativity. Pauling[1] has noted that it is possible to assign to each element a definite number x such that for any two elements, say, A and B, the relation $(x_A - x_B)^2 = \Delta_{AB}$ (where Δ_{AB} is the particular value of Δ from Table 12) is approximately fulfilled.[2] This makes it possible to assign a series of values of x from the Δ values, and this number x may be taken as the measure of the electronegativity; it will be seen then that the greater the distance between the electronegativities of two elements, the greater the corresponding value of Δ, as expected. We then proceed as follows: We first note, by inspection of the Δ values, that the elements can be arranged in the following order of increasing electronegativity: P, H, S, I, C, Br, N, Cl, O, F. We then find values for $x_A - x_B$, where A and B represent two adjacent elements in the list. Thus, for Cl and O, the value of $\sqrt{\Delta}$ for the Cl — O bond gives at once a value of[3] $x_O - x_{Cl}$. However, another estimate of $x_O - x_{Cl}$ can be

obtaining with oxygen and nitrogen. By comparison of the energies of formation of C_2H_4 and C_2H_2 with that of ethane, the C=C double-bond energy and C≡C triple-bond energy may be estimated to be 5.28 and 7.08 electron volts, respectively, assuming that the energy of the C—C bond is 3.00 volts. It thus appears that the energy of a double bond is less than twice, and the energy of a triple bond less than three times that of a single bond. (This conclusion could hardly be upset even though the value for the C—C bond should turn out to be appreciably different from 3.00). The dissociation energy of oxygen, which contains a bond that resembles a double bond, is much more than twice the single-bond energy, and the dissociation energy of nitrogen, which contains a triple bond, is much more than three times its single-bond energy. Although the single-bond energies of nitrogen and oxygen may not be exactly correct, it does not appear possible that the error is sufficiently great to account for the peculiarity; for *all* single-bond energies involving oxygen and nitrogen are smaller than the corresponding single-bond energies of carbon, and an appreciable increase in the values of the O—O and N—N energies would result in the appearance of many more large negative values of Δ in Table 12 than are there at present.

[1] Pauling, *loc. cit.*

[2] For a theoretical deduction of this relationship and further discussion of electronegativity scales, see Mulliken, *J. Chem. Phys.*, **3**, 573, 586 (1935).

[3] In the cases where Δ is negative, it is, of course, impossible to find $\sqrt{\Delta}$. Since, if the ideas back of the procedure used are correct, negative values of Δ are due to experimental error, we have in these cases set $\sqrt{\Delta} = 0$.

obtained from the values of $x_C - x_{Cl}$ and $x_C - x_O$, which are calculated from the respective Δ values. Thus a series of values for $x_O - x_{Cl}$ can be found and the average taken. If this is done for all adjacent pairs of elements in the list, and H is arbitrarily assigned the value $x = 0$, the values of x for all the elements can be tabulated as shown in Table 14. It will be observed that with respect to the halogens the order is the same, and even the relative spacing somewhat the same as in Table 9.

TABLE 14.—ELECTRONEGATIVITIES

P	−0.12	Br	0.88
H	0.0	N	1.02
S	0.30	Cl	1.09
C	0.47	O	1.46
I	0.47	F	1.97

In this table, no attempt has been made to include the electropositive elements, though values for the single-bond energies of the alkali halides and hydrides can be obtained. These substances, especially the halides, are so much more polar than nonpolar that no satisfactory results can be expected from treating the deviations from nonpolarity.[1] They are considered as polar compounds in Chap. XIV.

Table 14 can, however, be readily extended to a few other negative elements (see Exercise 7, page 204).

12.9. Dipole Moments.—In the case of an elementary molecule in which the two atoms are alike and in which there is perfect symmetry with respect to the center of gravity, there is no net average displacement of either positive or negative charge from the center of gravity (unless the molecule is placed in an electric field, in which case an electric moment will be developed due to the polarizability of the molecule); the situation is as described on page 163, so that on the average the center of positive charge

[1] Since this book went to press there has appeared a table of electronegativity values which includes the electropositive elements (Pauling, "The Nature of the Chemical Bond," Cornell University Press, 1939). This table is obtained on the assumption that in metals the bonds are essentially single covalent bonds. Although the values obtained fit in well with the periodic table and the table of ionization potentials, they appear to have a much more speculative basis than the electronegativities of the negative elements.

coincides with the center of negative charge. On the other hand, the situation will be different with a gaseous molecule of NaCl, for example; for chlorine is more electronegative than sodium, and the electrons will tend to be drawn over toward the chlorine. Thus, there will be a displacement of electric charge, the center of positive charge will no longer coincide with the center of negative charge, and the molecule will possess a *permanent* dipole moment. If it were strictly true that NaCl were composed of Na^+ and Cl^- ions, it would be possible to calculate the dipole moment from the distance between the ions. For this purpose, it could be assumed that the net charge of each ion was concentrated at its center. However, since actually the ions are polarizable, they might be expected to be distorted in each other's fields, resulting in a decrease in the electric moment. Such a polarizing effect would be especially marked in the hydrogen halides because of the

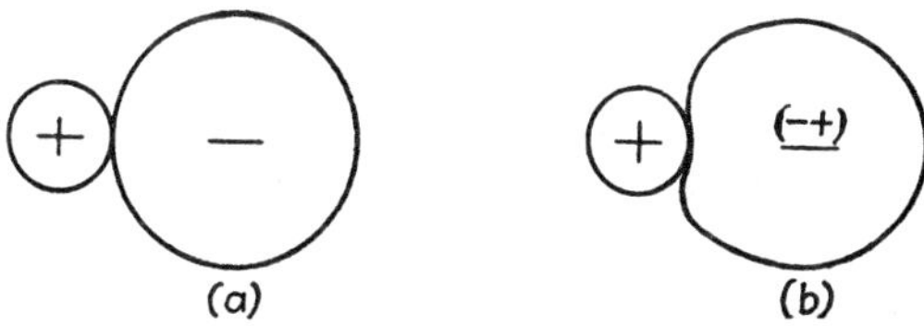

FIG. 48.—Model of a polar molecule; (*a*) ideal case; (*b*) showing distortion of anion by cation: the induced dipole partially cancels the original dipole.

large polarizing action of the small H^+ ion, and, as we have seen, HCl, HBr, and HI are to be considered as having nonpolar rather than polar binding.[1]

Enough has been said to make it evident that measurements of the dipole moment will give valuable information concerning the polarity of a bond. The magnitude of the moment can be obtained from measurements of the dielectric constant of a substance. It will be recalled that the dielectric constant D of a substance is defined in the following way. Suppose we have a parallel-plate condenser with a certain charge on the plates; let the potential between the plates in a vacuum (whose dielectric constant is 1) be Φ. Then if, without changing the charge on the plates, a substance whose dielectric constant is D is introduced between them, the potential will drop to Φ/D. The dielectric constant of a substance is thus a measure of its ability to counteract an electric field; it does this, of course, by producing an

[1] See footnote 2, p. 172.

electric field itself in the opposite direction, either by displacement of electric charge within itself, *i.e.*, in virtue of its polarizability, or by orientation of its permanent dipoles. There is one essential difference between these two effects, which enables us to separate them. The displacement effect is independent of temperature, but the orientation depends upon the temperature, because the greater the thermal agitation, the greater the tendency for the directions of the dipoles to be randomly distributed and the more difficult it is for the applied field to orient them. To find the dipole moment, the dielectric constant may be measured at various temperatures.[1]

The dielectric constant is closely related to the index of refraction n; in fact, according to the electromagnetic theory of light, for light of infinite wave length $n^2 = D$. The relation between dielectric constant, polarizability α, and permanent electric moment μ is given by a relation similar to Eq. (4),[2] page 175,

$$\frac{D-1}{D+2}V = \frac{4\pi}{3}N\left(\alpha + \frac{\mu^2}{3kT}\right), \tag{4a}$$

where k is Boltzmann constant (see page 461) and T is absolute temperature. It is thus seen how measurements at different temperatures give μ. μ can also be found by measuring the dielectric constant and then finding the index of refraction for visible light instead of light of infinite wave length. In this case, $n^2 \neq D$, because the frequency of the light is so great that the orientation of the molecule and hence of the permanent dipole does not have time to take place. Yet as far as the polarizability is concerned, the wave length can, at least in many cases, be considered practically infinite, because the *displacement* of charge in the molecule, as opposed to orientation, takes place very rapidly. Thus α can be found from Eq. (4), after which μ can be obtained from Eq. (4*a*).

In Table 15, we give, following Sidgwick,[3] some results for the hydrogen halides. The second column gives μ; the third column

[1] Debye, "Polar Molecules," Chap. III, Reinhold Publishing Corporation, 1929.

[2] Equations (4) and (4*a*) are the same except that Eq. (4) tacitly assumes $\mu = 0$.

[3] Sidgwick, "The Covalent Link in Chemistry," p. 144, Cornell University Press, 1933.

the distance δ obtained by setting $\mu = e\delta$ (where e is the electronic charge); the fourth column r_0, the actual distance between hydrogen and halogen nuclei, as obtained from band spectra; and the last column δ/r_0. It is seen that δ/r_0 is considerably less than 1, indicating a relatively small net displacement of charge within the molecule, or a great polarization of the halogen ion by H^+. It is seen, furthermore, that the moment μ as well as the ratio δ/r_0 decreases as the polarizability increases, in the order Cl^-, Br^-, I^-, as is to be expected.

TABLE 15.—SOME DIPOLE MOMENTS

	$\mu \times 10^{18}$, e.s.u.	δ, Å	r_0, Å	δ/r_0
HCl	1.03	0.216	1.27	0.17
HBr	0.78	0.163	1.41	0.12
HI	0.38	0.080	1.61	0.050
NaI	4.9	1.03	2.90	0.35
KCl	6.3	1.32	2.79	0.47
KI	6.8	1.42	3.23	0.44

To Table 15 have been added results for some gaseous alkali halide molecules, obtained by Scheffers[1] by means of a molecular-beam method, somewhat similar to the Stern-Gerlach experiment on magnetic moments described in Chap. VI. It is seen that even in molecules as polar as these there is considerable polarization of the anion. It might well be that there would be less distortion in a crystal, where the field of the cations is more or less evenly distributed around the anion.[2]

In the case of polyatomic molecules, the situation is more complicated. However, in certain cases, where the structure of the molecule is known, the bond moments can be readily determined from measurements of the moment of the molecule and the assumption that this moment is the resultant of bond moments directed along the bond directions. Thus water is known (see Chap. XV) to have a kinked structure, the two

[1] SCHEFFERS, *Phys. Zeit.*, **35**, 425 (1934). The values of r_0 in Table 15 are the r_0' (expt.) of Table 27 of Chap. XIV.

[2] In this connection, it is of interest to note that Ewing and Seitz, *Phys. Rev.*, **50**, 760 (1936), believe, on theoretical grounds, that even in the crystals the electrons are considerably displaced from the negative toward the positive ion.

O—H bonds being at an angle of 105°. If these bonds have equal moments directed at this angle to each other, the bond moment is readily calculated from the measured resultant. In carbon compounds, the problem is complicated on account of the tetrahedral symmetry of carbon. The bond moments in methane, for example, all cancel, and no information can be obtained regarding their magnitude from the fact that the electric moment of methane is zero. However, if the C—H moment is obtained by extrapolation from O—H and N—H (the N—H bond moment can be found because ammonia has a pyramidal structure, as is noted in Chap. XV, and hence the resultant moment is not zero), other carbon moments can be obtained from measurements of organic compounds. It is generally assumed that a bond moment depends only on the bond and is independent of the compound in which it is situated. It is not our purpose to go here into the details of the determination of bond moments, and we shall merely state that a considerable number have been determined with more or less accuracy and certainty. It has been found[1] that the hydrogen bond moment of quite a series of elements is roughly proportional to the electronegativity value given in Table 14. This would lead us to expect the moment of a single bond formed by *any* two elements, in general, to be proportional to the difference in their electronegativities. However, when neither of the elements involved is hydrogen, this is found to be at best only qualitatively true. It is, however, clear that there is a close connection between the dipole moment and the polar character of the bond, even though it does not appear possible to express it so satisfactorily as in the case of bond energies.

12.10. Moments and Bond Energies of the Hydrogen Halides, Considered as Ionic Molecules.—In the last section, we took the point of view that the hydrogen halides could be considered as polar molecules in which the anion was highly polarized by the field of the cation. In this section, a somewhat more quantitative development of this idea will be attempted,[2] though it will

[1] Smallwood, *Zeits. physik. Chem.*, **B19**, 242 (1932); Malone, *J. Chem. Phys.*, **1**, 197 (1933); Smyth, *J. Phys. Chem.*, **41**, 209 (1937), *J. Am. Chem. Soc.*, **60**, 183 (1938).

[2] See Heisenberg, *Zeits. Physik*, **26**, 196 (1924); Hund, *ibid.*, **31**, 81 (1925); **32**, 1 (1925); Debye, "Polar Molecules," Chap. IV; Kirkwood, *Physik. Zeits.*, **33**, 259 (1932).

perhaps appear that this is really approaching the problem from the wrong end, since the hydrogen halides are more nearly covalent than ionic. But there are eight electrons in the valence shell, and if the displacement, which cancels part of the electric moment expected for a purely ionic molecule, is shared among them all, the displacement of any one need not be great. This is consistent with the ideas of the Hund-Mulliken theory of valence, and by a detailed wave mechanical calculation on essentially this basis, Kirkwood has been able to give a very good description of these molecules. For our purposes, however, we shall use a much cruder picture.

A number of very rough assumptions will be made. We shall consider the measured dipole moment μ of the molecule to be composed of two parts (1) the moment er_0 to be expected if it really consists of two ions a distance r_0 apart and (2) the moment, say μ_0, which is induced in the anion by the proton and which is, of course, in the opposite direction. We have then

$$\mu = er_0 - \mu_0. \tag{10}$$

It is obvious that μ_0 is nearly equal to er_0, since μ is small compared with er_0. We shall suppose that, as noted, μ_0 is due to a displacement of a considerable amount of charge through a distance small compared with r_0, and that the dipole is located at the center of the halide ion. Now the field of force due to the proton at the center of the halide ion is equal to e/r_0^2, and we may set $\mu_0 = \alpha e/r_0^2$. However, if this expression for μ_0 is substituted into (10), it is found necessary to give α a value considerably smaller than that given in Table 10 in order to get the right value of μ. This discrepancy is not at all surprising, considering the closeness of the proton to the anion and the consequently great variation of its field over the volume occupied by the anion, as well as the fact that the proton has penetrated well into the electron shell of the anion.

Let us now consider the work necessary to remove the proton to infinity against the electrical attraction due to the charge on the negative ion and the induced dipole. The potential of the proton due to the charge of the anion is $-e^2/r_0$, and by page 467 the potential due to the induced dipole is $-\frac{1}{2}\alpha e^2/r_0^4 = -\frac{1}{2}\mu_0 e/r_0^2$. The total work of removal, therefore, is $\dfrac{e^2}{r_0}\left(1 + \dfrac{\frac{1}{2}\mu_0}{er_0}\right)$. From the values of $\frac{\delta}{r_0}\left(= \frac{\mu}{er_0} = 1 - \frac{\mu_0}{er_0}\right)$ in Table 15, it will be seen that the term due to the induced dipole is a very appreciable fraction of the whole work. Just how much of this work is to be referred to as the covalent part of the binding energy is probably an academic question, since covalent forces are also electrostatic in their ultimate origin.

Use of the values of δ/r_0 from Table 15 in the formula of the preceding paragraph to evaluate the energy of dissociation of the hydrogen halides into ions gives fairly good results. Still, because of the approximations involved, including in particular the neglect of the repulsive forces due to the penetration of the proton into the electron shell, the calculation must be considered to be quite rough. The discussion, however, should serve to indicate something of the nature of the forces involved in the transition between polar and covalent binding.

12.11. Resonance.—As has been explained in some detail, the considerations of Secs. 12.7 and 12.8 rest upon the hypothesis of constant bond energies. That is, it has been assumed that a definite energy is associated with a given bond between a given pair of atoms, no matter where this bond occurs. Of course, this is only an approximation in any event, but there are certain sets of circumstances under which this assumption can break down rather badly, and the subject should not be closed without some discussion of this matter. These exceptions are connected with a type of resonance of considerable importance, though most of the applications are in organic chemistry.

In Sec. 12.5, the phenomenon of resonance between covalent and ionic states of a molecule was considered in some detail, and a good deal of the material of this chapter is based directly, or indirectly, upon this discussion. It will be clear, however, that resonance will not necessarily be confined to this particular type of interaction. There can also be resonance between different possible electronic structures of a molecule, all of which may be essentially covalent. In fact, it has already been intimated that this is the case in the discussion of SO_2 in Sec. 11.1.

It is especially easy for such a situation to occur in compounds containing double bonds. Consider, for example, formic acid. The electronic structure of this substance might conceivably be written either

$$\mathrm{H{:}C}\begin{smallmatrix}\cdot\cdot\ddot{\mathrm{O}}{:}\mathrm{H}\\ \cdot\cdot\ddot{\mathrm{O}}{:}\end{smallmatrix} \qquad \text{or} \qquad \mathrm{H{:}C}\begin{smallmatrix}\ddot{\mathrm{O}}{:}\mathrm{H}\\ \vdots\ddot{\mathrm{O}}{:}\end{smallmatrix}$$

These two electronic configurations would probably have nearly enough the same energies and the same atomic positions so that they could interact or resonate with each other. That the energies would be nearly equal is expected from the fact that both structures contain just the same bonds, with but a slightly different arrangement. The molecule, then, would actually be in a state intermediate to those described by the two Lewis formulas, and its energy would be lowered by the resonance. A similar situation might be expected with any carboxylic acid or ester.[1]

In the case of carbon dioxide, resonance might be expected among the three structures

$$\underset{\cdot\cdot}{\overset{\cdot\cdot}{\mathrm{O}}}{::}\mathrm{C}{::}\underset{\cdot\cdot}{\overset{\cdot\cdot}{\mathrm{O}}}, \qquad {:}\underset{\cdot\cdot}{\overset{\cdot\cdot}{\mathrm{O}}}{:}\mathrm{C}{\vdots\vdots}\mathrm{O}{:}, \qquad \text{and} \qquad {:}\mathrm{O}{\vdots\vdots}\mathrm{C}{:}\underset{\cdot\cdot}{\overset{\cdot\cdot}{\mathrm{O}}}{:}$$

The first of these is the one usually written, but the two others, involving a triple bond to one or the other of the oxygens, are quite conceivable, and their energies may not be greatly different from that of the first.

It will be interesting, then, to find the normal C═O bond energy from compounds in which this resonance cannot occur, and compare it with C═O bond energies in which resonance does occur. Resonance cannot occur in aldehydes. We may find the C═O bond energy, knowing the C—H bond energy, either by comparing the heat of combustion or of formation of acetaldehyde or glyoxal, OHC—CHO, with that of ethane, or by comparing the heat of combustion or formation of formaldehyde with that of methane. Unfortunately, these do not give consistent results. From these three compounds, 7.23, 7.39, and 6.69 electron volts, respectively, are obtained. In any event, these values are all lower than the apparent value for C═O obtained from compounds with resonance. Methyl formate (compared with methyl ether), formic acid (compared with methyl alcohol), and carbon dioxide (compared with methane) give 7.90, 7.97,

[1] PAULING and J. SHERMAN, *J, Chem. Phys.*, **1**, 606 (1933).

and 7.88 electron volts, respectively, showing that the resonance energy has strengthened the bond considerably.

Resonance can also have an effect on the interatomic distance, as will be seen in Sec. 16.9. Thus the interatomic distances in CO_2 are abnormally small as compared with distances in compounds having C=O groups without having the possibility of resonance. Irregularities in the interatomic distances also occur in C_2H_3Cl, which can conceivably have either of the structures

```
H.   :Cl:              H   .Cl:
 .C::C.          or   :C:C.
H.    .H               H  .H
```

It must be said, however, that the energy of this compound shows no abnormality.

Resonance between different electronic structures has recently received many applications in organic chemistry, especially in the explanation of properties of aromatic compounds.[1]

Exercises

1. Calculate the energy of dissociation of IF into neutral atoms, assuming that the binding is polar. The IF distance may be estimated (see Table 33) at 2.0Å.; the electron affinity of F is 4.12 electron volts. Estimate the actual energy of dissociation from Table 12. Do you believe IF to be predominantly ionic?

2. Make a similar calculation for HF. The interatomic distance is 0.864Å. In this case, it will be necessary to allow for the repulsive forces in the assumed polar molecule. This may be done by multiplying the electrostatic potential by $\left(1 - \frac{1}{n}\right)$, where n may be taken as 6, which is probably an overestimate (see Sec. 14.5 and Table 25).

3. Check the assumption of constancy of bond energies by calculating the energy of the C—O bond from Eq. (24) and again from (25) of Table 13.

4. Check the value of the bond energy for the S—S bond, using Eq. (10), Table 13. Repeat, using Eq. (11).

5. Check the bond energy of the C—Cl bond; of the N—F, N—Cl, and N—Br bonds; of the S—Cl and the S—Br bonds.

6. Calculate the average difference of the electronegativities of S and I from the Δ values of Table 12.

7. Discuss the electronegativity table from the point of view of the periodic table. Estimate the positions of the following elements: As, Se, Si, B. For values, see Pauling "The Nature of the Chemical Bond," Chap. II.

[1] See PAULING, Chap. 22 of Gilman, "Organic Chemistry: An Advanced Treatise," vol. II, John Wiley & Sons, Inc., 1938.

CHAPTER XIII

THE NATURE OF THE SOLID STATE

The classification of bonds as polar or nonpolar, with intermediate cases, is adequate to cover the situation with practically all gaseous molecules. In the case of solids, matters are not quite so simple. In some solid substances, the forces cannot be classed as either polar or nonpolar, in the restricted sense in which we have hitherto used these terms; on the other hand, there are many examples of solids in which just these kinds of forces are entirely adequate to describe the situation.

Solid substances (*i.e.*, crystals) may be classified roughly into four classes,[1] as follows:

1. Ionic (typically polar) crystals.
2. Atomic (typically nonpolar) crystals.
3. Molecular crystals.
4. Metals.

There are of course many gradations and many kinds of gradations between these types; the differences between them can, however, be best understood by considering the properties of the "most typical" members of each class.

In the typical ionic crystals, the forces are of the type already described as polar. The crystal consists of positively and negatively charged ions, held together by electrostatic attraction; any given ion exerts comparable forces on all its neighbors, so that it is not possible to group the atoms into molecules within the crystal, but, as is well known, the whole crystal must be considered as one gigantic molecule. Typical examples of these substances are furnished by the salts, *e.g.*, the alkali halides, which in the gaseous state form molecules with polar bonds. Ionic crystals have in general certain characteristic properties;

[1] See Ruff, *Ber. deut. chem. ges.*, **52**, 1223 (1919). Kossel, *Zeits. Physik*, **1**, 395 (1920). Biltz and Klemm, *Zeits. anorg. allgem. Chem.*, **152**, 267 (1926). Grimm, *Zeits. Elektrochem.*, **34**, 430 (1928); *Naturwiss.*, **17**, 535, 557 (1929); *Angewandte Chemie*, **47**, 53 (1934).

they are hard, often easily soluble in water, in which they are ionized; they are nonvolatile, having boiling points usually above 1000°C., and practically always above 700°; and the molten salts, consisting of mixtures of positive and negative ions, conduct the electric current. The molecular volumes are determined by the ionic radii in a way which will presently be discussed in some detail.

In atomic crystals, the forces are of the nonpolar-valence-type variety, such as occur in the elementary diatomic gases. Again the whole crystal must be considered as one gigantic molecule. Typical examples are furnished by C (in the form of diamond) Si, SiC, AlN, BeO, ScN, TiC. They are formed mostly from elements near the central portion of the periodic system.[1] They are very hard, diamondlike bodies; in the most typical cases, they are extremely nonvolatile, indicating very strong forces. The molecular volumes are small, a property that goes with strong interatomic forces.

In the molecular crystals, in contrast to the two types just considered, the molecules retain their identity in the crystal complex. Thus, solid chlorine consists of Cl_2 molecules held together by intermolecular forces which are weak compared with the forces that hold together the two atoms in a chlorine molecule. The intermolecular forces may be of two types. In some cases, notably the solids of diatomic molecules of elements on the right-hand side of the periodic table, the solids of the rare gases (which, though the gases are monatomic, may best be classed with the molecular solids), CO_2, CO, SF_6, probably the hydrogen halides (except HF), CH_4 and most organic substances, the forces are of a type that we may designate as van der Waals forces; they are the forces that are chiefly responsible for the deviations of the so-called permanent gases from behavior as perfect gases. They are due to the mutual interaction and polarizability of the electron shells surrounding the molecules, and are very weak forces. They have the unusual property that the force between large molecules is greater than the force between small ones; thus the forces increase from top to bottom of a column in the periodic system. For example, the forces between

[1] In referring to the periodic table, we always have in mind the arrangement with the long rows (Thomsen's arrangement) presented in Table 1, p. 91.

molecules increase in the order HCl, HBr, HI, as is indicated by the boiling points, −85°C, −67°C., −35°C., respectively; and, in general, the higher the molecular weight of an organic compound in a homologous series, the higher its boiling point. In many of these substances, the forces are sufficiently small, or at least depend so little on the relative orientation of the molecules, that there is probably actual rotation of the molecules inside the solid except at very low temperatures. At the very lowest temperatures, of course, unless the intermolecular forces were absolutely symmetrical, the molecules would be rigidly fixed in position, and no such rotation would occur. As the temperature is raised, however, it will be anticipated that a few molecules will get enough energy to begin to rotate. Now, when a molecule is itself rotating, this will obviously make it easier for its neighbors to rotate.[1] Furthermore, as the temperature goes up, the volume increases and the molecules get less in each other's way.[2] The effect is cumulative; as the temperature increases, the number of rotating molecules increases very rapidly, and rather suddenly the crystal will become unstable, and they will all rotate. Since increased rotation is accompanied by absorption of heat, this can be followed by observing the specific heat, which will rise suddenly over a transition region of a few degrees and then drop, even more sharply, to near its normal value; or a regular transition with a latent heat of transition may occur. Transition points that have been interpreted in this way by Pauling[3] are actually known in a number of cases.

Another type of force which may be instrumental in certain cases in holding together the molecules of molecular crystals is the electrostatic attraction that occurs between electric dipoles when they are properly oriented (they tend, of course, to orient themselves naturally in such a way as to attract rather than repel each other). This type of force will be of most importance when the molecules have relatively large electric moments, and when the molecules are small, so that these electric moments can effectively get close to each other. Examples of crystals in which this type of force is probably predominant are H_2O, HF,

[1] Fowler, "Statistical Mechanics," 2d ed., pp. 810*ff*., Cambridge University Press, 1936.

[2] Rice, *J. Chem. Phys.*, **5**, 492 (1937).

[3] Pauling, *Phys. Rev.*, **36**, 430 (1930).

HNO_2, H_2CO, CH_3OH, NH_2OH, H_2O_2. In these cases, the dipole forces are larger than the van der Waals forces in the same compound, and the properties of the substances indicate, in general, a greater attraction between molecules than one would expect from the van der Waals forces alone. Thus the liquids tend to be fairly high boiling, compared with liquids of comparable molecular weight which have only van der Waals forces, and there is a tendency to polymerize in the liquid state, or for the molecules to associate in more complicated ways. If dipole forces predominate in holding a crystal together, there cannot be free rotation of molecules within the crystal.

Both types of molecular compounds tend to form rather soft crystals, an indication that the forces which come into play are relatively small compared with those in the ionic and atomic crystals; the compounds in which the chief forces are the van der Waals forces will be expected to be softer than similar compounds in which the dipole forces predominate, though not many data are available, and in any event, softness and hardness are qualities that are difficult to compare.

Finally, we have the metallic crystals, in which the forces in the typical cases, *e.g.*, the alkali metals, are principally the forces between "free" electrons in the metals and positive ions. These substances are solid, relatively involatile, and conduct electricity readily owing to the more or less free electrons which they contain. The alkali metals are soft, but the metals toward the center of the periodic table (*i.e.*, the iron and platinum metals and their neighbors) are hard; in these latter, however, and even more especially in substances like bismuth, forces of the nonpolar-valence-bond type undoubtedly come into play.

In the foregoing account, we have confined ourselves for the most part to a discussion of typical properties of the various kinds of crystals, and have illustrated our discussion with typical examples. There are numerous cases in which a compound has some of the characteristics of one extreme class and some of the characteristics of another. We have just mentioned some metals that have some of the characteristics of atomic crystals. There are intermediate cases between atomic and ionic crystals, for example, AgI. And there are examples of molecular compounds, such as HCl, in which both dipole and van der Waals forces must play some part. Then there are many cases in which forces of more than one type come into play in different parts of the same structure. Thus, there are numerous examples of complex anions and cations, such as SO_4^{--} or NH_4^+, in which a central atom, or ion (sulfur or nitrogen in the examples considered), holds others around it by forces intermediate between the ionic and covalent types; the complex ion then enters as a whole into the crystal structure, and is bound by ionic forces to the other constituents. The CdI_2 crystal furnishes a remarkable example of a mixture of molecular, nonpolar, and ionic forces. This crystal is composed of layers; each layer consists of three planes—one of cadmium atoms, or ions, surrounded on either side by a plane of an equal number of iodine atoms, or ions. These three planes are bound into a compact layer by forces that are presumably intermediate between atomic and ionic and form a gigantic two-dimensional molecule. These molecules are then held together largely, presumably, by van der Waals forces.

The classification, with respect to the periodic table, of the substances forming the various types of compounds will probably be evident to the reader from the examples given and from the previous discussion of polar and nonpolar compounds. The ionic-type crystals always contain elements that are widely separated in the periodic table, one of which tends to give up electrons, the other of which tends to take on electrons. Typical solids of the other three types are either elementary solids, or compounds the elements of which are close together in the periodic table, the metals coming from the left, the atomic compounds from the center, and the molecular compounds from the right of the periodic table.[1] Thus substances that tend to lose their electrons easily form metals, in which the electrons are more or less free; while substances intermediate in this respect form atomic compounds, in which there are several bonds to each atom; and in the case of substances that have a great attraction for electrons, there is such a tendency for localization of all the electrons that the presence of other molecules does not disturb the bonds already formed. Thus the electrons are held so tightly in a Cl_2 molecule that they are not greatly disturbed by the near presence of other Cl_2 molecules; but on the other hand, the electrons in a C_2 molecule would be affected by the presence

[1] See footnote, p. 206.

of other carbon atoms or molecules sufficiently to cause the C_2 to become part of the atomic crystal. (See also pages 368–369.)

The various types of solid compounds, briefly discussed above, will be treated in more detail in the succeeding chapters. Of course, it will not be possible to give a treatment of solids that is independent of the considerations of gases and liquids; some further properties of bonds in gaseous molecules will also be considered, and, occasionally, some discussion of the forces in liquids will be presented, though the situation in liquids is so complicated that no generally adequate treatment yet exists. In Chap. XII, although it dealt exclusively with gaseous substances, the general program for the study of the chemical bond has already been begun. The molecules treated in Chap. XII are of the type that form molecular solids, so that many of the properties of the bonds between atoms within the molecules of these solids have already been studied. And as already indicated, this type of force is not qualitatively different from that encountered in atomic crystals.

Although in the preceding discussion a variety of properties has been mentioned, two properties that will be stressed in the ensuing study are the energy of formation of compounds and interatomic distances. The utility of the study of the energy is already clear, and it will be seen that the interatomic distance also depends on the type of binding, and can throw light in turn on the nature of the binding. Interatomic distances are measured by means of X rays; the methods used are briefly described in Chap. XIV, in connection with some simple examples.

Chapter XIV deals with polar or ionic crystals. Ionic radii, giving interatomic distances in polar crystals, are first considered. Then the energies of crystal lattices and energies of formation of polar crystals are studied, with the aid of a certain type of thermochemical cycle, known as the "Born-Haber cycle." The effect upon the energy of the transition from ionic to covalent binding is considered. The Born-Haber cycle is a powerful tool, and a number of special applications are given. Chapter XIV also contains a brief section on the energies of polar compounds in the gaseous state.

In Chap. XV, we return to the study of covalent binding, and present a discussion of some properties common to covalent bonds in both gases and crystals, which have not hitherto been considered. In particular, the direction of the bonds in space is discussed. A new criterion for type of binding, the magnetic criterion, is introduced.

In Chap. XVI, the ideas developed in Chap. XV are applied, and a variety of properties of various complex compounds are considered. After a discussion of the methods of investigation and the chemistry of complex compounds, and some energetic considerations in a few cases where these are possible, there is a section on the stereochemistry of complex compounds. This is followed by a discussion of the arrangement of atoms in complex crystals. Next follows a discussion of covalent radii, which are compared with ionic radii, after which various experimental data that throw light on the type of binding and, in particular, the transition from ionic to covalent type in crystals are summarized. The chapter ends with two sections on special types of crystals and special types of bonds.

Chapter XVII deals with the nature of intermolecular forces in molecular crystals, and gives a discussion of the relevant experimental material, including material bearing on the transition between this type of binding and other types.

Chapter XVIII discusses metallic crystals. The energies of binding and the nature of the binding in the alkali metals are considered in some detail; there is a section on the transition between metallic and covalent binding, a brief discussion of intermetallic compounds, and finally, a brief discussion of the recently developed wave mechanical picture of metallic binding.

The general aim of these chapters is, by discussion and a wealth of illustration, to give the reader a reasonably accurate idea of the nature of chemical binding in the various types of compounds described in this chapter, and to show what can be done toward an understanding in relatively simple terms of the various properties of compounds which are generally described in treatises on inorganic chemistry.

In Chap. XIX, the ideas developed are applied to aqueous solutions.

CHAPTER XIV

IONIC CRYSTALS

As noted in the last chapter, the typical examples of ionic crystals are furnished by the alkali halides. These crystals may be considered to be composed of positive and negative ions, the forces that hold the ions together being almost entirely electrostatic and being balanced when the crystal is at equilibrium by the repulsive forces due to interpenetration of the electron shells. As the alkali halides are not only the most typical ionic crystals, but are also in many respects the simplest of all crystals, we shall treat them in some detail and illustrate with them the methods that may be used in the study of crystals and, in particular, ionic crystals. We shall show that most of their properties are satisfactorily explained by the assumption that the attractive forces are purely electrostatic.

14.1. The Crystal Structure of the Alkali Halides.—All the alkali halides except CsCl, CsBr, CsI have the structure shown in Fig. 49. It may be seen that the ions are arranged at the corners of cubes, positive and negative ions alternating. In such crystals, the ions are said to have a coordination number of six, the coordination number being the number of nearest neighbors. (The term is also used to designate the number of atoms surrounding a central atom. Thus in methane or in chloroform, for example, the carbon atom has a coordination number of four; in the ion $PtCl_6^{--}$, the platinum atom has a coordination number of six.)

Figure 49 represents a very simple arrangement, and it is probably the first guess that one would make for the structure of sodium chloride if he knew nothing about the structure of crystals. Its experimental verification by the help of X rays will be discussed in the subsequent pages of this chapter. It is by no means the only possible crystal structure for crystals of this valence type; the chloride, bromide, and iodide of cesium have, in fact, another arrangement of atoms, in which all the ions at the corners of a cube are alike and the other ions form a similar system of cubes, the vertices of which are at the middle of the first set of cubes, so that the ions have a coordination number of eight. This is known as a "body-centered cubic lattice"; for a figure, see Appendix IV.

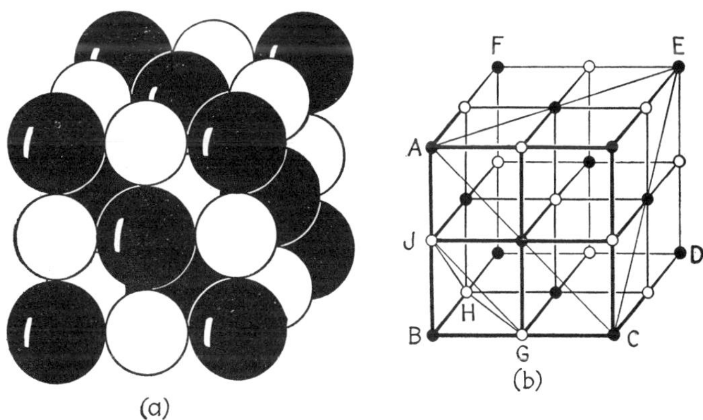

FIG. 49.—The sodium chloride lattice: (*a*) showing the ions as contacting spheres; (*b*) showing the positions of the centers of the ions.

14.2. The Use of the X Ray in the Study of Crystals.—X rays offer a convenient tool for the study of crystals.[1] The wave length of ordinary light is so great compared with atomic dimensions that even with the most powerful microscope we could never hope to see details of anything even several thousand times as large as an atom; but the wave lengths of X rays are of the order of atomic dimensions, and so they are admirably adapted to use in "seeing" the arrangement of atoms in crystals. The reflection of X rays from crystal faces has already been discussed in Sec. 3.2, and it was shown that from any crystal face the X rays will be reflected only if they are incident at certain angles θ such that

$$2d \sin \theta = n\lambda, \tag{1}$$

[1] For more detailed accounts of the use of X rays in crystal analysis and the results, the following books may be consulted: W. H. Bragg, "Introduction to Crystal Analysis," D. Van Nostrand Company, Inc., 1929; W. H. Bragg and W. L. Bragg, "The Crystalline State," The Macmillan Company, 1934; Compton and Allison, "X Rays in Theory and Experiment," D. Van Nostrand Company, Inc., 1935.

where d is the distance of successive atomic planes, λ the wave length of the X rays used, and n an integer called the order of the reflection. If X rays of known wave length are available and the various angles at which reflection occurs can be measured it is possible to determine d. This can easily be accomplished by having a crystal mounted so that it can be rotated with respect to the incident beam of X rays. If d is known but not λ, the method can be used to determine λ. When the method was first used, neither d nor λ was known. However, by working with a constant source of X rays, so that λ, though unknown, would remain the same from one experiment to another, it was possible to determine the *relative* distances between various planes and thus determine the positions of the ions with respect to one another. And if the density of the crystal and Avogadro's number are known, the number of ions in a given volume can be determined. The arrangement of the ions and the number per unit volume being known, it was possible to compute the actual distances in the crystal and then from the distances determine λ by use of Eq. (1). Having now a standard wave length, we can determine distances in other crystals, and thus determine their density by use of the X ray alone, obtaining values that may be compared with the density determined in the ordinary way. In the present state of the science, it is often possible to get more accurate values of the density of a substance by use of X rays than by direct measurements, accurate results by the ordinary methods often being, for various reasons, very difficult to obtain.

14.3. Application of X Rays to the Study of the Alkali Halides. If we examine Fig. 49 we see that there are many ways in which planes can be passed through the crystal so that the planes will go through centers of ions which lie close together. Three types of such planes are shown in the figure: (1), planes like ABC, AFE, etc., which (rather than use the usual nomenclature, which for our purposes is needlessly complicated) we shall call "straight planes"; (2) planes like BAE and FEC which we shall call "semidiagonal planes"; and (3) planes like AEC and GHJ which we shall call "diagonal planes." It may be readily seen from the geometry of the figure that if the distance between two adjacent straight planes is δ (equal, *e.g.*, to BG) then the distance. between two adjacent semidiagonal planes is $\frac{1}{2}\sqrt{2}\delta$ and between two diagonal planes is $\frac{1}{3}\sqrt{3}\delta$. If the structure represented by

Fig. 49 is correct, an examination of the crystal with X rays and application of Bragg's law should show that there are actually distances in the crystal which bear the ratio $1:\frac{1}{2}\sqrt{2}:\frac{1}{3}\sqrt{3}$. Furthermore, it is possible to determine, by observing the directions of the incident and the reflected beams of X rays with reference to the position of the crystal, the angles between the various kinds of planes, *e.g.*, the angle between various straight planes or between straight planes and semidiagonal planes.

There is another detail to which the attention of the reader must be directed. It will be noted that the straight planes and semidiagonal planes contain equal numbers of each of the two kinds of ions, but that the diagonal planes contain only one kind of ion. Furthermore, the diagonal planes alternate—if one contains one kind of ion, the next plane will contain the other kind. Therefore, alternate diagonal planes are different, and this has a special effect on the X-ray diagram, which we shall proceed to investigate.

Consider a set of planes, all alike, the distance between them being d. Then the angles at which reflection of X rays will take place are given by Eq. (1). Now insert another set of planes, just like the original set, halfway between the latter. The possible angles of reflection θ' will now be given by the modified expression

$$2d' \sin \theta' = n\lambda, \tag{2}$$

where $d' = d/2$. Solving for $\sin \theta$ from Eq. (1), we get

$$\sin \theta = \frac{n\lambda}{2d}, \tag{3}$$

and solving for $\sin \theta'$ from Eq. (2)

$$\sin \theta' = \frac{n\lambda}{2d'} = \frac{2n\lambda}{2d} = \frac{m\lambda}{2d}, \tag{4}$$

which is just the same as (3) except that $m = 2, 4, 6. \ . \ . \ .$ It is thus seen that inserting an extra set of planes halfway between the original ones is equivalent to suppressing the odd orders, and only half as many reflections occur. It is easy to understand why this occurs. The reflection from the new planes is in such phase that the odd orders are canceled out by the destructive interference; the reflection from the new planes, however, reinforces the even orders, and these will now have an increased intensity. If, however, the new planes which are inserted are different from the original planes, and so do not reflect the X rays with the same intensity, the odd orders of reflection will not be completely blotted out, but will still be there, though with diminished intensity. This is exactly what we should expect to happen with the diagonal planes in the crystal we have been considering. Actually it has been found in NaCl and other alkali halides of like structure that there are reflections at all the angles corresponding to the distance between two *like* diagonal planes, rather than merely the reflections corresponding to the distance between adjacent planes, but the odd orders are weak.

It should be noted that the straight, semidiagonal, and diagonal planes are not the only possible planes of atoms in the crystal. Many other such planes can of course be found, but they will contain fewer atoms per unit area, and the reflection will therefore be less intense. Furthermore, the distance between adjacent planes of this type will be small, as will be clear from the two-dimensional section shown in Fig. 50. For most of them, d is so small that Eq. (1) cannot be satisfied even for $n = 1$, since $\sin \theta$ cannot be greater than 1; therefore, from these planes there will be no reflection at all.

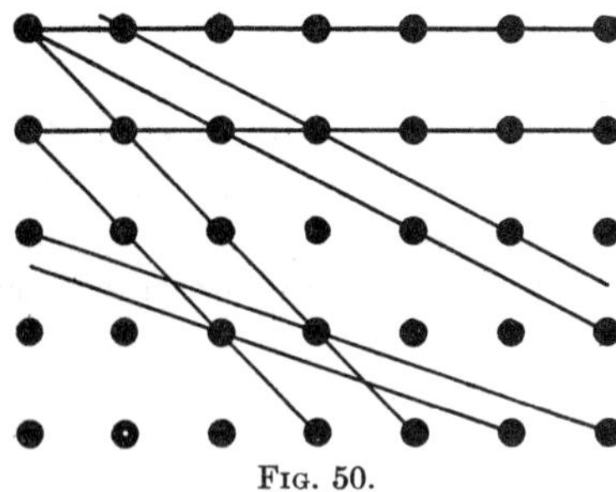

FIG. 50.

Before leaving the discussion of the structure of sodium chloride, it should be stated that there is at least one other possible arrangement of the ions which would have reflections at the same angles as the actual structure. This is the zinc-blende structure[1] with Zn replaced by Na and S replaced by Cl. However, the relative intensities of the various reflections would be very different for the two different structures, and a study of them leaves no doubt as to the one to be assigned to NaCl. In general, the intensities are of great importance in the determination of crystal structures.

[1] See Appendix IV.

The foregoing discussion has shown how it is possible to verify, by the use of X-ray analysis, the structure of the alkali halides that have the atomic arrangement shown in Fig. 49. The verification of the structure of cesium salts that do not have this arrangement proceeds just as simply. Planes of various kinds can be found in the crystal and their distances calculated. It is obvious that the distances involved will be different in this case, and the distances can be found and the postulated arrangement of the crystal verified by the use of X rays in the same manner as before.

Of course when dealing with a crystal of a more complicated type, one is often confronted with a problem that is very much more difficult than those just discussed. It will often be difficult to make a preliminary guess as to the structure of the crystal, and sometimes X rays alone will not enable one to distinguish among various possibilities. In many cases, however, it is possible to obtain definite results, and in other cases one can supplement the information given by X rays with other information. As far as the X-ray analysis goes, however, there is no new principle involved, and as it is not desired to go into details of crystallography we shall in the future merely report the results of the X-ray analysis.

It should be mentioned that there are other methods of using the X rays than the Bragg method described above. These are the von Laue method and the Debye-Scherrer-Hull powder method. The von Laue method has already been briefly described in Chap. III; it has been very generally employed.

The powder method is really a modification of the Bragg method. A beam of X rays of definite wave length is allowed to pass through a powder. These X rays strike crystals that are oriented in all possible directions. Bragg reflection takes place from those crystals which happen to be oriented properly with respect to the incident beam. Reflection thus occurs in cones the rays of which make definite angles with the incident beam. If a photographic plate is placed some distance below the powder (which is laid upon a thin plate of some material as transparent as possible to X rays), there appear upon it, in addition to a spot due to undeflected X rays, a series of rings where it intersects the cones of reflected X rays, and from the size of these rings and the geometry of the setup, information concerning the atomic distances within the crystal can be obtained.

14.4. The Ionic Radius.—The determination of interionic distances in crystals by means of X rays has brought out the fact that in many crystals these distances can be correlated by assuming that each ion has a fixed and definite radius.[1] The interionic distances, however, are always distances between the centers of two ions, and the actual radii of the separate ions cannot be found from a knowledge of these distances alone.

We may illustrate these statements with an example. Thus if the distances between positive and negative ions in NaCl, KCl, and KBr crystals are known, it is possible to calculate the distance in an NaBr crystal. Let an arbitrary radius be assigned to Na^+. Then the radius of Cl^- is found from the ionic distance in NaCl, on the assumption that two adjacent ions are touching and the distance between centers is the sum of the two radii. From the radius of Cl^- and the distance in KCl, the radius of K^+ is found. And similarly, from KBr, the radius of the Br^- is found. When the latter is added to the radius of Na^+, it gives the expected ionic distance in NaBr. This agrees with experiment, thus, in this instance, confirming the assumption that the distances in the crystal can be explained by assuming ions of fixed radii. But it is clear that any division of the distance in the NaCl crystal will give the same result. Taking a smaller value for the radius of Na^+, for example, eventually yields a correspondingly larger value for that of Br^-, the sum being the same.

Various methods for dividing the distances in crystals between the ions have been suggested. Perhaps the most satisfactory is a theoretical one introduced by Pauling.[2] The reason that the

[1] The additive rule was first proposed by W. L. Bragg, *Phil. Mag.*, **40,** 169 (1920), and the early development of the idea is due largely to Goldschmidt. See Goldschmidt, "Geochemische Verteilungsgesetze der Elemente," I Kommisjon Hos Jacob Dybwad, Oslo, 1923–1927, especially vols. 7 and 8.

[2] Pauling, *J. Am. Chem. Soc.*, **49,** 765 (1927). Earlier Wasastjerna, *Soc. Sci. Fenn. Comm. Phys. Math.*, **1,** No. 38 (1923); *Zeits physik. Chem.*, **101,** 193 (1922), and Goldschmidt had used other methods for dividing the distance which gave results in agreement with those obtained by Pauling. Wasastjerna's method was based on the assumption that the mole refraction of an ion depends upon its size. The mole refraction of an ion is closely related to its polarizability, and that this should be related to the size of an ion will be clear from the discussion of polarizability of molecules given in Sec. 12.4. Another method of getting the absolute size of ions, based upon the assumption of anion-anion contact in certain crystals (see end of Sec. 14.4), was proposed by Landé, *Zeits. Physik*, **1,** 191 (1920). Recently the matter has been discussed from still a different point of view by Jensen, Meyer-Gossler, and Rohde, *Zeits. Physik*, **110,** 277 (1938).

ions appear to have fairly definite radii lies in the nature of the repulsive forces which set in when the ions are brought close together. These forces are the same as in diatomic ionic-type molecules (see Sec. 9.1) and are due to the interpenetration of the electron clouds surrounding the ions. A relatively small interpenetration results in a very large increase of the repulsive force, so that the apparent size of the ion is very largely determined by the extension in space of the electron cloud.

In a crystal such as KCl, in which the positive and negative ions are, respectively, the ions of alkali and halogen atoms coming just after and just before a given rare gas in the periodic table, the electronic structures are exactly similar, the only difference being in the charge on the nucleus. Now obviously in the K^+ ion, which has a greater charge on its nucleus than the Cl^- ion, the electrons will be pulled in closer, so that K^+ will be smaller than Cl^-. The exact difference in size will depend upon the effective charge on the outer electrons, which, of course, depends on the shielding effect of the inner electrons, which in turn depends upon how much the outer electrons penetrate the inner shells and upon the difference in size of the inner shells themselves in the two ions. In general, it may be said that the size of an orbit with a definite quantum number is inversely proportional to the effective charge. This is seen by replacing[1] e^2 by $Z'e^2$, where $Z'e$ is the effective charge acting on the electron considered, in Eqs. (9) and (13) of Chap. V and eliminating E between them. It is then found that with a given value of n, the quantity a, which effectively determines the size of the orbit, is inversely proportional to Z'. The determination of the effective charge acting on the outer electron is a very complicated calculation, into the details of which we cannot enter here. It has been done in a number of cases by Pauling, who has thus divided the distance in[2] NaF, KCl, RbBr, and CsI.

[1] Compare Chap. VII, Eq. (1). In this equation, N is the number of inner electrons surrounding the nucleus, and the outer electron is revolving about them and is supposed not to penetrate. Hence it is moving under the influence of a charge $(Z - N)e$. Here the electron penetrates and moves under the influence of a charge whose *average effective value* is $Z'e$. Z' thus takes the place of $Z - N$ in Eq. (1), Chap. VII.

[2] Since CsI has a body-centered structure, and since the interatomic distances in such a structure are somewhat larger than in the sodium chloride structure (see Sec. 14.6), correction has been made for this.

Once the radii of these ions are calculated, they can be used to obtain the radii of other ions with the same number of electrons but with different effective charges acting on these electrons by again assuming that the extension of the electron atmosphere in space is inversely proportional to the effective charge. Thus the radius of O^{--} can be calculated from that of F^-, the radius of Ca^{++} from that of K^+, etc. The radius of Li^+ is obtained from the radius of O^{--} and distances in Li_2O, after applying the correction for the charge on the ions and the cordination number discussed below (Sec. 14.6). The results obtained are given in Table 16.

TABLE 16.—IONIC RADII
(In Angstroms)

			H^-	He	Li^+	Be^{++}	B^{3+}	C^{4+}	N^{5+}	O^{6+}	F^{7+}
			2.05	0.92	0.59	0.43	0.34	0.29	0.25	0.22	0.19
C^{4-}	N^{3-}	O^{--}	F^-	Ne	Na^+	Mg^{++}	Al^{3+}	Si^{4+}	P^{5+}	S^{6+}	Cl^{7+}
4.14	2.47	1.76	1.36	1.12	0.95	0.82	0.72	0.65	0.59	0.53	0.49
Si^{4-}	P^{3-}	S^{--}	Cl^-	A	K^+	Ca^{++}	Sc^{3+}	Ti^{4+}	V^{5+}	Cr^{6+}	Mn^{7+}
3.84	2.79	2.19	1.81	1.54	1.33	1.18	1.06	0.96	0.88	0.81	0.75
					Cu^+	Zn^{++}	Ga^{3+}	Ge^{4+}	As^{5+}	Se^{6+}	Br^{7+}
					0.96	0.88	0.81	0.76	0.71	0.66	0.62
Ge^{4-}	As^{3-}	Se^{--}	Br^-	Kr	Rb^+	Sr^{++}	Y^{3+}	Zr^{4+}	Cb^{5+}	Mo^{6+}	
3.71	2.85	2.32	1.95	1.69	1.48	1.32	1.20	1.09	1.00	0.93	
					Ag^+	Cd^{++}	In^{3+}	Sn^{4+}	Sb^{5+}	Te^{6+}	I^{7+}
					1.26	1.14	1.04	0.96	0.89	0.82	0.77
Sn^{4-}	Sb^{3-}	Te^{--}	I^-	Xe	Cs^+	Ba^{++}	La^{3+}	Ce^{4+}			
3.70	2.95	2.50	2.16	1.90	1.69	1.53	1.39	1.27			
					Au^+	Hg^{++}	Tl^{3+}	Pb^{4+}	Bi^{5+}		
					1.37	1.25	1.15	1.06	0.98		

NOTE: The radius of Li^+ and other radii derived therefrom (first row in the table) are slightly different from the values given by Pauling.

It will be noted that ions with an outer shell of eighteen electrons are also included in Table 16. These are treated as though they had the same structure as the ions of the rare-gas type. Although it might seem that this would not be justified, actually a wave mechanical calculation shows that the eighteen outer electrons of Cu^+, for example, would on the average be somewhat closer to the nuclei than the eight outer electrons of K^+, even if the effective charge acting on the outer electrons were the same in the two cases. It turns out that the electron density in the extreme outer portion of an eighteen-electron shell ion, calculated on the assumption that the effective charges are the same in the two cases, is about equal to the electron density for an ion with shell of eight. It is the extreme outer part of the electron shell that determines the effective radius. Hence, it is only necessary in computing the radius of Cu^+, for example, from that of K^+, to correct for the difference in the effective charge.[1]

We may now check the hypothesis of constant ionic radii by comparing experimental and calculated distances for the various combinations of ions in the alkali halide crystals which have the sodium chloride structure. This is done in Table 17. It will

TABLE 17.—CALCULATED AND OBSERVED DISTANCES IN ALKALI HALIDES WITH THE SODIUM CHLORIDE STRUCTURE
(In Angstroms)

	Li^+	Na^+	K^+	Rb^+	Cs^+
F^-, calc	1.95	(2.31)	2.69	2.84	3.05
obs	2.01	2.31	2.67	2.82	3.01
Cl^-, calc	2.40	2.76	(3.14)	3.29	
obs	2.57	2.81	3.14	3.29	
Br^-, calc	2.54	2.90	3.28	(3.43)	
obs	2.75	2.97	3.29	3.43	
I^-, calc	2.75	3.11	3.49	3.64	
obs	3.00	3.23	3.53	3.66	

From Pauling, reference 2, p. 218. Calculated values for lithium salts corrected to conform with Table 16.

be observed that the agreement is good, except in the case of lithium salts. The reason for the discrepancy with these salts is readily understood if we note the small size of the lithium ion. The anion is so large that there is anion-anion contact, the cation being left free to bounce around in the interstices. The situation is illustrated in Fig. 51, which shows the cross section of one plane

[1] Of course, the effective charge is different for the various electrons in the outer shell, depending largely upon the amount of penetration. In the case of K^+, it is the outer *p*-electrons that principally determine the size, and in the case of Cu^+ it is the outer *d*-electrons. It is, therefore, the effective charge on *p*-electrons in the one case, and on *d*-electrons in the other, which is of importance.

of a normal sodium chloride type crystal and also a similar section of a sodium chloride type crystal with anion-anion contact.

Anion-anion contact occurs if the ratio of the radius of the cation to the radius of the anion is less than $(\sqrt{2} - 1):1$. This condition is fulfilled in the case of all lithium halides except lithium fluoride. It would then be expected that one could obtain a direct measure of the anion radii by taking half the distance between anion centers in these crystals, and this is indeed the case. Half the distance between anion centers is 1.82Å., 1.95Å., and 2.12Å. for Cl^-, Br^-, and I^-, respectively, agreeing closely with the values given in Table 16. In the case of sodium iodide, and lithium fluoride the ratio of cation and anion radii is very close to the critical value $\sqrt{2} - 1$. This means that double repulsion will occur, *i.e.*, any given ion will be affected by the

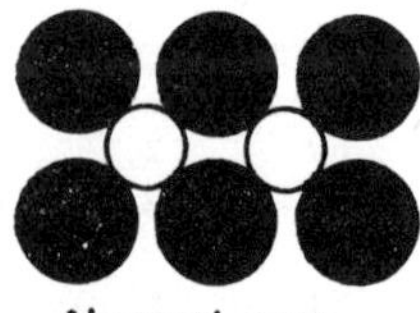

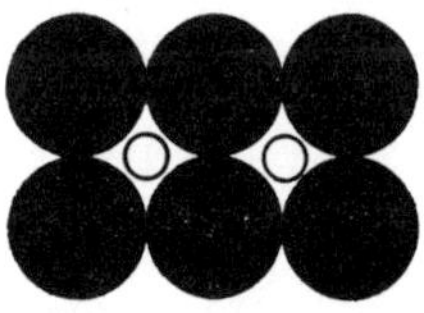

Fig. 51.

repulsive forces of ions of both signs instead of only ions of one sign. The effect of the extra repulsive force is to make the actual interionic distance slightly greater than the calculated distance.

14.5. Energy of Crystal Lattices.—Before proceeding further with the discussion of crystal radii, it will be advantageous to consider in somewhat more detail the forces operative in ionic crystals.[1] The nature of these forces will already be clear from the discussions at the beginning of this chapter and elsewhere (*e.g.*, Sec. 9.1). They consist of electrostatic forces and the repulsive forces due to interpenetration of electron shells.

Let us consider first the electrostatic forces. The potential, due to the electrostatic forces on some ion, say the ith ion, is given by the sum[2] over all other ions $\Sigma_j\, z_i z_j e^2/r_{ij}$, where $z_j e$ is the

[1] The first calculations of the type to be considered were made by Born. See, *e.g.*, Born: "Atomtheorie des festen Zustands," B. G. Teubner, Leipzig, 1923.

[2] We use the usual sign Σ to indicate summation. Σ_j means the sum is to be taken over all values of j. A double summation (one summation followed by another) is indicated by a double Σ.

charge on the jth ion (e is the charge on the electron), $z_i e$ being, of course, the charge on the ith ion,[1] and r_{ij} is the distance between the ith ion and the jth ion. The contributions to this sum come largely from the ions in the immediate neighborhood of the ith ion, so that if the number of ions is large enough the value of the sum is independent of their number. This is true because there is practically complete cancellation of the contribution from positive and negative distant ions; there is as much positive as negative charge in the crystal, and the positive and negative charges are close to each other; so a distant positive charge will be balanced in its effect on the ith ion by an almost equally distant negative charge. The sum is therefore equal to $-e^2/r$ (where r is the shortest anion-cation distance) multiplied by a factor of the order of magnitude of one, and where the sign is negative because the ions closest to a given ion are all, or at least predominantly, of opposite sign, giving a negative potential (attraction). In simple crystals like the alkali halides, the sum for any given ion is the same as that for any other ion, because they are all similarly placed in the crystal. To find the total energy of such a crystal, we can then multiply the sum $\Sigma_j z_i z_j e^2/r_{ij}$ by the number of ions and divide by 2 (it is necessary to divide by 2 because the potential of any ion due to a second ion is exactly the same thing as the potential of the second ion due to the first—by considering the potential of any ion due to all the other ions, then summing again over all ions, we include the mutual potential of all pairs of ions twice). In other more complicated types of crystal, there may be different kinds of ions placed in different ways, each one with its own potential; but in any case the electrostatic energy of the crystal will be equal to $\frac{1}{2}e^2\Sigma_i\Sigma_j z_i z_j/r_{ij}$, and this sum will again be proportional to N, the number of "molecules"[2] in the crystal. The sum can be expressed in terms of r, N, and a constant depending solely on the geometry of the crystal. Suppose we allow r to vary, keeping the crystal structure the same, *i.e.*, holding each ion in the same relative position, so that every distance r_{ij} changes in the same ratio as r; we shall then have

[1] The sign of the charge is given by the sign of z_i or z_j.

[2] Although the ions in ionic crystals are not arranged in the form of molecules, nevertheless we may speak without ambiguity of the number of molecules, it being understood that this refers to the number of neutral groups, *e.g.*, the number of pairs, NaCl, in a sodium chloride crystal.

$r_{ij} = l_{ij}r$ where l_{ij} is independent of r. Setting this into the sum, we have $\frac{1}{2}(e^2/r)\Sigma_i\Sigma_j z_iz_j/l_{ij} = -NAe^2/r$, where A is a quantity that depends only on the l_{ij}'s and the z's and hence only on the geometry of the crystal and the charges on the ions. The values of A, the so-called "Madelung constant" (after the investigator who made the first calculations of this constant), are given in Table 18 for several different types of crystal.[1] (Actually A/α^2 is given, where α is the largest common factor of the valences of the ions.) This then completes our description of the attractive forces in the crystal.

It has been usual to represent the repulsive energy between two ions i and j by an approximation of the type b_{ij}/r_{ij}^n, where b_{ij} and n are constants. The repulsive forces are known to be practically zero to fairly small values of r_{ij} and then to rise rapidly as r_{ij} decreases. This behavior is reproduced by making n fairly large. It varies for different crystals but is of the order of magnitude of 10; the method of evaluating it is discussed below. It is assumed to be the same for all ion pairs in a given crystal. The total repulsive energy of the crystal, like the electrostatic energy, is given by a double summation

$$\frac{1}{2}\sum_i\sum_j \frac{b_{ij}}{r_{ij}^n} = \frac{1}{2r^n}\sum_i\sum_j \frac{b_{ij}}{l_{ij}^n} = \frac{NB}{r^n}, \tag{5}$$

where B is again a constant depending on the geometry of the crystal. (B, however, also depends on the magnitude of the repulsive forces.) The total potential energy per "molecule" is thus given by

$$U = -\frac{Ae^2}{r} + \frac{B}{r^n} \tag{6}$$

if r is allowed to vary,[2] keeping the crystal structure the same. Equation (6) defines a kind of potential-energy curve in which U is defined as a function of r when the crystal is expanded or

[1] SHERMAN, J., *Chem. Rev.*, **11,** 107 (1932).

[2] We may assume that r varies arbitrarily, though an actual crystal naturally tends to an equilibrium condition in which r has a definite value. This equilibrium position may be changed by application of pressure. A process in which r changes, but in which departure from the condition of equilibrium is always infinitesimally small (*i.e.*, in which the pressure is varied and always has the proper value), is said to be *reversible.*

compressed without displacement of the relative positions of the atoms. U becomes zero when r is infinite. The curve is very similar both in nature and in origin to that shown in Fig. 34, page 106, the only difference being that here we deal with the interaction of many ions, whereas there we were considering but two.

TABLE 18.—VALUES OF THE MADELUNG CONSTANT FOR DIFFERENT TYPES OF CRYSTALS

Crystal type	Example	Coordination number	A/α^2	A_0
Sodium chloride	NaCl	6	1.748	1.748
Cesium chloride	CsCl	8	1.763	1.763
Sphalerite	ZnS	4	1.638	1.638
Wurtzite	ZnS	4	1.641	1.641
Fluorite	CaF_2	Ca, 8; F, 4	5.039	1.680
Rutile	TiO_2	Ti, 6; O, 3	4.816	1.606
Anatase	TiO_2	Ti, 6; O, 3	4.800	1.600
β-Quartz	SiO_2	Si, 4; O, 2	4.439	1.480
Cadmium iodide	CdI_2	Cd, 6	4.71	1.570

NOTE: The coordination numbers are given to give the reader some idea of the characteristics of the crystal. Some of the structures are discussed in Appendix IV, and for descriptions of the others the reader may see Sherman, reference 1, p. 224, or the "Strukturbericht," vol. I. In some cases, there is some leeway as to the positions of the atoms in the designated type of crystal. In this case, the maximum value of the Madelung constant is given; it is not to be expected that any existing crystal of that type will have a Madelung constant very different from the maximum. See Sherman, *loc. cit.*, p. 107. For A_0, see Sec. 14.6.

Equation (6) represents only the potential energy, and neglects any thermal motion of the ions. There is still the thermal energy to be considered. This will be of the order of, or less than, $3kT$ per ion, where k is the Boltzmann constant (see Appendix II), plus the zero-point energy, which will exist for vibrations of the crystal lattice just as it does for vibration of a molecule as seen in Secs. 9.2 and 9.4. This extra energy is in all cases extremely small compared with the potential energy and except in rather refined calculations can be neglected.

To the approximation that it is correct to neglect the thermal energy, no heat will flow in or out of the crystal when the distance r, and consequently the volume per molecule V, is changed. U, of course, changes for, by Eq. (6), it is a function of r. If the pressure on the crystal is P, and if the crystal is expanded reversibly by an amount dV, it does work $P\,dV$, per molecule. This work is wholly supplied from the energy of the crystal, under the assumptions made, and so is equal to $-dU$. Thus we have $P = -\frac{dU}{dV}$. Under ordinary conditions, the pressure is practically equal to zero. A pressure of 1 atm., for example, is entirely negligible compared with the pressure necessary to change V by an exceedingly small amount. If we set $dU/dV = 0$ (which, if we assume that the relative positions of the atoms remain the same,[1] is equivalent to $dU/dr = 0$ and, of course, means that the energy U has its minimum value), then this will be the ordinary condition of equilibrium for the crystal. That is, the crystal, if not under pressure, will naturally take the volume that makes this condition true. Let us call the value of r corresponding to this condition r_0. Then we have, differentiating Eq. (6) and setting the derivative equal to zero (multiplying through by r_0^2),

$$Ae^2 - \frac{nB}{r_0^{n-1}} = 0. \tag{7}$$

If we are interested in the energy of the crystal at equilibrium, as is usually the case, we may readily find it from (6) by setting $r = r_0$ in that equation; and evaluating B from Eq. (7), we get

$$U_0 = -\frac{Ae^2}{r_0}\left(1 - \frac{1}{n}\right). \tag{8}$$

It is thus seen that the energy of formation of the crystal from the ions (generally known as the "lattice energy" of the crystal[2]) can be found if n can be determined.

Since A and r_0 may be taken as known, Eq. (7) gives one relation between B and n. If we could find one more relation between them, they could both be evaluated "experimentally."

[1] We are certainly free to imagine and deal with expansions and contractions of the crystal in which this is true, whether it is always true in actual practice or not. We know that the most stable condition of the crystal must be stable with respect to changes in which the relative positions remain the same, as well as all others, and our considerations, therefore, give a necessary condition for the state of equilibrium. In other words, if we have equilibrium, $dU/dr = 0$ must hold.

[2] U_0 always has a negative value; the lattice energy is generally taken as the positive quantity, $-U_0$.

This second relation is furnished by the measurement of the compressibility of the crystal. The compressibility is defined as the relative decrease in volume per unit pressure applied. This may be written as

$$\kappa = -V^{-1}\frac{dV}{dP}, \tag{9}$$

where V is the molecular volume and P the applied pressure. Since $P = -\frac{dU}{dV}$ if all the energy of the crystal is potential energy, we also have $\frac{dP}{dV} = -\frac{d^2U}{dV^2}$. By Eq. (9), this gives

$$\frac{d^2U}{dV^2} = -\frac{1}{\frac{dV}{dP}} = \frac{1}{\kappa V}. \tag{10}$$

V is proportional to r^3, if the atoms remain in the same relative positions, and the constant of proportionality is readily found if the geometry of the crystal is known. Let us call this constant a. Furthermore, U is given as a function of r by Eq. (6). The various terms in Eq. (10) may, therefore, be evaluated; it is found that

$$-\frac{4Ae^2}{9a^2r^7} + \frac{n(n+3)B}{9a^2r^{n+6}} = \frac{1}{\kappa a r^3}. \tag{10a}$$

If κ is measured under conditions such that the volume of the crystal has its normal value, *i.e.*, when $r = r_0$, then, since r_0 and all other quantities in the equation except B and n are known, this gives the second relation between the latter quantities.

The compressibility of the alkali halides has been measured by Slater,[1] who extrapolated his results to absolute zero. The kinetic energy of the lattice, which in our consideration has been neglected, may well be expected to have a greater effect on the compressibility, which involves a second derivative, than on the energy of the crystal, so that in calculating values of r_0 and n it is obviously best to make the extrapolation to absolute zero.

Pauling[2] has used the following values for n for ions with electron structure of the type indicated: He (*e.g.*, Be^{++}, B^{+++}),

[1] Slater, *Phys. Rev.*, **23**, 488 (1924).

[2] Pauling, *J. Am. Chem. Soc.*, **49**, 772 (1927).

5; Ne, 7; A, Cu^+, 9; Kr, Ag^+, 10; Xe, Au^+, 12. In the case of compounds containing ions with different electron configurations, the average value was used for n. In the limited number of cases in which data are available, these values check fairly well with the experiments (see also end of Sec. 14.6). No data are available for the ions with an outer shell of eighteen, however.

The values of n are far from certain, but fortunately it is not necessary for n to be known very accurately in order to be able to calculate the energy of a crystal with a fair degree of accuracy. This is readily seen from the expression for the energy of a crystal, as given by Eq. (8). An error of 1 in n will make an error of about one-nth part in the term $1/n$ of Eq. (8), which in turn contributes about $1/n$ of the energy. Thus an error of 1 in n will make a fractional error of $1/n^2$ in the energy, which would vary from 4 per cent for $n = 5$ to less than 1 per cent for $n = 12$. Sherman believes that the error in actual cases will not usually exceed 3 per cent.

There is some probability of error due to the approximate form B/r^n which has been taken for the repulsive potential. Evaluating B and n by use of Eqs. (7) and (10a) amounts to determining B and n in such a way that the first and second derivatives of the repulsive part of the potential have the correct values at $r = r_0$. This, of course, does not necessarily mean that the repulsive potential itself will have the right value at $r = r_0$, and the correctness of the assumption will depend upon the validity of our form of approximation.

Another type of approximation, which is probably preferable from the theoretical point of view and which does not give very different results from the approximation already used, is one of the form $be^{-\frac{r}{\rho}}$, where b and ρ are constants, so that, instead of (6), we write

$$U = -\frac{Ae^2}{r} + be^{-\frac{r}{\rho}}. \tag{11}$$

This form of repulsive potential has certain theoretical advantages. It was first used in the actual calculation of crystal energies, a subject to which we shall revert later, by Born and Mayer.[1] The repulsive potential rises the more suddenly and

[1] Born and Mayer, *Zeits. Physik*, **75**, 1 (1932). Actually we have used a somewhat oversimplified form for the repulsive potential in Eq. (11). Born

rapidly the smaller ρ so that r_0/ρ (a dimensionless quantity) is in a certain sense the counterpart of n of Eq. (6) and it also can be evaluated by a consideration of the compressibility of the crystal. For details, the paper of Born and Mayer must be consulted. With a potential of the form (11), we get in place of Eq. (8)

$$U_0 = -\frac{Ae^2}{r_0}\left(1 - \frac{\rho}{r_0}\right) \tag{12}$$

in which it will be seen that the term $1/n$ is replaced just by ρ/r_0.

It is worthy of note that the value of ρ as determined by the compressibilities was found by Born and Mayer to have nearly the same value, namely, 0.345Å., for all the alkali halides. With the single exception of lithium iodide, in which case Born and Mayer believe the compressibility may be in error, the values range between 0.310 and 0.384Å. In finding the values of ρ, Born and Mayer used a somewhat more exact form for the repulsive potential[1] and took into account a number of small corrections to the formula (11) which also of course enter into the expression connecting the compressibility with ρ. Instead of attempting to estimate the compressibility at 0°K., they set up a thermodynamic expression (involving the coefficient of thermal expansion and the change of the compressibility with temperature and pressure—all at room temperature) by means of which it was possible to relate the desired quantity ρ with the compressibility at room temperature. Enough data were available for nine alkali halides to carry out the calculations, and estimates could be made for other alkali halides. In addition, they took into account the van der Waals forces between the ions, which cause further modifications in the expressions used in the determination of ρ. The van der Waals forces, as explained in Chap. XVII, result from mutual polarization of the ions. The motion of the electrons in an ion causes a temporary dipole moment in that ion, which in turn induces a dipole moment in neighboring ions in such a direction as to produce attraction between the ions. The *permanent charge* on an ion (say ion I—

and Mayer have made use of a more complicated form, arising because they took into account explicitly the repulsive forces of the more distant, as well as the nearest neighbors of a given ion.

[1] See preceding footnote.

see Fig. 52) will also induce a polarization in a neighboring ion (ion II) so as to cause attraction between them; however, there will be another ion (ion III) on the other side of ion II, and the electric moment produced in ion II by the permanent charge of ion III will just cancel that produced by ion I, so that no net attraction occurs. This, of course, assumes that the dipoles produced in ion II may be taken as located exactly at the center of this ion, so that the dipole produced in ion II by ion III has as much effect on ion I as the dipole produced in ion II by ion I, itself. If ion II is readily distorted, this will not be the case, and there will then be an extra attraction due to this cause between the ion pairs; such an attraction may be considered to be due to an incipient covalent binding force.

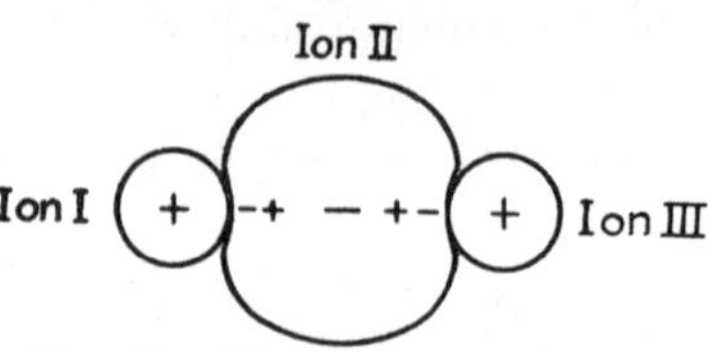

FIG. 52.—Illustrating the distortion of a negative ion by two positive ions. The induced dipoles cancel as far as their moments are concerned but may not be located at the center of the negative ion.

14.6. Interionic Distances in Real Crystals.—Since an ion is not simply a hard sphere like a billiard ball, but has a certain compressibility, the actual distances between ions in a crystal will depend not only upon the extension in space of the electron atmospheres of the ions, but also upon the forces the ions exert on each other,[1] as expressed, for example, by Eq. (7). Since the radii of the alkali and halide ions were determined from distances in uni-univalent crystals with sodium chloride structures, and since the radii of the other ions in Table 16 really represent extensions in space of the electron cloud about the ion, determined by comparing with alkali and halide ions, it is seen that the distances in a crystal should be given by such radii only if the crystal were a uni-univalent crystal with sodium chloride structure. But, of course, the polyvalent ions cannot form uni-univalent crystals and they may not have the sodium chloride structure. Let us, therefore, define what may be called a "reduced" Madelung constant A_0 by the equation

$$A = \frac{z_c z_a A_0 f}{2}, \tag{13}$$

[1] ZACHARIASEN, *Zeits. Krist.*, **80**, 137 (1931). See, also, Goldschmidt, reference 1, p. 218. We have departed somewhat from Zachariasen's procedure.

where z_c and z_a are the absolute values of the charges (expressed as multiples of e) of cation and anion, respectively, f is the number of ions in the molecule (*i.e.*, $f = 2$ for NaCl, $f = 3$ for $CaCl_2$, etc.), and A_0 depends solely upon the geometry of the crystal and not on the valence. Thus it will be clear from the expressions for the attractive potentials at the beginning of Sec. 14.5, taken together with Eq. (13), that A_0 will be the same for NaCl and MgO, both of which have the sodium chloride structure, and for both of which, of course, $f = 2$. It will be observed from Table 18 that A_0 does not depend very strongly on the geometry of the crystal, and the value of A is determined mostly by z_c, z_a, and f. As a first approximation, we shall consider A_0 to be independent of the form of the crystal.

In the case of the repulsive term, we can write approximately

$$B = f_c C_N B_0, \tag{14}$$

where f_c is the number of cations per molecule, C_N the coordination number of the cation, and B_0/r^n the repulsive potential between a single pair of ions. The total number of contacts between anion and cation in a crystal having but one kind of cation is $f_c C_N N$, where N is the number of molecules. Therefore, if we take into account only the potential between adjacent atoms and assume that the anion-cation distances for all C_N contacts of a given ion have the same value, namely, r, then the total repulsive potential of the crystal is equal to $f_c N C_N B_0/r^n$, so that Eq. (14) follows from Eq. (5). Substituting Eqs. (13) and (14) into Eq. (7) and solving for r_0 give

$$r_0 = \left(\frac{2f_c n C_N B_0}{z_c z_a A_0 f e^2}\right)^{\frac{1}{n-1}}. \tag{15}$$

The radii of Table 16 give the correct value of r_0 for crystals of the sodium chloride type, which are crystals with $z_a = z_c = 1$, with $2f_c/f = 1$, and with $C_N = 6$. We now assume on the basis of Eq. (15) that if these quantities do not have these values, the value of r_0 obtained from Table 16 may be corrected by multiplying by $(2f_c C_N/6z_c z_a f)^{\frac{1}{n-1}}$. Pauling's set of values of n (see Sec. 14.5) can be used. The calculation is easily made by means of tables constructed by Zachariasen.

In Table 19 are presented the results for the alkaline earth oxides, sulfides, and selenides, which have the sodium chloride

TABLE 19.—DISTANCES FOR ALKALINE EARTH OXIDES, ETC.
(In Angstroms)

	Mg^{++}	Ca^{++}	Sr^{++}	Ba^{++}
O^{--}, calc.	2.05	2.41	2.56	2.79
obs.	2.10	2.40	2.58	2.77
S^{--}, calc.	2.54	2.83	2.98	3.21
obs.	2.60	2.84	3.01	3.19
Se^{--}, calc.	2.72	2.97	3.12	3.34
obs.	2.73	2.96	3.12	3.30
Te^{--}, calc.	. . . .	3.18	3.32	3.55
obs.	. . . .	3.17	3.33	3.50

Data from "Strukturbericht." Calculated values for MgS and MgSe corrected for anion-anion contact.

structure. The excellent agreement is perhaps the best justification for the use of Pauling's values of n in attempting to make a rough estimate of the expected interionic distance in an ionic crystal, which can be compared with experimental distances in connection with the study of the transition between ionic and covalent binding. The discussion of other cases is deferred until Chap. XVI, after the consideration of covalent crystals.

Pauling's values of n do not give so completely satisfactory a result when an attempt is made to find the contraction on going from a coordination number of six to a coordination number of one (vaporization of a sodium chloride crystal). This is treated in Sec. 14.9.

14.7. The Born-Haber Cycle.—In the preceding pages of this chapter, we have considered the energy of formation of a crystal from the ions which compose it. In most cases, it is not possible to measure this energy directly, but the energy of formation of the crystal from the elements is known. The latter can be decomposed into the energies of a number of processes, which constitute hypothetical intermediate steps in the formation of the crystal. Let us consider, for example, the formation of sodium chloride from elementary solid sodium and elementary gaseous chlorine. The reaction may be supposed to take place either directly or through the following steps:[1]

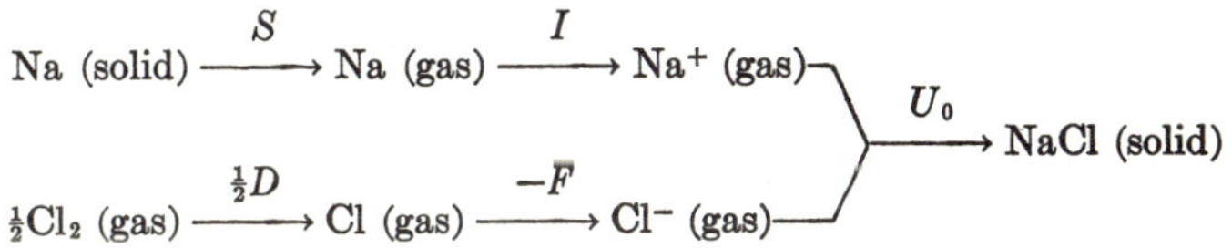

$$\left.\begin{array}{l} \text{Na (solid)} \xrightarrow{S} \text{Na (gas)} \xrightarrow{I} \text{Na}^+ \text{ (gas)} \\ \tfrac{1}{2}\text{Cl}_2 \text{ (gas)} \xrightarrow{\frac{1}{2}D} \text{Cl (gas)} \xrightarrow{-F} \text{Cl}^- \text{ (gas)} \end{array}\right\} \xrightarrow{U_0} \text{NaCl (solid)}$$

The energy placed above each arrow indicates the energy absorbed by the system in the particular step. If the cycle takes place at the absolute zero of temperature, S is the energy of sublimation of sodium at absolute zero, I is the ionization potential, D is the energy of dissociation of chlorine at absolute zero, F is the electron affinity, and $-U_0$ is the lattice energy. S and I represent the energy absorbed when the process involved goes in the direction indicated, and D is the energy absorbed when the reaction Cl_2 (gas) $\rightarrow$ 2Cl (gas) takes place. As energy is actually absorbed in these processes, they are all positive quantities. F is the energy absorbed when an electron is removed from Cl^-; it is also a positive quantity. When the reverse reaction takes place, the energy absorbed is $-F$ (*i.e.*, energy is actually evolved). U_0 is given, at least in good approximation, by Eq. (8) or (12). In Eqs. (8) and (12), the energy of the crystal is assumed to be zero when the ions are completely separated, and U_0 is a negative quantity. The positive quantity $-U_0$, the lattice energy, is the energy absorbed when the ions are separated. In the reverse reaction, energy U_0 is absorbed; since U_0 is negative, energy is actually evolved. The total energy of formation of the crystal from the elementary substances Q [energy absorbed in the reaction Na (solid) $+ \frac{1}{2}Cl_2$ (gas) $\rightarrow$ NaCl (solid)] is given by

$$Q = S + I + \tfrac{1}{2}D - F + U_0. \tag{16}$$

Q is accurately known experimentally for many substances, as are all the quantities in Eq. (16) except F and U_0. The latter can be calculated as indicated in Sec. 14.5, and the equation may thus be used to obtain F. As a matter of fact, there have been recently several direct experimental determinations of F, but these are probably not to be regarded as being so reliable as the

[1] See, *e.g.*, BORN: "Atomtheorie des festen Zustands," B. G. Teubner, Leipzig, 1923.

determination from Eq. (16). We shall therefore use this equation for the purpose of calculating F for the various halogens. The reliability of the results may then be estimated from the consistency of the values F obtains from different compounds of a given halogen. (Concerning energy units, see note, page 461.)

Calculations of this type were made by Mayer and Helmholz[1] for the alkali halides and have recently been improved by Huggins.[2] In calculating the lattice energy, Eq. (12) was used, but with appropriate corrections for van der Waals forces,[3] etc., as discussed in Sec. 14.5. As already noted, ρ was found not to vary greatly from crystal to crystal, and in calculating U_0 the same (mean) value was used for all the salts. Born and Mayer took $\rho = 0.345$Å., and Huggins took $\rho = 0.333$. The zero-point energy of vibration of the crystals was also corrected for. The values for U_0 are given in Table 20. These follow Huggins but have been corrected to room temperature by taking into account the (relatively very small) thermal energy of the crystal[4] and the gaseous ions. The tabulated values also include the heat absorbed in order to furnish energy for the work done against the atmosphere when the process is carried out at constant pressure; in other words, they are the changes in heat content (see Appendix II), which are the values obtained directly experi-

[1] Mayer and Helmholz, *Zeits. Physik*, **75**, 19 (1932).

[2] Huggins, *J. Chem. Phys.*, **5**, 143 (1937). Verwey and de Boer [*Rec. trav. chim. Pays-Bas*, **55**, 431 (1936)] have also made a recalculation, but they merely corrected the van der Waals potential used by Mayer and Helmholz without considering how this change affected the equilibrium condition, and hence ρ and other dependent quantities; the latter changes partially cancel the original change in the van der Waals potential.

[3] The effect of the van der Waals forces is small but not entirely negligible. Their contribution to the attractive potential ranges from around 6 (Li salts) to 12 kg.-cal. (Cs salts). Recent work of May, *Phys. Rev.*, **52**, 339 (1937), and **54**, 629 (1938), indicates that the contribution may possibly be larger.

If the van der Waals forces are neglected altogether, this will be partially compensated for by using in Eq. (12) the different value of ρ which is obtained from the compressibility by neglecting the van der Waals forces. Some idea of how much the van der Waals force affects ρ may be obtained by calculating U_0 as Huggins has done, and then reversing the process, calculating an effective value of ρ directly from Eq. (12) without making the corrections. Such values of ρ are given for the alkali halides in Table 27.

[4] Following Mayer and Helmholz. Energies (except I and F) in all tables throughout the chapter are for room temperature and, unless otherwise noted, contain the work done against the atmosphere.

mentally. The other quantities also include the work done against the atmosphere and are for room temperature, except for the ionization potentials. The average kinetic energies of a gaseous atom and a gaseous ion at any temperature will be equal; so taking the ionization potential at 0°K. merely neglects the

TABLE 20.—THE BORN-HABER CYCLE FOR THE ALKALI HALIDES
(All energies in kilogram-calories)

Salt	$-Q$	S	I	$\frac{1}{2}D$	$-U_0$	F	AvF
LiF	145.6	39.0	123.8	31.8	245.1	95.1	
NaF	136.0	25.9	118.0	31.8	216.4	95.3	
KF	134.5	19.8	99.7	31.8	193.2	92.6	95.0
RbF	133.2	18.9	95.9	31.8	183.4	96.4	
CsF	131.7	18.8	89.4	31.8	175.9	95.8	
LiCl	97.6	39.0	123.8	28.9	201.1	88.2	
NaCl	98.3	25.9	118.0	28.9	184.0	87.1	
KCl	104.4	19.8	99.7	28.9	168.3	84.5	87.3
RbCl	105.1	18.9	95.9	28.9	162.1	86.7	
CsCl	106.3	18.8	89.4	28.9	153.2	90.2	
LiBr	87.6	39.0	123.8	23.1	189.9	83.6	
NaBr	90.6	25.9	118.0	23.1	175.9	81.7	
KBr	97.9	19.8	99.7	23.1	161.5	79.0	81.8
RbBr	99.6	18.9	95.9	23.1	156.1	81.4	
CsBr	101.5	18.8	89.4	23.1	149.6	83.2	
LiI	72.5	39.0	123.8	18.1	176.2	77.2	
NaI	76.7	25.9	118.0	18.1	164.4	74.3	
KI	86.3	19.8	99.7	18.1	152.5	71.4	74.3
RbI	88.5	18.9	95.9	18.1	147.9	73.5	
CsI	91.4	18.8	89.4	18.1	142.4	75.3	

Experimental quantities from Bichowsky and Rossini, "Thermochemistry of the Chemical Substances," Reinhold Publishing Corporation, 1936.

kinetic energy (and the work done against the atmosphere) of the free electrons. The resulting electron affinity then fails to include the same quantity, so that the electron affinity obtained is actually that for 0°K. This is presented in the last column of the table, together with the average for each of the halogens.

The values given for the electron affinities are probably correct to within 5 kg.-cal. and are seen to decrease, as expected, from fluorine to iodine. The fact that, within rather small limits, the same electron affinity is obtained for a given halogen regardless of the salt used is evidence in favor of the correctness of the assumptions made concerning the nature of the forces operating in these crystals, and we can feel fairly certain that they are chiefly electrostatic and repulsive forces as supposed.[1]

These conclusions are further supported by some experiments in which the equilibrium between ions and a solid salt has been measured directly. Knowing the equilibrium constant, for example, for the reaction KI (solid) $\rightleftarrows K^+$ (gas) + I^- (gas), together with already known thermodynamic data (entropies of the salt and of the ions—see footnote, page 406) enables one to obtain the energy of the reaction by means of a well-known thermodynamic relation [see Eq. (6) of Appendix II].

The experiment consists of heating the salt to a high temperature in a graphite furnace with a small opening, the whole being in a vacuum. The number of ions issuing from the hole in the furnace is determined by measuring the current they carry due to their electrical charges. From the laws of efflux of gases at low pressures, the concentration of ions in the furnace may then be calculated. [As a matter of fact, the ions are not actually in equilibrium with the salt, but with the gaseous molecules produced by heating the salt which were themselves heated to a higher temperature, so that actually the equilibrium constant for the reaction KI (gas) $\rightleftarrows K^+$ (gas) + I^- (gas), for example, was measured and the value for the reaction KI (solid) $\rightleftarrows$ KI (gas) taken from data of Wartenberg and others.[2]] In this way,

[1] This is probably a fair statement, though the lattice may be rather insensitive to deviations from the assumed electrostatic character of the forces in the crystal; see, *e.g.*, Ewing and Seitz, *Phys. Rev.*, **50,** 760 (1936), who believe that the electrons are considerably displaced from the negative toward the positive ion. Pauling ("The Nature of the Chemical Bond, pp. 69*ff*., Cornell University Press, 1939) has attempted to estimate quantitatively the degree to which these bonds are ionic, but the estimate is only tentative.

We may mention at this point other papers which attempt to penetrate more deeply into the fundamentals of the crystal lattice energy calculations, for example, Jensen, *Zeits. Physik*, **101,** 164 (1936), Neugebauer and Gombás, *ibid.*, **89,** 480 (1934), Landshoff, *ibid.*, **102,** 201 (1936). See also Wasastjerna, *Phil. Trans. Roy. Soc. London*, **A237,** 105 (1938).

[2] v. Wartenberg and Albrecht, *Zeits. Elektrochem.*, **27,** 164 (1921); v. Wartenberg and Schulz, *ibid.*, **27,** 570 (1921); Sommermeyer, *Zeits. Physik*, **56,** 554 (1929).

Mayer and Helmholz[1] found the following values of the lattice energy for several salts (in kilogram-calories at 0°K.):

KI	CsI	RbBr	NaCl
153.8	141.5	151.3	181.3

Tandon,[2] using a similar method, has obtained the following results:

NaI	KI	NaBr	KBr
166.4	150.6	176.3	159.7

These results are in satisfactory agreement with the values in Table 20, and this agreement lends further support to the belief that the forces in these crystals are of the ionic type.

Further support comes to the theory from the direct measurement of the electron affinity of iodine. Sutton and Mayer[3] have measured the rates of emission of electrons and negative ions from a tungsten filament heated to a high temperature in the presence of iodine. Electrons may be separated from negative ions by taking advantage of the much greater deflection of electrons in a magnetic field, due to their small mass. It is assumed that practically all iodine molecules striking the surface (their number per second can be calculated from the pressure of iodine and the area of the wire) are dissociated. Some appear as atoms, but others are converted to negative ions by the electrons coming from the wire, and it is assumed that equilibrium is established at the surface of the wire between atoms, ions, and electrons. There are then enough data to calculate the equilibrium constant for the reaction I (gas) + electron (gas) $\rightleftarrows$ I^- (gas). This gives the free energy change in the reaction; the entropy change can be calculated theoretically, and so the energy of the reaction can be calculated from the considerations of Appendix II. In this way, Sutton and Mayer obtain 72.4 kg.-cal. for the electron affinity of iodine. The uncertainty in this result may be several kilogram-calories, and it would seem to be in satisfactory agreement with the value obtained from Table 20. More

[1] MAYER, *Zeits. Physik,* **61,** 798 (1930); MAYER and HELMHOLZ, *ibid.,* **75,** 22 (1932); HELMHOLZ and MAYER, *J. Chem. Phys.,* **2,** 245 (1934).

[2] TANDON, *Proc. Nat. Acad. Sci. India,* **7,** 102 (1937); *Indian J. Phys.,* **11,** 99 (1937).

[3] SUTTON and MAYER, *J. Chem. Phys.,* **3,** 20 (1935).

recently Glockler and Calvin[1] have, by means of an experiment based on the same principles but different in details, obtained a value of 74.6 kg.-cal.

Similar direct determinations for bromine have led to estimates of the electron affinity of[2] 88 and[3] 80 kg.-cal.

Once reliable values for the electron affinities are available, it is possible to use them, in conjunction with Eq. (16) (or the appropriate modification of it if other than a uni-univalent salt is being considered), to find an "experimental" value of U_0, provided the other quantities in the equation are known experi-

TABLE 21.—THE BORN-HABER CYCLE FOR CUPROUS, SILVER, AND THALLOUS HALIDES

(All energies in kilogram-calories)

Salt	$-Q$	S	I	$\frac{1}{2}D$	F	$-U_0$ (expt.)	$-U_0$ (theor.)	$U_0 - U_0$ (theor.) (expt.)	Van der Waals contribution to $-U_0$, theor.
CuCl........	34.3	81.2	177.4	28.9	87.3	234.5	216	18	(15)
CuBr.......	30.5	81.2	177.4	23.1	81.8	230.4	208	22	(15)
CuI.........	25.2	81.2	177.4	18.1	74.3	227.6	199	29	(17)
AgF........	48.7	68.0	174.0	31.8	95.0	227.5	219	9	(24)
AgCl........	30.3	68.0	174.0	28.9	87.3	213.9	203	11	(29)
AgBr.......	27.6	68.0	174.0	23.1	81.8	210.9	197	14	(27)
AgI.........	22.4	68.0	174.0	18.1	74.3	208.2	190	18	(31)
TlCl.........	48.6	40	140.2	28.9	87.3	170	167	3	(28)
TlBr........	45.3	40	140.2	23.1	81.8	167	164	3	(28)
TlI..........	38.5	40	140.2	18.1	74.3	162	159	3	(30)

Experimental values from Bichowsky and Rossini, "Thermochemistry of the Chemical Substances," Reinhold Publishing Corporation, 1936.

mentally. U_0 may then be calculated theoretically, as in the case of the alkali halides, and the theoretical and experimental values compared. This has been done by Mayer and Mayer and Levy[4] for a number of the silver, thallous, and cuprous

[1] GLOCKLER and CALVIN, *J. Chem. Phys.*, **3**, 771 (1935).

[2] GLOCKLER and CALVIN, *ibid.*, **4**, 492 (1936).

[3] WEISBLATT and MAYER, Baltimore meeting of the American Chemical Society, April, 1939.

[4] MAYER, *J. Chem. Phys.*, **1**, 327 (1933); MAYER and LEVY, *ibid.*, **1**, 647 (1933). These authors have corrected for van der Waals forces, following Mayer, *J. Chem. Phys.*, **1**, 270 (1933). The contribution of the van der Waals potential to $-U_0$ (theor.) is given in Table 21, in parentheses, so that the reader may form some judgment as to its importance.

halides. In these calculations the important constants in the repulsive part of the potential were obtained from the compressibilities, and allowance was made for the van der Waals forces. The results are presented in Table 21. It is seen that in the case of the silver halides the agreement is good for the fluoride, but that the discrepancy between experimental and theoretical values increases with the size of the halide ion and is rather considerable in the case of silver iodide. It seems probable that this is due to the silver iodide bond being partly covalent. The copper salts appear to be more covalent than the silver

TABLE 22.—RESULTS OF THE APPLICATION OF THE BORN-HABER CYCLE TO HALIDE CRYSTALS

Crystal	$-U_0$ (expt.)	$-U_0$ (theor.)	U_0 (theor.) $-U_0$ (expt.)
MgF_2	688.8	696.8	−8.0
CaF_2	617.2	617.7	−0.5
SrF_2	587.5	587.5	0.0
BaF_2	553.4	556.4	−3.0
MnF_2	645.0	656.3	−11.3
FeF_2	668.6	657.7	10.9
NiF_2	703.2	697.1	6.1
CdF_2	661.9	628.7	33.2
PbF_2	589.7	580.7	9.0
$SrCl_2$	501.3	493.6	7.7
CdI_2	563.1	473.6	89.5
PbI_2	497.1	457.7	39.4

salts, which is expected since copper ion is smaller. In the case of the thallous salts, the theoretical and experimental values of U_0 agree within the limits of error, and it seems unlikely that the thallous halide bonds are appreciably covalent. Inasmuch as thallium has three valence electrons, it would hardly be expected to have a covalent bond in a compound in which it has lost only one valence electron.

It will be seen, nevertheless, that even in the cases in which discrepancy is greatest, the assumption of ionic binding gives a fairly good value for the binding energy. This is true for a large number of salts of metals in the transition region of the periodic table, as will be seen from the accompanying Table 22 of experimental and calculated values of U_0, which is taken from the work of Sherman.[1] These calculated values of U_0 are probably not so accurate as those of Mayer and his collaborators. Equation (6) rather than Eq. (11) was used to approximate the potential, for one thing, and no corrections were made for van der Waals forces; nevertheless, these values of Sherman's are

TABLE 23.—THE BORN-HABER CYCLE FOR THE ALKALINE EARTH OXIDES, SULFIDES, AND SELENIDES

Crystal	n	$-Q$	S	I	$\frac{1}{2}D$	$-U_0$	F	Av. F
MgO	7.0	146	36.5	520.6	59.2	940.1	−178	
CaO	8.0	151.7	47.5	412.9	59.2	842.1	−171	−166.5
SrO	8.5	141	47	383.8	59.2	790.9	−160	
BaO	9.5	133	49.1	349.0	59.2	747.0	−157	
MgS	8.0	82.2	36.5	520.6	53	778.3	−86	
CaS	9.0	114	47.5	412.9	53	721.8	−95	−92
SrS	9.5	113	47	383.8	53	687.4	−91	
BaS	10.5	111	49.1	349.0	53	655.9	−94	
CaSe	9.5	81.8	47.5	412.9	51	698.8	−106	
SrSe	10.0	83.4	47	383.8	51	667.1	−102	−105
BaSe	11.0	81.3	49.1	349.0	51	637.1	−107	

In the case of S and Se, the quantity $\frac{1}{2}D$ includes the heat of sublimation and Q is the energy of formation from solid S or Se; $\frac{1}{2}D$ has been corrected to conform to Table 7. $-U_0$ was calculated by Sherman from Eq. (8), using the values of n given. It contains the work done against the atmosphere, so it is a ΔH, rather than a ΔE, term.

NOTE: A recent recalculation of F for oxygen [de Boer and Verwey, *Rec. trav. chim. Pays-Bas*, **55**, 443 (1936)] yields a value of −173. An earlier calculation by Mayer and Maltbie, *Zeits. Physik*, **75**, 748 (1932), using the method of Born and Mayer, gave less consistent results for the oxides and sulfides than the calculation of Sherman.

sufficiently good to illustrate the point.[2] The largest discrepancy occurs in the case of cadmium iodide, which is undoubtedly a bad example of an ionic crystal (see Chap. XIII). The approximate agreement between theory and experiment in the

[1] SHERMAN, *Chem. Rev.*, **11**, 93 (1932).

[2] The experimental value of U_0 in Tables 22 and 24 is obtained from the available data by the use of Eq. (16) (properly modified if the number of anions or cations in the molecule is greater than 1). The values of heats of formation and sublimation given by Sherman have been compared with the more recent tabulation of Bichowsky and Rossini (footnote, Table 20) and a recalculation made wherever the differences were more than a few kilogram-calories. In Table 22, no allowance was made for the slightly different value of F obtained here, because Sherman's values of U_0 were all calculated from Eq. (8) rather than Eq. (12), and it seemed more consistent to use his values of F. Energies, Tables 22 to 25 are in kilogram-calories.

case of all these salts (which is to be contrasted with the results obtained when we attempted in Sec. 12.6 to calculate the binding energy of chlorine fluoride on the assumption that the bond was ionic[1]) may be an indication that the bonds in all these compounds have only a relatively small amount of covalent character. We cannot be certain of this conclusion, however, for we do not know just what results might be reasonably expected on the basis of a purely covalent bond, and it seems quite possible that they would not be very different.

The Born-Haber cycle has been applied[2] to the alkaline earth oxides, sulfides, and selenides, yielding values for the double electron affinities of oxygen, sulfur, and selenium. The results are given in Table 23. It will be noted that the electron affinities in this case are negative because it requires energy in order to put on the second electron after the atom is already negatively charged. The values of the electron affinities obtained from Table 23 have been used, as in the case of the halides, to obtain theoretical and experimental values of U_0 for oxides, sulfides, and selenides, the results being given in Table 24. It will be observed again that although some discrepancies occur they are not exceedingly large.

Equation (16) may also be used to find a value of U_0 for the alkali hydrides, in which hydrogen plays the role of a negative ion,[3] for the electron affinity of hydrogen has been quite accurately determined to be 16.4 kg.-cal. by a wave mechanical calculation.[4]

[1] It is to be remarked that in Sec. 12.6 we were considering the reaction $ClF \rightarrow Cl + F$, whose energy is not strictly comparable with the lattice energy, which is the energy to decompose a crystal into *ions*. However, the energy of the comparable reaction, $ClF \rightarrow Cl^+ + F^-$, is 12.58 electron volts, whereas the calculated energy (Sec. 12.6) is 8.78. It is seen that the percentage error is considerably greater than in any of the cases of Table 22.

[2] Sherman, *Chem. Rev.*, **11**, 149 (1932).

[3] Sherman, *Chem. Rev.*, **11**, 159 (1932). These substances are entirely saltlike in their properties, and there is every reason to believe that they consist of positively charged alkali ions and H^- held together by electrostatic forces. When molten they conduct electricity, H_2 appearing at the anode. All the crystals have the sodium chloride structure.

Early attempts to get the electron affinity of H from the Born-Haber cycle and an estimate of U_0 were made by Kasarnowsky, *Zeits. Physik*, **38**, 12 (1926), and *Zeits. anorg. allgem. Chem.*, **170**, 311 (1928).

[4] Bethe, *Zeits. Physik*, **57**, 815 (1929); Hylleraas, *Zeits. Physik*, **60**, 624; **63**, 291 (1930).

The results are given in Table 25. In the case of these salts, it is not possible to compute U_0 very exactly directly from Eq. (8)

TABLE 24.—RESULTS OF THE APPLICATION OF THE BORN-HABER CYCLE TO OXIDES, SULFIDES, AND SELENIDES

Crystal	$-U_0$ (expt.)	$-U_0$ (theor.)	U_0 (theor.) $- U_0$ (expt.)
Li_2O	692	695	−3
Cu_2O	786	644	142
Ag_2O	714	585	129
MnO	929	912	17
FeO	937	944	−7
CoO	963	950	13
NiO	965	968	−3
ZnO	964	977	−13
CdO	913	867	46
SnO_2	2812	2734	78
PbO_2	2829	2620	209
Al_2O_3	3613	3708	−95
Cr_2O_3	3447(?)	3655	−208(?)
Na_2S	523	516	7
Cu_2S	682	612	70
MnS	798	788	10
ZnS	846	818	28
CdS	801	770	31
HgS	841	774	67
PbS	731	705	26
Cu_2Se	689	599	90
MnSe	789	757	32
ZnSe	845	790	55
CdSe	803	745	58
HgSe	848	749	99
PbSe	739	684	55

U_0 (theor.) from Sherman, *Chem. Rev.*, **11,** 154 (1932), (some corrected). U_0 (expt.) from Sherman, but corrected when necessary to conform to Table 4 and Bichowsky and Rossini, "Thermochemistry of the Chemical Substances." $-U_0$ contains the work done against the atmosphere, so is a ΔH, rather than a ΔE, term.

or (12), because the compressibilities are not known. The calculation can be turned around, however, and Eq. (8) can be used to calculate a value of n from U_0. The values of n so obtained are also included in Table 25. The results appear very reasonable, and Hylleraas[1] has succeeded in treating the many

TABLE 25.—APPLICATION OF THE BORN-HABER CYCLE TO THE ALKALI HYDRIDES

Salt	$-Q$	$-U_0$ (expt.)	r_0	n
LiH	21.6	219.9	2.042	4.45
NaH	14	193.4	2.440	5.39
KH	10	165.0	2.850	5.28
RbH	12	162.3	3.018	6.42
CsH	12	155.7	3.188	6.92

Recalculated from data in Bichowsky and Rossini, "Thermochemistry of the Chemical Substances." $-U_0$ contains the work done against the atmosphere, so it is a ΔH, rather than a ΔE, term, but n is calculated from the ΔE term.

electron problem of the lithium hydride lattice quantum mechanically, obtaining for the lattice energy a value of 219 kg.-cal. as compared with 220 from Table 25; he obtained a value for the interionic distance of 2.21Å. as compared with the experimental value of 2.042Å.

14.8. The Stability of Salts of Different Valence Type.—Grimm and Herzfeld[2] have made some interesting calculations intended to show why certain conceivable salts with "odd" formulas, like NaF_2 and CaCl, either do not exist or are unstable. Let us consider, for example, the reaction

$$NaF_2 \rightarrow NaF + \tfrac{1}{2}F_2 \quad (17)$$

If it can be shown that the tendency of this reaction to go is very great, then NaF_2 will be unstable. The reaction can be considered broken up into steps as follows:[3]

$$NaF_2 \xrightarrow{-U_2} Na^{++} + 2F^- \xrightarrow{-I_2 + F} Na^+ + F^- + F \xrightarrow{U_1 - \frac{1}{2}D} NaF + \tfrac{1}{2}F_2 \quad (18)$$

The energy absorbed in each step is indicated above the arrow, $-U_2$ and $-U_1$ being the lattice energies for the reactions $NaF_2 \rightarrow Na^{++} + 2F^-$ and $NaF \rightarrow Na^+ + F^-$, respectively, I_2 the second ionization potential of sodium, and D the dissociation energy and F the electron affinity of fluorine. Grimm and

[1] HYLLERAAS, *ibid.*, **63**, 771 (1930). His value neglects thermal and zero-point energy, etc.

[2] GRIMM and HERZFELD, *Zeits. Physik*, **19**, 141 (1923).

[3] All substances in gaseous state except NaF_2 and NaF.

Herzfeld make the assumption that the energy of the reaction $NaF_2 \rightarrow Na^{++} + 2F^-$ is the same as that of the reaction $MgF_2 \rightarrow Mg^{++} + 2F^-$, magnesium being the most nearly similar alkaline earth metal. This enables us to calculate the over-all energy for the reaction (17). We get for the energy absorbed

$$Q = -U_2 - I_2 + F + U_1 - \tfrac{1}{2}D = 689 - 1084 + 95 - 216 - 32 = -548 \text{ kg.-cal.} \quad (19)$$

Thus an enormous amount of energy is *given out* when reaction (17) goes to the right, and NaF_2 will be very unstable. The fact that a gas is given off increases still more the tendency of (17) to go to the right, since the motion of gas molecules is less restricted than that of solids (see Appendix II). The energy is so great principally because of the high second ionization potential of the alkali metal.

It is seen that it is extremely unlikely that sufficient error could arise from the approximations involved in the estimate of U_2 to affect the result. As a matter of fact, our value of $-U_2$ is probably an overestimate, for Na^{++} would probably have a larger radius than Mg^{++} since the nucleus of the latter has a greater positive charge. This would naturally result in the lattice energy of NaF_2 being less than that of MgF_2, giving Q a still more negative value.

Sufficient data are given in the immediately preceding pages for the reader to convince himself that a similar result will hold for all alkali fluorides (though for CsF the value of $-Q$ has gone down to about 100) and, since the fluorides have the greatest lattice energies, the other halides of "odd" formula will be even less stable.

We turn now to a consideration of the alkaline earth halides. For example, let us consider $SrCl_2$. If the reaction

$$SrCl_2 \rightarrow SrCl + \tfrac{1}{2}Cl_2 \quad (20)$$

does not tend to go, $SrCl_2$ will be stable toward decomposition. This reaction may be analyzed into steps, as follows:[1]

$$SrCl_2 \rightarrow Sr^{++} + 2Cl^- \rightarrow Sr^+ + Cl^- + Cl \rightarrow SrCl + \tfrac{1}{2}Cl_2 \quad (21)$$

The only part of this scheme whose energy is unknown is $Sr^+ + Cl^- \rightarrow SrCl$. In this case, it is assumed that the reaction

[1] All substances in gaseous state except $SrCl_2$ and SrCl.

has the same energy as $Rb^+ + Cl^- \rightarrow RbCl$. This would be expected to give an *overestimate* of the energy evolved, or an underestimate of the energy absorbed, since Sr^+ would be expected to have a larger radius than Rb^+ on account of the valence electron, though it is true that it is not known just how the presence of this valence electron may affect the result otherwise. We find the energy absorbed when (20) goes to the right to be about 155 kg.-cal. This is so high that, in spite of the fact that gas is formed when the reaction goes to the right, there will be no tendency for this to occur. [The reaction (20) in which a gas is formed is somewhat, though not exactly, analogous to the boiling of a substance, and, when it is considered that substances with half that energy of vaporization boil at well over 1000°, it is readily understood that there will be no tendency for (20) to go.]

It is also of interest to consider the reaction

$$Sr + SrCl_2 \rightarrow 2SrCl \qquad (22)$$

The function of the metal may be thought of as the removal of the Cl_2 in (20) and thus the displacement of the equilibrium to the right, but (22) may be handled by itself and does not involve the complication resulting from the formation of gas in the reaction. It may also be analyzed into steps, as follows:

$$\begin{aligned} Sr\ (\text{solid}) + SrCl_2\ (\text{solid}) &\rightarrow Sr^{++}\ (\text{gas}) + 2Cl^-\ (\text{gas}) + Sr\ (\text{gas}) \\ &\rightarrow 2Sr^+\ (\text{gas}) + 2Cl^-\ (\text{gas}) \rightarrow 2SrCl\ (\text{solid}) \qquad (23) \end{aligned}$$

and it is found that the reaction requires the absorption of approximately 112 kg.-cal.[1] The heats absorbed in the formation of other subhalides have been calculated in similar fashion, and are presented in Table 26. In spite of the large energies of reaction indicated by the calculation, it has often been claimed that subhalides of alkalies and silver, as well as of alkaline earths,

[1] None of the reactions which we have considered can be taken as indicating that such a substance as SrCl is unstable with respect to decomposition into its elements, according to the reaction $SrCl \rightarrow Sr + \frac{1}{2}Cl_2$. It is undoubtedly stable with respect to this reaction, but the question of practical importance is whether the reverse of (22) will go or not, *i.e.*, whether the subhalide is stable with respect to the decomposition $2SrCl \rightarrow Sr + SrCl_2$. (This may be considered to be double the reaction $SrCl \rightarrow Sr + \frac{1}{2}Cl_2$, followed by $Sr + Cl_2 \rightarrow SrCl_2$. Because of the great tendency of the latter reaction to go, SrCl is not stable with respect to decomposition into Sr and $SrCl_2$.) NOTE: All U's recalculated from F of Table 20 and Q (expt.).

may be formed by heating the metal in the presence of the normal halide, or by electrolysis of the latter in the molten state. Some authors, however, have claimed that these subhalides are merely solutions of the metal in the normal halide. A careful investigation was made by L. Wöhler and Rodewald,[1] who came to the conclusion that compounds with the formulas Ag_2F, CaF, CaCl, and CaI do actually exist. Ag_2F reverts to the metal and normal halide above 90°C., whereas CaCl and CaI are unstable with respect to the metal and normal halide *below* about 800°, and CaF is unstable below about 1000°; by rapid cooling, the calcium salts may be brought down to room temperature without decomposition. Guntz and Benoit[2] claim to have prepared all the subhalides (as well as suboxides and subsulfides) of strontium and barium, though not in pure form. These do not tend to become unstable at low temperatures because of the higher heats of formation from metal and normal halide. [In the case of the calcium subsalts the heat of formation probably becomes negative at high temperatures, and their behavior is then explained by application of Eq. (8) of Appendix II.] The following heats of formation at room temperature are given by Guntz and Benoit:

TABLE 26.—CALCULATED HEAT ABSORBED ON FORMATION OF SUBHALIDES
(In kilogram-calories)

Reaction	
$Mg + MgF_2 \rightarrow 2MgF$	128
$Mg + MgI_2 \rightarrow 2MgI$	84
$Ca + CaF_2 \rightarrow 2CaF$	154
$Ca + CaCl_2 \rightarrow 2CaCl$	114
$Sr + SrCl_2 \rightarrow 2SrCl$	112
$Ba + BaF_2 \rightarrow 2BaF$	147
$Ba + BaI_2 \rightarrow 2BaI$	100

$Ca + CaCl_2 \rightarrow 2CaCl$, 2.7 kg. cal. evolved
$Sr + SrCl_2 \rightarrow 2SrCl$, 14.6 kg. cal. evolved
$Ba + BaCl_2 \rightarrow 2BaCl$, 16.5 kg. cal. evolved

These values, it will be noted, are in marked disagreement with Table 26.

[1] WÖHLER and RODEWALD, *Zeits. anorg. Chem.*, **61**, 54 (1909). The existence of Ag_2F has been often confirmed, and its structure determined by X-ray analysis. It has been reported that it has a structure similar to that of CdI_2, described on page 209. See the "Strukturbericht," vol. II.

[2] GUNTZ and BENOIT, *Bull. Soc. Chim.*, **35**, 709 (1924).

On the other hand, Bichowsky and Rossini[1] have listed heats of formation of gaseous MgF and CaF (from spectroscopic data). If it is assumed that the heats of sublimation of these substances are the same as those of NaF and KF, respectively, we can find the energy of formation of the solid substances and so get an "experimental" value of the heat of reaction for $Mg + MgF_2 \rightarrow 2MgF$ and $Ca + CaF_2 \rightarrow 2CaF$ (involving only solid substances, of course). We find, respectively, 310 and 196 kg.-cal. absorbed, again not in agreement with Table 26, the discrepancy this time being in the opposite direction. It is obvious that our knowledge of the subhalides is not in a very satisfactory state.

There may, therefore, be some grounds for the doubts of some authors that the existence of the subhalides has been unequivocably proved. However, the chemical evidence appears to be in favor of their existence, and they seem to be much more stable, or less unstable, than expected.[2] The existence of Ag_2F suggests the possibility that the subhalides contain a doubly charged halogen and a normal cation. This suggestion is hard to test because of the difficulty of estimating the electron affinity of such an ion as Cl^{--}. If it is assumed that the difference in energy between the reactions $Cl^{--} \rightarrow Cl + 2E^-$ and $S^{--} \rightarrow S + 2E^-$ is the same as the difference between $K \rightarrow K^{++} + 2E^-$ and $A \rightarrow A^{++} + 2E^-$ (K and A having the same electronic structure as Cl^{--} and S^{--}, respectively), then sufficient data are available to estimate the energy for $Cl^{--} \rightarrow Cl + 2E^-$, and a Born-Haber cycle can be

[1] Bichowsky and Rossini, "Thermochemistry of the Chemical Substances."

[2] It is, of course, often quite possible to prepare, indirectly, substances that are quite unstable, but do exist by virtue of the fact that their decompositions are very slow. If equilibrium were established rapidly, such substances could not exist at all. The considerations of this section exclude certain conceivable substances only on the assumption that equilibrium is established rapidly. The occurrence of the subhalides cannot be explained, however, by supposing that it is due to the slow establishment of equilibrium. It is true that the subhalides of calcium are, according to Wöhler and Rodewald, unstable at room temperature, and exist at that temperature only by virtue of the slowness with which equilibrium is attained, but the fact that they are made by heating the metal with the normal halide indicates that the equilibrium represented by Eq. (22) is actually toward the subhalide at these temperatures. The heats of reaction given in Table 26 are so high, however, that this reaction should not have the slightest tendency to occur at any attainable temperature.

set up (E^- = electron). On this basis, however, even the reaction $CaCl \rightarrow Ca + \frac{1}{2}Cl_2$ evolves energy, so the possibility does not seem very promising. We can only conclude that the alkaline earth subhalides do not have an ionic lattice. Their unexpected stability must have something to do with the extra valence electron of the metal atom, and it is possible that this electron gives the binding a partly metallic character. Pauling has suggested that in Ag_2F, in which there are layers of silver and fluorine atoms, the binding between adjacent silver atoms is metallic.

In spite of the fact that calculations based on ionic lattices do not work well for the subhalides, it will be of interest to consider from this point of view the possibility of existence of alkaline earth trihalides. The reaction

$$MgF_3 \rightarrow MgF_2 + \tfrac{1}{2}F_2$$

for example, will proceed with the production of about 1100 kg.-cal., if we assume that the lattice energy of MgF_3 is the same as that of AlF_3 and obtain the latter from available experimental data in the same way that U_0 (expt.) was obtained in Table 22. We may also estimate that $BaCl_3$ will decompose in similar fashion with the evolution of about 270 kg.-cal.

In quite a similar manner, it may be shown that rare gas monohalides should be expected to be unstable. Similar considerations have also been used by Grimm and Herzfeld in discussing the stability of oxides and sulfides. In connection with the latter calculations, the existence of peroxides, such as Na_2O_2, is of interest. The calculations definitely indicate that a lattice composed of Na^{++} and O^{--} ions should not be stable, but if the lattice is composed of Na^+ and O^-, the energy evolved in the reaction $2\ Na + O_2 \rightarrow Na_2O_2$ (found by using the value for the energy of O^- given by Bichowsky and Rossini, and assuming the lattice energy to be the same as NaF) is found to be about 80 kg.-cal. as compared with an experimental value of 119. These probably agree within the limits of error of the calculation. It is usually assumed, however, that in sodium peroxide the ions are Na^+ and O_2^{--}, the latter having the electron structure $:\ddot{\underset{..}{O}}:\ddot{\underset{..}{O}}:$ (See footnote, page 191.)

There is, of course, no doubt that in general the "normal" formulas for the halides and oxides of the alkalies, alkaline earths, and metals of the aluminum group represent the most stable possible combinations, and the preceding discussion will help to show why that is so. This discussion has not included consideration of many possible formulas in which the halogen or oxygen has an abnormal valence, because of the difficulty of estimating the energy of a halide or oxide ion with abnormal charge. The reason the metal ions generally exhibit a normal charge in their compounds may perhaps be summarized in general terms. The lattice energies increase very rapidly with increase in charge. This tends to make compounds in which the metal exhibits a large valence more stable. On the other hand, the work necessary to overcome the ionization potential tends to make it difficult to get an ion of large charge, and each successive ionization potential is larger. The increase in lattice energy, however, more than counterbalances the extra ionization potential required to increase the charge on the ion until all the valence electrons are removed. But once the last valence electron is removed, the work necessary to remove another electron becomes so much greater that it now more than counterbalances the extra lattice energy. This, of course, is only a rough picture of the situation as other factors, *e.g.*, the energy of sublimation of the metal, play a role, but it is accurate in a general way. In order to verify these statements, the reader should work through some of the calculations given in the preceding paragraphs in detail (see, for example, Exercises 3, 4, 6, 7, and 8 at the end of this chapter). In the case of copper, the second ionization potential is small enough and the deviations from ionic binding great enough to make cupric compounds stable (see especially Exercise 8, page 261). It may be remarked that deviations from the ionic type of binding can only make a compound *more* stable than one would calculate from the assumption that it is purely ionic, just as deviations from the pure covalent type make a predominantly covalent compound more stable, and for the same reason (see Sec. 12.5). If, therefore, it is found that a compound actually exists when the calculations indicate that it should be unstable, it may simply be because the calculations do not allow for a mixed type of binding, though, of course, other approximations may cause appreciable errors also.

14.9. The energy of the alkali-halide and hydride molecules in the gaseous state may be readily calculated from the crystalline energy, provided we assume that the attractive and repulsive forces are of the same nature.[1] Since experimental data are available for comparison, it seems worth while to make this calculation. The calculation is of interest because of the light it throws on the repulsive forces, and we shall accordingly carry out the computation by three methods, using different expressions for the repulsive potential.

Method I. In this case, the calculation is essentially like that of Sec. 14.6. If the crystal is of the sodium chloride type, vaporization involves a change of coordination number from 6 to 1 and a change of the Madelung constant A from 1.748 to 1. It is seen from Eq. (15), therefore, that the value of the interionic distance r_0' of the gaseous molecule will be given by

$$r_0' = r_0\left(\frac{1.748}{6}\right)^{\frac{1}{n-1}}, \tag{24}$$

where r_0' is the distance in the crystal and n has the value assigned by Pauling.

The energy of the gaseous molecule is determined in terms of its equilibrium distance by an equation exactly analogous to Eq. (8), but with $A = 1$ and with r_0 replaced by r_0'. Thus

$$U_0' = -\frac{e^2}{r_0'}\left(1 - \frac{1}{n}\right). \tag{25}$$

The value of $-U_0'$ thus obtained corresponds to the energy absorbed in a reaction of the type

$$NaCl(g) \rightarrow Na^+(g) + Cl^-(g)$$

An appropriate modification can be made for salts of the cesium chloride type, but it is just as satisfactory to calculate a value of r_0 for a salt of the sodium chloride type from Table 16, and then use this value in Eq. (24) without modifying the latter.

Method II is the same as Method I, except that we use the value of n necessary to make Eq. (8) hold when U_0 is given the value from Table 20 or, in the case of other salts, the experimen-

[1] See Reis, *Zeits. Physik*, **1**, 294 (1920); Born and Heisenberg, *ibid.*, **23**, 403 (1924); Verwey and de Boer, reference 2, p. 234. Since our calculation was made, another by May, *Phys. Rev.*, **54**, 629 (1938), has appeared.

tal value. It must be remembered, however, that $-U_0$ of Table 20 is the energy of a reaction of the type

$$\text{NaCl (solid)} \rightarrow \text{Na}^+(\text{g}) + \text{Cl}^-(\text{g})$$

including the work done against the atmosphere, *i.e.*, it is the ΔH of such a reaction, while the ΔE of this reaction should be used in Eq. (8) (see Appendix II).

Method III uses the Born-Mayer approximation for the repulsive force. The energy of a pair of ions forming a gaseous molecule will be given by the expression

$$U' = -\frac{e^2}{r'} + b'e^{-\frac{r'}{\rho}} \tag{26}$$

The first term is the electrostatic energy and differs from the electrostatic energy of the crystal only in that the factor A does not appear. In the last term, giving the repulsive energy, b of Eq. (11) is replaced by b'; if b is the coefficient for a crystal with sodium chloride structure and without anion-anion contact, then

$$b' = \frac{b}{6}, \tag{27}$$

because each ion has one neighbor instead of six.[1]

We may determine the equilibrium distance r_0' in the gaseous molecule by setting dU'/dr' equal to zero, getting

$$\frac{e^2}{r_0'^2} - \frac{b'}{\rho}e^{-\frac{r_0'}{\rho}} = 0. \tag{28}$$

If we solve for b' in terms of r_0' and substitute into Eq. (26), we get an equation like Eq. (12):

$$U_0' = -\frac{e^2}{r_0'}\left(1 - \frac{\rho}{r_0'}\right). \tag{29}$$

Substituting for b' from Eq. (27) into Eq. (28) and rearranging slightly, we get

$$\frac{6\rho e^2}{r_0'^2 b} = e^{-\frac{r_0'}{\rho}} \tag{30}$$

[1] In considering the repulsive forces we take into account only nearest neighbors.

or, taking logarithms of both sides and again rearranging the terms,

$$r_0' = \rho \ln \frac{b}{e^2 \rho} - \rho \ln 6 + 2\rho \ln r_0'. \tag{31}$$

If we differentiated Eq. (11) and solved for r_0, the equilibrium distance in the crystal, we would get

$$r_0 = \rho \ln \frac{b}{e^2 \rho} - \rho \ln A + 2\rho \ln r_0. \tag{32}$$

From (31) and (32),

$$r_0 - r_0' = \rho(\ln 6 - \ln A) + 2\rho \ln \frac{r_0}{r_0'}. \tag{33}$$

In this development, van der Waals forces, thermal energy, and zero-point energy have been neglected. It seems best to assume that these are all included in the repulsive term, and then to determine ρ in such a way that Eq. (12) gives a value of U_0 equal to that found in Table 20. This is similar to the way n is determined in Method II. Born and Mayer and Huggins took the van der Waals force, thermal energy, and zero-point energy into account separately, and their calculated values of U_0 contain special corrections for these factors, and are based on a value of ρ obtained from compressibilities by taking these factors into account. The difference between our value of ρ and theirs will, to a certain extent, correct for the fact that we have not considered these quantities explicitly.[1] However, it is clear that a term which involves the sum of a repulsive potential, a van der Waals potential, and a correction for kinetic energies will be only approximately represented by an exponential function; furthermore, these may differ in crystal and gas.

The results obtained by the various methods for the alkali halides and hydrides are given in Table 27, where they may be compared with each other and with the experimental results. The "experimental" values of $-U_0'$ are obtained from the heat of sublimation $U_0 - U_0'$ and from $-U_0$, taken from Table 20 or 25. In Table 27, none of the energy quantities include work done against the atmosphere in a constant-pressure process—they are ΔE's rather than ΔH's.

[1] See footnote 2, p. 234.

On the whole, the agreement between the calculated and experimental values of r_0' and U_0' is fair, but the discrepancies are appreciable. Also the agreement between the different

TABLE 27.—PROPERTIES OF GASEOUS ALKALI HALIDES AND HYDRIDES

Salt	r_0, Å	n I	n II	ρ III	r_0' I	r_0' II	r_0' III	r_0' expt.	$U_0' - U_0$ expt.	$-U_0'$ I	$-U_0'$ II	$-U_0'$ III	$-U_0'$ "expt."
LiF....	(1.95)	6	5.90	0.33	1.52	1.52	1.25	....	63	181	181	194	181
LiCl....	(2.40)	7	7.75	0.31	1.95	2.00	1.86	....	46	145	143	148	154
LiBr...	(2.54)	7.5	8.46	0.30	2.10	2.15	2.04	....	44.2	136	135	138	145
LiI.....	(2.75)	8.5	9.49	0.29	2.33	2.38	2.28	....	41	125	124	126	134
NaF...	2.310	7	7.32	0.316	1.88	1.90	1.74	....	71	150	150	155	144
NaCl...	2.814	8	9.38	0.300	2.36	2.43	2.33	2.51	56.5	122	121	123	126
NaBr...	2.981	8.5	10.47	0.285	2.53	2.62	2.54	2.64	53.2	115	114	115	122
NaI....	(3.11)	9.5	11.51	0.27	2.69	2.76	2.70	2.90	49.8	110	109	110	113
KF.....	2.665	8	8.97	0.297	2.23	2.28	2.18	....	49.6	129	128	131	142
KCl....	3.139	9	11.21	0.280	2.69	2.78	2.71	2.79	51.4	109	108	109	116
KBr....	3.293	9.5	12.00	0.274	2.85	2.94	2.88	2.94	49.5	103	103	103	111
KI.....	3.526	10.5	13.60	0.259	3.10	3.20	3.15	3.23	48.3	96	96	96	103
RbF ...	2.815	8.5	9.15	0.308	2.39	2.42	2.32	....	52.8	122	121	123	129
RbCl...	3.270	9.5	11.59	0.282	2.83	2.91	2.84	2.89	50.9	104	104	104	110
RbBr..	3.427	10	12.81	0.268	2.99	3.09	3.03	3.06	50.2	99	98	99	105
RbI....	3.663	11	14.98	0.245	3.24	3.35	3.11	3.26	49.2	93	92	92	98
CsF....	3.004	9.5	11.29	0.266	2.60	2.66	2.60	....	50	113	113	114	125
CsCl...	3.566	10.5	14.98	0.238	3.08	3.20	3.15	3.06	50.3	97	96	97	102
CsBr...	3.713	11	19.47	0.190	3.22	3.40	3.38	3.14	49.3	93	92	92	99
CsI....	3.95	12	25.1	0.157	3.45	3.66	3.64	3.41	45.8	88	86	87	95
LiH....	2.042	....	4.45	.....	....	1.43	....	1.6	54.4	...	179	...	164
NaH...	2.440	....	5.39	.....	....	1.84	....	1.88	39.0	...	146	...	153
KH....	2.850	....	5.27	.....	....	2.14	....	2.24	36.5	...	125	...	127

NOTE: When anion-anion contact occurs in the crystal, values of ρ and n (Method II) are estimated by extrapolation.

Values of r_0 in parentheses are calculated from Table 16 (cases of anion-anion contact). The other values are those given by Huggins, *J. Chem. Phys.*, **5**, 146 (1937).

Values of $U_0' - U_0$ for halides from Bichowsky and Rossini, "Thermochemistry of Chemical Substances"; for hydrides calculated directly from dissociation energy from Sponer, "Molekülspektren," Tables 4 and 25, and the electron affinity of H.

Expt. values of r_0' for halides from Maxwell, Hendricks, and Mosley, *Phys. Rev.*, **52**, 968 (1937); for hydrides, from Sponer, *loc. cit.*

methods of calculation is not bad, except in a few cases; in particular, Method III cannot be used at all for the hydrides. The value of U_0' seems rather insensitive to the method of calculation. Since the differences between the calculated and experimental values of U_0' are in almost all cases fairly small compared with $U_0' - U_0$, we may be said to have made a reasonably good calculation of the energy of sublimation of the alkali halides. Yet it is seen that the calculated value of $-U_0'$ is rather consistently low, and the very fact that it is rather insensitive to the exact method of calculation makes this a little difficult to understand. If the greater experimental value of $-U_0'$ is due to greater polarization of the anion by the cation in the gaseous molecule where the force fields are very unsymmetrical, it would be expected that the observed values of r_0' would be lower than the calculated values, whereas in a number of cases the reverse is true. One might account for the observed results by supposing that there is considerable polarization, and hence an appreciable contribution of covalent forces, in the gas molecule combined with a rather sudden increase of the repulsive force at distances near the observed values of r_0'.

14.10. The Proton Affinity of Ammonia.—Another interesting application of the Born-Haber cycle has been made by Grimm,[1] whereby he has determined the proton affinity of ammonia, namely, the energy absorbed in the reaction

$$NH_4^+ \rightarrow NH_3 + H^+$$

It is possible to analyze the reaction

$$\tfrac{1}{2}N_2 \text{ (gas)} + 2H_2 \text{ (gas)} + \tfrac{1}{2}X_2 \text{ (gas)} \rightarrow NH_4X \text{ (crystal)}$$

(where X is any halogen), which involves a total energy change Q, in the following manner:

$$\begin{array}{l}
\tfrac{1}{2}N_2 + \tfrac{3}{2}H_2 \xrightarrow{-M} NH_3 \\
\tfrac{1}{2}H_2 \xrightarrow{D_H} H \xrightarrow{I} H^+ \\
NH_3 + H^+ \xrightarrow{-P} NH_4^+ \\
\tfrac{1}{2}X_2 \xrightarrow{D_X} X \xrightarrow{-F} X^- \\
NH_4^+ + X^- \xrightarrow{U_0} NH_4X
\end{array}$$

(all substances in gaseous state except NH_4X), where the energies absorbed in the various reactions are designated by the symbols written above the arrows. We have, then,

$$Q = -M + D_H + D_X + I - F - P + U_0. \tag{34}$$

[1] See SHERMAN, *Chem. Rev.*, **11**, 150 (1932).

All these quantities have been determined experimentally or otherwise except U_0, which may be calculated in the same manner as for other salts, on the assumption that the NH_4^+ ion is a spherically symmetrical ion much like the alkali ions. This certainly should be a good approximation if the ammonium ion

TABLE 28.—PROPERTIES OF AMMONIUM HALIDES

Salt	Ionic distance, Å.	Q, kg.-cal.	$-U_0$, kg.-cal.
NH_4F	2.63	111.9	177.5
NH_4Cl	3.35	75.1	153.3
NH_4Br	3.51	64.7	147.4
NH_4I	3.62	48.6	143.6

is rotating in the crystal. The values for the proton affinity thus obtained from ammonium fluoride, chloride, bromide, and iodide, respectively, are, according to Sherman, 219.6, 209.0, 208.6, and 202.7 kg.-cal. per mole. These values are corrected to 0°K.

The high value obtained from the fluoride (which means that the calculated value of $-U_0$ is less than the true value, provided the values for the other salts are approximately correct) has been interpreted by Sherman on the basis of the assumption that the ammonium ion is *not* rotating in the fluoride crystal. If this is the case, then the calculation of the Madelung constant A in the expression for U [Eq. (6) or (11), Sec. 14.5], will require the consideration of the structure of the ammonium ion and distribution of charge within it; a value of A obtained by supposing the ammonium ion to be a spherically symmetrical positive ion will clearly give an incorrect value for U_0.

The question as to whether the ammonium ion will be freely rotating in the crystal is largely one of the relative energies of the form in which it is rotating and the form in which it is not. In general, the crystalline form in which rotation does not occur will have the lower energy. As will be seen in later chapters, the hydrogens in an ammonium ion are in all probability placed around the nitrogen at the corners of a regular tetrahedron. The charge distribution in such an ion cannot be spherically symmetrical, and there is undoubtedly a preponderance of positive charge near the hydrogens. A negative ion which is near one of these positive centers has a lower potential than would be the case if all the charge were symmetrically placed about the center of the ammonium ion. In order to make the most of this arrangement in a crystal, the ammonium ion should not be rotating and there should be a negative ion near each hydrogen, so that each ammonium ion would have four negative neighbors. This, however, means that the crystal would have the wurtzite, sphalerite, or some similar structure (see Appendix IV), with a lower value of the constant A (see Table 18), and hence a smaller lattice energy, assuming that the interatomic distance is not greatly changed. We may consider the potential of an ammonium halide crystal to be that of a crystal with a symmetrical cation, plus a contribution due to the tetrahedral structure of the ammonium ion; unless the latter contribution is sufficiently great, the crystal will not tend to have the form in which the ammonium ion has a coordination number of four. In the case of ammonium fluoride, the crystal has the wurtzite structure; this is probably because the fluoride ion is so small that it can get close enough to the ammonium ion to make the energy due to the lack of symmetry appreciable. On the other hand, the chloride and bromide have the cesium chloride structure, and the iodide has the sodium chloride structure (cesium chloride below about $-15°C.$), and it is probable that the ammonium ion rotates freely at room temperature; at any rate, the effect of lack of spherical symmetry is not pronounced.

Since the nonrotating state of the ammonium ion is the one of lower energy, it would be expected to be the stable condition at low enough temperatures. A further factor would favor it at low temperatures. If the electron atmosphere of the ammonium ion extends farther out in some directions than in others, then when the ions are closely packed, there might be hindrance to the rotation due to a purely steric effect; *i.e.*, when the ammonium starts to rotate the parts that bulge out will tend to strike other ions, and so the rotation will be stopped. As the salt is warmed up, a few of the ammonium ions will get enough energy to start rotating. Since energy is required for this, the specific heat of the crystal will begin to increase over its normal value. As the temperature is raised still further, it will become easier for ions to rotate, not only because it is easier to get the requisite energy, but because the fact that a number of its neighbors are already rotating will mean that less energy will probably be required for a given ion to begin rotating, and further, the thermal expansion of the crystal will allow rotation to occur more readily. It is thus expected that the specific heat will first gradually then rapidly rise over a range of a few degrees, as the setting in of rotation becomes more noticeable, and then rather suddenly fall off when practically all the ions are rotating. Transitions of just this type, occurring over a range of about 10°, have been observed in ammonium chloride, bromide, and iodide[1] at about −30, −38, and −42°C., respectively, and have been ascribed to the setting in of rotation in the crystal.[2] The temperature at which the transition occurs decreases from chloride to iodide, indicating decreasing forces, as expected with increasing ionic radii. The force exerted by the very small fluorine ion must be considerably greater. It should, however, be noted that the fluoride also shows an anomaly in the specific heat at −31°C, but it is much less than those shown by the other salts. Its interpretation is doubtful, as the crystal structure is certainly best interpreted on the assumption that no rotation occurs at room temperature.

It is worth noting that the ammonium chloride crystal does not change from the cesium chloride structure below −30°. It thus does not make fullest use of the possible energy lowering due to the lack of spherical symmetry (probably because of the smaller *A* value for the wurtzite structure, as explained above). It is still possible for four chlorides to be especially near the hydrogens of a given ammonium ion, but there are four other chlorides nearby which are not especially near the hydrogens. In the wurtzite structure, each anion is near four hydrogens, one from the ammonium in each direction; in the cesium chloride structure, it is also near four hydrogens, but only half the eight anions near any cation are favorably placed. In the cases of NH_4Br and NH_4I, the low temperature forms are neither wurtzite nor cesium chloride, and it is possible that the ammonium ion is still rotating but only about one axis.

[1] See Simon, Simson, and Ruhemann, *Zeits. physik. Chem.*, **129**, 339 (1927).

[2] Pauling, *Phys. Rev.*, **36**, 430 (1930).

In the gaseous state, the effect of the structure of the ammonium ion becomes much more pronounced. In the crystalline state, a rotating ammonium ion strongly resembles the rubidium ion; it has just about the same apparent size, and it is seen from Tables 20 and 28 that NH_4I, which has the sodium chloride structure, has about the same lattice energy as RbI. The lattice energies of the other ammonium salts, which do not have the sodium chloride structure, are, except for the fluoride, still not far from those of the rubidium salts. The heat of sublimation of NH_4Cl is, however, more than 20 kg.-cal. less than that of RbCl, indicating that gaseous NH_4Cl has a relatively very low energy (*i.e.*, its dissociation energy into NH_4^+ and Cl^- is higher by 20 kg.-cal.[1] than it would be if NH_4^+ were spherically symmetrical). This is probably due to the fact that in the gas phase, as we have seen in the last section, the distance between anion and cation is relatively small, so that the asymmetry of the cation becomes more important. Further, there is undoubtedly a displacement of one of the hydrogens[2] toward the negative chloride ion in the gaseous molecule of NH_4Cl which will still further lower the energy; this effect would also be present in the crystal, but it would probably be less marked, since the force is applied on all the hydrogens, and a displacement of one of them would probably make the displacement of another harder. It is still a little difficult to see why the lowering of the energy of a gaseous molecule of NH_4Cl should be so much greater than the lowering of the energy in a crystal of NH_4F (about 12 to 15 kg.-cal. as judged by the abnormality of the proton affinity of NH_3 calculated from NH_4F), considering that the fluoride ion is so small and that there are four fluoride contacts per ammonium ion in the crystal. It is seen that the apparent anomaly of NH_4F crystal is certainly not greater than would be expected from the anomaly in the heat of sublimation of NH_4Cl.

14.11. Some Properties of OH and SH.—Estimates of the electron affinity of OH and SH have been made by calculation of lattice energies of crystals in which they occur by Goubeau

[1] That this is a legitimate conclusion is evident from inspection of Table 27, where it is seen that the heats of sublimation of all the potassium, rubidium, and cesium halides are practically alike.

[2] See FAJANS, *Zeits. physik. Chem.*, **A137**, 361 (1928).

and Klemm,[1] and West,[2] respectively. OH has the same electronic structure as F, whereas SH has the same electronic structure as Cl. The difference between OH and F is that in F all the positive charge is in the nucleus, whereas in OH part of it is near the edge. A positive charge that is near the edge of an ion does not exert so strong a force on an electron as one that is at the center. This difference shows up in OH having a much smaller electron affinity than F. It also shows up in other ways. Thus F^- is smaller than OH^-, and the polarizability[3] of F^- is less than that of OH^-. In a similar way, the electron in SH^- is less strongly held than that in Cl^-. A comparison of the various properties mentioned follows:

Electron affinity: F, 95.0; OH, 48; Cl, 87.3; SH, 61
Polarizability: F^-, 0.96; OH^-, 1.88; Cl^-, 3.57; SH^-, 5.23
Radius in crystals: F^-, 1.36; OH^-, 1.53; Cl^-, 1.81; SH^-, 2.00

It will be observed that the difference between the electron affinities of F and OH is much larger than the corresponding difference between Cl and SH. This peculiarity is very likely connected with a change in the force on the proton in OH or SH when the electron is added. The proton probably moves to a new position of equilibrium, with a resultant decrease in the energy of the system. This effect must be greater in the case of SH than in the case of OH, and this may be possible because the proton is originally very loosely held in SH, and because this structure is larger and "softer" than OH. In any event, the proton is finally quite deeply buried in the electron shell in the SH^- ion, which appears to be spherically symmetrical in crystals and does not have marked dipole properties as does OH^-.

The process of separating the two OH groups in H_2O_2 bears much the same relation to the dissociation of an F_2 molecule as the separation of an electron from OH^- bears to the separation of an electron from F^-. In H_2O_2 or F_2, half the molecule shares an electron from the other half, and, as in F^- or OH^-, the shell of eight of each part is filled. It is therefore of interest to note

[1] Goubeau and Klemm, *Zeits. physik. Chem.*, **B36**, 362 (1937).
[2] West, *J. Phys. Chem.*, **39**, 493 (1935).
[3] Fajans, *Zeits. physik. Chem.*, **B24**, 134 (1934).

that according to Bichowsky and Rossini, "Thermochemistry of Chemical Substances," the reaction

$$H_2O_2 \rightarrow 2OH$$

requires 21.7 kg.-cal. This is to be compared with

$$F_2 \rightarrow 2F$$

which, as is to be expected from the electron affinities, requires much more energy, namely, 63.5 kg.-cal.

Another closely related quantity is the bond energy. The O—O bond energy (determined from H_2O_2) is 34.6 kg.-cal., and the S—S bond energy (determined from H_2S_2) is 38.3 kg.-cal. In Sec. 12.8, we commented upon the fact that the S—S bond energy is greater than the O—O bond energy. This now seems less surprising when considered in the light of the electron affinities of OH and SH, in view of the analogy between breaking an electron-pair bond and pulling off an electron. The surprising difference between the O—O single-bond energy and the C—C bond energy is, however, still not explained.

West has also considered the proton affinity[1] of SH^-, *i.e.*, the energy given out by the reaction

$$SH^- + H^+ \rightarrow H_2S$$

and found that this is 338 kg.-cal. The proton affinity of Cl^- is 328 kg.-cal. It is interesting that the electron affinity of Cl is greater than that of SH and the proton affinity less than that of SH^-. The large proton affinity of SH^-, in spite of its greater size, is due to the combination of its large polarizability, the possibility of internal adjustment by shift in position of the other proton, and the dipole moment, which, though small, makes it easier for a proton to be added since it means that some of the positive charge already there can be at a relatively great distance from the added proton.

West has pointed out that in some respects SH^- resembles Br^- more than it does Cl^-. Thus the size of Br^- in sodium chloride type crystals is 1.95Å., very close to that of SH^-, and

[1] The proton affinities of a number of different ions and molecules have been tabulated by Juza, *Zeits. anorg. allgem. Chem.*, **231**, 134 (1937). Although many of these figures represent only rough estimates, they are of interest for comparison.

the polarizability of Br^- is 4.99, also close to that of SH^-. The electron affinity of Br, 81.8 kg.-cal., is also closer, but the proton affinity of Br^-, 318 kg.-cal, is not so close as that of Cl^-.

Exercises

(s = solid; g = gas)

1. Show the results to be expected from an X-ray analysis by the Bragg method of a crystal of the CsCl type.

2. Calculate the energies of the following reactions: $Li^+(g) + NaCl(s) \rightarrow Na^+(g) + LiCl(s)$; $Na^+(g) + KCl(s) \rightarrow K^+(g) + NaCl(s)$; $K^+(g) + RbCl(s) \rightarrow Rb^+(g) + KCl(s)$; $Rb^+(g) + CsCl(s) \rightarrow Cs^+(g) + RbCl(s)$. State in each case whether the reaction will be expected to go spontaneously in the direction written or not when the concentration of the gaseous ions are unity.

3. Calculate the energy of the following reactions: $RbF_2(s) \rightarrow RbF(s) + \frac{1}{2}F_2(g)$; $RbBr_2(s) \rightarrow RbBr(s) + \frac{1}{2}Br_2(g)$. (Assume $RbBr_2$ like $SrCl_2$.)

4. Calculate the energy of the reaction $CaBr_2(s) + Ca(s) \rightarrow 2CaBr(s)$.

5. From the data given in Table 25 and elsewhere in this chapter, check the values of U_0 (expt.) and n for the alkali hydrides. The alkali hydrides have the sodium chloride structure.

6. $CdCl_2$ crystallizes in a form somewhat similar to the cadmium iodide structure. Assuming that the Madelung constant A is the same as for CdI_2, calculate U_0 (see Table 36, page 328). Find the experimental value for U_0 from data given in tables in this and earlier chapters, together with the following data: heat of formation from the elements (heat evolved in kilogram-calories per mole of salt), 93.0; heat of sublimation of Cd, 26.8. Compare the theoretical and experimental values of U_0, and discuss the result.

7. Find the experimental value of U_0 for $ZnCl_2$ and $CuCl_2$ from data given in tables in the book and the following: heats of formation, $ZnCl_2$, 99.5; $CuCl_2$, 53.4 (in both cases heat evolved in kilogram-calories per mole of salt); heat of sublimation of Zn, 27.4.

8. Calculate the energy absorbed in the reaction

$$2CuCl(s) \rightarrow CuCl_2(s) + Cu(s)$$

from the lattice energies of CuCl, $CuCl_2$, the ionization potentials of copper and the heat of sublimation of copper. Similarly, calculate the energy absorbed in the reaction

$$2AgCl(s) \rightarrow AgCl_2(s) + Ag$$

assuming the lattice energy of $AgCl_2$ to be the same as that of $CdCl_2$ (see Exercise 6). Repeat the latter calculation, assuming that the lattice energy of $AgCl_2$ differs from that of $CdCl_2$ by the same amount that the lattice energy of $CuCl_2$ differs from that of $ZnCl_2$. Analyze the principal reasons for the instability of $AgCl_2$ as compared with $CuCl_2$.

9. Calculate and compare the proton affinities of the halide ions.

CHAPTER XV

FURTHER PROPERTIES OF COVALENT BONDS

Before proceeding further in our study of the nature of chemical compounds, it will be necessary to extend somewhat our considerations concerning atomic binding in general; in particular, we shall have to discuss the phenomenon of directed valence. It has long been realized, or at least suspected, that valence bonds (of the nonpolar type) have direction; for example, it has long been customary in organic chemistry to speak of the tetrahedral carbon atom, implying that the valence bonds extend out toward the corners of a regular tetrahedron. To be sure, in this case, one might claim that, since carbon generally has four other atoms attached to it, the corners of a regular tetrahedron are the most natural positions for these atoms to take, regardless of whether the bonds have any specific orientative influence. A better example of the effect of directed valence is therefore afforded by a molecule like the water molecule. It is definitely known that this molecule is kinked, having the form

```
H     H
 \   /
   O
```

Such an unsymmetric configuration is best explained as a property of the OH bonds; the two bonds tend to form at a certain definite angle to each other.

15.1. Methods of Determining Molecular Structure.—We shall here list and briefly discuss some of the principal methods by which the structure of molecules—involving such questions as the distances between atoms and the angles between bonds—can be determined experimentally.

1. *Measurement of electric moment* (see Sec. 12.9) will often give considerable qualitative information about the structure of a molecule,[1] and from such measurements the kinked structure of the water molecule may be at once inferred. It is to be expected, of course, that the hydrogen atoms would be positively charged with respect to the oxygen atoms, thus producing a dipole.

[1] DEBYE, "Polar Molecules," Reinhold Publishing Corporation, 1929.

However, if the molecule is a straight structure the dipoles would cancel, because of the hydrogen's being oriented in exactly opposite directions. On the other hand, with the kinked structure this would not occur. Thus the finding of a finite dipole moment was definite evidence in favor of the kinked structure.

2. *Analysis of the Vibration and Rotation Spectra.*—The vibrational and rotational energy levels of a polyatomic molecule depend upon its structure, as has been shown in Chap. IX. Any means by which the various energy levels of a molecule may be inferred will be of value in the determination of the structure of the molecule.

3. *Stereochemical Considerations.*—The inferences that may be drawn from stereoisomerism in organic chemistry are well known. Similar considerations applied to the so-called inorganic complex molecules also yield much information. This subject will be treated more fully later.

4. *Diffraction of X Rays and Electrons.*—It was shown in Chap. XIV that information regarding the arrangement of atoms in crystals may be obtained from X-ray analysis. It is possible from such measurements to learn about the arrangements of atoms in stable ions which generally enter into the crystal as units that are the same in all crystals. Also the diffraction of X rays by gases may be made to yield information about the structure of the gas molecules.[1] This may be considered to be a limiting case of the powder method of X-ray analysis, the powder being so finely divided that it consists of single molecules. There will still be a diffraction pattern but the diffraction rings will be broad instead of narrow. With a crystal, the reflection of X rays from a surface takes place at the Bragg angles[2] and practically at no other angle because of the reflection from many planes of the crystal, which ensures that, except at the Bragg angles, there will be waves in practically every phase, guaranteeing complete destructive interference. Of course, this does not occur when the reflection is from a single molecule, and instead of a sharp maximum of intensity at a definite angle there is a band with a broad maximum. Nevertheless, these diffraction patterns can be utilized, by the aid of

[1] See Debye, *Zeits. Elektrochem.*, **36**, 612 (1930); Bragg, "The Crystalline State," p. 191, The Macmillan Company, 1934.

[2] See Sec. 3.2, especially Eq. (1).

Fourier analysis, to give very directly the atomic distances in the molecule.

Instead of X rays, electrons may be used to give the diffraction pattern,[1] since, as we have seen, they are diffracted in much the same way as X rays. In the study of gases, electrons are in some respects better than X rays, because it is easy (by speeding up the electrons) to get shorter wave lengths and thus more rings in the diffraction pattern and because the scattering is more intense.

15.2. Directional Properties of Chemical Valence.—The physical reasons for the phenomenon of directed valence will now be briefly considered. Probably the simplest formulation of this problem is that given by Pauling,[2] and though it involves many arbitrary assumptions, it will give a sufficiently good picture of the physical reality for our purposes. It has been remarked before that if a number of different wave functions correspond to the same value of the energy then any linear combination of those wave functions will also be an acceptable solution of the wave equation (see Sec. 10.1). Furthermore, we have seen that if wave functions are obtained on the basis of certain approximations, then, even though these wave functions do not correspond to exactly the same energy, the true wave function after the approximations are removed will be a linear function of the approximate wave functions.[3]

Let us now consider an atom in the fields produced by the atoms with which it is combining. These fields produce a strong perturbing force, the potential of which will be great compared with the energy differences of the outer electron orbits. The result will be that the wave functions of these outer orbits will be rearranged so to speak; there will be a special set of linear combinations (equal in number to the original wave functions) which will give the resulting molecule the lowest energy, and which will, therefore, be involved in the formation of the ground state of the molecule.[3]

[1] For a summary and general discussion with references see Brockway, *Rev. Mod. Phys.*, **8**, 231 (1936); see also Bragg, p. 264 (*loc. cit.*). The application of electron diffraction to gases was originally due to Mark and Wierl.

[2] Pauling, *J. Am. Chem. Soc.*, **53**, 1367 (1931). See also Slater, *Phys. Rev.*, **37**, 481; **38**, 1109 (1931). For a critical review and detailed discussion see Van Vleck and Sherman, *Rev. Mod. Phys.*, **7**, 167 (1935).

[3] The situation is similar to that described in Sec. 12.5.

Some indication of the type of linear combination to be expected may be obtained from the hydrogen molecule. In this molecule, a concentration of electrons between the nuclei causes the attraction between the atoms. It is, therefore, natural to assume that the set of linear combinations which is to be favored is one that will cause the wave functions each to project out farthest in some specific direction, the direction of the bond it is to form, and to give a large electron density along the bond. It is on this assumption that the considerations of Pauling rest.

We may illustrate these considerations by means of a simple example, the methane molecule. The carbon atom has four outer electrons. In the ground state, there are two 2*s*-electrons and two 2*p*-electrons. According to the theory of London, all the electrons must be unpaired before combination with the four hydrogens can take place. This assumption has been discussed in Chap. XI. In the present case, the unpairing is helped by the fact that "good" bond-eigenfunctions (*i.e.*, wave functions that "stick out" markedly in certain directions) can be formed from a linear combination of one *s*-wave function and three *p*-wave functions (with the quantum number $m_l = -1, 0, 1$, respectively) better than from other combinations. It is found that the "best" four independent linear combinations of these waves have, respectively, high electron densities in the directions of the corners of a regular tetrahedron. Thus in place of the original 2*s*- and 2*p*-states, we have these bond-eigenfunctions, or bond-forming orbits, and into each of these orbits goes one of the outer electrons. Each outer electron then "pairs" with the electron of a hydrogen atom. The calculations show that the pairing is most effective, *i.e.*, the bond energy lowest and the bond consequently most stable, if there is as much overlapping as possible between the wave function of a given electron and that of the electron with which it is paired and as little as possible between it and those of all other electrons on other atoms. This is realized if the hydrogens are situated at the corners of the tetrahedron toward which the bond-eigenfunctions point, and the stable form of the methane molecule is therefore tetrahedral.

The theory is not able, at the present time, to give more than the roughest estimate of the energy involved in the bond formation. Even the estimate of the energy of the preliminary unpairing of the electrons offers some difficulty, for any one of the unpaired electrons may have its spin oriented in either direction (for an unpaired electron does not have to share its orbital quantum state with another electron, and so its spin state is not restricted by the Pauli exclusion principle). There are, then, various possible spin combinations of the unpaired electrons. Although the energy due to electron spin is generally neglected in our calculations, it is nevertheless sufficient to cause a considerable difference between the energies of some of these states with different spin combinations; and since it is not known just which combination is to be used in bond formation, the energy of unpairing is left in some doubt. The energy to bring about the unpairing in carbon has, however, been estimated by Van Vleck,[1] with what he believes to be about 10 per cent accuracy, to be around 7 electron volts or 160 kg.-cal. per mole. It has this rather large value on account of the necessity of exciting one of the *s*-electrons to a *p*-state. It is clear that unless there were a considerable gain from the formation of the tetrahedral bonds, as is the case with carbon compounds, the unpairing would not occur. In cases where there are fewer than four groups or atoms attached to a central atom which has a completed octet, the energy gained by tetrahedral bond formation may not be great enough to counterbalance the disturbance of the low energy *s*-electrons, which in such a case as the water molecule need not take part in the bond formation. Under these circumstances the *s*-wave function is said to be unavailable for bond formation. The case of the water molecule will be discussed below, and in this case it is not possible to say unequivocally whether the *s*-wave function takes part in bond formation or not.

A tetrahedral arrangement of atoms about the central atom is not the only possibility. The arrangement which will occur in any given case depends upon the original wave functions involved in forming the bond-eigenfunctions. In general, of course, the wave functions which are concerned in the formation of the bond-eigenfunction cannot differ too greatly in energy, even though the effect of this difference in energy is largely erased by the effect of the other atoms. Thus one might expect the *s*- and *p*-wave functions of a given shell to combine to form bond-eigenfunctions, but *s*- and *p*-wave functions of different

[1] Van Vleck, *J. Chem. Phys.*, **2**, 297 (1934).

shells would not be expected to so combine. A *d*-wave function might be expected to combine with *s*- and *p*-wave functions of a shell just outside, however, since the corresponding energies are not very different; *e.g.*, 4*s*- and 4*p*-wave functions may combine with 3*d*-functions. It is possible, on the other hand, for such a *d*-state to be occupied by an electron or electrons not taking part in bond formation (*i.e.*, not shared by an adjacent atom), in which case the *d*-wave function is not available for formation of a bond-eigenfunction. Although such a *d*-state can be occupied by two electrons with opposite spins, one such electron is sufficient to render it unavailable, inasmuch as a bond-eigenfunction, after being further transformed by interaction with the other combining atom, is generally occupied by a pair of shared electrons. On the other hand, there may be some cases in which the *d*-wave function is used in bond formation, and then one of the bond eigenfunctions is occupied by an electron or pair of electrons which is not shared. This situation, however, seems to be fairly rare.

The type of structure that will be produced by various wave functions may be inferred by a study of the nature of the bond-eigenfunctions formed. Such a study has been made by Pauling, and although it involves a number of approximations which are not entirely satisfactory, it yields rules that appear to be fairly well confirmed by the experimental facts. These rules are as follows:

1. If only *p*-wave functions are involved in formation of the bond-eigenfunction, the valence bonds are at right angles to each other.

2. If all *s*- and *p*-wave functions of a given shell are involved, the structure will be a tetrahedron. Expressing the fact that one *s*-wave function and three *p*-wave functions are involved, these are designated as sp^3 bonds.

3. If all *s*- and *p*-wave functions of a given shell and one *d*-wave function (usually from the shell just inside) are involved, the structure is a square. In this case, there are five wave functions that form five linear combinations four of which give electron distributions projecting out each in a specific direction, the four directions thus determined being toward the corners of a square. The fifth combination, which does not involve either the *d*- or the *s*-eigenfunction at all, does not give a large electron density

except relatively near the central atom, and so is not a good bond-eigenfunction. Thus, only four of the five possible eigenfunctions are used in bond formation, and the fifth possible state will generally not be filled by an electron pair at all. In fact, an electron pair that has no other place to go than this fifth state will tend, instead, to render the *d*-eigenfunction unavailable. It is to be noted that it would be possible, by not using the *d*-wave function, to form a tetrahedron of the type (2), but the square here described is more stable, and is formed when a *d*-state of sufficiently low energy is available. These are known as sp^2d bonds (p^2 because the equivalent of one *p*-eigenfunction is used in forming the nonbonding combination).

4. If all *s*- and *p*-wave functions of a given shell and three or more *d*-wave functions (presumably from an inner shell) are available, then a tetrahedron may be formed, only four of the seven resulting eigenfunctions being good bond-eigenfunctions. These are sp^3d^3 bonds. This is a more stable arrangement than the square, and if there are three or more *d*-spaces available, the resulting molecule or complex will have a tetrahedral structure (at least if its coordination number is 4). If only one or two *d*-spaces are available, it is not possible to have this type of bond-eigenfunction, and if the coordination number is 4, only one of the *d*-wave functions is used, the square structure being formed.

5. If all *s*- and *p*-wave functions of a given shell and two or more *d*-wave functions are available, six bond-eigenfunctions may be formed, giving a regular octahedron. These are sp^3d^2 bonds.

There are still other possibilities, but the five given are the most important.

These rules will be illustrated by means of specific examples. As has already been pointed out, if the atoms surrounding a central atom are held to it by ionic forces, they will tend to take up symmetrical positions about the central atom. Since most of the configurations mentioned above as resulting from atomic binding are also symmetrical, it will be seen that the atomic arrangement will not always suffice by itself to distinguish between binding which is predominantly polar and binding predominantly nonpolar, and in general it is not possible to get a simple criterion for the degree of polarity of the bonds in complex compounds. In illustrating the rules for configuration, we shall select compounds that are believed to be predominantly nonpolar, but it will be necessary to defer until later the discussion of the reasons why they are believed to be nonpolar. For the most part, the discussion of the determination of the structure of the molecules will also be deferred to later chapters. With this understanding, we shall proceed to give examples of the various types of atomic configuration.

1. It would be natural to suppose that the water molecule would furnish a good example of compound formation involving only p-wave functions. However, the angle between OH bonds is nearer the tetrahedral angle than it is to a right angle. This may be, indeed, due in part to the repulsion between the hydrogen atoms, but it may also be due to the bond-eigenfunctions actually being of the sp^3 tetrahedral type.[1] The eight valence electrons would then be housed in those states, but two pairs would belong only to the oxygen, and only two pairs would actually be shared between the oxygen and a hydrogen. On the other hand, H_2S seems certainly to offer an example of p bond-eigenfunction, as do also the crystals of arsenic, antimony, and bismuth. In these crystals, the atoms are so arranged that each atom has three closest neighbors, and, therefore, three of the five outer electrons are shared. There will be three p bond-eigenfunctions belonging to each atom, each one of which can interact with a similar bond-eigenfunction of another atom; two atoms share each pair of electrons, and each atom shares a pair with three other atoms altogether. However, even here there may be some effect of the s-wave function, for the angles between the bonds are not exactly 90°, but are as follows: As, 97°; Sb, 96°; Bi, 94°.

[1] It has been argued by Stuart, *Zeits. physik. Chem.*, **B36**, 155 (1937), that the normal valence angle for N, O, and F, and the heavier elements in the same columns of the periodic table is 90°, but that in many cases the mutual repulsion of the attached group causes these angles to be larger. It is observed that the angles are greater, the smaller the central atom, and the larger the group attached to it, as would be expected. The attached groups can be assigned an effective radius which checks fairly closely with the effective radius estimated from interatomic or intergroup distances between atoms or groups attached to different (neighboring) molecules. Although this seems to be evidence in favor of the supposition that N, O, and F do have a normal valence angle of 90°, it may still be true that when the angle is forced to a larger value the wave functions approach the tetrahedral type.

TABLE 29.—BOND DISTANCES AND ANGLES IN UNSYMMETRICAL MOLECULES OF VALENCE TYPES (1) AND (2)

Molecule	Distance, Å	Angle	Authority
$HCCl_3$	(C—Cl) 1.77	112°	B
H_2CCl_2	(C—Cl) 1.77	112°	B
NH_3	1.01	109°	S
PF_3	1.52	104°	B; S
PCl_3	2.00	101°	B; S
AsF_3	1.72	....	B; S
$AsCl_3$	2.16	103°	B; S
OH_2	0.955	104°40′	S
OCl_2	1.68	115°	B

TABLE 29.—BOND DISTANCES AND ANGLES IN UNSYMMETRICAL MOLECULES OF VALENCE TYPES (1) AND (2).—(*Continued*)

Molecule	Distance, Å	Angle	Authority
$O(CH_3)_2$ (O bonded to two CH_3 groups)	1.42	111°	B
H_2S (S bonded to two H atoms)	1.35	92°20′	S

NOTE: There has been some controversy as to whether the bond angle in H_2S should be 92° or 85°. See Nielsen and Nielsen, *J. Chem. Phys.*, **5**, 277 (1937), and Cross and Crawford, *ibid.*, **5**, 370, 371 (1937).

2. Perhaps one of the simplest examples of the sp^3 tetrahedral bond-eigenfunction is that furnished by the carbon atom. Here the four wave functions of the L-shell are combined to form four tetrahedral bond-eigenfunctions. Each one of these interacts with the bond-eigenfunction of another atom, and an electron pair is shared between them. In the diamond crystal, for example, each carbon atom is surrounded by four carbon atoms tetrahedrally arranged about it, and each bond-eigenfunction interacts with a bond-eigenfunction of one of the surrounding atoms. In some compounds, two of the bond-eigenfunctions of a given atom interact with two from another atom, forming a double bond. Similarly, interaction between three eigenfunctions gives a triple bond. Another example of tetrahedral bond-eigenfunction is furnished by the atoms in the AlN crystal, which has an arrangement much like the diamond, but with each atom surrounded by four others of a different kind.

As has already been noted, those cases are particularly interesting in which less than four atoms are arranged about a central atom with the angles between the bonds approximately equal to the tetrahedral angle 109°28′, or smaller. If there are less than four atoms, the structure cannot be explained by the tendency of the molecule to take a symmetrical form, but must be due to the properties of the bond. It is also an indication that the binding is nonpolar rather than ionic, since ions of the same sign surrounding the central ion would tend to get as far away from each other as possible.

It is true that it has been shown that the effects considered in Secs. 12.9 and 12.10 are capable of producing the appearance of directed valence (see reference 2, page 200). We have seen that a cation induces a dipole moment in an anion. This dipole moment, if it is large enough, may exert sufficient force on another cation to change very appreciably its equilibrium position with respect to the first cation. A cation induces in the anion a dipole that is oriented in such a direction as to attract the cation or any other positive ion very close to it; therefore the induced dipole will have the effect of partially overcoming the repulsion between any positive ions that may be near the anion. The first attempt to consider the structure of the water molecule was made in this way, and although an incorrect valence angle was deduced, a definitely kinked structure was predicted (in fact, the angle was *too* acute). It may be that recently these calculations have not been given the consideration they deserve, but in any event, since the polarization of one atom by another is a step in the transition from polar to nonpolar binding, this cannot alter our conclusion that a kinked structure for such a molecule as the water molecule is evidence for a predominance of covalent character in the bonds.

In view of the special interest of these unsymmetrical molecules, a table of some whose structures have been determined, either by the study of molecule spectra or by electric diffraction, has been included. Data for molecules whose structures have been determined by the former method are taken from Sponer, "Molekülspektren," vol. I (indicated by S in Table 29), and data for those studied by means of electron diffraction are taken from the article by Brockway[1] (indicated by B in the table). The structure of the molecule is indicated, as well as may be in two dimensions, and the column headed "Distance" gives the distance between the central atom and the one next to it (if there is more than one distance, the particular distance is specified). The column headed "Angle" gives the angle between bonds. (If there is more than one kind of bond, it is the angle between two bonds of the kind for which the distance is given.) Other unsymmetrical molecules will be found listed in Brockway's article.

[1] Brockway, *Rev. Mod. Phys.*, **8**, 260 (1936).

A slight digression about the properties of double and triple bonds may not be amiss at this point. The reason that it is possible to have free relative rotation of the two methyl groups in ethane but not of the two methylene groups in ethylene may now be made clear. The tetrahedral bond-eigenfunction is symmetrical about an axis through the line of centers of the two carbon atoms in ethane. Relative orientation of the two methyl groups does not affect the relation between the two interacting bond-eigenfunctions. The only forces tending to prevent free rotation are presumably those between the hydrogen atoms

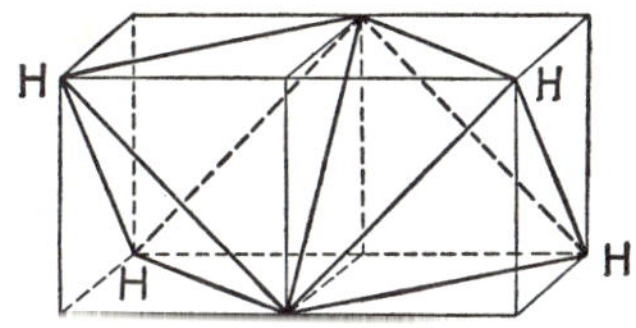

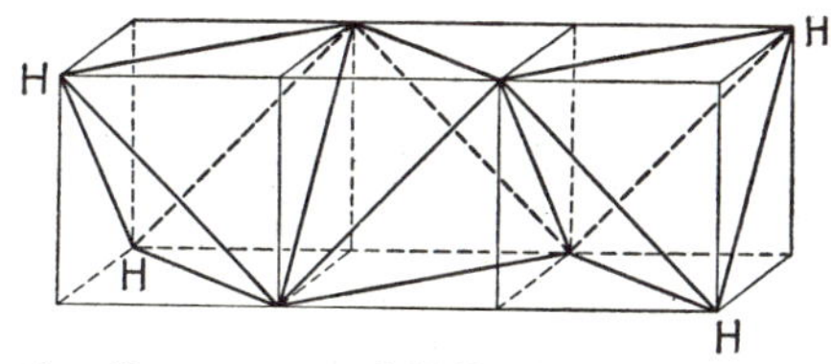

FIG. 53.—Illustrating the structure of $H_2C{=}CH_2$ and $H_2C{=}C{=}CH_2$, showing the carbon atoms as tetrahedra inscribed in cubes (see Appendix IV).

themselves. In the case of ethane, for example, these forces are probably great enough to prevent continuous rotation of the methyl groups,[1] but not sufficiently great to prevent frequent rotation, as the methyls must frequently acquire sufficient energy to turn around with respect to each other against these forces. This means that, as far as chemical phenomena are concerned, the rotation may be considered as free, since the rotation occurs much more rapidly than most chemical reactions. On the other hand, with two interacting bond-eigenfunctions in each group, as in the case of ethylene, for example, relative rotation of groups cannot take place without affecting the interaction between at

[1] HOWARD, *J. Chem. Phys.*, **5**, 451 (1937); KEMP and PITZER, *J. Chem. Phys.* **4**, 749 (1936); KISTIAKOWSKY and WILSON, *J. Am. Chem. Soc.*, **60**, 494 (1938).

least one pair of bond-eigenfunctions, and this requires a very great energy. Thus practically no rotation occurs around a double bond.

Elementary geometrical considerations show that the centers of gravity of all the atoms in an ethylene molecule are in a plane. This follows if we assume that the carbon atom has a tetrahedral structure and that a double bond involves sharing an edge, whereas with a single bond only a corner is shared. Similarly, in an arrangement like C═C═C, all the carbon atoms lie in a straight line. Assuming that a triple bond means that the tetrahedra involved share a face, we see that all the atoms in acetylene lie in a straight line. (See Figs. 53 and 54.)

When a central atom is attached to one atom by a single bond and to another atom by a double bond, the three atoms should not be in a straight line, as is seen from the geometry of the tetrahedron (Fig. 53). The angle would be expected to be somewhat larger than the normal tetrahedral angle of 109°28′. An example is furnished by SO_2 in which the two oxygens are attached to a central sulfur atom, one presumably by a single bond, the other by a double bond (else the octet of an atom is not complete). There is a possibility of resonance in this case, as was shown in Sec. 11.1, but all the possible electronic states would result in a kinked molecule. It is, therefore, not surprising that the molecule is actually found to be bent,[1] the O—S—O angle being in the neighborhood of 125°.

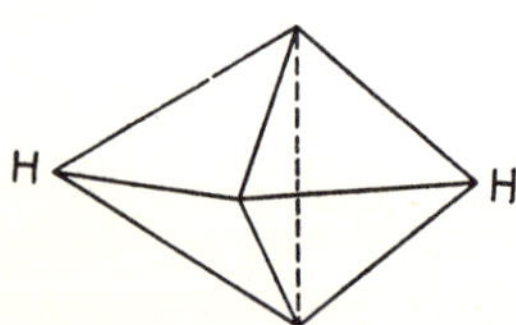

FIG. 54.—Illustrating the structure of HC≡CH, showing the carbon atoms as tetrahedra.

3. An example of the square structure is furnished by the nickel cyanide ion, $Ni(CN)_4^{--}$. Nickel in its normal state has eight 3*d*-electrons and two 4*s*-electrons. Now the nickel atom shares an electron with each of the cyanide radicals. A cyanide radical is a neutral radical with an odd electron, resembling a halogen atom. The two extra electrons, which give the negative charge to the complex ion, may be assigned to the nickel (for the purpose of discussion only). This gives an ion Ni^{--} with eight 3*d*-electrons and four electrons in the $n = 4$ shell, only the latter

[1] CROSS and BROCKWAY, *J. Chem. Phys.*, **3**, 821 (1935); GIAUQUE and STEVENSON, *J. Am. Chem. Soc.*, **60**, 1389 (1938).

of which are shared. The states occupied by the eight $3d$-electrons do not, in this case, share in the formation of bond-eigenfunctions, as the electrons are paired among themselves. The unoccupied $3d$-state is free to interact with the $4s$- and $4p$-states to form bond eigenfunctions, into which the four electrons, which have been roughly described as having $n = 4$, can go. With one $3d$-state and all the $4s$- and $4p$-states available, we have a situation corresponding to that required for the formation of a square.

It will be seen that nickel has the maximum number of electrons with which a square configuration for the divalent state would be expected. Nevertheless, copper, which has one more electron, has a square structure in such compounds as $CuCl_2(H_2O)_2$. In this case, the extra electron is evidently housed in the fifth eigenfunction, noted in the description of sp^2d bonds, and does not render the last $3d$-state unavailable. A single electron is apparently not so effective in rendering this state unavailable as a pair.[1]

4. An example of the tetrahedron of the more stable type is CrO_4^{--}. This ion may conveniently be considered to be formed from Cr^{++} and O^- ions. The Cr^{++} ion has four $3d$- and no $4s$- or $4p$-electrons. All these electrons are shared, and it is obvious, since all of the $3d$-wave functions are available, that the conditions for forming the tetrahedron of the more stable type are fulfilled.

[1] It should be noted that Cox, Wardlaw, Webster, and coworkers [*J. Chem. Soc.*, **1935**, 459, 731, 1475, **1936**, 775, **1937**, 1556; *Nature*, **139**, 71 (1937); *Sci. Prog.*, **32**, 463 (1938)], consider that the plane structure is a characteristic of a bivalent central element. Thus they claim from X-ray examination that not only nickel, copper, platinum, and palladium show the plane structure, but also bivalent silver, cobalt, manganese, tin, and lead. Such a structure for tin and lead involves *outer* d-states, or contravenes Pauling's criteria. Also the square structure for the cobalt compounds they examined (isomers of $CoCl_2Py_2$, where Py = pyridine) would be rather surprising since the magnetic criterion (Sec. 15.3) shows the binding to be ionic (or sp^3) [Barkworth and Sugden, *Nature* **139**, 374 (1937)], and $CoCl_4^{--}$ has a tetrahedral structure. Cox and Webster also state that $Pt(CH_3)_3Cl$ is tetrahedral, whereas platinic platinum with a coordination number of four might be expected to be square. If, in this compound, the bonds are of the sp^3d^3 tetrahedral type, one of the nonbinding eigenfunctions must contain a pair.

Two elements not in their bivalent states, namely, trivalent gold and manganese, are also said to show square coordination.

5. The ion $PtCl_6^{--}$ is an example of the octahedron. This may be considered to be made up of neutral Cl's and a Pt^{--} ion with ten $5d$- and two $6s$-electrons; six of these twelve electrons are not shared. These six unshared electrons occupy three d-states, leaving just enough d-states of the $n = 5$ shell available to satisfy the necessary conditions for an octahedron.

15.3. Magnetic Criterion for Type of Binding.—As remarked in the opening paragraph of this chapter, the mere fact that a compound or complex ion has one of the structures noted above is not a sufficient criterion to show that the binding is covalent if the structure is symmetrical. A kinked structure, such as exhibited by water, may be regarded as evidence that the bond is predominantly covalent, for if the bonding were ionic the two hydrogen ions would tend to get as far away from each other as possible. Similarly, the square structure of $Ni(CN)_4^{--}$ would appear to indicate that the binding within the ion is covalent.

Another property that is of some value in distinguishing between ionic and covalent binding in such compounds or ions is the magnetic moment.[1] As we have seen in Chap. VI, an electron in an atom contributes to the magnetic moment of the atom in two ways, by its orbital motion and by its spin. The contribution of an electron in a *molecule* depends largely on whether it is shared between atoms in the molecule or whether it may be regarded as belonging essentially to one atom only. An electron that is not shared, and is well shielded from the action of other atoms in the molecule, acts much like an electron in an atom. The inner, unshared electrons usually occur in closed shells or closed subshells; in such cases, the magnetic moments of the different electrons in the shell or subshell just cancel each other, so that the net value of the magnetic moment is zero. We shall in the following discussion be most interested in the valence electrons and in the uncompleted subshells of d-electrons, which are appreciably affected by surrounding atoms.

It was shown in Chap. VI that the orbital magnetic momentum of an electron is proportional to the angular momentum. This

[1] PAULING, *J. Am. Chem. Soc.*, **53**, 1391 (1931). The matter is discussed by Stoner, "Magnetism and Matter," Methuen & Co., Ltd., 1934, and Klemm, "Magnetochemie," Akademische Verlagsgesellschaft, Leipzig, 1936. For an account of magnetic theory, see also Van Vleck, "Electric and Magnetic Susceptibilities," Oxford University Press, 1932.

holds for molecules as well as atoms, and the fact that angular momentum and orbital magnetic moment parallel each other is of some value in our exposition, since it is often easier to discuss the angular momentum than it would be to discuss the magnetic moment directly. We shall see that such a discussion will show that in a *molecule* there is usually no orbital magnetic moment, and that the measured magnetic moment, being thus entirely due to spin, makes it possible to tell how many paired and how many unpaired electrons are present, thus throwing much light on the valence phenomena.

In previous discussions, much stress has been laid on the fact that there is conservation of the angular momentum of an electron moving in a field of force, such as is found in an atom, directed toward a fixed center. The proof of the conservation of angular momentum given in Appendix I depends directly upon the assumption that the field of force is of this type. In a molecule, however, the force on an electron is not necessarily at all times directly toward the center of gravity of the system, nor toward any single point. The angular momentum of an electron in a molecule will not remain constant. This does not mean, of course, that the angular momentum of the molecule as a whole is not conserved: the electrons merely exchange angular momentum with the nuclei. This exchange takes place continuously (though not necessarily uniformly). Since the nuclei are heavy bodies, so that a relatively small change in their velocity results in an appreciable change in angular momentum, the angular momentum of an electron is not only not conserved, but changes sign, and will on the average be as often in one direction as in the opposite. The average angular momentum with respect to the nuclei will be zero;[1] therefore, since the motion of the nuclei is very slow compared with that of the electrons, the average angular momentum of the electrons with respect to fixed space will also be practically zero.[2] This null value of the

[1] In the case of diatomic molecules, or molecules all of whose atoms are arranged in a straight line, the component of the angular momentum of electrons *along this line of centers* is conserved, because there will be no force on the electrons perpendicular to such a line. The component of the angular momentum of an electron along the line of centers need not, therefore, be zero.

[2] There will be some contribution to the magnetic moment from the motion of the nuclei. This, however, will be small compared, for example,

angular momentum implies a null value of the orbital magnetic moment. The orbital magnetic moment about such centers is said to be quenched.

The angular momentum due to spin, however, is not readily exchanged with the angular momentum of the nuclei. The electrons in a molecule, therefore, contribute a spin magnetic moment just as in atoms. Often the magnetic moment of a molecule is contributed almost entirely by spins.

This statement might be thought to be incorrect because of the inner electrons which are so well shielded from the fields of the surrounding atoms that they move in the essentially spherical symmetrical field of a single atom; the relative motion of the atoms disturbs them but little, and their angular momentum is conserved (at least for a time that is long compared with the time necessary for them to become orientated in a magnetic field.)[1] But, as we have seen, such inner electrons ordinarily occur in closed subshells, and the magnetic moments of the different electrons in such a subshell cancel, so that the net value is zero. The only case in which electrons in an uncompleted subshell are well shielded from external fields is that of the rare earths, in which the 4*f*-electrons are in such a subshell. In this case, the 4*f*-electrons contribute to the magnetic moment through both their orbits and their spins. However, such electrons have little or nothing to do with valence phenomena, and so the problem of their magnetic moments will not be considered in this book.

The uncompleted *d*-shells of the transition element do enter into valence phenomena and are sufficiently affected by outside fields so that in many cases only their spins contribute to the

---

with the magnetic moment of the electron in the lowest state of the hydrogen atom (*i.e.*, to a Bohr magneton), for the nuclei have the same charge as an electron, and move in paths of similar dimensions, but have a much smaller velocity. That the contribution due to the motion of the nuclei will be small may also be seen from Eq. (2) of Chap. VI, which holds for the nuclear motion if we substitute the nuclear rotational quantum number j for l and the reduced mass of the nuclei for m. Although j may be on the average larger than l, the reduced mass is so much larger than the mass of an electron that the magnetic moment arising from nuclear motion is always small.

[1] Since the process of measuring the magnetic moment consists in applying a magnetic field and observing the effects of orientation, such electrons would be expected to behave experimentally as though their orbital magnetic moments were unquenched.

magnetic moment, although sometimes there is some residual effect due to the orbital moment, which, however, we shall neglect in the discussion. It is in the case of these d-electrons that the measurement of magnetic moments can give information concerning the type of binding in a compound. It is a general rule, empirically observed to be practically always obeyed, that in the lowest energy level of an atom the unpaired electrons in the atom have their spins in the same direction. Thus if there are n unpaired electrons, these are lined up so the magnetic moments add, and if, in a magnetic field, the complex is lined up as a whole with or opposed to the field, the component of the angular momentum along the field will be $\pm\frac{\frac{1}{2}nh}{2\pi}$ since each electron contributes $\pm\frac{\frac{1}{2}h}{2\pi}$. Or the complex of electron spins may be oriented in such a way that the component of the angular momentum in the direction of the field differs by an integral number of units from $\pm\frac{\frac{1}{2}nh}{2\pi}$, thus giving the series of values $\frac{\frac{1}{2}nh}{2\pi}, \frac{(\frac{1}{2}n - 1)h}{2\pi}, \ldots, \frac{(-\frac{1}{2}n + 1)h}{2\pi}, -\frac{\frac{1}{2}nh}{2\pi}$, the corresponding components of magnetic moment being obtained simply by multiplying these values by[1] e/mc. It is usual to designate $n/2$ as S. It will be noted (see Sec. 5.4) that the values of the component of angular momentum are the same as would be contributed by an orbit with quantum number $l = S$, whose total angular momentum would be $\frac{\sqrt{S(S+1)}\,h}{2\pi}$. The values of the component of magnetic moment are twice those resulting from such an orbit, and such as would result from a magnetic moment of $2\sqrt{S(S+1)}(he/4\pi mc)$. As has been noted in Sec. 6.2, the fundamental moment $he/4\pi mc$ is known as a Bohr magneton, and the quantity $2\sqrt{S(S+1)}$, which is commonly used as a measure of the magnetic moment, is consequently the number of Bohr magnetons. It is seen that the number of Bohr magnetons is determined by, and is itself a measure of, the number of unpaired electrons.[2]

[1] See p. 79.

[2] This, however, may be considered as only the roughest sort of approximation. Very often the orbital motion and the spin affect each other in such a way that the magnetic moment is different from what would be

In view of the tendency of electrons that are not shared between atoms to be unpaired if possible (a tendency that presumably persists even if the atom is combined in a compound), measurement of the magnetic moment may throw light on the number of shared electrons. Since in ionic binding, electrons are not shared, whereas in covalent binding they are, something may thus be learned about the type of binding.

The way that the determination of the number of unpaired electrons can throw light upon the nature of the binding in a chemical compound may be illustrated by noting the difference between FeF_6^{---} and $Fe(CN)_6^{---}$. The experimental values[1] of $2\sqrt{S(S+1)}$ for these ions are ~ 5.9 and 2.6, indicating five and one unpaired electrons, respectively $[2\sqrt{\frac{5}{2}(\frac{5}{2}+1)} = 5.92$ and $2\sqrt{\frac{1}{2}(\frac{1}{2}+1)} = 1.73]$. If the binding in such an ion is ionic, *i.e.*, if the ion is composed of an Fe^{+++} and six X^- ions (where $X^- = F^-$ or CN^-), then the X^- ions have only closed shells and contribute no magnetic moment, and the Fe^{+++} has five unpaired d-electrons whose spins tend to add, giving $2\sqrt{\frac{5}{2}(\frac{5}{2}+1)}$ Bohr magnetons. On the other hand, if octahedral bond-eigenfunctions are formed, there will be six pairs of shared electrons, each X^- ion furnishing one pair (or one of the electrons of each pair may be thought of as coming from an Fe^{---} ion, and one from a neutral X). The sp^3d^2 octahedral bond-eigenfunctions, containing the shared electrons, use up two d-wave functions, so that only three d-places are left to hold the five unshared d-electrons of the Fe. Therefore, only one of these electrons may be unpaired. From this discussion, and from the respective values of the magnetic moments, it would be inferred that the FeF_6^{---} ion is itself composed of ions, whereas the $Fe(CN)_6^{---}$ ion, containing shared electron pairs, has predominantly covalent bonds.

---

expected from spin alone. Usually it is greater, and, roughly speaking, one may say that the angular momentum acts as though it were not fully quenched, but were contributing to the measured number of Bohr magnetons.

[1] Calculated from data of Cotton-Feytis, *Ann. Chim.*, **4**, 9 (1925), and Jackson, *Proc. Roy. Soc.* (*London*), **A140**, 695 (1933), respectively. The actual salts used were $(NH_4)_3FeF_6$ and $K_3Fe(CN)_6$; NH_4^+ and K^+ have no magnetic moments. The number of Bohr magnetons found for the cyanide is actually closer to the value for two unpaired electrons but probably agrees with the value for the expected one unpaired electron as closely as may be hoped for, considering the uncertainties involved.

Recent calculations of Van Vleck and Howard[1] appear to indicate a possibility that the conclusion that the forces in $Fe(CN)_6^{---}$ are predominantly covalent is not necessarily correct, for it is shown that the forces of surrounding ions might conceivably be sufficient to change the order of the energy levels of the Fe^{+++}. However, since this does not occur in the FeF_6 ion, in which, on account of the small size of the F^- ion, the ionic forces are especially great, it seems reasonable to suppose that the magnetic criterion for distinguishing between polar and nonpolar bindings is correct. In any event, such a change in its energy levels implies a distortion of the Fe^{+++} and hence may reasonably be assumed to be a step in the direction of covalency, although it is the wrong ion that is being distorted. There being a lack of other criteria applicable to compounds involving elements of the transition group, we shall use this one where possible. It will be obvious that if there are less than four *d*-electrons on the central element or ion this criterion cannot be applied.

In the foregoing discussion, we have assumed that the shared electrons in a covalent bond of the square or octahedral type necessarily involve the inner *d*-eigenfunctions, *i.e.*, the 3*d*-eigenfunctions in the case of elements of the iron group. It is possible on the other hand that there are some cases in which the *outer* *d*-eigenfunctions (4*d* in the iron group) are involved.[2] Such a bond would obviously be more polar in its properties than a bond involving the inner *d*-eigenfunctions, and its magnetic moment would be the same as for an ionic bond, and hence could not be distinguished from the latter by the magnetic criterion.[3] Since such a bond has properties intermediate between truly ionic and truly covalent bonds, we shall designate it as semi-covalent. It will be understood that this term will be applied only in cases in which use of the inner *d*-eigenfunctions is possible. If these are filled up, so that a bond involving the outer eigenfunctions is the only possible one, then that type will be designated as covalent, as usual.

We shall conclude this section with a few words on the experimental method for determining the number of Bohr magnetons, though a full discussion of this subject is beyond the scope of this

[1] Van Vleck and Howard, *J. Chem. Phys.*, **3**, 807, 813 (1935).

[2] Huggins, *J. Chem. Phys.*, **5**, 527 (1937).

[3] True of any case involving no inner *d*-electrons, *e.g.*, sp^3 bonds.

book. The measurement is based on observing what happens when the substance under consideration is introduced into a magnetic field. The action of a magnetic field upon a magnetizable body is somewhat similar to the action of an electric field on a dielectric. Magnetization of a body by a magnetic field occurs (1) on account of the effect of the magnetic field on the orbits of the electrons in the body (by changing their size or the speed of the electrons) and (2) owing to the lining up of molecules or ions having permanent magnetic moments (so-called "elementary magnets") under the influence of the field. On account of these effects, the mutual potential of any system of magnetic poles will be obtained from the expression for their potential in a vacuum by dividing by a factor μ called the magnetic permeability of the material. It is more usual to deal with the susceptibility χ, which is related to μ by the equation $\mu = 1 + 4\pi\chi$.

When elementary magnets are oriented in a field, they turn in such a way as to decrease the energy of the system. The effect, therefore, is to draw the magnetized body into the field. A body that is thus drawn into the field is said to be "paramagnetic." One method of measuring the susceptibility is based upon the measurement of the force with which such a body of known volume is drawn into a field of known intensity.

The first type of action of the magnetic field on a body, namely, its action on the electron orbits, usually results (unlike anything in the electrical case) in a force tending to push the body out of the field. This force is generally very small compared with, and hence more than counterbalanced by, the attraction of the field for the body, provided there are molecules present that do have permanent magnetic moments. If, however, there are no permanent magnetic moments present, there will generally be a slight residual force tending to push the magnetized body out of the field. Such a body is said to be "diamagnetic."

If a paramagnetic substance is diluted by mixing with it some nonmagnetic substance (*i.e.*, a substance with no permanent magnets), χ is approximately inversely proportional to the volume containing one mole of paramagnetic substance (at least after allowing for the diamagnetism of the diluent). Thus the susceptibility multiplied by the molal volume, which is called the "molal susceptibility" χ_M, is a constant characteristic of the substance.

In the electrical case, it is the *temperature coefficient* of the dielectric constant that is directly connected with the electric moment of the molecules. Similarly, in the magnetic case, the temperature coefficient of the molal susceptibility is the quantity directly connected with the molecular magnetic moment. The situation is somewhat complicated by the interaction of the elementary magnets with each other. However, it is generally found experimentally that the molal susceptibility χ_M is given as a function of the absolute temperature T by an equation of simple form, involving two constants, C_M and θ:

$$\chi_M = \frac{C_M}{T - \theta}$$

The number of Bohr magnetons per molecule of the substance is given, at least in simple cases, by $2.84 \sqrt{C_M}$. It is beyond the scope of this book to go into the theory that lies back of this expression, and the doubts and difficulties that may arise in attempting to apply it in special cases; for these matters, the student is referred to treatises on magnetism.[1] Often measurements are available only at room temperature, and in this case C_M is usually calculated on the assumption that θ is zero. θ is usually, but not always, small, and sometimes this may lead to considerable errors. Large values of θ may be connected with strong interaction between neighboring atoms of the substance being investigated, or other disturbing features, and if θ is large, the calculated value of the number of Bohr magnetons is less likely to be of significance.

There is a considerable amount of experimental work available on the magnetic susceptibilities of compounds of the transition metals.[2] In general, in these compounds, it is only the transition

[1] See footnote, p. 276.

[2] The data have been summarized by Pauling "The Nature of the Chemical Bond," pp. 107, 109, Cornell University Press, 1939; Pauling and Huggins, *Zeit. Krist.*, **87**, 215 (1934); in the treatises mentioned in the footnote, p. 276; by Gorter, *Arch. du Musée Teyler*, **7**, 183 (1932); by Cabrera, "Le Magnétisme," Rapports et Discussions du Sixième Conseil de Physique, pp. 104*ff*. 1932; in Landolt-Börnstein, "Tabellen," and in "International Critical Tables."

metals themselves that have any magnetic moment, and these measurements, therefore, may be considered to give the magnetic moments of ions of these metals; or, in compounds in which the metal is bound covalently, the measurements give the moments of ions or groups containing the atom of the transition metal. Measurements have been made on simple compounds, on complex compounds, and on aqueous solutions. Some of the results on simple compounds are summarized below:

Oxides of the First Row of Transition Elements.—The temperature dependence of the susceptibilities indicates large negative values of θ for some oxides of chromium and manganese. The resulting values of C_M give values of the Bohr magneton number agreeing fairly well with what is to be expected for ionic compounds. With other oxides susceptibilities that have been measured only at room temperature are smaller than calculated for the given ion (knowing the number of Bohr magnetons expected for that ion) assuming $\theta = 0$; but since θ is probably not near zero this is not surprising, and it seems fair to conclude that the results are at least consistent with the assumption that oxides are ionic.

Sulfides of the First Row of Transition Elements.—Data on vanadium, chromium, and manganese compounds inconclusive. FeS_2 is probably covalent.[1]

Halides and sulfates of the first row of transition elements are almost certainly ionic, or, at least, semicovalent.

Exercises

1. Why would the PCl_3 molecule be expected to have either p or sp^3 bond-eigenfunctions?

2. Discuss the electronic structure of $Cu(CN)_4^{---}$. Do you expect it to have a tetrahedral or square structure? For the experimental result, see Table 31.

3. Complex compounds of platinic platinum generally have an octahedral structure, whereas platinous compounds generally have the square structure. Assuming that they are covalent, give an explanation in terms of their electronic structure.

4. Assuming covalent binding and assuming ionic binding, calculate the number of Bohr magnetons for the elements V to Cu in the first transition series in their common valence states. Assuming $\theta = 0$, calculate the susceptibility at 300°K.

[1] According to Pauling and Huggins, ref. 2, p. 283.

5. Write a Lewis electronic structure for nitrous oxide, NNO. Assuming N and O to be tetrahedra, predict the form of the molecule. Do the same for thiocyanate ion, SCN^-, hydrogen peroxide, HOOH, and nitrite ion, ONO^-. Name a number of other ions you would expect to have the same structure as NNO.

6. Suggest two possible electronic structures for the molecule ONCl. Make a prediction as to the shape of this molecule [see Ketelaar and Palmer, *J. Am. Chem. Soc.*, **59**, 2629 (1937)].

CHAPTER XVI

COMPLEX COMPOUNDS AND COMPLEX CRYSTALS, INCLUDING ATOMIC CRYSTALS

Although there are simple atomic crystals, such as diamond and AlN, many of the properties of covalent bonds, especially those discussed in the last chapter, are better illustrated by the complex compounds and complex ions of various types, and this interesting branch of chemistry is much illuminated by the general principles developed. Further, the questions that arise in connection with the transition to other types of binding are by no means confined to atomic crystals of the simple type, but present much the same problem in the case of complex compounds. As we have seen in the preceding pages, there are complex ions of various types, which generally represent quite stable configurations. In some respects, the study of these ions yields more information than the study of the simpler atomic crystals, and it seems convenient to begin this chapter with a consideration of ions rather than crystals.

In the common type of ion, there is a central atom surrounded by a number of more negative atoms (or sometimes groups or molecules) in a more or less regular configuration. Often these negative atoms are all alike; occasionally more than one kind are present. Sometimes the central atom is the more negative, though this type is less common. Sometimes there is no single central atom and the geometrical arrangement is more complicated than indicated above. Complex ions are usually stable in aqueous solutions, and generally enter as units in chemical reactions. If this were not true of at least some chemical reactions with any given ion, it would scarcely be recognized as a distinct entity, though a compact grouping in a crystal is often used to define an ion.

16.1. Methods of Investigation.—In our consideration of a complex ion, we are interested in knowing something of its composition and its structure. There are many methods available that throw light on the composition of complex ions.[1] Most elementary, of course, is the method of chemical analysis. The fact that a certain combination of atoms occurs again and again in a series of different compounds leads one to infer that these atoms are bound together in a kind of complex, though it does not, of course (at any rate without other considerations being, at least, implied), lead to a determination of the charge on the ion, which is one of its important properties. Again, a salt may be soluble in water, and it may react with other substances to form various precipitates. If all these precipitates contain certain elements (which were present in the same proportions in the original salt) in a certain combination, it seems likely that this group exists also as an entity in aqueous solution.

It is often possible to infer the composition of a complex ion by its chemical behavior and the behavior of substances from which it is formed in aqueous solution. This may perhaps best be illustrated by an example. Suppose we have a solution that is known to contain one mole of silver ion and suppose it has previously been ascertained that cyanide ion has a definite composition. Now we add one mole of potassium cyanide to the silver solution and obtain a precipitate containing 1 gram atom of silver and 1 gram equivalent of cyanide. If one more mole of potassium cyanide is added, all the precipitate goes into solution. We therefore infer that there is now present in solution a complex ion having the formula $Ag(CN)_2^-$. (If the charges on silver and cyanide ions are known, we infer therefrom the charge on the complex ion, and the result must be correct if no other ions are involved.) The fact that many sets of experimental observation can be explained in similar fashion, and the further fact that observations on related substances give mutually consistent results, lend some confidence in the conclusions reached in this manner.

[1] For more detailed accounts of these methods see Jacques, "Complex Ions," Longmans, Green & Company, 1914; Schwarz, "Inorganic Complex Compounds," English trans., John Wiley & Sons, Inc., 1923; Werner, "New Ideas on Inorganic Chemistry," English trans., Longmans, Green & Company, 1911; and Weinland, "Einführung in die Chemie der Komplexverbindungen," Ferdinand Enke Verlag, Stuttgart, 1924. The book by Werner is a classic by the man principally responsible for the development of this field of chemistry; the work by Weinland is a comprehensive account of the experimental results. See also footnote 2, p. 297.

It must be borne in mind, however, that cases will arise in which the foregoing method of investigation will not be applicable. It will be seen that its success in the preceding case depends upon a particular set of circumstances. Solid AgCN reacts with CN^- to form $Ag(CN)_2^-$, and it requires only an exceedingly small concentration of CN^- to bring the precipitate into solution. However, solid AgCN is also in equilibrium with Ag^+ and CN^- As long as there is an excess of Ag^+ present, it keeps the concentration of CN^- down sufficiently so that the AgCN is not brought into solution as $Ag(CN)_2^-$. If these equilibria did not bear the right relation to each other, the observations would not have the simplicity observed. However, it might be possible to measure the concentration of the various substances in solution, either directly or indirectly, and so obtain the values for the equilibrium constants involved. This method of investigation of the properties of complex ions is illustrated by Exercise 2 at the end of the chapter.

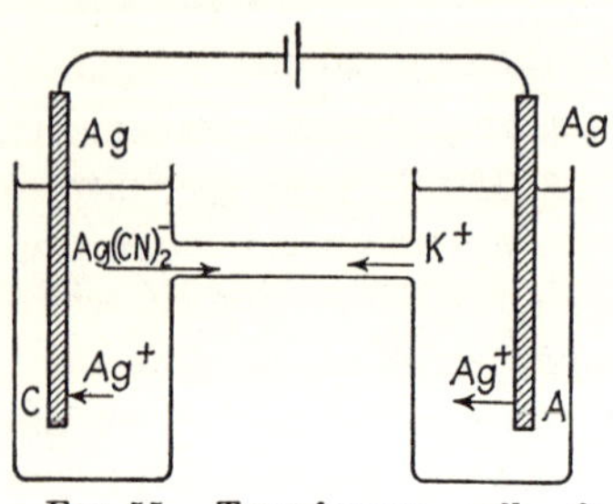

FIG. 55.—Transference cell, for the study of the complex ion, $Ag(CN)_2^-$. The ionic carriers of electricity are $Ag(CN)_2^-$ and K^+. Though $Ag(CN)_2^-$ moves toward the anode, silver is deposited on the cathode, the mechanism undoubtedly involving $Ag(CN)_2^- \rightleftarrows Ag^+ + 2CN^-$, though Ag^+ is present in very small concentration.

Observations of the type described above may be confirmed by transference experiments. Consider, for example, the experimental setup shown in Fig. 55 in which we have a silver anode at A and cathode at C immersed in a solution of the complex salt $KAg(CN)_2$ (obtained, say, by filtering the AgCN precipitate, washing and dissolving in an equivalent amount of KCN). Suppose one equivalent of electricity to be passed through the cell. One gram atom of silver will then be found to be deposited on the cathode, and 1 gram atom is removed from the anode. The current will have been carried by the ions, part of it by K^+, part by $Ag(CN)_2^-$. Suppose that the fraction of the current carried by K^+ is x. Then after the passage of the current, x equivalents of K^+ will have been transferred from the region near A to the region near C and $1 - x$ equivalents of $Ag(CN)_2^-$ will have been transferred from the region near C to that near A.

There would then be one equivalent more of K^+ than of $Ag(CN)_2^-$ near C, provided nothing else had happened. But, since one equivalent of $Ag(CN)_2^-$ will have been deposited on the cathode as silver, two equivalents of CN^- will have been released, and the net effect will be that the number of equivalents of K^+ will exceed the number of equivalents of $Ag(CN)_2^-$ by two, and two equivalents of $(CN)^-$ will be present. In the anode region, the amounts of K^+ and $Ag(CN)_2^-$ will be equal, but two moles of AgCN will precipitate. These predictions can be verified by analysis of the anode and cathode portions of the cell for silver, cyanide radical, and potassium after the electricity has passed through. It will be observed that the results predicted depend upon the assumptions made about the ions. It is, namely, assumed that the silver and cyanide move together so that silver goes to the anode (Ag^+ would go to the cathode) and that one atom of silver is tranferred for each two cyanide radicals. Furthermore, it is assumed that each $Ag(CN)_2^-$ ion has a single negative charge (this, of course, is not independent of the assumption that the salt of the formula $KAg(CN)_2$ gives potassium ions and complex ions in solution). The verification of these assumptions may be considered to be a verification of the formula of the complex ion. The transference method is a very powerful one, and has been much used in the study of complex ions.

Another method that has been used in the study of complex ions is the measurement of the molal electrical conductivity of the salt. In the case of simple salts and complex salts with ions of known composition, this quantity has been found to depend chiefly on the valence type of the salt, being greater the higher the charge on the ions. Conversely, the conductivity can be used to determine the valence type of the salt, which may be sufficient to determine the formula of the complex ion or, at least, throw light on it.[1] Other methods, such as the freezing point lowering, can also be used to find the number of ions present, and hence to make possible inferences concerning the formula of the complex ion.

In the case of complex ions in which various groups of different character are attached to the central atom, the isomerism, particularly stereoisomerism, can give much information con-

[1] For an example, see Schwarz, *op. cit.*, p. 32.

cerning the structure of the ion, as already remarked in Sec. 15.1. This will be discussed later in the chapter.

Finally, as already seen in preceding chapters, the use of X rays can give much information concerning the arrangement of atoms in complex ions as they exist in crystals.

16.2. The Nature and Properties of Complex Ions.—A glance at Weinland's "Chemie der Komplexverbindungen" will indicate the vast variety of complex ions that may be formed. The best known of these are, of course, the anions in which a central atom is surrounded by oxygen atoms, the sulfate ion being a typical example. A few cations containing oxygen, such as VO^{++}, are also known. There are numerous anions and cations that can be thought of as being formed by the combination of a positive ion, generally of an element in or near the transition region of the periodic table, either with negative ions or with such neutral molecules as water or ammonia. The negative ions or neutral molecules thus combined are invariably groups that are themselves capable of separate existence, and the variety of ions that may occur in such combinations is very great indeed, among them being F^-, Cl^-, Br^-, CN^-, CNS^-, NO_2^-, SH^-, SO_3^{--}, OH^-, a variety of organic ions, etc. The hydrates, which will be studied in Chap. XIX, may be classed among the complex ions. It is interesting to note that certain organic substances having two free ends, like oxalate ion and ethylene diamine, can occupy two coordination places[1] in a complex ion. CO_3^{--} and SO_3^{--} are also included among these.

Since most of our chemistry is the chemistry of aqueous solutions, the hydrates play a special role among the complex ions. Water in crystals may occur either as a "filler" occupying interstices in the crystal or it may be a coordination molecule more or less firmly bound to a central positive ion and thus forming part of a complex ion. The actual strength of the binding is not always a good criterion as to the type of binding, for some filler molecules are more tightly bound than coordination mole-

[1] Following a common usage, we shall speak of a molecule that is attached to a central ion to form part of a complex ion as a "coordination molecule," and say that it occupies a place in the "coordination sphere." The number of molecules or groups surrounding a central atom is the coordination number of that central atom. (If a group occupies two coordination places it is counted twice.) For the definition of coordination number when no complex ion or molecule is involved see Sec. 14.1.

cules in other crystals.[1] However, when a water molecule in the coordination sphere of a complex ion is driven out, the properties of the ion change in a characteristic way. This may best be illustrated by an example.[2] $[Co(NH_3)_5H_2O]Cl_3$ is a brick-red crystalline powder which is soluble in water, giving a solution that has none of the characteristic properties of a solution containing ammonia. This indicates that the ammonia is not free. On the other hand, the chlorine may all be precipitated by silver ion without apparently affecting the ammonia. The conclusion is that the ammonia is bound in the cation whereas the chlorine is ionized. It is then inferred that the water molecule is also in the cation, since experience shows that trivalent cobalt has a coordination number of six. This, however, is mere inference, but it can be verified, or at least the inference can be strengthened, by heating the crystal to around 100°C. at which temperature the water is driven out. It is then found that only two of the three chlorines can be precipitated by silver ion, the third having presumably taken the place of the water in the coordination sphere. By heating this substance in aqueous solution, the chlorine is again driven out by a water molecule and the original substance recovered.

There are cases in which isomers are known, the difference in the compounds consisting entirely in the mode of linkage of the various constituents. For example, there are three salts with the empirical formula $CrCl_3 \cdot 6H_2O$, one of them being violet, and two green.[3] The violet salt gives a solution in which all three chlorines are ionized, as indicated by its behavior with silver ion. In the two green forms, two chlorines and one chlorine, respectively, are ionized. The salts, therefore, presumably have the formulas

$$[Cr(H_2O)_6]Cl_3, \quad [Cr(H_2O)_5Cl]Cl_2 \cdot H_2O$$

and

$$[Cr(H_2O)_4\, Cl_2]Cl \cdot 2H_2O.$$

The violet salt is the most stable, and in aqueous solutions the other salts change to the violet form in a short time.

[1] See the discussion in Sec. 19.9.

[2] See Weinland, *op. cit.*, pp. 15–17, 37

[3] See Werner and Gubser, *Ber. deut. chem. Ges.*, **34,** 1579 (1901); Bjerrum, *ibid.*, **39,** 1599 (1906); *Zeits. physik. Chem.*, **59,** 596 (1907).

16.3. The binding force within complex ions may be due to the electrostatic attraction of the central ion for the surrounding ions or dipoles, or it may be of the nonpolar type, or it may be any combination of the two. In most ions it appears from stoichiometrical considerations that at least part of the binding force must be ionic. Thus we might imagine an ion like SO_4^{--} to be made up of S^{6+} ions and O^{--} ions, in which case the binding force would be purely electrostatic, or of S^{++} and O^{-} ions, in which case there would be the binding force due to four shared pairs of electrons, but still some electrostatic force due to the attraction of the positive and negative ions, even though these are not so highly charged. Another possibility is that the sulfur is practically neutral and is combined with two neutral and two singly charged oxygens. (This would not mean that there would be any observable difference between the oxygens, as the charge would undoubtedly move around from one oxygen to the other, so each would have an average charge of one half.) The true state of SO_4^{--} probably lies between the extremes mentioned, and indeed, it seems reasonable to suppose that the combination of S^{++} and four O^{-} is a fairly good approximation.[1] (See also Sec. 16.14.)

In special instances, the arrangement of the atoms in an ion can give information as to the nature of the binding. In Chap. XV, we discussed the arrangements of the groups within the complex resulting from nonpolar binding involving various kinds of wave functions. The expectation for purely electrostatic binding is that the arrangement should be the most symmetrical possible with the given number of groups. Unfortunately, this coincides in most cases with the grouping predicted by the theory of the covalent forces. Only case (3) of those considered in Chap. XV, the arrangement of four groups about a central atom in the form of a square, is different from what would be expected of polar binding. However, as seen in Chap. XV, an unsymmetrical arrangement can often persist when less than four atoms are attached to the central atom. This happens, for example, in the case of the sulfite ion, which has the

[1] This assumption seems very plausible and is supported to some extent by the considerations of Sec. 19.10. It at first appeared that the analysis of the spectra of such ions yielded evidence for it [Urey and Bradley, *Phys. Rev.*, **38**, 1969 (1931)], but more recently [Rosenthal, *Phys. Rev.*, **49**, 535 (1936)] objections have been raised to this interpretation.

electronic formula

$$\begin{array}{l} \;\;\,\cdot\cdot \\ \;\;:\!\mathrm{O}\!: \\ :\!\mathrm{S}\!:\!\mathrm{O}\!: \\ \;\;:\!\mathrm{O}\!: \\ \;\;\,\cdot\cdot \end{array}$$

The oxygens are actually at three of the corners of a nearly regular tetrahedron about the sulfur.[1] The only difference between this ion and an ion in which four oxygens are attached is that in the present case one of the pairs of electrons occupying the bond-eigenfunctions of the sulfur is not shared with another atom. Other ions of this type are ClO_3^- and BrO_3^-. Such an unsymmetrical arrangement of the atoms about the central atom is a clear indication of the covalent character of the bonds. Since they differ only by having one more atom and one more shared electron pair, it is natural to infer that SO_4^{--} and ClO_4^- ions have covalent binding also.

Another criterion, which can give some information about the character of the bonds in an ion, has also been discussed in Sec. 15.3. This is the magnetic moment of the ion. This criterion is applicable only to elements of the transition series, and then only if the number of *d*-electrons in the uncompleted shell lies within certain limits. It is, unfortunately, not applicable to such ions as CrO_4^{--} and MnO_4^-, which are possibly examples of the sp^3d^3 tetrahedron, since the same magnetic moment would be predicted for ionic as for nonpolar binding. The magnetic criterion is usually applicable to the elements in the iron and platinum groups, and a considerable amount of data exists. An attempt has been made to summarize this in the accompanying table. In order to indicate the valence of the metal in any given complex ion, the symbol for the metal which is given at the top is the symbol that would be written if the binding were ionic. The nature of the various types of binding is indicated either by the symbol *i*, meaning ionic (or semicovalent), or, in the case of covalent binding, by giving the type of binding. This will indicate the kind of ion formed, its coordination number, etc.

It will be understood that in some cases a certain amount of judgment has been used in compiling this list with respect to the actual constitution to be assumed for a compound with a given formula. In a similar manner, measurements of susceptibilities of ions in solution have been assumed to be measurements of the

[1] Zachariasen, *J. Am. Chem. Soc.*, **53**, 2123 (1931).

TABLE 30.—TYPES OF BINDING IN COMPLEX IONS FROM THE MAGNETIC CRITERION

	Cr^{++}	Mn^{++}	Mn^{+++}	Fe^{++}	Fe^{+++}	Co^{++}	Co^{+++}	Ni^{++}	Mo^{4+}	W^{4+}	Ir^{+++}	Pt^{++}	Pt^{4+}
Hydrates	*i*	*i*	*i*	*i*	*i*	*i*	....	*i*					
Fluorides	....	....	.....	....	*i*	....	*i*						
Ammonia complexes	....	....	.....	*i*	....	*i*	d^2sp^3	*i*	....	....	....	dsp^2	d^2sp^3
Amminochlorides	....	....	.....	....	....	....	d^2sp^3	....	....	....	....	....	d^2sp^3
Amminonitrites	....	....	.....	....	....	....	d^2sp^3	....	....	....	d^2sp^3		
Chlorides	....	....	.....	....	....	....	....	....	....	....	....	dsp^2	d^2sp^3
Chlorohydrates	....	....	.....	....	*i*								
Cyanides	d^2sp^3	d^2sp^3	d^2sp^3	d^2sp^3	d^2sp^3	....	d^2sp^3	dsp^2	d^4sp^3	d^4sp^3			
Nitrites	....	....	.....	....	....	d^2sp^3	d^2sp^3						

See footnote 2, p. 283, for sources of data. Molybdenum and tungsten have a coordination number of 8 in the cyanide complexes, and the bond type is presumably that given.

susceptibilities of the hydrates, unless definite evidence to the contrary is available. In general, no attempt has been made to go beyond the compilations of magnetic data assembled by various investigators on the subject, and no exhaustive analysis of the crystal structure and magnetic data with respect to the light they might throw upon each other has been made. It is believed, however, that in its general aspect the table cannot be greatly in error; at the same time it seems obvious that further work on the subject, both with respect to the gathering of new data and the analysis of the old, would be highly desirable.

In using Table 30 it should be borne in mind that the magnetic criterion says a bond is either covalent or ionic and gives no information about the transition between the two extreme types. It is, therefore, quite possible that a different criterion would give a different result in some cases.

16.4. Some Energy Relations among Iron and Cobalt Complexes.—There are not many easily correlated data for estimating the relative stability of various complex ions. Data do happen to be at hand, however, that make it possible to find the effect of forming the cyanide complex on the oxidizability of ferrous iron to ferric and to compare it with the same effect in the case of cobalt. The matter will be considered in some detail as it illustrates the utility of the rules set forth in Chap. XV, and at the same time may be correlated with the magnetic data.

The effect of the formation of the cyanide may be compactly expressed in the following manner. The energy (actually the free energy—see Appendix II) given out by the reaction (in aqueous solution)

$$Fe^{+++} + Fe(CN)_6^{4-} \rightarrow Fe^{++} + Fe(CN)_6^{---}$$

is 9.5 kg.-cal., and that given out by the analogous reaction

$$Co^{+++} + Co(CN)_6^{4-} \rightarrow Co^{++} + Co(CN)_6^{---}$$

is 60 kg.-cal.[1] The latter reaction, therefore, has a much greater tendency to go as written. The great difference between the reactions is explained on the supposition that $Co(CN)_6^{4-}$ is very unstable, which is evidenced by the fact that it decomposes water

[1] From the tables of electrode potentials in Latimer, "Oxidation States of the Elements and their Potentials in Aqueous Solutions," pp. 201, 294–298, Prentice-Hall, Inc., 1938.

with the evolution of hydrogen. This instability is expected from its electron structure.[1] Let us suppose that $Co(CN)_6^{4-}$ is built up from neutral CN's and a Co^{4-} ion, which has seven $3d$-electrons, two $4s$-electrons, and four $4p$-electrons, and that the binding within the ion is covalent. The seven $3d$-electrons are not shared, and in order that two d-states may be available for combination into octahedral sp^3d^2 bond-eigenfunctions it is necessary for one of these d-electrons to be promoted to a $5s$-state. This would naturally cause the ion to have a high energy and hence be unstable. It is very likely that this circumstance causes the binding in the $Co(CN)_6^{4-}$ to be ionic (or at least semicovalent). In general, CN^- tends to form covalent complexes, so the fact that the complex is divalent and is forced into the relatively unstable ionic form makes the complex cobalt easy to oxidize. In the case of the iron complex, the magnetic measurements show that both the ferro- and ferricyanides have covalent binding.[2]

The instability of complex cobaltous compounds is a general phenomenon, though quantitative data are not available in most cases. In the case of the ammonia complexes, however, it is known that the cobaltous compound $[Co(NH_3)_6]Cl_2$ has a much greater ammonia pressure than the cobaltic[3] $[Co(NH_3)_6]Cl_3$. Furthermore, magnetic data are available, and these indicate that the binding in $[Co(NH_3)_6]Cl_2$ is ionic, but that it is covalent in $[Co(NH_3)_6]Cl_3$ (see Table 30). In the case of the iron compounds, the magnetic data indicate that $[Fe(NH_3)_6]Cl_2$ is ionic,

[1] HOARD, quoted by Pauling, *J. Am. Chem. Soc.*, **54**, 994 (1932).

[2] It must be remembered that in aqueous solution Co^{++}, Co^{+++}, Fe^{++}, and Fe^{+++} probably represent hydrate complexes (see Chap. XIX). The relative stability of the simple bivalent and trivalent aqueous ions should, however, not be affected by factors similar to those which influence the stability of the cyanides, since in the hydrates ionic forces undoubtedly predominate.

[3] MELLOR, "Comprehensive Treatise on Inorganic and Theoretical Chemistry," vol. 14, pp. 631, 654, Longmans, Green & Company, 1935. The free energy made available in the aqueous reaction

$$Co^{+++} + Co(NH_3)_6^{++} \rightarrow Co^{++} + Co(NH_3)_6^{+++}$$

has been given by Latimer (Ref. 1, p. 295) as 40 kg.-cal. $Co(NH_3)_6^{++}$ is therefore not quite so unstable as the cobaltous cyanide, probably because of the somewhat greater tendency of ammonia complexes to have ionic bonds.

but unfortunately the data appear not to be available for the ferric compound. The indications seem to be that iron has a greater tendency than cobalt to form ionic compounds. $Fe(NH_3)_6^{++}$ has a marked tendency to oxidize but this seems to be little if any greater than that of Fe^{++}, and the solid chloride $[Fe(NH_3)_6]Cl_3$ has a greater ammonia pressure[1] than the solid $[Fe(NH_3)_6]Cl_2$. As far as these observations go, they are what would be expected from the results with the cyanides. More experimental work on the ammonia complexes, which would show to what extent their properties parallel those of the cyanides, would be of considerable interest.

16.5. Stereoisomerism.—Much can be learned about the structure of the ions or compounds formed by a given atom by studying the various types of isomerism exhibited by these compounds.[2] Two types of isomerism may be distinguished, geometrical isomerism and optical isomerism. The first type of isomerism may be illustrated by dichloroethylene, which occurs in the so-called cis- and trans-forms:

```
Cl       H          H       H
  \     /             \     /
   C=C                 C=C
  /     \             /     \
H       Cl          Cl       Cl
 Trans-form          Cis-form
```

In this particular instance, the isomerism is possible because there is no rotation about a double bond. These two compounds have different physical properties and, in particular, the trans-form has no resultant electric moment, because it has a plane of symmetry, whereas the cis-form has an electric moment.[3] This

[1] ABEGG, "Handbuch der anorganischen Chemie," IV Band, 3 Abt., 2 Teil, pp. B91, B390, S. Hirzel Verlag, Leipzig, 1930.

[2] A much more extended account of this material than can be given here will be found in Freudenberg's compendium, "Stereochemie," Franz Deuticke, Leipzig and Vienna, 1933, which has been largely used as a source book. Especially useful for our purposes are articles by Meisenheimer and Theilacker, Ziegler, and Pfeiffer. Another useful work is Wittig, "Stereochemie," Akademische Verlagsgesellschaft, Leipzig, 1930. For a brief account including some more recent work, see Bailar, *Chem. Rev.*, **19,** 67 (1936). See these reviews for references to the literature for the work discussed in this section.

[3] See DEBYE, "Polar Molecules," p. 53, Reinhold Publishing Corporation, 1929.

property thus serves to tell which is which, and the resolution has been confirmed by X-ray and electron-diffraction experiments.[1] Optical isomerism, as is well known, is exhibited by a carbon atom to which four different groups are attached, being due, in this case, to the tetrahedral structure of the carbon atom. When four different groups are attached, the compound has no plane of symmetry and cannot be made to coincide with its mirror image. The compound, therefore, occurs in right- and left-handed forms, which have in general identical physical properties, but which react in different ways toward anything that has similar right- or left-handed properties. For example, the plane of polarized light will be rotated to either the right or left, depending upon which form the light passes through, whence the optical activity. If allowed to combine with another optically active substance of definitely determined right- or left-handed configuration, the two forms of a substance which it may be desired to investigate yield compounds which differ more fundamentally than an object and its mirror image and which can be separated from a mixture by taking advantage of their difference in physical properties.[2] After separation, the original substance (or rather, in each case, an optically active isomer of it) can often be regenerated. This method of separating optical isomers has been successfully applied in many cases.[3]

A complex ion in which the central atom has a coordination number of four and in which the four substituents are arranged in a plane in the form of a square can exhibit geometrical isomerism similar to that of dichloroethylene. An example of this sort of isomerism is furnished by the compound[4] $Pt(NH_3)_2Cl_2$

[1] See Brockway, *Rev. Mod. Phys.*, **8**, 260*ff.* (1936).

[2] It is readily seen that the combination "right-left" bears an essentially different relation to the combination "right-right" than "right" bears to "left." On the other hand "right-left" bears a relation to "left-right," which is similar to the relation "right" bears to "left." The first relationship is the one of which advantage is taken in the separation of optical isotopes.

[3] It has been applied to many optically active inorganic compounds, such as are discussed in the succeeding paragraphs, as well as to organic compounds. See, *e.g.*, Werner, Ref. 1, p. 287.

[4] This classical case of stereoisomerism has recently been questioned, but in view of the behavior of other platinum compounds it is probably a valid case. For discussion and references see Bailar, Ref. 2, p. 297. For a dis-

(which is, of course, a neutral substance and not really an ion). The structures of the two forms are as follows:

```
NH3      NH3     NH3      Cl
   \    /           \    /
    Pt                Pt
   /    \           /    \
Cl        Cl     Cl        NH3
  Cis-form         Trans-form
```

This type of isomerism, in which the isomers have different physical properties, does not occur if the compound has a tetrahedral structure, and hence can serve to distinguish between the two types. However, the same isomerism would occur with a square in which the central atom was not in the plane of the other groups, and so it does not serve to eliminate this possibility. In cases in which X-ray examinations have been made of this type of compound (*e.g.*, $K_2[PtCl_4]$ and $K_2[PdCl_4]$), the central atom has been found to lie in the plane of the others (Table 31, p. 308). If the central atom were not in the plane, compounds of the type $MeABC_2$ (cis) or MeABCD (where Me stands for the central atom and A, B, C, and D for different atoms or groups in the coordination sphere) should show optical activity.[1] Such optical activity has not been observed, but it does not appear to have been particularly looked for. There is thus not very much evidence on this point, but it seems reasonably safe to assume that when the square configuration occurs the central atom lies in the plane.

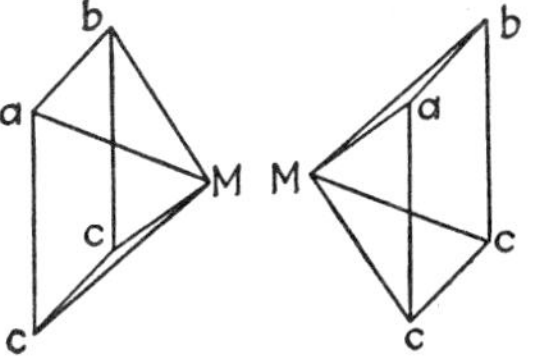

Fig. 56.—Optical isomers of square coordination compound with central atom not in plane.

In discussing the tetrahedral and square configurations, it will be realized that we have considered only the most symmetrical possible arrangements. If the arrangement were not symmetrical (if, for example, the four substituents in the "square" configuration were not equally distant from the central atom), the number

cussion of the evidence furnished by dipole moments see Jensen, *Zeits. anorg. allgem. Chem.*, **225**, 97 (1935).

[1] For a more detailed discussion of possibilities of stereoisomerism see Pfeiffer in Freudenberg's "Stereochemie," pp. 1200*ff*. The reader should also note the discussion, pages 304*f*, below, of the optical isomerism of diamine complexes.

of isomers would be greatly increased; such large numbers of isomers have not been found.[1] It will be understood that this does not exclude the possibility of distortions from these most symmetrical forms, provided that the distortion is conditioned by the substituents and not by the central atom. Thus, two optical isomers could be slightly distorted out of the tetrahedral shape owing to attractions and repulsions between the substit-

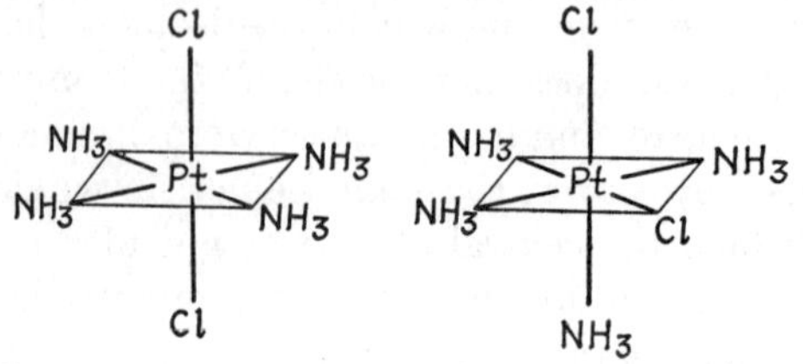

FIG. 57.—Cis-trans isomerism of an octahedral coordination compound.

uents, but the distortions of the right- and left-handed forms would be so related that they would still be mirror images, and they could not differ more fundamentally. Similarly, the cis- and trans-forms might be distorted in different but definite ways, depending upon the relative positions of the substituents, and this would not give rise to any new possibility, since it depends on the substituents and not on any peculiarity of the valence bonds of the central atoms.

In the octahedral type of complex ion, geometrical isomerism can also occur.[1] For example, there exist two ions with the formula $[PtCl_2(NH_3)_4]^{++}$ with slightly different properties. The possibility of geometrical isomerism is readily seen, for the two chlorines can be either in adjacent positions in the octahedron (cis-compound) or at opposite vertices (trans-compound), as will be clear from Fig. 57. The fact that in compounds of this type more than two isomers have never been found is evidence against a plane configuration, similar to benzene, which would allow the possibility of three isomers.

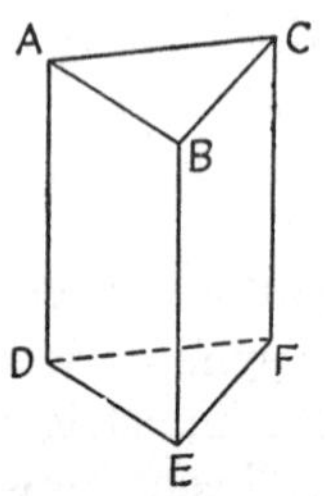

FIG. 58.—Trigonal prism.

Other less symmetrical arrangements also would allow the existence of more isomers, just as in the case where the coordination number is four. This is true, for example, for the trigonal prism, as will be seen from Fig. 58, since the edge AB

[1] See footnote on preceding page.

is different from the edge AD. The trigonal prism is indeed, not unknown; it has been discovered, as noted in Sec. 16.12, not from isomerism, but (in cases in which all the atoms surrounding the central atom are the same and in which the crystal cannot really be said to consist of complex ions) by the use of X rays. It is, however, apparently quite rare.

The isomerism of compounds with coordination number six but containing groups or radicals that occupy two coordination places is of particular interest. Let us consider, for example, the ion

$$[Co(NH_3)_4(C_2O_4)]^+,$$

in which the oxalate radical occupies two coordination places. Now it is natural to assume that these two places are *adjacent* places and not opposite places in the octahedron; in other words, the compound must be a cis-compound. As a matter of fact, in cases of this kind isomerism does not occur. If there are two oxalate groups, it is possible for stereoisomerism to occur. Let us consider, for example, the ions of the formula $[Co(C_2O_4)_2(NH_3)_2]^-$. The trans-compound (Fig. 59) has a plane of symmetry and is identical with its

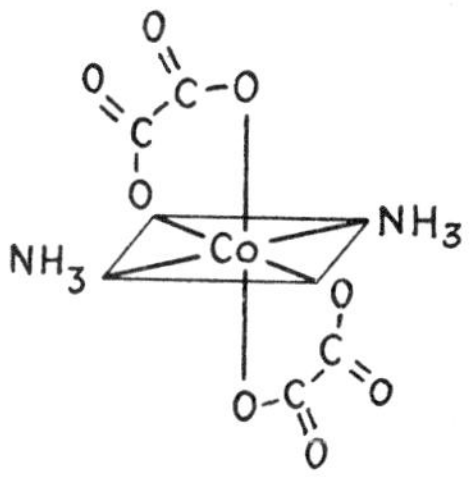

FIG. 59.—Trans-oxalato compound.

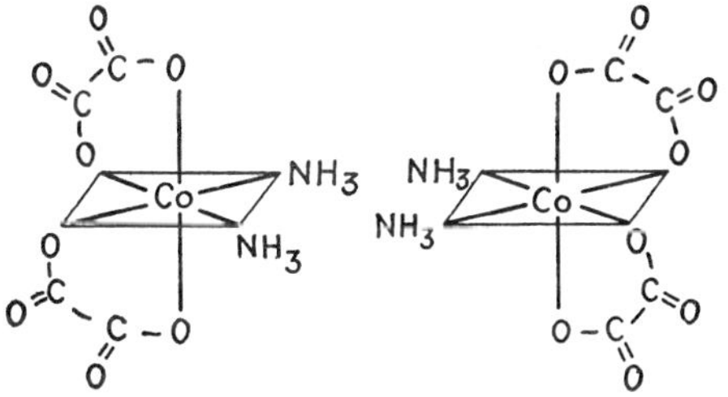

FIG. 60.—Optical isomers, cis-oxalato compound.

mirror image. This is not true of the cis-compound (Fig. 60). The cis- and trans-compounds may therefore be identified by finding which one can be resolved into optical isomers. This type of consideration, together with the fact that the two places occupied by an ion such as the oxalate ion are cis-places, has proved invaluable in the study of these compounds. By use of substitution reactions, in which it can be assumed that substitution takes place without a shift in position, a group like the oxalate group may be replaced by two others, which are then known to be in the cis-position, and so the cis- and trans-forms of other compounds can be identified. It is true that sometimes shifts in position do occur, so that due caution must be used in drawing conclusions from such experiments; but the conditions under which such shifts occur are believed to be known in many cases. These shifts can often be studied by use of the compounds with two oxalate groups, for in these compounds the configuration can be determined if it is known whether optical isomers can be separated or not. There is at present a large body of experimental material bearing on the question of the determination of the configuration, especially of the cobalt compounds, and the configuration is believed to be known in many cases.[1]

An excellent summary of the results of experimental work in this field is comprised in Freudenberg's "Stereochemie." The following elements have been shown by study of isomerism to have a tetrahedral valence structure: Be, B, C, Si, Ge, Zn, Sn, N, P, As, S, Se, Te. In all compounds in which this tetrahedral structure occurs, the elements in question have (presumably) an octet of electrons, some shared and some unshared (or often all shared), in their outer shells.

The compounds by means of which these structures have been determined are for the most part complex organic compounds. Usually it is optical isomerism of a compound with four different substituents (or with three different constituents and with one corner of the tetrahedron empty, which offers the possibility of the right- and left-handed forms just as well, since all four corners of the tetrahedron are different) that have been observed. In the case of nitrogen, structural isomerism of the dichloroethylene type is quite common. A typical example is offered by the two compounds

$$\begin{array}{ccc} C_6H_5\text{—}\underset{\displaystyle \underset{\displaystyle HO\text{—}N}{\|}}{C}\text{—}CH_2Br & \quad\text{and}\quad & C_6H_5\text{—}\underset{\displaystyle \underset{\displaystyle N\text{—}OH}{\|}}{C}\text{—}CH_2Br \end{array}$$

which have different physical properties. Quaternary ammonium bases having four different groups attached to the nitrogen have been separated into optical isomers, but in most cases

[1] See WERNER, Ref. 1, p. 287.

attempts to separate tertiary amines in which there are three different substituents with one corner of the tetrahedron blank have failed. This is probably due to the ease with which right- and left-handed isomers change into each other in this case. In one instance in which separation was effected, racemization was found to take place rather easily.

In the cases of beryllium, boron, and zinc optical isomerism has apparently been found only in compounds having two substituents, each one occupying two coordination places.

In the case of sulfur, selenium, and tellurium, optically active substances in which there are three groups attached to the central ion have been separated. Typical of these compounds is the tellurium salt

$$\left[CH_3-C_6H_4-Te\begin{array}{l} \diagup CH_3 \\ \\ \diagdown C_6H_5 \end{array} \right] I$$

The tellurium has a completed octet without sharing electrons from the iodine ion, and the latter is apparently not part of the complex, but held merely by ionic binding. In certain cases, *e.g.*, $Te(CH_3)_2I_2$, geometrical isomerism was found to occur, which at first led to the supposition that the tellurium in this compound had a square configuration. In this case, also, ionization occurs in aqueous solution, and it was eventually found that the isomerism was due to the formation of more complex ions.[1] This example illustrates the care that must be exercised before definite conclusions can be drawn from examples of isomerism.

It has been claimed that in some of their compounds the divalent elements Cu, Pd, Pt, and Ni also have tetrahedral structures. This is based in part on the existence, in the case of nickel, platinum, and palladium, of salts of the type

$$\left[\begin{array}{c} \quad NH_2CH_2CH_2 \quad \\ \diagup \qquad\qquad \diagdown \\ Ni-NH_2CH_2CH_2-N \\ \diagdown \qquad\qquad \diagup \\ \quad NH_2CH_2CH_2 \quad \end{array} \right] X_2$$

[1] DREW, *J. Chem. Soc.* (*London*), 1929, p. 560; LOWRY and GILBERT, *ibid.*, 1929, p. 2076.

where X is univalent. In this salt, the nickel is in contact with four nitrogens,[1] and when the spatial arrangement is investigated it is seen that this would be practically impossible without great distortion and strain if all the bonds lay in a plane.[2] The compounds could be octahedral, however, with the halogens occupying places in the coordination sphere.

There are other cases in which these elements are believed to be tetrahedral. In these, the central atom is surrounded not by four different substituents, but by two substituents, each of which occupies two coordination places. The arrangement is such that if the valences of the central atom are in a plane cis-trans isomerism would result, but not optical isomerism; whereas optical isomerism and not cis-trans isomerism would be expected if the arrangement is tetrahedral. In divalent salts of copper, platinum, and palladium, such optical isomerism has been observed. An example is the palladium ion

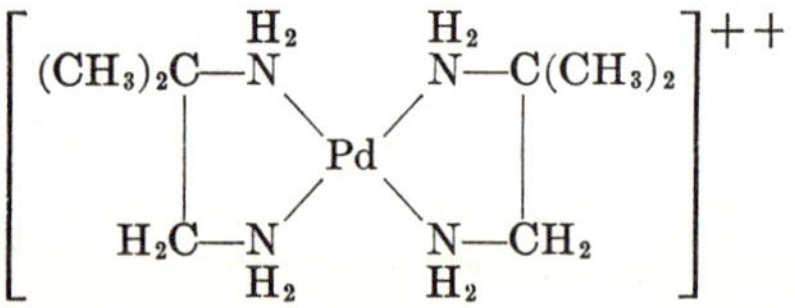

$$\left[\begin{array}{ccccccc}
 & & \mathrm{H_2} & & \mathrm{H_2} & & \\
(\mathrm{CH_3})_2\mathrm{C} & - & \mathrm{N} & & \mathrm{N} & - & \mathrm{C(CH_3)_2} \\
| & & & \diagdown \quad \diagup & & & | \\
| & & & \mathrm{Pd} & & & | \\
| & & & \diagup \quad \diagdown & & & | \\
\mathrm{H_2C} & - & \mathrm{N} & & \mathrm{N} & - & \mathrm{CH_2} \\
 & & \mathrm{H_2} & & \mathrm{H_2} & &
\end{array}\right]^{++}$$

On the other hand, the compound shown in Fig. 61 should have optical isomerism if the valences of the central atom lie in a plane, but not if they are arranged as a regular tetrahedron. It was found that the substance was optically active. In other platinum salts, cis-trans isomerism has been observed, which would also indicate a plane rather than a tetrahedral configuration.

[1] Each bond represents a pair of electrons shared between nickel and nitrogen. We have drawn them as four equivalent bonds though this gives nitrogen an apparent valence of four. Bonds that give an element an apparently abnormal valence are frequently indicated by dotted lines, but as this carries with it the implication that these bonds have a special character, which actually they do not possess, it seems preferable to represent them in the same way as other bonds.

[2] The plane structure is expected for the elements under consideration in their divalent states, and either a tetrahedral or an octahedral structure is contrary to what would be expected from the valence rule, especially for palladium and platinum, which usually form covalent bonds. However, we must not be surprised if the valence rules are broken when their fulfilment would cause great strain in the attached organic radicals, and other cases of this type are known.

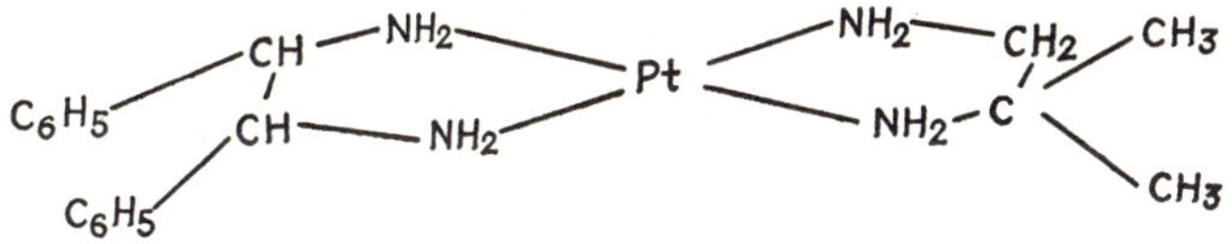

FIG. 61.—Optically active ion (doubly positively charged) with plane configuration. The two plane rings are both horizontal. If one were horizontal and the other vertical, as would be demanded by a tetrahedral valence structure of the central atom, there would be a plane of symmetry and, hence, no optical activity.

In the case of the ion

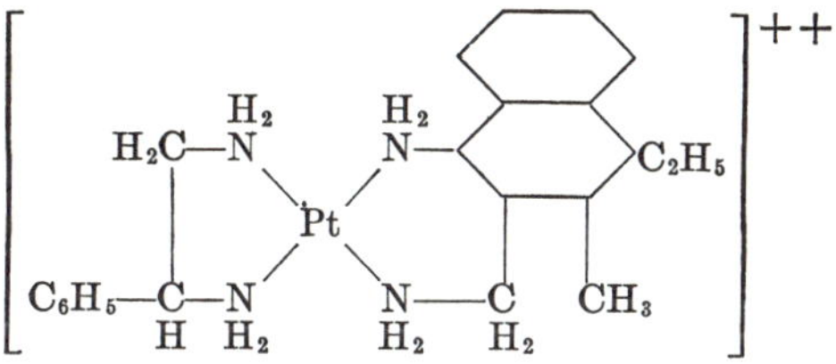

both cis-trans isomerism and optical activity are found.[1] The former would be expected from a plane and the latter from a tetrahedral structure in this case, and so the compound seems to have some of the properties of both kinds of structure.

The observations on these compounds thus seem to be self-contradictory; however, they could be explained by supposing that the valence bonds of the central atom form a tetrahedron, but an irregular one, or that the valence bonds form a pyramid, or that the diamine rings themselves are not planar. However, with respect to the last possibility, it must be said that on account of the large size of the central atom a nonplanar form of the ring would result in a distortion of the bond angles in the ring from the tetrahedral angle and so would cause strain (see Fig. 62). In any event, the problem presented by these salts does not seem to be finally cleared up.

In connection with this discussion, it is interesting to note that the nickel compound[2] (in which the nickel is divalent, *i.e.*, contributes two valence electrons to the electron structure)

[1] REIHLEN and HÜHN, *Liebigs Ann. Chem.*, **519**, 80 (1935). It was demonstrated that optical activity was due to the platinum, as well as to the active carbon atom present.

[2] In the formula shown each valence line represents, as usual, a pair of shared electrons, and all carbon, nitrogen, and oxygen atoms have a completed octet.

```
                     O         O
                     |         |
CH3—CH2—C=N         N=C—CH2—CH3
              |    \     /    |
              |      Ni       |
              |    /     \    |
       CH3—C=N         N=C—CH3
                     |         |
                    OH        OH
```

in which there is a five-membered ring with double bonds has a cis- and trans-form (the cis-form, with the OH groups adjacent, is shown). It has, therefore, presumably, a plane configuration. In other cases (with zinc and copper), six-membered rings involving double bonds give an apparent tetrahedron.

Whenever the elements Co, Pd, and Pt, in their divalent forms, are not members of a ring, the isomerism observed indicates a plane configuration. We have already noted that such an arrangement is not the one that would be anticipated were the forces ionic, so the formation of such a compound is indication that the bond is covalent. It is, therefore, gratifying that the magnetic criterion indicates covalent bonds for $PtCl_4^{--}$, $Pt(NH_3)_4^{++}$, and $Pt(C_2O_4)_2^{--}$. It may be noted here that X-ray experiments on copper compounds, in particular $CuCl_2(H_2O)_2$, indicate a square configuration. Divalent Ni also tends to have a square configuration, as is to be inferred, *e.g.*, from the isomorphism of $K_2[Ni(CN)_4]H_2O$ and $K_2[Pd(CN)_4]H_2O$. ($PdCl_4^{--}$ has been found to be a plane square by X-ray examination, and so it is probable that $Pd(CN)_4^{--}$ is a square also.) $Ni(CN)_4^{--}$ has been found by the magnetic criterion to be covalent. On the other hand, compounds of Ni^{++} with ammonia have ionic binding, and there are many compounds with the octahedral ion $Ni(NH_3)_6^{++}$. The bond between Ni and amines has also been found in some cases to be ionic, and, undoubtedly, this is true in general. This being the case, it would not be at all surprising if further stereochemical or X-ray research on the complex nickel salts containing diamine rings should show that the nickel is either tetrahedral or octahedral. But if a tetrahedral structure is finally confirmed in the case of the similar platinum and palladium salts, it would be much more unexpected from the theoretical point of view, as the binding in these salts usually seems to be covalent.[1]

[1] The reader should compare footnote 1, p. 275.

An octahedral arrangement of six surrounding groups forming a complex ion has been found[1] by observation of isomerism for the following divalent elements: Fe, Ni, Ru, Pt, Zn, Cd; for the following trivalent: Cr, Fe, Co, Ru, Rh, Ir, Al; for the following tetravalent: Pd, Ir, Pt, Ti; and for pentavalent As. In each case, there are six pairs of shared electrons.

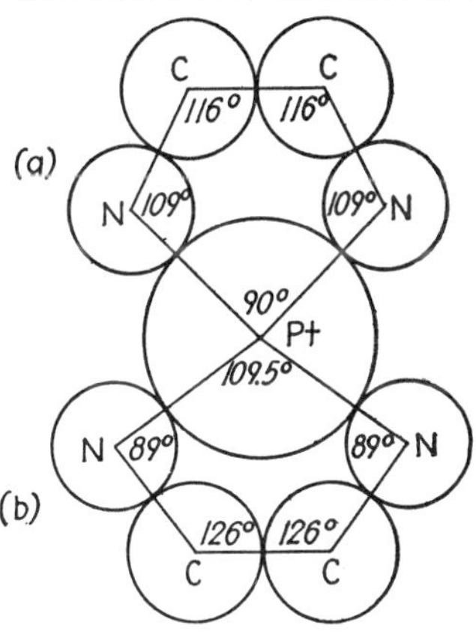

FIG. 62.—Showing the angles in a diamine ring attached to a platinum atom, assuming (*a*) that the platinum has a plane square configuration, (*b*) that the platinum has a tetrahedral configuration; and assuming in each case that the ring is in a plane. Radius of divalent platinum taken as 1.32 Å., of carbon, as 0.77Å., of nitrogen, as 0.70 Å. [Table 33 below, and Pauling and Huggins, *Zeits. Krist.*, **87**, 232 (1934)]. It is seen that if the normal valence angles of carbon and nitrogen are tetrahedral (109.5°), and if the radius of platinum is the same for both configurations, there is much more strain for a tetrahedral arrangement than for a plane square. On this basis, then, as well as from the point of view of the electronic structure of platinum, one would expect the plane square to be favored. (Figure after Mills and Quibell, *J. Chem. Soc. (London)*, **1935**, 842).

Some of these cases call for special mention. Divalent nickel forms octahedral complexes with some substituents and square complexes with others. It was seen earlier in the chapter (Sec. 16.4) that the complex, six-coordinated cyanide of divalent cobalt is unstable. A covalent six-coordinated complex of nickel should be unstable for the same reason; in fact it should be more unstable, for there are more electrons left in Ni^{++}. It is thus interesting that with cyanide nickel forms the ion $Ni(CN)_4^{--}$ with a square configuration. On the other hand, the ions such as $Ni(NH_3)_6^{++}$ and others with six substituents are not particularly unstable, and the binding is probably ionic or semicovalent in all these cases. Optical activity was observed in a tridipyridine compound, each dipyridine molecule occupying two coordination places.

Again, divalent platinum would not be expected to have stable complexes with coordination number six, and mostly it has the

[1] PFEIFFER in Freudenberg, "Stereochemie," pp. 1200*ff;* BURSTALL, *J. Chem. Soc.*, **1936**, 173. In the case of Zn and Cd isomers were not actually isolated, but certain optical effects in solution were attributed to optically active forms which rapidly racemize.

square configuration. There are a few compounds with coordination number of six, however, and cis-trans isomerism has been observed in $Pt[(CH_3CN)_2(NH_3)_4]Cl_2$. The tetravalent ion Pt^{4+} has only six 5*d*-electrons, and is therefore capable of forming covalent octahedral bonds involving two 5*d*-places without difficulty.

16.6. Results of X-ray Analysis.—It appears to be generally agreed by crystallographers that the ions that are commonly observed in aqueous solution occur as units in crystals. For example, many crystals of which the sulfate ion is a part have been examined, and it is invariably found that the sulfur atom is surrounded by four oxygens at the corners of a regular tetrahedron, the distance between oxygen and sulfur centers being about 1.50Å., and the distance between adjacent oxygen centers being about 2.45Å.

The size and shape of a given ion are not, however, entirely invariable from crystal to crystal but depend upon the forces

TABLE 31.—SHAPES OF CERTAIN IONS IN CRYSTALS
(In all cases, the central ion appears first in the formula)

Straight lines:
HF_2^-, N_3^-, NCO^-, $Ag(NH_3)_2^+$, $Cd(NH_3)_2^{++}$, $Hg(NH_3)_2^{++}$, I_3^-, ICl_2^-, $Ag(CN)_2^-$

Kinked lines:
NO_2^-, ClO_2^-

Triangles, central atom in plane:
BO_3^{---}, CO_3^{--}, $C(NH_2)_3^+$, $CO(NH_2)_2$, NO_3^-

Triangles, central atom out of plane (pyramids, or decapitated tetrahedra):
SO_3^{--}, ClO_3^-, BrO_3^-

Tetrahedra:
SiO_4^{----}, CrO_4^{--}, WO_4^{--}, MoO_4^{--}, MnO_4^-, PO_4^{---}, SO_4^{--}, SeO_4^{--}, ClO_4^-, IO_4^-, ReO_4^-, VS_4^{---}, AsS_4^{---}, SnS_4^{----}, BeF_4^{--}, BF_4^-, $Be(H_2O)_4^{++}$, $CoCl_4^{--}$, $Cu(CN)_4^{---}$, $Zn(CN)_4^{--}$, $Cd(CN)_4^{--}$, $Hg(CN)_4^{--}$, $N(CH_3)_4^+$, $N(CH_3)_2H_2^+$, $PH_2O_2^-$, OsO_3N^-, $Zn(NH_3)_2Cl_2$

Squares, central atom in plane:
$PtCl_4^{--}$, $PdCl_4^{--}$, $Pd(NH_3)_4^{++}$, $CuCl_2(H_2O)_2$, $AuBr_4^-$

Octahedra:
AlF_6^{---}, SiF_6^{--}, GeF_6^{--}, FeF_6^{---}, $MoO_3F_3^{---}$, $TiCl_6^{--}$, $ZrCl_6^{--}$, $PtCl_6^{--}$, $PdCl_6^{--}$, $SnCl_6^{--}$, $PbCl_6^{--}$, $SeCl_6^{--}$, $TeCl_6^{--}$, $OsO_2Cl_4^{--}$, $SeBr_6^{--}$, $Sn(OH)_6^{--}$, $Te(OH)_6$, $Co(NH_3)_6^{++}$, $Co(NH_3)_6^{+++}$, $Co(NH_3)_5H_2O^{+++}$, $NH_4(H_2O)_6^+$, $Rb(H_2O)_6^+$, $Cs(H_2O)_6^+$, $Mg(H_2O)_6^{++}$, $Al(H_2O)_6^{+++}$, $Cr(H_2O)_6^{+++}$, $Zn(H_2O)_6^{++}$, $Ni(H_2O)_6^{++}$, $Tl(H_2O)_6^+$, $Cr(NH_3)_5Cl^{++}$, $Co(NH_3)_5Cl^{++}$, $Rh(NH_3)_5Cl^{++}$, $Co(NO_2)_6^{+++}$, $Pt(SCN)_6^{--}$

exerted by the other ions present. Distortions and changes in size involving variation in the interatomic distances of as much as 0.2Å. appear to be by no means uncommon. It is not altogether impossible, however, that some of these differences are due to experimental error in determining the positions of the atoms in the complex crystals involved.

In Table 31 are given the structures of a number of ions, as obtained by X-ray analysis of crystals.[1] This table is meant to be suggestive rather than exhaustive.

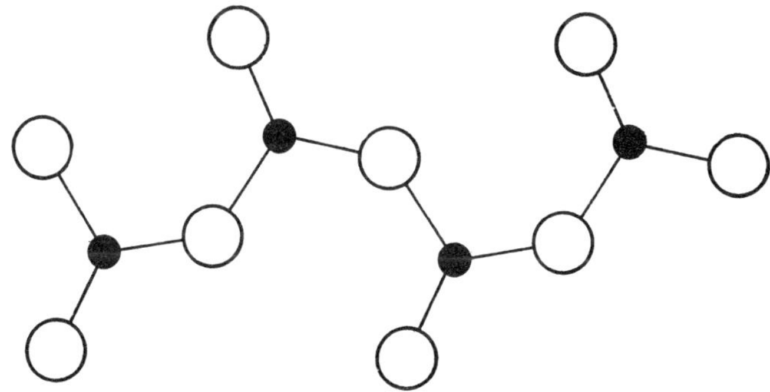

FIG. 63.—Extended BO_2^- ion. [*After Zachariasen, Proc. Nat. Acad. Sci.*, **17**, 619 (1931).]

16.7. Complex Crystals.—There are many crystals of rather complex composition that are better considered as a whole than as composed of ions. For example, in CaB_2O_4 the borons are surrounded by oxygens which form an almost equilateral triangle about the borons. These triangles are extended indefinitely, forming chains of the composition BO_2^-, as shown in Fig. 63. These chains form what Bragg has characterized as an extended acid radical. These radicals are held together by the positive calcium ions.

The silicates, however, are the compounds that furnish the examples of complex crystals, par excellence.[2] In fact, among the silicates, practically every gradation of complexity is realized in some one or another of the known forms. Every crystal

[1] For more detailed descriptions of crystal structures and references to the original literature, the reader should consult the "Strukturbericht," *Zeits. Krist.* For $Cu(CN)_4^{---}$ see Cox, Wardlaw, and Webster, *J. Chem. Soc. (London)*, **1936**, 775.

[2] W. L. BRAGG, *Zeits. Krist.*, **74**, 237 (1930).

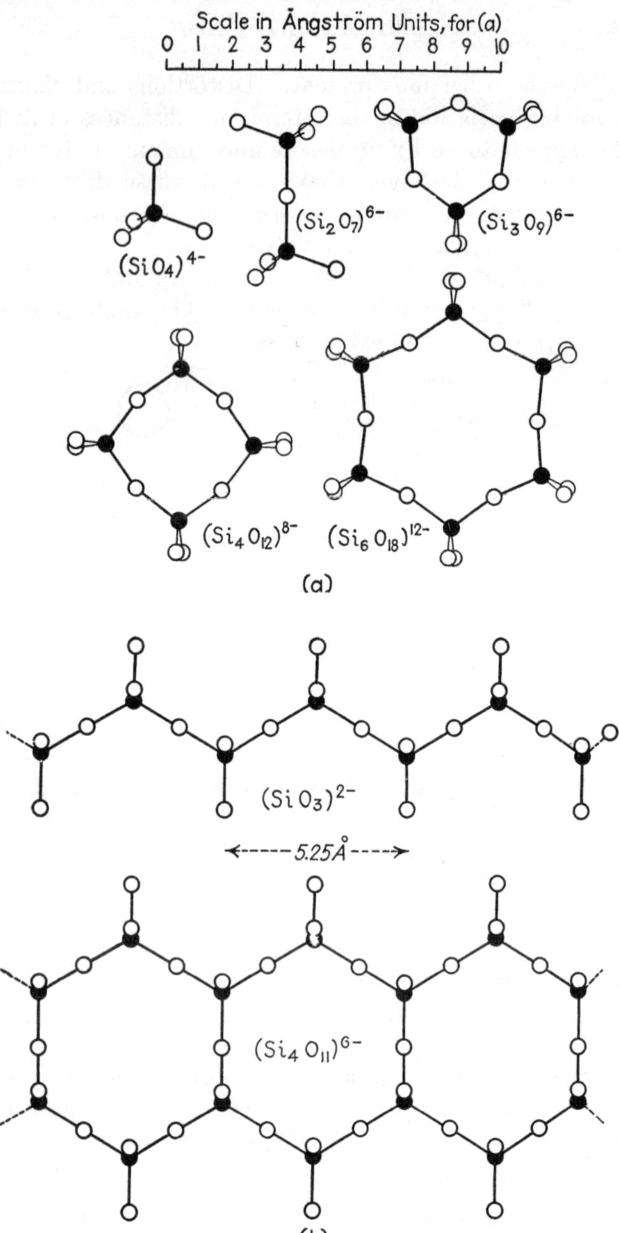

FIG. 64*A*.—Types of silicon-oxygen groupings: (*a*) closed groups, (*b*) chains and bands. Black circles represent silicon, white circles oxygen. [*From Bragg, Zeits. Krist.*, **74**, 237 (1930), *and Bragg and Bragg*, "*The Crystalline State*," *vol.* I, *p.* 134, *The Macmillan Company*, 1934.]

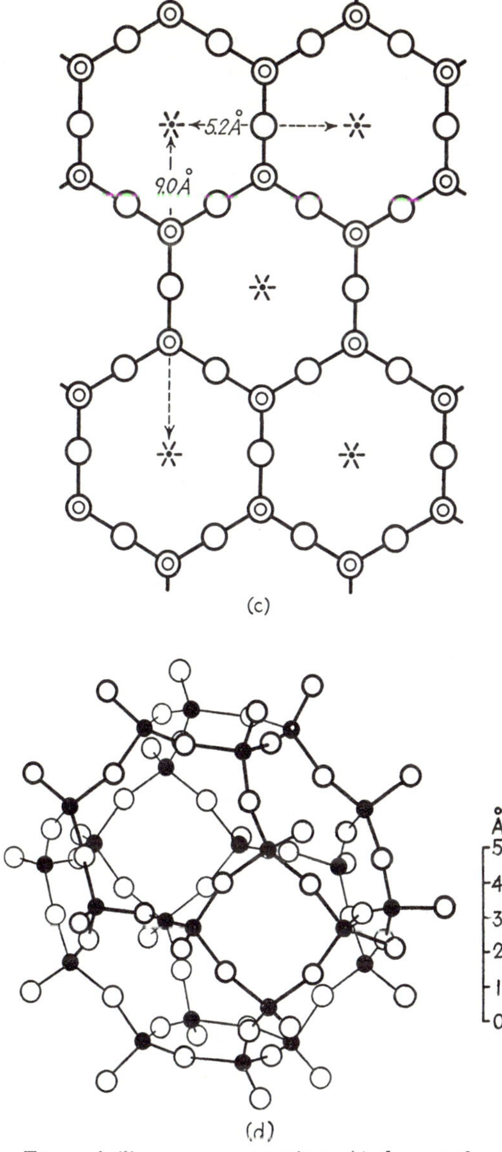

FIG. 64*B*.—Types of silicon-oxygen groupings: (*c*) sheet, such as found in talc and mica; all tetrahedra may point the same way, or alternate tetrahedra may point in opposite directions; a superimposed silicon and oxygen are shown by a double circle, oxygen by a single circle; (*d*) three-dimensional net (ultramarine); black circles represent aluminum or silicon, white circles oxygen. [*From Bragg, Zeits. Krist.*, **74,** 237 (1930), *and Bragg and Bragg,* "*The Crystalline State,*" *vol.* I, p. 135, *The Macmillan Company,* 1934.]

known, however, is characterized by the fact that silicon is surrounded by four oxygens in the form of a nearly regular tetrahedron,[1] the silicon-oxygen distance being in the neighborhood of 1.6Å., varying slightly from crystal to crystal. In the orthosilicates, in which the ratio of oxygen to silicon is at least 4:1, simple anions of the formula SiO_4^{4-} are formed. In crystals in which there is relatively less oxygen, various types of complex structure occur. In some crystals, there are finite silicon-oxygen groups of varying complexity, often forming a ring structure. In others, there are twisted silicon-oxygen chains, each link composed of an oxygen tetrahedron about a silicon and each tetrahedron having one atom in common with the preceding and following one. These chains have the composition SiO_3^{--}. There also occur silicon-oxygen sheets, which extend indefinitely in two dimensions and for which a typical formula would be $Si_2O_5^{--}$. Such a structure occurs in micas, and is responsible for their peculiar property of splitting and slipping easily in one direction. Finally, silicon and oxygen may form a network in three dimensions; this occurs in silicon dioxide (quartz) itself, the composition being SiO_2 (neutral). In some minerals in which part of the silicon is replaced by aluminum, three-dimensional network anions can occur, since Al^{+++} has a smaller charge than Si^{4+}. In all cases, the extended anions are held together by positive ions dispersed among them in such numbers that the whole is neutral. In its ability to form either simple[2] or extended anions, silicon is intermediate in its properties between the alkali and alkaline earth metals and the negative elements like chlorine and sulfur. The alkali and alkaline earth metals form oxides which (though actually, of course, neutral) might perhaps be thought of as extended anions, but they do not

[1] It is of interest to note that the angle between the bonds to a given oxygen atom is by no means invariable in the silicates, but seems to be determined by the geometrical exigencies of the situation. There are even different forms of SiO_2 in which the Si—O—Si bond angle is slightly different, though never far from 180° in SiO_2. This may indicate that the binding is predominantly ionic rather than covalent.

[2] It will be noted that the word "simple" has been used to characterize ions that are elsewhere called "complex." The terms, of course, are relative, and an ion that is simple compared with an extended acid radical is complex when compared, for example, with Cl^-

form simple anions. The negative elements, on the other hand, tend[1] to form simple anions like ClO_4^- and SO_4^{--}.

If sulfur-oxygen chains tended to exist, then we should expect to have neutral chains of composition SO_3. There is an undoubted tendency for the SO_3 molecule to double up[2] according to the formula

$$\begin{array}{cccc} & :\ddot{O}: & :\ddot{O}: & \\ :\ddot{O}: & \ddot{S} & :\ddot{O}: & \ddot{S}, \\ & :\ddot{O}: & :\ddot{O}: & \end{array}$$

but combinations of this sort, in particular long chains, cannot be extremely stable, as is evidenced by the relatively low boiling point of SO_3. A possible explanation of the difference between SO_3 and SiO_2 in this respect is the following. Sulfur is a much more negative element than silicon, and tends to draw the oxygen electrons in toward it to a much greater extent, so that they are less free to be shared with another sulfur or another element in general. It may be this same tendency which shows up in the fact that sulfuric acid is a strong acid (see Sec. 19.10).

It is a very interesting fact that in complex crystals containing oxygen, and in crystals with oxygen anions, the more electronegative elements tend to show a constant coordination number, being surrounded by a definite number of oxygen atoms, whereas the electropositive elements have coordination numbers that vary from crystal to crystal, and the oxygen atoms are often arranged about them in a rather irregular way; *i.e.*, the more electronegative elements determine the structure, the more electropositive ones "take what they can get." The reason for this becomes clear if we think of the crystal as formed from ions of all the elements. Take, for example, a BPO_4 crystal, and think of it as composed of B^{+++}, P^{5+}, and O^{--} ions. The force exerted by the P^{5+} ion on the O^{--} ions will be greater than that exerted by the B^{+++} ion, both because of the greater charge and the greater electronegativity of the phosphorus. The fact that P^{5+} is highly charged and phosphorus is highly electronegative means that the

[1] The following heats absorbed are relevant:

$$Ca_2SiO_4 \rightarrow CaSiO_3 + CaO, \text{ 8 kg.-cal.};$$
$$2Na_3PO_4 \rightarrow Na_4P_2O_7 + Na_2O, \text{ 59 kg.-cal.};$$
$$2K_2SO_4 \rightarrow K_2S_2O_7 + K_2O, \text{ 124 kg.-cal.}$$

[2] By Trouton's rule (see p. 362).

P—O bond will certainly not be a strictly ionic bond, but this does not affect the conclusion, for it is really only a result of the fact that the phosphorus exerts a large force on the oxygen. This particular crystal may be used as an example of the fact that the most highly charged ion, here P^{5+}, determines the structure of the crystal, for the phosphorus atoms are surrounded by four oxygens in the form of a nearly regular tetrahedron, whereas the boron is also surrounded by four oxygens, forming a less regular tetrahedron. The normal position for boron, as indicated by the borate ion, is in the center of three oxygens. This dominance of the highly charged ions is one of the determining rules of complex crystal structure.

Other rules have been formulated by Bragg and by Pauling.[1] Of these, the most important is the rule that electrical charge is locally neutralized (electrostatic-valency rule). Consider an anion, which we designate by a subscript 0, with charge $-z_0$ (the unit being taken as the charge on the electron). This ion will be close to a number of cations, which will be designated by the running subscript i. Let the ith ion have a coordination number of n_i and a charge of z_i; then it may be said to share an n_ith part of its charge, namely, z_i/n_i, with any one anion. The principle of local neutralization then says that as nearly as may be possible the charge on the original ion just balances the charge shared with it by the surrounding cations, *i.e.*, $z_0 \cong \sum_i \frac{z_i}{n_i}$. This assumes that every charge is surrounded as closely as possible by charges of the opposite sign, which means, of course, a low potential energy. In applying this rule, it is assumed that each ion has a charge corresponding to its valence, as though the binding were purely ionic. The fact that in the actual compound the binding may be more nearly covalent means only that there is a displacement of charge, due to distortion of the electron atmosphere of the anion toward the cation, and does not affect the applicability of the rule at all. This rule applies to single (monatomic) ions, which may themselves be constituents of complex ions.

Another rule generally followed by complex crystals is that highly charged ions are far away from each other. This is another expression of the fact that the electrostatic potential

[1] W. L. Bragg, *Zeits. Krist.*, **74,** 287–304 (1930); Pauling. *J. Am. Chem. Soc.*, **51,** 1010 (1929).

must be low. The explanation of the relative instability of chains with the formula SO_3 may be given an alternative statement. Since S^{6+} is a highly charged ion and since these ions are brought fairly close to each other in such a chain its instability is explained. This explanation may be compared with the one previously given. It illustrates the fact that similar results are often obtained, regardless of the type of binding assumed. (Again compare Sec. 19.10.)

16.8. The coordination number in complex ions and crystals depends primarily upon the ratio of the radius of the central atom or ion to that of the atoms and ions surrounding it. This applies in a general way whether the binding is ionic or covalent, and something can be learned by assuming that it is purely ionic, so that within the complex ion itself the central atom is a cation which has lost all of its outer electrons and the outer atoms are anions which have completed their outer shell of electrons. This assumption is useful despite the fact that it is often not too good an approximation to the truth, and even though the actual interatomic distances may not be reproduced too well by it. The rule usually cited for the determination of coordination number is that there will be as many anions surrounding a cation as can be arranged in symmetrical fashion about it, maintaining contact between the cation and the anions surrounding it, *i.e.*, avoiding anion-anion contact;[1] this we shall call the "rule of close-packing." Thus, we have a coordination number of three if the space within three closely packed anions is less than the size of the cation, but the space available between four of the anions closely packed in the form of a tetrahedron would be more than needed to accommodate the cation, etc. If we attempt to

[1] This rule may be generally applied whether the cation and surrounding anions are part of a complex ion or not; indeed it probably holds better when they are a part of a complex ion.

The "rule of close-packing" is not to be confused with another kind of close-packing often mentioned in the literature. In many complex crystals with multiply charged cations, even relatively small anions such as O^{--} and F^- are large compared with the cations. These anions then generally have an arrangement in which they are approximately close-packed among themselves. That is to say, if we look at the anions alone, we find that they simulate one of the close-packing arrangements discussed in Appendix IV. This fact has been of great assistance in crystal analysis. Approximate close-packing of this kind is in no way incompatible with the rule of close-packing discussed in the text.

calculate the electrostatic energy of the various arrangements, there appears to be no reason why the rule of close-packing should hold. It would naturally be expected that the electrostatic energy would be lower the more anions there were about a given cation; hence an arrangement with a high coordination number would be expected to persist, even though anion-anion contact were occurring, unless a rearrangement would result in a considerable lowering of the anion-cation distance, sufficient to more than compensate the effect of decreasing the number of anion neighbors near a given cation. The matter is further complicated by the existence of repulsive forces and van der Waals forces, which undoubtedly often play an important part in determining the crystal form. It is not surprising that exceptions to the rule of close-packing, *e.g.*, cases where anion-anion contact occurs, are known.

Indeed, it could hardly be supposed that in crystals of fixed composition, such as the alkali or alkaline earth halides, the rule of close-packing would necessarily give the crystal form correctly. But in case the crystal contains more than two constituent elements (as is always the case when complex ions are present), and the composition has some possibility of variation, without upsetting the balance of positive and negative ions, then the close-packing rule may well be an important factor in determining just what composition the crystal will assume. It may, for example, determine how many molecules of water will enter into the reaction when the crystal is formed by evaporation of an aqueous solution.

Goldschmidt and Pauling have applied the rule of close-packing to a study of the coordination number in crystals. Pauling[1] has applied it, in particular, to the oxygen anions of a number of the more negative elements. Purely geometric considerations, then, show that the triangular form of an ion will be stable if the cation-anion ratio is less than 0.225; the tetrahedron is stable if the cation-anion ratio is between 0.225 and 0.414; the octahedron, if it is between 0.414 and 0.732; and the cube, if it is above this. The results of Pauling are shown in the accompanying table and, in spite of the rough character of the assumptions made, are in good accord with experience. Some of the ions listed are too electropositive in character and not of sufficiently

[1] Pauling, *J. Am. Chem. Soc.*, **55**, 1895 (1933).

high valence to form stable complex anions; the considerations, however, should be equally applicable to complex cations formed with water. Thus we have the ions $Be(H_2O)_4^{++}$ occurring in $Be(H_2O)_4SO_4$, the ion $Al(H_2O)_6^{+++}$ occurring in many crystals, and other ions as noted in Table 31. B^{+++}, C^{4+}, and N^{5+} have a normal coordination number of three. The other ions, according to Pauling's summary and available data in the "Strukturbericht," have normal coordination numbers as indicated by

TABLE 32.—RATIOS OF RADII: CATION/OXYGEN
COORDINATION NUMBERS IN COMPLEX CRYSTALS AND, IN PARTICULAR, IN OXYGEN IONS

Be^{++}	B^{+++}	C^{4+}	N^{5+}			
0.24	0.19	0.16	0.14	C.N. = 3 region		
Mg^{++}	Al^{+++}	Si^{4+}	P^{5+}	S^{6+}	Cl^{7+}	C.N. = 4 region
0.47	0.41	0.37	0.34	0.30	0.28	
Zn^{++}	Ga^{+++}	Ge^{4+}	As^{5+}	Se^{6+}	Br^{7+}	
0.50	0.46	0.43	0.40	0.37	0.35	
Cd^{++}	In^{+++}	Sn^{4+}	Sb^{5+}	Te^{6+}	I^{7+}	
0.65	0.59	0.55	0.51	0.47	0.44	
		C.N. = 6 region				

The regions shown on the table give a rough indication of the experimental results.

Table 32. There are many salts in which this might not be indicated by the stoichiometrical formula, but it must be remembered that polymerization, *i.e.*, formation of chains, rings, or sheets, as described for the silicates, can occur. For example, this occurs with salts of HPO_3. Iodine, near the borderline, can have a coordination number of four or six (IO_4^-, IO_6^{5-}). Tin, antimony,[1] and tellurium form the ions $Sn(OH)_6^{--}$, $Sb(OH)_6^-$, and TeO_6^{6-}.

[1] It was suggested by Hammett, "Solutions of Electrolytes," p. 108, McGraw-Hill Book Company, Inc., 1929, and by Pauling (Ref. 1, p. 316) that in the antimonates the ion $Sb(OH)_6^-$ is present. This is supported by the fact that in such a substance as $K_2H_2Sb_2O_7$ the amount of water of crystallization is generally such as to make it possible to write the formula $KSb(OH)_6$, and by the fact that a solution of the salt has an acid reaction and (as shown by measurements of conductivity) a singly charged anion. However in crystals of $Ca_2Sb_2O_7$, $Ca_2Sb_2O_7{\cdot}4H_2O$, and a number of other antimonates, the antimony is surrounded by eight oxygens, all at about the same distance.

Since F^- is slightly smaller than O^{--}, the central atom in a fluoride often has a larger coordination number. This is illustrated by BF_4^-, AlF_6^{---}, SiF_6^{--}, PF_6^-, SF_6, and probably SnF_8^{4-} Of course, some of the ions mentioned exist only in crystals, not in aqueous solution.

Though very good results are obtained by assuming that the binding in all these ions is ionic, it is worth while noting that in every case the observed coordination number is quite consistent with the electronic structure if covalent binding is assumed. It is true that the binding in the ions of coordination number six must involve outer *d*-electrons, having the same value of the quantum number *n* as the *s*- and *p*-electrons. The cases where the coordination number is three are of particular interest. In these cases, the central atom lies in the plane with the surrounding ones. The carbonate ion may be taken as typical. There are three possibilities regarding its electronic structure. (1) It may be ionic, being a combination of C^{4+} with three O^{--}'s. (2) It may have the structure

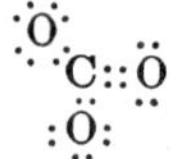

In this situation, all the atoms would lie in a plane. Since there would be resonance between the three possibilities (double bond to any one of the three oxygens), the actual structure would presumably be symmetrical. (3) It may have the structure

```
   ..
  :O:
 ..  ..  ..
:O:C:O:
 ..      ..
```

That is to say, it is not inconceivable that three of the four levels of the carbon atom could be combined to give bond-eigenfunctions of a different type from the tetrahedral bond-eigenfunctions;[1] however, in view of the fact that carbon appears in general to maintain its octet, this alternative seems somewhat less probable. But it is not possible to say definitely which of the alternatives gives the best description of the carbonate ion. It may resonate among all the possibilities.

[1] A combination can be formed of one *s*- and two *p*-levels, giving coplanar bonds making angles of 120° with each other.

16.9. Covalent Radii.—X-ray investigations have been made of a number of crystals in which the binding is almost certainly covalent and the atomic distances obtained. Pauling and Huggins[1] have made estimates, based on these investigations, of the radii of various atoms in compounds in which the binding is covalent; their values are purely empirical, but are of use in correcting the experimental data. These radii will obviously depend upon the valence state of the atom and upon which of the types of bonds mentioned in Sec. 15.2 is involved. Pauling and Huggins first consider the atomic radii for atoms with sp^3 tetrahedral bonds. In designating a bond as an sp^3 tetrahedral bond, we do not necessarily mean that the coordination number of the atom is four. Thus, as explained in Sec. 15.2, it appears that the oxygen bonds in water may be of this type, though only two hydrogens are attached to the oxygen. In Table 33, we give the

TABLE 33.—COVALENT RADII

	Be	B	C	N	O	F
	1.07	0.89	0.77	0.70	0.66	0.64
	Mg	Al	Si	P	S	Cl
	1.40	1.26	1.17	1.10	1.04	0.99
Cu	Zn	Ga	Ge	As	Se	Br
1.35	1.31	1.26	1.22	1.18	1.14	1.11
				(1.21)	(1.17)	(1.14)
Ag	Cd	In	Sn	Sb	Te	I
1.53	1.48	1.44	1.40	1.36	1.32	1.28
				(1.41)	(1.37)	(1.33)
Au	Hg	Tl	Pb	Bi		
1.50	1.48	1.47	1.46	1.46		
				(1.51)		

tetrahedral radii of Pauling and Huggins. These were obtained in the following way. For C, Si, Ge, and Sn the radii were taken as half the observed interatomic distances in the diamond-type crystals in which each atom is surrounded by four other atoms of the same kind at the corners of a regular tetrahedron. The radius for sulfur was assumed to be half of the distance between sulfurs in pyrite, FeS_2, or hauerite, MnS_2. [In these crystals, each sulfur is surrounded by four other atoms, one sulfur and three iron or manganese atoms, and presumably shares an electron pair with each of these atoms. Since all its inner shells are full and since there are thus just eight outer (shared)

[1] PAULING and HUGGINS, *Zeits. Krist.*, **87**, 205 (1934).

electrons, the sulfur is presumably in a tetrahedral state.] The radii of Zn, Cd, and Hg were then found from the interatomic distances in their sulfides, and the selenides and tellurides of these metals then furnished data for the determination of the Se and Te radii. The radius of O was obtained from ZnO. Other radii were then obtained through interpolation and extrapolation in the periodic table, together with some other approximate considerations described in the paper. The extrapolations and interpolations are based on the relationship between the elements of a given row of the periodic table rather than that between elements of a given column. It will be noted that the radius increases from top to bottom of a column, as might be expected, but the relation is rather irregular; however, as one goes across a row from left to right, the radius decreases regularly and gradually. This gradual decrease is readily understood as being due to the increase of the charge on the positive nucleus, if it is borne in mind that the electronic structure for tetrahedral valence is the same for all atoms of a given row. In every case, there are four pairs of outer shared electrons and the inner electronic structure is the same. If one of the electrons of a shared pair should be arbitrarily assigned to each of the sharing atoms, then the charge on the atom would vary in a regular way across a given row of the table. Thus tetrahedral magnesium would be Mg^{--} and tetrahedral Cl would be Cl^{+++}. Of course these formulas are far from representing the true state of affairs, but they illustrate the point.[1]

In a compound in which an element exhibits its normal valence, but in which the coordination number is less than four so that tetrahedral bond-eigenfunctions are not necessarily formed, a different radius may well be anticipated. Pauling and Huggins have given a set of such normal radii, based chiefly on a comparison of half the interatomic distance in the diatomic halogen molecules with the corresponding tetrahedral radius and on the radii of the elements of the carbon group, in which the normal valence radius must be the same as the tetrahedral radius. Where the normal radii differ from the tetrahedral, they are shown in parentheses in Table 33.

[1] In connection with the statements of this paragraph, the reader should see Sec. 16.10.

The method of knowing the cases in which normal valence radii are to be used and those in which tetrahedral radii are to be used may well be brought out by means of examples. Let us consider the iodine atom, for instance. Obviously in I_2 the normal radius is to be expected. The same is true in the tetraiodides, such as SiI_4. In these compounds, the crystals are composed of molecules loosely held together. Each iodine is held by valence forces to only one other atom, namely, the central atom. In the AgI crystal, on the other hand, each iodine atom is equidistant from four silver atoms (at least at a sufficiently low temperature—for further discussion see Sec. 16.12) and may be considered to share an electron pair with each of the four, so in this case we use the tetrahedral radius. It may be remarked that in this case each electron pair is assumed to be equally shared between a silver and an iodine atom, and the crystal might be thought of as composed of Ag^{---} and I^{+++} ions. Such a picture would be far from the truth, however, and the crystal no doubt much more closely resembles a complex of Ag^+ and I^- ions, though this is going too far in the other direction.

In the case of elements of the nitrogen-phosphorus group, the normal radii apply when the element has a valence of three, with three atoms attached directly, as in $AsCl_3$; in the case of the oxygen-sulfur group, the normal radii apply when the element has a valence of two, with two atoms attached, as in SCl_2. In these molecules, each atom has its completed octet.

The normal and tetrahedral radii apply to molecules in the gas phase as well as to crystals. Gas molecules have recently been extensively investigated by the electron-diffraction methods, and in Tables 34 and 35 we give some of the results[1] and compare them with the calculated covalent distances. The values given in Table 29 may also be compared with values calculated from Table 33.

It will be seen that the agreement between calculated and observed values is generally very good, in the case of Table 34. In Table 35, some discrepancies appear; these are greatest in

[1] The results are taken from the following sources: Brockway, *J. Phys. Chem.*, **41**, 185 (1937); Springall and Brockway, *J. Am. Chem. Soc.*, **60**, 996 (1938); Brockway, *Rev. Mod. Phys.*, **8**, 260–261 (1936); Gregg, Hampson, Jenkins, Jones, and Sutton, *Trans. Faraday Soc.*, **33**, 856 (1937).

just those cases in which the bond might be expected to have some ionic character.

Perhaps most interesting are the compounds of boron.[1] In all cases, the bond distances are smaller than calculated from the covalent radii. This is not unexpected, for the binding in these compounds is certainly not of the simple tetrahedral sp^3 type. BF_3 has exactly the same number of electrons as CO_3^{--}, and the remarks anent the latter made at the end of

TABLE 34.—BOND DISTANCES AND RADIUS SUMS IN METHYL COMPOUNDS, FROM ELECTRON-DIFFRACTION MEASUREMENTS
(In Angstroms)

	C—C	N—C	O—C	F—C
Calc.	1.54	1.47	1.43	1.41
Obs.	1.55	1.47	1.42	1.42
				(1.385)[a]
	Si—C	P—C	S—C	Cl—C
Calc.	1.94	1.87	1.81	1.76
Obs.	1.93	1.87	1.82	1.77
				(1.66)[a]
	Ge—C	As—C		Br—C
Calc.	1.99	1.98	....	1.91
Obs.	1.98	1.98	....	1.91
				(1.88)[a]
	Sn—C			I—C
Calc.	2.17	....	....	2.10
Obs.	2.18			
				(2.07)[a]

[a] These are spectroscopic values derived by Sutherland, reference 3, p. 325.

Note on configurations: Compounds of C, S, Ge, Sn, tetrahedra; compounds of N, P, As, pyramids; compounds of O, S, bent lines.

Sec. 16.8 will hold equally well for BF_3, though the binding in BF_3 may well be supposed to have more of an ionic character than that in CO_3^{--} because boron is less electronegative than carbon, and fluorine is more electronegative than oxygen. Also the double-bonded structures would be relatively less likely in BF_3 than in CO_3^{--}, as compared with the single-bonded structure, since the single-bonded structure, having fewer shared electrons,

[1] LÉVY and BROCKWAY, *J. Am. Chem. Soc.*, **59**, 2085 (1937).

TABLE 35.—BOND DISTANCES IN CERTAIN HALIDES FROM ELECTRON-DIFFRACTION MEASUREMENTS
(In Angstroms)

Central atom	Compound	Distances for				
			F	Cl	Br	I
B	BX_3	Calc.	1.53	1.88	2.03	
		Obs.	1.31	1.75	1.87	
C	CX_4	Calc.	1.41	1.76	1.91	
		Obs.	1.36	1.755	1.93	
Si	SiX_4	Calc.	1.81	2.16		
		Obs.	1.54	2.00		
Ge	GeX_4	Calc.	....	2.21		
		Obs.	....	2.08		
Sn	SnX_4	Calc.	....	2.39		
		Obs.	....	2.30		
P	PX_3	Calc.	1.74	2.09	2.24	2.43
		Obs.	1.52	2.00	2.23	2.52
As	AsX_3	Calc.	1.85	2.20	2.35	2.54
		Obs.	1.72	2.16	2.36	2.58
Sb	SbX_3	Calc.	....	2.40	2.55	2.74
		Obs.	....	2.37	2.52	2.75
O	OX_2	Calc.	1.30	1.65		
		Obs.	1.41	1.68		
Te	TeX_2	Calc.	....	2.36	2.51	
		Obs.	....	2.36	2.49	
Cl	ClX	Calc.	....	1.98		2.32
		Obs.	....	1.983		2.315

Note on configurations: BX_3, plane triangles; CX_4, SiX_4, GeX_4, SnX_4, tetrahedra; PX_3, AsX_3, SbX_3, pyramids; OX_2, bent lines; TeX_2, probably straight lines.

is of a more ionic character. All the halides of boron have, of course, the same number of valence electrons. The distances in BCl_3 and BBr_3 suggest an effective radius of 0.75Å. for boron in compounds of this type. This agrees fairly well for $B(CH_3)_3$, also.[1] The observed B—C distance in this compound is 1.56Å.; calculated distance, assuming boron radius of 0.75Å., is 1.52Å. The observed B—F distance, however, is rather lower than the calculated distance of 1.39Å. But this is not unexpected if the binding in this compound has an appreciably greater ionic character than the other compounds.[2]

Cases in which the observed distances are greater than the calculated covalent distances are found in[3] ONCl and ONBr. The N—Cl distance is 1.95Å., calculated 1.69Å.; the N—Br distance is 2.14Å., calculated 1.84Å. It has been suggested that this is due to some ionic character of the bond on the supposition that the sum of the ionic radii is larger than the sum of the covalent radii in this case; however, actually it seems difficult to predict what the ionic distance should be.

The F—F distance in F_2 is also unexpectedly large (1.45Å.—see Table 8), and here no ionic contribution is possible. This may indicate some peculiarity in the binding in fluorine, but in view of the empirical nature of the covalent radii, it may be that too much stress should not be laid on discrepancies of this kind.

It is perhaps not without interest to note that in $TeCl_2$ the Cl—Te—Cl valence angle is not a tetrahedral or a right angle; instead, the atoms are probably in a straight line.[4] A similar statement holds for $TeBr_2$. This must indicate some peculiarity in the electron structure of these compounds; it is clear that the bonds are neither of the usual sp^3 type nor of the pure p type. In spite of this, the agreement between observed and calculated distances is very good as may be seen from Table 35.

$HgCl_2$, $HgBr_2$, HgI_2, ZnI_2, and CdI_2 are other examples of linear molecules.[4,5] Here there is considerable discrepancy between observed and calculated distances. The observed distances are: 2.34, 2.44, 2.61, 2.42, 2.60Å.; calculated: 2.47, 2.62, 2.81, 2.64, 2.81Å., respectively. The observed distances in the

[1] It will, of course, be observed that the double-bonded structure of CO_3^{--} cannot be written for $B(CH_3)_3$ unless a CH_3 group shares electrons as a unit. This may possibly occur.

[2] For some remarks on $B(OH)_3$ see Sec. 19.10, p. 438.

[3] Ketelaar and Palmer, *J. Am. Chem. Soc.*, **59**, 2629 (1937).

[4] Brockway, *Rev. Mod. Phys.*, **8**, 260–261 (1936).

[5] Hassel and Strömme, *Zeits. physik. Chem.*, **B38**, 466 (1938); Gregg, *et al.*, reference 1, p. 321.

gaseous molecules differ considerably from the distances in the crystals (see Table 36).

Other interesting substances are P_4 and As_4. In these molecules, the four atoms occupy the corners of a tetrahedron,[1] and the bond angles are, therefore, equal to the angle of an equilateral triangle, 60°. This is far from the normal tetrahedral angle (which is, of course, the angle made by the lines joining the center of gravity of the tetrahedron with two vertices), or from a right angle, but in spite of this the interatomic distances (2.21Å. in P_4 and 2.44Å. in As_4) are quite normal.

There are some cases where the interatomic distance between a given pair of atoms apparently varies in different molecules. According to the electron-diffraction data, the C—F distances in the various fluorochloromethanes vary by as much as 0.06Å. This has been discussed by Brockway.[2] Recently, however, Sutherland[3] has made estimates of carbon-halogen distances in methyl halides, with the use of spectroscopic data; these are included in Table 34 for comparison. In the cases of CH_3F and CH_3Cl, these depend chiefly upon analysis of the rotational spectrum; in the cases of CH_3Br and CH_3I, the rotational spectrum cannot be resolved, and recourse is had to a discussion of the vibrational levels, which involves certain theoretical guesses. The distances thus derived for C—Br and C—I are subject to considerable uncertainty, but in general there appears to be enough discrepancy between the results obtained from the two types of data to leave room for some doubt in the whole matter. For further details, the reader must be referred to the original papers of Brockway and of Sutherland.

Brockway[2] has also called attention to the fact that in the chlorine derivatives of ethylene, containing a double bond, the C—Cl distance, as obtained from electron-diffraction data, is considerably smaller (about 0.06 or more) than in saturated compounds. This is supposed to be due to resonance (see Sec. 12.5) between two different states of the molecule.[4] For exam-

[1] Brockway, *Rev. Mod. Phys.*, **8**, 260–261 (1936).

[2] Brockway, *J. Phys. Chem.*, **41**, 185 (1937).

[3] Sutherland, *Trans. Faraday Soc.*, **34**, 325 (1938).

[4] The fact that the electric moment of $H_3C—CH_2Cl$ is 2.03×10^{-18} e.s.u. whereas that of $H_2C═CHCl$ is only 1.66×10^{-18} has been cited as evidence in favor of the occurrence of resonance in the latter compound. The

ple, the compound C_2H_3Cl can conceivably have either of the two following electron structures:

```
H.    .Cl        H  .C̈l:
 :C::C:       H:C̈:C̈˙˙
H˙    ˙H          ˙˙ Ḧ
```

The first is the one usually assumed, but if the compound actually has properties intermediate between those suggested by the two formulas, then it is clear that the C—Cl bond will have some of the properties of a double bond, and hence the C—Cl distance should be somewhat shortened. For it is found in general that double-bond distances are about 0.20Å. less, and triple-bond distances are about 0.34Å. less, than single-bond distances. It may be remarked that a somewhat similar situation occurs with CO_2. Here resonance could take place between the following structures:

```
:O⋮⋮C:Ö:     Ö::C::Ö     :Ö:C⋮⋮O:
```

The middle one is the one usually written, but the others, one with a triple bond on one oxygen, the other with a triple bond on the other oxygen, are also possible. The distance between carbon and oxygen in this molecule, 1.13–1.16Å., is definitely smaller than one would expect from a double bond and more like what would be expected for a triple bond. It appears that resonance, which is in general accompanied by a lowering of the energy of the molecule, also generally results in a shortening of the intermolecular distance, so that in this case the tendency is to make the distance resemble that expected for a triple bond, rather than to increase the double-bond distance toward that expected for a single bond.

Although the results of electron diffraction by gases are very interesting, and the number of researches is continually increasing, it is probably still true that the main source of material available for comparison with the atom radii is the result of X-ray examination of crystals. We shall now proceed to a discussion of this material. Instead of merely reproducing

---

resonance should result in the chlorine sometimes sharing two electrons with the neighboring carbon atom. The chlorine in the unsaturated compound should thus be less negative than the chlorine in the saturated compound where no resonance can occur. The hypothesis of resonance therefore offers an explanation of the lower electric moment in the unsaturated compound.

TABLE 36.—DISTANCES IN CRYSTALS

Compound	Obs.	Ionic	Covalent	R_c/R_a	C.N.
Li_2O	2.00	2.00	....	0.335	4
Li_2S	2.47	2.43	....	0.269	4
Li_2Se	2.59	2.57	....	0.254	4
Li_2Te	2.82	2.77	....	0.236	4
Na_2O	2.40	2.37	....	0.540	4
Na_2S	2.83	2.80	....	0.433	4
Na_2Se	2.95	2.93	....	0.409	4
Na_2Te	3.17	3.14	....	0.380	4
K_2O	2.79	2.75	....	0.756	4
K_2S	3.19	3.18	....	0.606	4
K_2Se	3.32	3.31	....	0.573	4
K_2Te	3.53	3.52	....	0.531	4
CsCl	3.56	3.61	....	0.933	8
CsBr	3.71	3.75	....	0.865	8
CsI	3.95	3.95	....	0.781	8
CuF	1.84	2.19	1.99	0.705	4
CuCl	2.35	2.63	2.34	0.530	4
CuBr	2.46	2.77	2.46	0.492	4
CuI	2.63	2.99	2.63	0.444	4
Cu_2O	1.84	2.19	(2.01)	0.545	2
Cu_2S	2.42	2.85	(2.39)	0.438	4
Cu_2Se	2.49	2.98	(2.49)	0.413	4
AgF	2.46	2.62	(2.17)	0.926	6
AgCl	2.77	3.07	(2.52)	0.696	6
AgBr	2.88	3.21	(2.64)	0.646	6
AgI	2.99	3.42	(2.81)	0.584	6
AgI	2.80	3.28	2.81	0.584	4
Ag_2O	2.05	2.48	(2.19)	0.715	2
BeO	1.65	1.53	1.73	0.244	4
BeS	2.10	(1.99)	2.11	0.106	4
BeSe	2.22	(2.15)	2.21	0.185	4
BeTe	2.43	(2.40)	2.39	0.172	4
Be_2C	1.81	(3.27)	(1.84)	0.104	4
MgF_2	1.99	1.82	(2.04)	0.603	6
$MgCl_2$	2.69(?)	2.25	(2.39)	0.453	6
MgI_2	2.94	(2.68)	(2.68)	0.380	6
MgTe	2.77	2.68	2.72	0.328	4
Mg_2Si	2.77	(3.43)	(2.57)	0.214	4
Mg_2Ge	2.76	(3.39)	(2.62)	0.221	4
Mg_2Sn	2.93	(3.48)	(2.80)	0.222	4
CaF_2	2.36	2.26	....	0.868	8
$CaCl_2$	2.74	2.60	....	0.652	6
CaI_2	3.12	2.97	....	0.545	6

TABLE 36.—DISTANCES IN CRYSTALS.—(*Continued*)

Compound	Obs.	Ionic	Covalent	R_c/R_a	C.N.
SrF_2	2.50	2.41	....	0.970	8
$SrCl_2$	3.02	(2.84)	....	0.729	8
BaF_2	2.68	2.62	....	1.13	8
ZnF_2	2.03	1.92	....	0.646	6
$ZnCl_2$	2.73(?)	2.34	(2.30)	0.486	6
ZnO	1.98	2.04	[1.97]	0.500	4
ZnS	2.35	2.45	[2.35]	0.401	4
ZnSe	2.45	2.59	[2.45]	0.379	4
ZnTe	2.64	2.80	[2.63]	0.352	4
CdF_2	2.34	2.24	(2.12)	0.839	8
$CdCl_2$	2.66	2.59	(2.47)	0.630	6
CdI_2	2.99	2.95	(2.76)	0.528	6
CdO	2.36	2.41	(2.14)	0.647	6
CdS	2.52	2.69	[2.52]	0.520	4
CdSe	2.62	2.83	2.62	0.491	4
CdTe	2.80	3.04	2.80	0.456	4
HgF_2	2.40	2.37	(2.12)	0.919	8
$HgCl_2$	2.25	2.43	2.47	0.690	2
$HgBr_2$	2.48	2.56	2.62	0.640	2
HgI_2	2.78	2.98	(2.76)	0.578	4
HgS	2.53	2.85	[2.52]	0.570	4
HgSe	2.62	2.98	2.62	0.539	4
HgTe	2.78	3.19	2.80	0.500	4
BN	1.45	1.61	(1.59)	0.138	3
AlF_3	1.70–1.89	1.54	(1.90)	0.530	6
Al_2O_3	1.90	(1.78)	(1.92)	0.410	6
Al_2O_3	1.62–1.78	1.66	(1.92)	0.410	4
AlN	1.91	2.07	1.96	0.292	4
AlP	2.36	2.42	2.36	0.258	4
AlAs	2.44	2.52	2.44	0.253	4
AlSb	2.64	2.70	2.62	0.244	4
Sc_2O_3	2.09	2.11	....	0.602	6
ScN	2.22	2.58	....	0.430	6
Y_2O_3	2.27	2.26	....	0.681	6
LaF_3	2.36–2.70	2.39	....	1.021	11
GaP	2.35	2.60	2.36	0.290	4
GaAs	2.44	2.69	2.44	0.284	4
GaSb	2.64	2.86	2.62	0.275	4
InSb	2.79	3.07	2.80	0.352	4
TlSb	3.33	(4.28)	(2.83)	0.390	8
C	1.54	(2.69)	[1.54]	0.070	4
SiC	1.89	(2.99)	1.94	0.157	4
SiF_4	1.60	1.35	1.81	0.478	4

TABLE 36.—DISTANCES IN CRYSTALS.—(*Continued*)

Compound	Obs.	Ionic	Covalent	R_c/R_a	C.N.
SiI_4	2.46	2.12	2.50	0.301	4
SiO_2	1.58	1.49	1.83	0.369	4
SiS_2	2.14	1.88	2.21	0.296	4
Si	2.34	(2.99)	[2.34]	0.169	4
TiO_2	1.96	1.91	2.02[a]	0.545	6
TiS_2	2.42	2.31	2.40[a]	0.438	6
$TiSe_2$	2.53	(2.44)	2.50[a]	0.414	6
$TiTe_2$	2.72	(2.72)	2.68[a]	0.383	6
TiC	2.16	(3.94)	2.13[a]	0.232	6
ZrO_2	2.20	(2.27)	(2.17)[b]	0.620	8
ZrS_2	2.58	2.44	2.55[b]	0.498	6
$ZrSe_2$	2.68	2.58	2.65[b]	0.470	6
ZrC	2.34	(4.0)	2.28[b]	0.264	6
CeO_2	2.34	(2.35)	....	0.722	8
GeI_4	2.57	2.28	2.55	0.352	4
GeO_2	1.86	1.77	(1.88)	0.431	6
Ge	2.44	(3.11)	[2.44]	0.205	4
SnI_4	2.63	2.46	2.73	0.445	4
SnO_2	2.06	1.95	(2.11)[c]	0.545	6
SnS_2	2.55	2.35	[2.49][c]	0.438	6
Sn (gray)	2.80	3.39	[2.80]	0.259	4
PbO_2	2.16	2.10	(2.16)[d]	0.602	6
$SiCl_4$	2.00	1.76	2.16	0.359	4
$GeCl_4$	2.08	1.91	2.21	0.420	4
$SnCl_4$	2.30	2.10	2.39	0.530	4
$TiCl_4$	2.21	2.06	....	0.530	4
PO_4^{---}	1.5	1.41	1.76	0.335	4
SO_4^{--}	1.5	1.31	1.70	0.302	4
ClO_4^-	1.5	1.24	1.65	0.279	4
IO_4^-	1.8	1.57	1.94	0.437	4

[a] From the octahedral radius, 1.36Å. for Ti, of Pauling and Huggins.
[b] From the octahedral radius, 1.51Å. for Zr, of Pauling and Huggins.
[c] From octahedral radius of Sn, 1.45Å., as modified by Pauling.
[d] From octahedral radius of Pb, 1.50Å., as modified by Pauling.

tables of experimentally observed and theoretically calculated distances in atomic crystals, it has seemed best to prepare and insert at this point a fairly comprehensive table of interatomic distances, both ionic and covalent. In Table 36, therefore, we present the experimental anion-cation distances for a large number of binary crystals; a few elementary crystals are also included. In a few cases, the data given in the literature indicate the possibility that the atoms surrounding a cation are not all at the same distance. Unless there is a large variation in these distances, however, a rough mean is given. Also, in some cases, different crystal forms with the same coordination number are known. In these cases, a mean distance is also used.

The experimental distances are compared with the calculated values obtained from the ionic radii according to the method[1] described in Sec. 14.6. When necessary, the calculated values are corrected for anion-anion contact.[2] In the case of elementary crystals, like diamond, it is assumed that half the atoms are anions and half cations. Where available, covalent radii from the tabulation of Pauling and Huggins are also included. Values that were originally used in obtaining the set of covalent radii are in brackets. In some cases, the covalent radii used are not applicable to the actual crystal form of the substance considered, and in such cases the values are enclosed in parentheses. The fact that a substance exhibits its normal valence in a crystal does not mean that one is justified in using the normal valence radius, as will be clear from the description above of when the normal radius is to be used. On the other hand, if the crystal is not tetrahedral, the tetrahedral radius is not directly applicable,

[1] The method of Sec. 14.6 was modified somewhat in the case of tetrahedral molecules. It seemed that in these substances it would be best to neglect forces between the separate molecules altogether. The electrostatic forces are due to the attraction of the central cation for the surrounding anions and the anions' mutual repulsion. If r is the anion-cation distance, then it is readily proved from the geometry of the tetrahedron (see Appendix IV) that the potential per molecule is $-12.33e^2/r$, so that the Madelung constant A in this case is 12.33, which is to be compared with the value of 1.748 for NaCl. Taking into account the fact that there are four repulsive contacts per molecule as compared with six in NaCl, we see, by the same reasoning as in Sec. 14.6, that the radius sum obtained from Table 16 should be multiplied by $(1.748/12.33)^{\frac{1}{n-1}} (4/6)^{\frac{1}{n-1}}$. This correction should apply equally well to a gaseous molecule, and at the end of Table 36 we have added some results for tetrahedral molecules obtained by the electron-diffraction method.

[2] Correction is made only for the unoccupied space. No correction is made for possible change in the number of contacts between ions, or for change in the value of n. The correction can, therefore, be only partially satisfactory. Where anion-anion contact occurs or where the radius ratio is very close to the limit so that it is very nearly, but not quite occurring, the results for the calculation of the ionic distances have been placed in parentheses.

either; however, unless normal valence radii are clearly indicated, tetrahedral radii are used. We have included in the table the ratio of the radius of the cation to that of the anion (R_c/R_a) and the coordination number (C.N.) of the cation from which it may be judged to what extent the rule of close-packing about the cation, discussed in Sec. 16.8, is obeyed. It will be observed that anion-anion contact does occur, but is relatively infrequent; on the other hand, sometimes not enough anions surround a cation to fill up the space around it.

A comparison of the experimental distance in a crystal with the results obtained from ionic and covalent radii may serve in some cases to distinguish between ionic and covalent binding, though often no certain conclusion may be drawn.[1] It is rather remarkable, in fact, how often the distances calculated from ionic radii and covalent radii check each other rather closely. Large discrepancies occur only when the elements involved are very close to each other in the periodic table, and the compounds are pretty definitely covalent. The way the distance is divided between adjacent atoms is very different, however, depending upon whether the ionic or covalent set of radii is used.

A special word should be said about BN. In this case the crystal is arranged in sheets, with the following type of structure (each bond line indicating an electron pair):

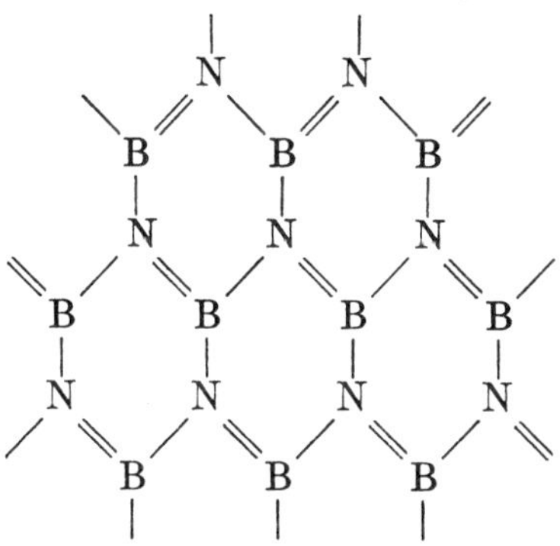

[1] It should also be borne in mind that when $z_c z_a$ [see Eq. (15) Chap. XIV] is greater than 4, the difference between the calculated value of r_0 and the radius sum of Table 16 becomes great. Under these circumstances, owing to the uncertainty in n, the value of r_0 can be considered only a rough approximation, and may be in error by 10 or, in extreme cases, 20 per cent. Assuming the "reduced" Madelung constant A_0 to be the same for all crystals may cause the calculated ionic distances in crystals having three atoms per stoichiometric molecule to be about 0.05Å. too low in the worst cases.

This particular structure can, of course, resonate with many other structures having the double bonds in other positions. The small interatomic distance is no doubt due to the double-bond character of the linkages and the resonance. Carbon, in the form of graphite, has the same type of structure and has a very similar interatomic distance, 1.42Å. In both these crystals the distance between the sheets of atoms is relatively very large.

To Table 36, we append the distance from the central atom to one of the surrounding atoms in some of the common complex ions. These distances vary somewhat from crystal to crystal, as we have noted, and the values given are "average" or typical values. The experimental distances have again been compared with covalent distances and with ionic distances, calculated much the same as for tetrahedral molecules, but allowing for the double charge of the oxygen and for the particular charge on the cation. It will be seen that the ionic distances agree remarkably well with the observed distances, though it seems doubtful that the forces can be purely ionic. The oxygen-oxygen distances in these crystals are very small, much smaller in all cases than twice the radius of oxygen in Table 16. In fact, they are so small that it seems probable that there is anion-anion (*i.e.*, oxygen-oxygen) contact even though the cation-anion ratio would not indicate that this should be the case. At least this is true unless the distortion of the oxygen ions is quite considerable (as would be the case if the binding were largely covalent).

Pauling and Huggins[1] have also given a table of octahedral radii, which they have used to correlate interatomic distances in crystals such as FeS_2, in which metal atoms are surrounded by six others in the form of a regular octahedron, the binding being of the sp^3d^2 type. The interatomic distances are obtained by combining the octahedral radius with the appropriate radius for the other atom involved. Where possible, distances calculated from these octahedral radii are included in Table 36.

16.10. Distances in Isosteres.—If two molecules have the same electronic structure, they are said to be isosteres. Examples are furnished by the pairs CO and N_2 and OCO and NNO. In each pair the molecules have the same number of valence electrons and the same number of inner electrons on corresponding

[1] Pauling and Huggins, *Zeits. Krist.*, **87**, 227*ff*. (1934); these have been slightly modified by Pauling, "The Nature of the Chemical Bond," pp. 168*ff*., Cornell University Press, 1939.

atoms. They differ only in the charges on the nuclei. The molecules in each pair have remarkably similar physical properties and also similar structures. Thus the interatomic distance[1] in CO is 1.13Å., in N_2 it is 1.09Å.; in OCO the distance between end atoms is 2.32Å., in NNO it is 2.38Å.

One may in the same way speak of isosteric crystals. Thus KF and CaO are isosteres. In this case, however, the interatomic distances are noticeably different, being 2.67Å. for KF and 2.40Å. for CaO. If we compare the radii sums directly from Table 16, the order of size is reversed, and we get 2.69Å. for KF and 2.94Å. for CaO. The small observed size for CaO is due to the higher charges on the ions. Such results are typical of ionic compounds.

On the other hand, Goldschmidt[2] has observed that in covalent crystals with tetrahedral structure, all of which have eight val-

TABLE 37.—INTERATOMIC DISTANCES IN ISOSTERIC CRYSTALS WITH COVALENT BINDING

Number of inner electrons[a]	Isosteres	Interatomic distances
2, 2	BeO, CC	1.65, 1.54
2, 10	BeS, CSi, NAl	2.10, 1.89, 1.91
2, 28	BeSe, OZn, FCu	2.22, 1.98, 1.84
10, 10	AlP, SiSi	2.36, 2.34
10, 28	AlAs, PGa, SZn, ClCu	2.44, 2.35, 2.35, 2.35
10, 46	MgTe, AlSb, SCd	2.77, 2.64, 2.52
28, 28	CuBr, ZnSe, GaAs, GeGe	2.46, 2.45, 2.44, 2.44
28, 46	CuI, ZnTe, GaSb, SeCd	2.63, 2.64, 2.64, 2.62
46, 46	AgI, CdTe, InSb, SnSn	2.80, 2.80, 2.79, 2.80

[a] Of the elements in the compounds, in the order in which they are given in the formulas of the compounds.

ence electrons per stoichiometric molecule, the interatomic distances are practically the same for all of a series of isosteres. This is brought out in Table 37, where it is seen that Gold-

[1] These are the data from band spectra, see Sponer, "Molekulspektren," Vol. I, Julius Springer, Berlin, 1935. Data from electron diffraction agree within the limit of error, but indicate a greater difference between CO_2 and N_2O.

[2] See GRIMM and WOLFF, "Handbuch der Physik," 2d ed., vol. XXIV/2, p. 996, Julius Springer, Berlin, 1933.

schmidt's rule holds better the more covalent the set of compounds would be expected to be. The table is so arranged as to bring out the way in which the interatomic distances gradually increase with increasing complexity of the electron structure. This method of exhibiting the results seems to bring out certain properties of the covalent bond better than merely finding a set of covalent radii, but is, of course, in no way inconsistent with the existence of the set of covalent radii. Table 37 suggests, in fact, a simplified set of covalent radii, as follows: all elements with 2 inner electrons, 0.77; with 10 inner electrons, 1.17; with 28 inner electrons, 1.22; with 46 inner electrons, 1.40. These values give radius sums that agree with the experimental data with the same accuracy as that with which the distances in isosteric crystals agree with each other. They will not reproduce the experimental data as well as the values of Table 33. It may be that the radii in Table 33 are not to be taken as pure covalent radii, but as containing an unconscious correction for the tendency of certain elements to form compounds with a partially ionic character. However, this is a matter that can scarcely be decided definitely.

16.11. The Transition between Ionic and Covalent Binding.—The halides of silver and copper and the oxides, sulfides, selenides, and tellurides of zinc, cadmium, and mercury are substances of simple composition in which the binding is undoubtedly intermediate in character between ionic and covalent. It should, therefore, be of interest to collect in one table the various properties that may be of use in studying the transition between ionic and covalent binding. This has been done as far as is possible in Table 38.

First is given the radius of the cation from Table 16 and the polarizability of the anion from Table 10. As seen in Sec. 12.4, these quantities should help us form a judgment as to the relative deviations from pure ionic type to be expected of the various compounds in the table. It is important, however, to notice that Cu^+ is about the same size as Na^+ and Ag^+ is only slightly smaller than K^+. It is, therefore, clear that if the silver and copper halides show greater departure from a purely ionic character than do the alkali halides, this cannot be due to the small size of the cation. Indeed, it appears to be generally true, as has been stressed by Fajans, that an ion with a shell of eighteen electrons exerts a greater distorting force on a given anion than does an otherwise similar ion with a shell of eight. This is undoubtedly connected with the greater force exerted by an eighteen-shell ion on a penetrating electron, as the outer electrons of the anion probably penetrate to some extent beneath the outer shell of the cation.

Table 38.—Properties Illustrating the Transition between Ionic and Covalent Binding

Substance	r (cation)	$\alpha \times 10^{24}$ (anion)	Ionic −Obs. dist.	Pure covalent[a] −Obs. dist.	U_0 (theor.) $-U_0$ (expt.)	C.N.
CuF	0.96	0.96	0.35	0.15	...	4
CuCl		3.57	0.28	0.04	18	4
CuBr		4.99	0.28	−0.02	22	4
CuI		7.57	0.36	−0.01	29	4
AgF	1.26	0.96	0.16	......	9	6
AgCl		3.57	0.30	......	11	6
AgBr		4.99	0.33	......	14	6
AgI		7.57	0.48	0.00	18	4
ZnO	0.88	2.74	0.06	0.01	−13	4
ZnS		8.94	0.10	0.04	28	4
ZnSe		11.4	0.14	−0.01	55	4
ZnTe		16.1	0.16	−0.02	...	4
CdO	1.14	2.74	0.05	......	46	6
CdS		8.94	0.17	0.05	31	4
CdSe		11.4	0.21	0.00	58	4
CdTe		16.1	0.24	0.00	...	4
HgS	1.25	8.94	0.32	......	67	4
HgSe		11.4	0.36	......	99	4
HgTe		16.1	0.41	......	...	4

[a] The covalent distances are calculated from the radii given in Sec. 16.10 rather than those given in Table 33, as it is believed that the former represent more nearly the radii to be expected from purely covalent bonds.

The series of silver halides have often been cited as exhibiting very beautifully the transition between ionic and covalent bonds. It will be seen that the deviation between the observed internuclear distance and the calculated ionic distance becomes progressively more pronounced going from AgF to AgI. Likewise, the difference between the experimental and theoretical lattice energies becomes progressively greater. Finally AgF, AgCl, and AgBr have the sodium chloride type of lattice, whereas AgI has a tetrahedral type of lattice[1] (with, to be sure, some complications—see Sec. 16.12). A tetrahedral type of lattice is, of course, characteristic of sp^3 binding, as we have seen, but long before the development of the wave mechanical theories, it was held to be characteristic of covalent binding. This type of lattice is, of course, to be expected with ionic binding if the cation-anion radius ratio is small enough; but the ratio is not sufficiently small in the case of AgI, and alkali iodides having cations smaller than Ag^+ have the sodium chloride structure. In addition to the properties thus far considered, it is believed that the increasing insolubility of the silver salts from the fluoride to the iodide is directly connected with the increasingly covalent character of the binding. This is discussed in Sec. 19.7. All the properties of these substances thus show trends consistent with the trend in the character of the bond.

The beautifully consistent picture that is thus obtained with the silver halides may, however, be to some extent deceptive, for as soon as an attempt is made to extend it to the other substances in Table 38 difficulties arise.

Consider, for example, the cuprous halides. One gains a general impression that these salts are more covalent than the silver salts. This appears from the larger values of U_0 (theor.) $-U_0$ (expt.) as well as the fact that the coordination number in all the salts is four, and a greater degree of covalency in the cuprous salts is quite to be expected since Cu^+ is a smaller ion than Ag^+. The values of U_0 (theor.) $-U_0$ (expt.) are not only larger than those for the silver salts but they have the expected trend. Yet when we look at the difference between ionic and observed distances, we see that the trend is entirely lacking: the discrepancy could be explained if it were supposed that, for some reason the calculations that resulted in the radii of Table 16 gave too high a value for Cu^+. However, the lack of a trend in this instance may be to some extent accidental, for it is seen that, except for the fluoride, which is the most ionic cuprous salt,

[1] The data in Table 36 for AgI with silver having a coordination number of 6 were obtained from solid solutions with AgBr, not pure AgI.

the agreement is good between observed distances and those obtained from *covalent* radii.

In the case of the bivalent elements, it is not always easy to find marked trends from oxygen to tellurium. This is perhaps due, in that instance, to uncertainty in the values of U_0 (theor.) $-U_0$ (expt.) for CdO, etc., and it may be fair to say from the small value of this quantity in ZnO, from the close agreements of ionic and observed distances in the oxides (which may, however, be accidental since the covalent distances also agree), and from the coordination number of six in CdO that the oxides are more ionic than the other compounds. Such a conclusion, nevertheless, does not seem very certain. There seems to be some evidence, from the large values of U_0 (theor.) $-U_0$ (expt.), that the mercury compounds are more covalent than the others, which is just what would *not* be expected from the sizes of the ions.

It will thus be seen that, although these compounds most probably do exhibit the transition between ionic and covalent binding, the simple regularities, for which one is sometimes inclined to hope, cannot always be found.

The trends that occur in the nature of the binding among the metals of the group under discussion are of interest in connection with the trends in certain other related properties. It is generally recognized that silver is a more noble metal than copper. It is often stated or implied that the more noble a metal is, the less electropositive it is. However, the quality of nobleness in a metal is connected with nonreactivity, whereas the quality of electropositiveness is reflected in the type of bonds that it forms. It is true that there is often a parallelism between the two properties. Thus cesium is more electropositive than lithium, and also less noble, for in general its compounds have a higher heat of formation (*i.e.*, more heat is evolved when they are formed) and hence are more stable. See, for example, the values of Q in Table 20. And, of course, it is well known that the true noble metals are among the least electropositive.

Nevertheless, we should not be surprised to find that occasionally a less noble metal is also less electropositive, as occurs in the case of copper and silver. The halides of copper have only slightly greater heats of formation[1] than those of silver, and the

[1] It is of interest to note (see Table 21) that this slight difference is actually due to the greater degree of covalency of the copper bonds, and, hence, with

halides of univalent copper are stable with respect to decomposition into copper and the halides of bivalent copper. The real difference between copper and silver lies in the stability of the oxides and sulfides. The heats of formation (evolved) are as follows: Cu_2O, 42.5 kg.-cal.; Ag_2O, 6.95; Cu_2S, 19.0; Ag_2S, 5.5. Since the stability of such compounds is not a simple function of any one property of either of the elements involved, it is seen that the nobleness of a metal cannot be easily related to any other single property.

In the case of the zinc, cadmium, mercury family, the stability of the oxides and sulfides decreases in the order in which the members of the family are written. Zinc and cadmium halides are not exceedingly different, but mercury halides are much less stable. We have seen, however, that mercury also appears to be less electropositive, so here there is the usual relation between nobleness and electropositivity. (Cf. also Cd and Hg in Table 4.)

16.12. Special Types of Molecular and Crystal Structure.—Though many complex ions, molecules, and crystals have simple symmetrical structures, such as regular triangles, tetrahedra, squares, octahedra, and cubes, this is by no means true in all cases. There are molecules whose form corresponds to none of the expected types, and whose structure is difficult to explain on the basis of electron theory or electrostatic forces. Also there are numerous crystals that have exceedingly complicated structures, frequently involving highly distorted tetrahedra and octahedra. Although distortions may sometimes be due to the tendency of certain atoms or ions to form as closely packed a structure as possible, there are certainly many cases in which the arrangement may, to say the least, be characterized as unexpected.

An illustration of a rather extreme type, which shows the complexities that may be involved, is furnished[1] by AgI. As noted in Sec. 16.9, at sufficiently low temperatures this substance has a tetrahedral type of structure in which each silver is surrounded by four iodine atoms, and each iodine by four silver atoms. In

---

respect to the halides, copper is actually less noble than silver because it is less electropositive.

[1] Helmholz, *J. Chem. Phys.*, **3**, 740 (1935); Strock, *Zeits. physik. Chem.*, **B25**, 441 (1934). The discussion has to do with the behavior of AgI having the wurzite structure (see Appendix IV). Another form having the sphalerite structure is also known.

order to have complete electron sharing, the crystal would have to be composed of Ag^{---} and I^{+++} ions; in other words, three electrons from iodine would have to be among those shared. That they would be very reluctantly shared is evident, and the softness of the crystal would indicate that the atoms do not have four strong tetrahedral bonds. So it is, perhaps, not very surprising that at a higher temperature (room temperature) irregularities begin to appear. At room temperature, many of the silver atoms find a position of equilibrium in which they are closer to three of the surrounding iodines than to the remaining one. Above 146°C., the sharing of electrons becomes still less, there is a transition to a cubic body-centered lattice of iodine ions, and the silver ions seem to move freely, like a fluid, in the interstices; this is in spite of the fact that the density of the high temperature form is greater. Owing to the free mobility of the silver ions, the high temperature form readily conducts electricity. No very satisfactory explanation of this rather remarkable behavior can be given, but it is certainly not typical of either the purely ionic or the purely covalent type of crystal. Although, as stated above, the case of AgI is extreme, it is to be noted that not many other crystals have been investigated so thoroughly. Whereas there are few, or no other, cases in which some of the atoms or ions have alternative positions, in other respects some other crystals are about as complicated. In a complicated crystal of this sort, there is usually present at least one transition element or, at any rate, one from the central portion of the periodic table. Generally the elements involved are such as would be expected to have a binding of type intermediate between covalent and ionic, and frequently, though by no means always, some of the elements are not in their maximum valence states. It may be of interest to note here that also in the case of the transition between metallic and covalent binding complicated structures appear (see Sec. 18.2).

It would be impossible to attempt here to give a detailed description of the various types of crystal structure that have been discovered, and little would be gained, since at present few unifying principles have been found. However, there are some types of molecular and crystal structure which, though not conforming, at least apparently, to the ordinary rules, nevertheless are of fairly frequent occurrence. It seems worth while, without presenting an exhaustive account, to give a brief discussion of some of these. They will be classified according to the coordination number.

Coordination Number Two.—There are a number of ions and molecules containing a total of three atoms, in which all the atoms lie in a straight line, and in which the central atom may, therefore, be said to have a coordination number of two. Most of these compounds have already been mentioned in Table 31 and Sec. 16.9. A partial list follows: $Ag(NH_3)_2^+$, $Cd(NH_3)_2^{++}$, $Hg(NH_3)_2^{++}$, $HgCl_2$, $HgBr_2$, HgI_2, ZnI_2, CdI_2, $TeCl_2$, $TeBr_2$, I_3^-. The ions, of course, are known from crystal measurements[1]; $HgCl_2$ and $HgBr_2$ are known as linear molecules in both gas and crystal. The others are known from electron diffraction in gases. The linear form with coordination number two does not persist in the crystal in the case of HgI_2, CdI_2, and probably ZnI_2. To the list of substances having linear configurations should be added Cu_2O, in the form of cuprite, in which copper has a coordination number of two and oxygen of four.

It will be noted that here is a large group of substances which may be described as composed of a central eighteen-electron shell ion, to which are attached two negative ions or groups with octets. This is a convenient description but should not be taken as necessarily implying ionic binding. Ionic binding would give a ready explanation of this type of molecule in the gas phase, but in many cases this type of structure is found in the crystal, and in any event, HgI_2 must certainly have largely covalent binding. It has been shown[2] that by combination of the s-wave function with one of the p-wave functions it is possible to get two bond-eigenfunctions which are oppositely directed. These are not quite so much concentrated in one direction as are the tetrahedral sp^3 functions, but when there are only two pairs of electrons in the valence shell they may tend to be formed, because they involve less excitation in the unpairing of the electrons of the central atom. The excitation is expected to be less because the low-energy s-wave function is less diluted, so to speak, by the

[1] The cases of $Cd(NH_3)_2^{++}$ and $Hg(NH_3)_2^{++}$ depend upon measurements of the corresponding halides. Since there are also halogen ions fairly close to the mercury it is to some extent a question of judgment as to whether halogens are also coordinated with the mercury or cadmium or not.

[2] Hultgren, *Phys. Rev.*, **40**, 891 (1932).

higher energy *p*-states than in the case of the sp^3 bond-eigenfunctions. Another possibility in the case of the mercury and cadmium halides is that sp^3 bonds are formed with the double-bonded structure $:\ddot{Cl}::Hg::\ddot{Cl}:$ Such a possibility does not exist with the ammonia compounds, however, unless ammonia can share its electrons as a unit.

The tendency to form compounds with a linear structure is not confined to the elements of the silver and zinc groups, however, as is seen from $TeCl_2$ and $TeBr_2$. Tellurium, in these compounds, has a completed octet outside the shell of eighteen. In these cases it is difficult to see why *sp* linear bonds should be formed as this requires unpairing of *s*-electrons which do not need to be shared. Most probably an outer *d*-state is involved.

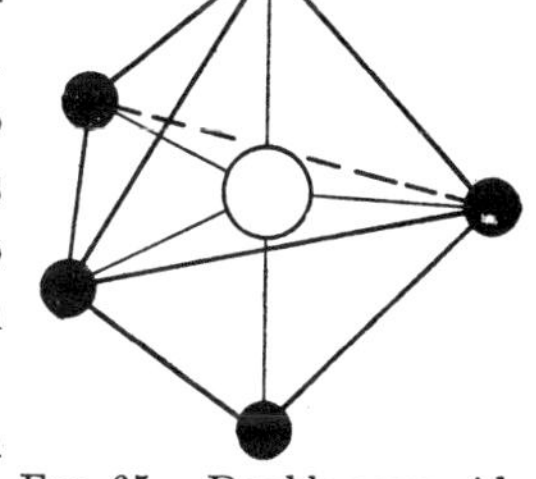

Fig. 65.—Double pyramid.

In the case of I_3^-, there are more than enough electrons to form octets on all the atoms, so one atom, presumably the central one, must have electrons in at least one *d*-orbit.

Coordination Number Four.—The ion ICl_4^-, occurring, for example, in $KICl_4$, deserves mention as an ion in which iodine has a peculiar type of coordination.[1] The chlorines are arranged in the form of a square, with the iodine in the middle. There are too many electrons for this to be binding of the regular square sp^2d type. It may be a case of octahedral binding, in which two of the corners of the octahedron have electron pairs but no substituent; outer *d*-orbits would be involved.

Coordination Number Five.—In PF_5, which has been investigated by electron diffraction,[2] the fluorines are arranged in the form of a double pyramid. Three of the fluorines form a plane triangle with the phosphorus in the middle. The other fluorines are directly above and below the phosphorus, forming the apices of the double pyramid. All fluorines are at the same distance 1.54Å. from the phosphorus. IF_5 probably has a similar structure, with an I—F distance of 2.57Å. PF_3Cl_2 and PCl_5 are similar molecules. In PF_3Cl_2 the distances[3] are, P—F, 1.59Å., and

[1] Mooney, *Zeits. Krist.*, **98**, 377 (1938).

[2] Braune and Pinnow, *Zeits. physik. Chem.*, **B35**, 239 (1937).

[3] Brockway and Beach, *J. Am. Chem. Soc.*, **60**, 1836 (1938).

P—Cl, 2.05Å. In PCl_5 the two chlorines at the apices are said to be at a slightly greater distance than the others; the distances[1] are given as 2.11Å. and 2.04Å.

Essentially the same structure occurs in the compounds[2] $Sb(CH_3)_3X_2$, where X may be Cl, Br, or I. The CH_3's form a triangle with the antimony in the middle, the halogens being directly above and below the antimony. The halogens, being attached directly to the antimony, will not ionize in acetonitrile solution, though hydrolysis occurs in water. $Sb(CH_3)_3(OH)_2$ is also a weak electrolyte, and probably has a similar structure, but $Sb(CH_3)_3XOH$ is a strong electrolyte. In this case, the antimony is probably tetrahedral, forming an ion $Sb(CH_3)_3X^+$. The tetrahedral form is what naturally would be expected, but it is seen that actually in this case it possesses no special stability.

The crystal covellite, CuS, offers another interesting example of an atom with a coordination number of five. There are two types of sulfur in this crystal; one of them is surrounded by five copper atoms in the form of a double pyramid. The copper atoms in the triangle are in this case somewhat nearer to the sulfur (2.19Å.) than those forming the apices of the double pyramid (2.35Å.).

Coordination Number Six.—A certain number of instances are known in which an atom with coordination number of six is surrounded by atoms at the corners of a trigonal prism rather than at the corners of a regular octahedron. Thus in NiAs an As atom is surrounded by six Ni's at the corners of a trigonal prism.[3] Other compounds of transition metals with S, Se, Te, and Sb crystallize in the same way. In MoS_2 and WS_2, the metal atom lies at the center of a trigonal prism formed by sulfur atoms. According to a wave mechanical treatment by Hultgren,[4] a trigonal prism can be formed if five *d*-eigenfunctions are available. Theoretically, the ratio of altitude to the edge of the base of the prism should be unity, a condition very nearly fulfilled by MoS_2 and WS_2.

[1] SCHOMAKER, quoted by Pauling, "The Nature of the Chemical Bond," p. 103, Cornell University Press, 1939.

[2] WELLS, *Zeits. Krist.*, **99**, 367 (1938).

[3] A trigonal prism is a figure whose base is a triangle (in this case an equilateral triangle) and whose sides are perpendicular to the base (see Fig. 58, p. 300).

[4] HULTGREN, *Phys. Rev.*, **40**, 891 (1932).

Electron-diffraction investigations have been made of the structures of the molecules[1] of UF_6, WF_6, and MoF_6. In contrast to results previously obtained with the hexafluorides of the sulfur group, it was found that these molecules were not regular octahedra, but that the octahedra were distorted. The fluorines can be divided into three groups of two each. Each group contains the pair of opposite fluorines, which form a straight line with the central atom. The three straight lines thus formed are at right angles to each other, just as they would be in a regular octahedron, but the distance between fluorine and the central atom is different in the different groups. In UF_6 and WF_6, the three different U—F distances are in the ratio 1:1.12:1.22. In MoF_6, the ratios may be the same, but the results are less certain. It is thus seen that uranium, tungsten, and molybdenum certainly do not tend to form regular octahedra, but the actual structure differs in the sulfides and fluorides.

16.13. One-electron and Three-electron Bonds.—There are some compounds of light elements, some of which have already been mentioned in passing, which, though hardly to be considered with the group discussed in Sec. 16.12, have rather unusual properties. These properties can be explained on the basis of certain special assumptions. It was seen in Chap. X that there is a very stable ion of the formula H_2^+ in which the two protons are held together by the action of one electron. This, of course, suggests that there may be other compounds in which some of the bonds are one-electron bonds. It is necessary for the formation of a one-electron bond that the two atoms (or ions) which are to be joined should exert forces of somewhat similar magnitude on the electron. Otherwise the electron will tend to be attached to one of the atoms, and the electron distribution about this atom will be fairly symmetrical. Since the electron is effective in producing a cementing action only when there is a tendency for it to remain between the atoms it is joining, it is seen that an overpowering attraction of one or the other of these atoms for the electron produces a weak one-electron bond; but, of course, a polar bond may be formed, if this leaves the atoms charged, and in this respect the one-electron bond is no different than the electron-pair bond.

[1] Braune and Pinnow, *Zeits. physik. Chem.*, **B35**, 239 (1937).

The structure of diborane has long been one of the outstanding puzzles of inorganic chemistry, for there appear not to be enough valence electrons to make such a compound stable.[1] Sidgwick has suggested[2] that it contains one-electron bonds. The possibilities are

$$\begin{array}{c} \mathrm{H\ H} \\ \mathrm{H\cdot\ddot{B}:\ddot{B}\cdot H} \\ \mathrm{\ddot{H}\ \ddot{H}} \end{array} \qquad \begin{array}{c} \mathrm{H\ H} \\ \mathrm{H:\ddot{B}:\dot{B}\cdot H} \\ \mathrm{\ddot{H}\ \ddot{H}} \end{array} \qquad \begin{array}{c} \mathrm{H\ H} \\ \mathrm{H:\ddot{B}\cdot\ddot{B}\cdot H} \\ \mathrm{\ddot{H}\ \ddot{H}} \end{array}$$

In the first two formulas, it is not important just which hydrogens have one-electron bonds, but only whether they are both on the same boron atom. The last arrangement, with a one-electron bond between the borons, is a distinct possibility, and it makes no essential difference which hydrogen has the remaining one-electron bond.

Lewis[3] has made a somewhat different suggestion. He supposes that at any given moment there is one point of linkage at which there are no electrons. Thus we may have

$$\begin{array}{c} \mathrm{H\ H} \\ \mathrm{H:\ddot{B}:\ddot{B}:H} \\ \mathrm{\ddot{H}\ H} \end{array} \qquad \text{or} \qquad \begin{array}{c} \mathrm{H\ H} \\ \mathrm{H:\ddot{B}\ \ddot{B}:H} \\ \mathrm{\ddot{H}\ \ddot{H}} \end{array}$$

It is supposed that there is resonance between all possible states. Since boron and hydrogen have about the same electronegativity, as will be evident from extrapolation of the electronegativities of the elements of the first row of the periodic table (see Table 14), the electrons will have no choice as between a B—B bond or a B—H bond. Hence, there being seven linkages and six pairs of electrons, each bond will have a pair six-sevenths of the time. Essentially the same result should follow from Sidgwick's hypothesis, and either Sidgwick's or Lewis's structure simply amounts to saying that the twelve electrons are uniformly distributed over the seven bonds. Since resonance produces an energy level which is lower than that of any of the resonating states, the uniform

[1] It is, as a matter of fact, very reactive. For the chemistry of boron see Stock, "Hydrides of Boron and Silicon," Cornell University Press, 1933.

[2] Sidgwick, "Electronic Theory of Valency," p. 103, Oxford University Press, 1927.

[3] Lewis, *J. Chem. Phys.*, **1**, 17 (1933).

distribution of all the electrons may well produce a stable B_2H_6, even though any given electron structure would be expected to result in an unstable molecule.

The literature has not been lacking in other suggestions as to the structure of the B_2H_6 molecule. Böeseken and Wahl[1] suggested that it is an acid of the anion $\left[\begin{array}{c}\text{H}\ \text{H}\\ \text{H:}\ddot{\text{B}}\text{:}\ddot{\text{B}}\text{:H}\\ \ddot{\text{H}}\end{array}\right]^-$, and Wiberg has suggested that it is an acid of the anion $\left[\begin{array}{c}\text{H}\quad\text{H}\\ \ddot{\text{B}}\text{::}\ddot{\text{B}}\\ \ddot{\text{H}}\quad\ddot{\text{H}}\end{array}\right]^{--}$.
Wiberg[2] has recently reviewed the evidence in favor of his structure. He cites the formation in liquid ammonia solution of the saltlike electrolytically conducting $B_2H_6 \cdot 2NH_3$, which he interprets as an ammonium salt of the acid. He states that the ethylenelike character of the anion is indicated by the fact that B_2H_6 readily adds two atoms of alkali metal. Further, he supports the opinion that four hydrogens of diborane are different from the others by citing the fact that four hydrogens, only, may be replaced by methyls.

However, the arguments of Wiberg can hardly be said to be conclusive. In the first place, Schlesinger and Burg[3] have brought forth evidence which indicates that $B_2H_6 \cdot 2NH_3$ is rather the ammonium salt of the anion $\left[\begin{array}{c}\text{H}\ \text{H}\ \text{H}\\ \text{H:}\ddot{\text{B}}\text{:}\ddot{\text{N}}\text{:}\ddot{\text{B}}\text{:H}\\ \ddot{\text{H}}\ \ddot{\text{H}}\ \ddot{\text{H}}\end{array}\right]^-$ which they believe to be formed in ammonia solution because in such solution B_2H_6 has a certain tendency to split into BH_3's, a tendency also supported by other evidence [*i.e.*, the reaction at low temperatures with trimethylamine to form $(CH_3)_3NBH_3$]. The fact that B_2H_6 adds alkali may indeed be an evidence of unsaturation, but this unsaturation is present according to any formulation. The alkali-addition compounds may be alkali salts of the anion $\left[\begin{array}{c}\text{H}\ \text{H}\\ \text{H:}\ddot{\text{B}}\text{:}\ddot{\text{B}}\text{:H}\\ \ddot{\text{H}}\ \ddot{\text{H}}\end{array}\right]^{--}$. Finally, the fact that only four hydro-

[1] BOESEKEN, *Verslag Akad. Wetenschappen Amsterdam*, **31**, 591 (1922); WAHL, *Zeits. anorg. allgem. Chem.*, **146**, 230 (1925).

[2] WIBERG, *Ber. deut. chem. Ges.*, **69**, 2816 (1936).

[3] SCHLESINGER and BURG, *J. Am. Chem. Soc.*, **60**, 290 (1938).

gens can be replaced by alkyls might be evidence that an anion of the type $[(CH_3)_2B{=}B(CH_3)_2]^{--}$ may be formed without it necessarily being true that an anion $[H_2B = BH_2]^{--}$ is formed. But more likely it simply shows that addition of a more negative group than H draws the electrons from the B—B linkage, so that further addition of methyl groups would result rather in splitting up the compound. As a matter of fact, boron trimethyl, $B(CH_3)_3$, and $B(C_2H_5)_3$ are not associated (see also the discussion in Sec. 17.3). It should, however, be stated that attempts to introduce a fifth methyl into $B_2H_2(CH_3)_4$ have been unsuccessful.[1]

Electron diffraction[2] and crystal structure measurements of B_2H_6 have shown that the B—B distance is about 1.86Å. This is larger than expected from Table 33 for a covalent single bond, and certainly is larger than would be expected for a double bond, and so offers further argument against Wiberg's structure. Furthermore, the electron-diffraction measurements indicate that all the hydrogens are equivalent, rather than that two of them are different, though, on account of the difficulty of determining positions of very light elements by electron diffraction, this conclusion may not be entirely certain.

Before concluding the consideration of diborane, we must mention one physical property that has frequently played an important role in discussions of its structure; namely, it is found to have no magnetic moment.[3] As we have seen, this means that there are as many electrons with spin in one direction as in the other. One might, perhaps, suppose that in the one-electron formulas, in which there are two unpaired electrons, there would be a chance of these electrons both having their spins in the same direction. If, however, resonance occurs between the one-electron state and states in which all of the electrons are paired, this cannot be the case. On account of the difficulty of turning over the spin of an electron, it is not possible for the molecule to change from one state to another in which the spins of some of the electrons are different, and hence no resonance occurs between such states.

[1] A. B. Burg, private communication.

[2] Bauer, *J. Am. Chem. Soc.*, **59**, 1096 (1937).

[3] See footnote 4 of Schlesinger and Walker, *J. Am. Chem. Soc.*, **57**, 621 (1935).

The compounds of boron appear to be the only important ones, outside of H_2^+, in which the idea of a one-electron bond offers promise of being useful. The situation is different, however, with respect to the three-electron bond,[1] which differs only slightly from a one-electron bond. Suppose we have two atoms, or ions, one with one outer electron and one with two, thus, $A\cdot$ and $B:$ We suppose the two electrons on B to be paired. Suppose the state in which there are two electrons on A and only one on B has a comparable energy. Then we can have one of the electrons shared between A and B and alternately paired with the electron on A and that on B. In certain energy states, the shared electron will tend to lie between A and B, and thus will form a bond between them. Since the shared electron must be alternately paired with an electron on A and one on B, its spin must be the opposite of that of each of these electrons, each of which must, accordingly, have the same spin. The spin of one of these electrons will be unbalanced, and the three-electron bond thus must contribute to the magnetic moment of the molecule.

Now the oxygen molecule is a molecule that spectroscopic analysis shows to have no resultant orbital angular momentum, yet it has a magnetic moment. This shows that it must possess unbalanced spins, which would not occur if all the electrons were paired in the usual way. Pauling suggests that it has two three-electron bonds and one electron-pair bond, its structure being $:O\overset{\cdots}{\underset{\cdots}{:}}O:$ In this structure, as in the structure one would get from the assumption of a double bond, four electrons are shared by the oxygen atoms, and it might well be expected that these two possibilities would result in molecules with somewhat similar energies. The empirical fact, apparently, is that the three-electron-bond structure has the lower energy, but this could hardly be predicted offhand.

Another molecule that Pauling postulates has a three-electron bond is nitric oxide, written $:N\overset{\cdots}{::}O:$ The reason for this assumption is the fact that nitric oxide, though possessing an odd electron, appears to be relatively little unsaturated. It does combine with itself to form a double molecule, but the energy of dissociation is low, only about 3 kg. cal.[2] The dissociation

[1] Pauling, *J. Am. Chem. Soc.*, **53**, 3225 (1931).

[2] For a discussion and references see Rice, *J. Chem. Phys.*, **4**, 367 (1936).

energy of N_2O_4, also rather low, about 13 kg.-cal., is nevertheless much larger than this.[1] Further, NO tends to combine with organic radicals having an unpaired electron, but these appear to be about as unsaturated as NO itself. Pauling writes for these a formula similar to the one he suggested for oxygen $\mathrm{R}:\overset{\cdots}{\underset{\cdots}{\mathrm{N}}}:\mathrm{O}:$

Pauling has also discussed magnetic moments of certain complex compounds containing NO, namely, $[Co(NH_3)_5NO]Cl_2$ which has a moment corresponding to two electrons with spin in the same direction and

$$Na_2[Fe(CN)_5NO]2H_2O, \qquad [Ru(NH_3)_4NO\ H_2O]Cl_3,$$

and

$$[Ru(NH_3)_4NO\ Cl]Br_2$$

which have zero moment. In these cases NO is assumed to have a different electronic structure than when it is uncombined. If we assume that in the first compound cobalt is trivalent and the NO is in the form of the ion NO^- with a structure and magnetic moment like oxygen, then, if the bonds between cobalt and the surrounding groups are covalent, the compound should have the magnetic moment observed, this coming entirely from the electrons on the NO group. If, in the iron and ruthenium compounds, these metals are in the divalent state and the NO is in the form of a positive ion NO^+ with structure like nitrogen (having, hence, no three-electron bonds and no magnetic moment) and if the bonds are covalent, the observed zero magnetic moment is explained.

It might be well to discuss the phenomena that we have explained in terms of one-electron and three-electron bonds from the viewpoint of the general theories which we have discussed in Chap. XI. A one-electron or three-electron bond is of course invoked precisely to account for phenomena that do not fit into the electron-pair scheme, and it would appear that the idea of the stability of the shell of eight loses a good deal of its significance in these cases. However, the Hund-Mulliken theory is general enough to explain the phenomena we have noted. Thus the magnetic moment of O_2 is explained simply by supposing that

[1] Pauling, however, also writes a structure for NO_2 involving three-electron bonds. See "The Nature of the Chemical Bond," p. 250.

the two promoted electrons are not in the same energy level, and that they have parallel spins, something about the interaction of the spins and the orbital motion stabilizing this arrangement. The assumption that one electron in NO is promoted and that there are six bonding electrons, giving the equivalent of two and a half bonds, corresponds well with the structure given in the preceding paragraphs.

There does, however, seem to be a difference between the Hund-Mulliken theory and the three-electron bond theory, which may be definitely stated, though we shall not attempt to say which one is a better description of the actual state of affairs in the molecule. In the case of a three-electron bond, only one of the three electrons involved may be properly called a bonding electron, as will be clear from our original description of a three-electron bond. Thus in O_2 there are four bonding electrons two in the pair and one in each of the three-electron bonds. The existence of the three-electron bonds is conditional upon the assumption that the oxygens are sufficiently far apart so that two distinct quantum states are involved, one belonging to one oxygen atom and the other to the other oxygen atom, in each three-electron bond. If the oxygens were close enough so that the electrons in these two quantum states began to interfere with each other, *i.e.*, if the united atom state of affairs were actually being approached, one of the three electrons would have to be promoted and two of them would constitute a bonding pair. From the three electrons, the net number of bonding electrons would still be one, but the three-electron bond appears essentially to require that the promotion of electrons shall not yet have occurred. In this way it may be said to differ from the Hund-Mulliken picture of promoted electrons. We may postulate a three-electron bond in cases like O_2 if we look at the combination from the separated atom point of view, and we get the Hund-Mulliken picture if we use the united atom point of view, just as in more ordinary instances the separated atom point of view leads us to electron pairs, whereas the united atom point of view leads to consideration of promoted electrons. Probably the truth lies somewhere in between.

In Chap. XI, it was suggested that the combination of two BH_3's to form B_2H_6 might be like the combination of two oxygen atoms to form O_2. Nothing that has been said about B_2H_6

would exclude the possibility that there is at least a certain amount of truth in this suggestion. However, since B_2H_6 has no magnetic moment and O_2 has, it would appear likely that the lowest state of B_2H_6 resembles the excited double-bonded state of oxygen with all electrons paired, rather than the normal lowest state with the three-electron bonds.

16.14. Double Bonds in Complex Compounds.—After the preceding sections of this chapter went to press, Pauling[1] published the suggestion that many of the oxygen ions, and complex compounds in general, have some double-bond character. For example, in SO_4^{--} he assumes that the oxygens tend to share more than one electron pair with the central atom, taking advantage of the ability of an atom like sulfur, in which relatively low-lying *d*-states are available, to expand its valence shell beyond the octet. The principal reason for this assumption is the fact that the interatomic distance is smaller than calculated for single bonds (see Table 36).

While such double-bonded structures are possibilities which cannot be definitely excluded, and although it is not easy to estimate just how much they may contribute, it is the opinion of the author of the present work that there is little reason to believe that they are usually of great importance. In the first place, it does not appear that the evidence from interatomic distances is very conclusive. The distance calculated from covalent single-bond radii is undoubtedly an overestimate, since a single-bonded structure must consist of S^{++} and O^- ions, and the superimposed electrostatic forces would undoubtedly shorten the bond. It is to be further noted that the small interatomic distance calculated on the basis of S^{6+} and O^{--} ions is an underestimate (for it does not take into account the forces due to the positive ions immediately outside the sulfate ion in a crystal), and, if it could be properly corrected, might well be expected to give something near the correct interatomic distance in spite of the improbable character of the assumptions on which it is based.

In SO_3^{--} the observed interatomic distance is also less than the sum of the covalent radii, but the form of the ion (regular tetrahedron with one corner missing) is as expected from a single-bonded structure. Similar remarks apply to ClO_3^-. In $H_2PO_2^-$

[1] PAULING, "The Nature of the Chemical Bond," Chap. VII, Cornell University Press, 1939.

the hydrogens are bound by single bonds to the phosphorus, but the oxygens may have double bonds. In this case the tetrahedron is rather noticeably distorted.

It is, perhaps, not possible to draw conclusions concerning bonds which involve *d*-states from bonds in which only *s*- and *p*-states play a role. It is of interest, however, that when only *s*- and *p*-states are involved, sulfur shows only a relatively weak tendency to form double bonds. The C—S single-bond energy is about 21 kg.-cal. less than the C—O single-bond energy, but by comparing CS_2 and OCS with CO_2 it is found that the C=S double-bond energy is about 65 kg.-cal. less than the C=O double-bond energy. It is, to be sure, hard to see just how the resonance with other bond structures may affect the results with these molecules, but since practically the same result is obtained from either CS_2 or OCS it is probable that the conclusions are not invalidated.

Pauling has also considered a possible type of double-bond structure in complex ions, in which the central atom furnishes the extra electrons. As a typical example we consider $Fe(CN)_6^{4-}$. The possible covalent electron-bond structures (neglecting electrons in the iron which are not in the particular bond) are Fe:C:::N: and Fe::C::N̤: The extra pair in the latter structure is contributed from the *d*-levels in the iron, and there are enough *d*-electrons in the iron to provide pairs for three double bonds. Pauling points out that if the electron pair between iron and cyanide in the first of the above formulas were equally shared between them, it would leave the iron with a charge of -4, which, he believes, could be only partly removed by the partially ionic character of the bond. Resonance with the second formula would obviate the difficulty. In the case of $Cr(CN)_6^{4-}$, however, another difficulty appears when the magnetic moment is considered. For if the double bonds have the usual character they should involve paired electrons, in which case all of the electrons in $Cr(CN)_6^{4-}$ should be paired, provided, at least, that all of them share in double-bond formation. But the magnetic moment indicates that there are two unpaired electrons, so at most two of the four[1] *d*-electrons of chromium can be used in

[1] Two electrons from chromium are surely shared, leaving thus four *d*-electrons which are shared only if there is double-bond formation.

the double bonds. If one pair of d-electrons were equally shared between chromium and cyanide it would reduce the negative charge on chromium by only one electronic charge, which would reduce the negative charge on the chromium by only one-fourth of its amount. It may nevertheless be possible that the great stability of numerous cyanide complexes (and nitrite complexes, too) is due in part to resonance with these double-bonded structures. However, at the present time it seems that most of our knowledge of these compounds is reasonably well expressed by the single-bond formula, allowing for partial ionic character of the bonds.

As Pauling points out, it is not possible to write double-bond formulas for complexes with NH_3, H_2O, etc. In most such cases the binding is more or less ionic, so the large negative charge is effectively removed from the central atom.

Exercises

1. A solution is known to contain cobalt, chlorine, and ammonia, and no other substance except water. It is known that for every atom of cobalt, in whatever form it may be, there are 3 atoms of chlorine in some form or other, and 1 molecule of NH_3. The total concentration of all cobalt is 1 gram atom per l. To 100 cc. of this solution, 0.3 mol of $AgNO_3$ are added. A precipitate is separated and found to consist of 0.2 mol of AgCl. To another 100-cc. portion of the solution, a 100-cc. portion of 0.1 N HCl is added, and the concentration of H^+ in the resulting solution is found to be 0.05 N. Give the formulas (including the charges) of the ions, and other substances, if any, present in the original solution in appreciable concentration.

2. The solubility of AgBr is about 7×10^{-7} mol per liter. It dissolves in 1 per cent $Na_2S_2O_3$ solution to the extent of 0.35 g. per 100 g. of solution. Calculate the equilibrium constant for the reaction between Ag^+ and $S_2O_3^{--}$ on the assumption (a) that the reaction is $Ag^+ + 2S_2O_3^{--} \rightleftarrows Ag(S_2O_3)_2^{---}$, ($b$) that the reaction is $Ag^+ + S_2O_3^{--} \rightleftarrows Ag(S_2O_3)^-$. In 10 per cent $Na_2S_2O_3$ solution, the solubility of AgBr is 3.50 g. per 100 g. of solution. Is assumption (a) or (b) correct? (Rough calculations, assuming perfect solutions, will be adequate.)

3. A solution was made up of a substance having the empirical formula K_3CrCl_5 (not counting possible water of hydration) and containing 1 g. of chromium per 100 g. of water. Some of this solution was electrolyzed between platinum electrodes, under conditions such that no chlorine was evolved. The portion of solution near the anode, containing 100 g. of water, was now analyzed, and was found to contain 1.252 g. of chromium and 5.13 g. of chlorine. From this, what do you infer respecting the formula of the compound? If the substance were really a mixture of $KCrCl_4(H_2O)_2 + 2KCl$, how would you expect the result to differ from that obtained?

4. Describe the possible stereoisomers of $[Co(NO_2)_3(OH)_3]Na_3$. Describe the possible stereoisomers, including optical isomers, of $[Co(NO_3)_2Cl_2(OH)_2]Na_3$. Can $[Co(C_2O_4)_3]^{---}$ exhibit optical isomerism?

5. Discuss the expected isomerism of the first two compounds in Exercise 4, if the configurations were trigonal prisms instead of regular octahedra.

6. Give Lewis electron structures for all the compounds mentioned in Sec. 16.5 whose formulas are set out in separate lines.

7. Look up the table of coordination numbers in silicate crystals given by Bragg (see, *e.g.*, Wyckoff, "Structure of Crystals," 2d ed., p. 194) (compare with Table 32), and show how it illustrates the rule discussed in Sec. 16.7 that the more electronegative cations have the greatest role in determining the crystal structure.

8. Check the electrostatic-valency rule (Sec. 16.7) in the case of BPO_4 for B and P.

9. The crystal NH_4HF_2 is presumably composed of NH_4^+ and FHF^- ions. Each N is surrounded by four H's, each one of which is also in contact with one F. Each F is in contact with three H's, two of which are parts of (two separate) NH_4^+ ions and one of which is in the FHF^- ion. Check the electrostatic-valency rule for N and F.

10. In CaB_2O_4, the boron and oxygen form chains as shown in Fig. 63. There are two types of oxygen atoms. The first type is linked to one boron and three calciums, the second type to two borons and one calcium. Each boron is surrounded by three oxygens, one of the first type and two of the second. Each calcium is surrounded by eight oxygens, six of the first type and two of the second. Show that the statement regarding the surroundings of a calcium follows from the other statements, and check the electrostatic-valency principle.

11. Write the electron formulas for the possible states among which B_4H_{10} can resonate. Assume that this is a straight chain compound, in this way resembling the straight chain hydrocarbons.

12. Give the electron structure for the complex compounds containing NO mentioned in Sec. 16.13. Show that the statements made regarding the magnetic moments follow from these electron structures.

13. Discuss the electron structures of NO_2^- and ClO_2^-, and show that these ions should have the form noted in Table 31.

14. Discuss electron structures and predict the shapes of SO_3, $AuCl_2^-$, and $CuCl_3^-$.

CHAPTER XVII

MOLECULAR CRYSTALS

In Chap. XIII, we saw that the forces in molecular crystals may be either van der Waals forces or dipole-dipole forces, and learned something of the nature of these forces. In the present chapter, we shall consider the magnitude of these forces from the point of view both of theory and of the effects they produce.

17.1. Van der Waals Forces.—Any molecule or atom will always have a dipole moment that continually fluctuates in magnitude and direction. This will be true even though the average moment of the molecule is zero, and is due to the fact that the electrons are continually in motion, so that the center of negative electricity is not always located exactly at the nucleus or center of positive electricity. It is clear, for example, that a hydrogen atom always has a dipole moment because the electron does not coincide with the positively charged nucleus. However, the orientation of this dipole, in an electric field of the order of magnitude that can be readily applied, is so much slower than the natural rate of motion of the electron around the nucleus that the measured moment is simply the average moment. The dipole is oriented in one direction as often as in the opposite direction, and so the average in any direction is zero, and the hydrogen atom appears to have no dipole moment; this is expressed by saying that it has no *permanent* dipole moment. Nevertheless, the temporary moment can exert a force on another atom that is close enough to it; it produces an electric field in the neighborhood of the other atom, and since the latter is polarizable a dipole moment is induced in it. The interaction between the temporary dipole in the one atom and the induced dipole in the other produces an attraction between them. (This is in addition to, and quite apart from, any attraction due to valence forces, which are not considered in this discussion, though, except for the rare gases, they are always large compared with the van der Waals forces if atoms are under consideration.)

We see from Eq. (21), page 467 that the expression for the potential energy associated with this attraction is $-\frac{\alpha_2 M_1^2(3\cos^2\theta_1 + 1)}{2r^6}$. M_1 is the temporary dipole of the first atom, θ_1 is the angle it makes with the line joining the centers of the two atoms, and α_2 is the polarizability of the second atom in which the dipole is induced. Since the temporary dipole is continually changing in magnitude and direction, to get an average value of the potential we must take the average value of M_1^2, which is called $\overline{M_1^2}$, and the average value of $\cos^2\theta_1$ over all angles in space, which is equal to $\frac{1}{3}$. The expression then becomes $-\frac{\alpha_2\overline{M_1^2}}{r^6}$. To get the total potential between two atoms, it is necessary to add the potential due to the dipole induced in the first atom by the temporary dipole of the second atom, which gives

$$U = -\frac{\alpha_2\overline{M_1^2}}{r^6} - \frac{\alpha_1\overline{M_2^2}}{r^6} \tag{1}$$

To find the value of U, we need to be able to estimate the α's and M^2's. We shall attempt to find only orders of magnitude. As seen in Sec. 12.4, in order to find the order of magnitude of α for an atom, it may be supposed, as a rough approximation, that the atom consists of a positive nucleus of charge Ze, surrounded by a sphere of negative electricity of uniform density and radius R, and that the electric field causes displacement of the sphere of negative electricity without distortion. Then [Eq. (3) of Chap. XII]

$$\alpha = R^3. \tag{2}$$

The major axis of the orbit of the outermost electron must also be of the order of R. We should naturally expect M to be of the order of magnitude of the distance from the nucleus of the outermost electron times its charge, so that

$$\overline{M^2} = e^2R^2 \tag{3}$$

approximately. It seems reasonable to suppose that the effect is essentially that due to *one* electron, as for the most part the electric moments of the various electrons will cancel each other.

As long as the outermost electron is outside all the other electrons, the net charge acting upon it is the same as that exerted by the proton of a hydrogen atom on the electron. On the assumption that the orbit of the outermost electron is reasonably hydrogenlike, we may write as an approximation for its energy

$$E = -\frac{e^2}{2a}, \tag{4}$$

where a is the semimajor axis of the ellipse and naturally of the order of R. Now $-E$ is just the energy necessary to remove this electron from the atom, *i.e.*, the ionization potential I. So

$$I \cong \frac{e^2}{2R}. \tag{5}$$

Using Eqs. (2) and (3) in Eq. (1), for the case that the two atoms are alike, so that $\alpha_1 = \alpha_2 = \alpha$ and $\overline{M_1^2} = \overline{M_2^2} = \overline{M^2}$, we get

$$U = -\frac{2\alpha\overline{M^2}}{r^6} = -\frac{2e^2R^5}{r^6}. \tag{6}$$

By the use of Eqs. (5) and (2), Eq. (6) may also be put into the form

$$U = -\frac{4\alpha^2 I}{r^6}, \tag{7}$$

which is useful if α and I are known from independent measurements. Except for the constant factor (which should be $\frac{3}{4}$), this is actually the approximate formula obtained by quantum mechanical calculations.[1] In view of the rough assumptions made in the preceding calculations, the agreement is all that could be expected; in fact, it may even be to some extent fortuitous.

The formula thus derived is confined in its applicability to atoms; however, it is actually reasonably good for molecules that are not too complex in structure, and especially molecules whose outer electron configurations are reasonably symmetrical. The

[1] EISENSCHITZ and LONDON, *Zeits. Physik*, **60**, 491 (1930); LONDON, *Zeits. phys. Chem.*, **B11**, 222–226 (1930). Or, see SLATER and FRANK, "Introduction to Theoretical Physics," pp. 439–442, 545–553, McGraw-Hill Book Company, Inc., 1933.

potential energy represented by the formula is, of course, of importance only in the case of saturated molecules or rare-gas atoms between which there are no valence forces.

This formula has been applied to the calculation of heats of sublimation of molecular crystals by London.[1] In doing this, it is necessary to take into account the forces exerted on any one atom or molecule by all its neighbors.[2] Forces of this type are additive, the force (and the potential) between two atoms or molecules being independent of others in the immediate neighbor-

TABLE 39.—THEORETICAL AND EMPIRICAL HEATS OF SUBLIMATION

Subs.	I Kg.-cal. per mole	Density, g./cc. at 0°K.	Molal volume cc. at 0°K.	$\alpha \times 10^{24}$, cc.	Heat of sublimation, kg.-cal. per mole	
					Theor.	Expt., 0°K.
Ne	495	1.46	13.8	0.39	0.39	0.59
A	361	1.70	23.5	1.65	1.77	2.03
Kr	321	3.2	26.	2.54	3.0	2.80
N_2	391	1.03	27.2	1.74	1.59	1.86
O_2	299	1.43	22.4	1.57	1.46	2.06
CO	329	1.05	26.7	1.99	1.82	2.09
CH_4	334	0.53	30.2	2.58	2.43	2.70
Cl_2	419	2.00	35.4	4.60	7.0	7.43
HCl	315	1.56	23.4	2.63	4.0	5.05
HBr	306	2.73	29.6	3.58	4.5	5.52
HI	292	3.58	35.7	5.4	6.7	6.21
NO	235	1.58	19.0	1.76	2.01	4.29

[1] LONDON, *op. cit.*, pp. 236–242; for a review, see LONDON, *Trans. Faraday Soc.*, **33**, 8 (1937).

[2] In a close-packed crystal (the usual type for simple molecular lattices), each molecule has 12 nearest neighbors. The potential of an atom in such a crystal, therefore, would be roughly twelve times that of the mutual potential of two atoms. However, each molecular contact is a contact of *two* molecules, and the corresponding intermolecular potential would be counted twice if we were to multiply the mutual potential of two molecules by twelve times the total number of molecules. The total energy of the crystal, therefore, is six times the mutual energy of a pair of molecules. Actually, however, the more distant molecules contribute something, so that the total attractive energy of a crystal in which the attractive energy follows the inverse-sixth-power law is 7.23 times that of a single pair of molecules with the same intermolecular distance.

hood. The distance between molecules is calculated from the density of the crystals, and the repulsive force [as well as appreciable corrections which really ought to be made in the attractive forces, because the distance of approach of the molecules is too close for Eq. (21) of Appendix III to hold exactly] is neglected. Both the experimental value of the heat of sublimation and the density are corrected to 0°K., in order to avoid complications due to thermal motions of the atoms. I was evaluated from spectroscopic data.

The results of the comparison between theory and experiment are presented in Table 39, taken chiefly from London. It will be seen that the agreement between theory and experiment is reasonably good, especially in view of the approximations involved, except for nitric oxide, which consists of double molecules in the liquid or solid phase.[1] Associated solids and liquids in general have high heats of vaporation, since the heat of vaporization also includes the heat of dissociation.

Of especial interest is the series of hydrogen halides, in which the heat of sublimation increases with the molecular weight. As already noted in Chap. XIII, this behavior is typical of van der Waals forces. On first glance at Eq. (6), assuming it to give a good approximation for the potential, this may seem a little strange; for one's first inclination would be to set the intermolecular distance r in the crystal equal, or at least proportional, to $2R$, which would make U inversely proportional to R and hence U would be expected to decrease with the molecular weight. If one may assume that the value of R for a hydrogen halide is somewhere near the radius of the corresponding halogen ion, then the intermolecular distances calculated from the respective densities (either at 0°K. or the boiling point) in solid or liquid HCl, HBr, and HI certainly parallel closely the values of R. On the other hand, it will be seen from the table that α actually varies more rapidly with R than R^3 (R^3 being proportional to the molal volume), so the fact that Eq. (6) does not appear to give an increase in the van der Waals potential with the molecular weight may be laid in part to the approximations in Eq. (2).

17.2. Dipole Forces.—The nature of the dipole forces, resulting from the interaction between molecules with permanent electric

[1] See Rice, *J. Chem. Phys.*, **4**, 367 (1936) for a discussion and references.

moments, will be sufficiently clear from the discussions in Chap. XIII and Appendix III. When permanent dipoles are present, the potential due to their interaction is simply superimposed upon the van der Waals potential; these two effects are independent of each other. (There will be an additional attraction due to the interaction between the permanent dipole and the dipole induced by it in another molecule. This, however, should be relatively unimportant compared with the forces due to the permanent dipoles, unless two different molecules, one with a permanent dipole and one without, are involved.)

The hydrogen halides offer a good opportunity to study the circumstances under which the van der Waals forces predominate, and those under which the dipole forces are more important. The calculation which was discussed in Sec. 17.1 would indicate that the cohesion properties of HCl, HBr, and HI are reasonably well accounted for by the consideration of van der Waals forces only. It will be of interest, however, to make a rough calculation to show that the dipole forces are actually small. For this purpose, we choose HCl, in which the dipole forces are most likely to be of importance.

At low temperatures, solid HCl has a somewhat distorted close-packed (face-centered cubic) lattice. Each molecule has 12 nearest neighbors, and from the density given in Table 39, the average distance between nearest neighbors is about 3.79 Å. Using, then, the expression developed in Appendix III for the potential between two dipoles, setting $\sin \theta_1 = \sin \theta_2 = 0$ and $\cos \theta_1 = -\cos \theta_2 = 1$ (*i.e.*, assuming the most favorable orientation for attraction), and taking $M_1 = M_2 = M$ from Table 40, we obtain a value of 3.88×10^{-14} erg for the potential of the two molecules. Multiplying by Avogadro's number and reducing to calories, we get 563. Suppose now that of the molecules near any one the equivalent of six are favorably oriented (which is impossibly large). Each of these contacts is a contact between *two* molecules; these forces will contribute, under the assumptions made, about $563 \times \frac{6}{2} = 1689$ cal. per mole to the heat of sublimation. This is certainly an overestimate, but even so, it is relatively small compared with the heat of sublimation given in Table 39. It might possibly be objected that some of the molecules are appreciably closer than this average distance;

these might exert greater mutual forces, but the difference must be small. The density of the high-temperature modification of HCl, in which, presumably, the molecules are rotating freely[1] and which is of the cubic close-packed variety, so that all near neighbors are equally distant, is only about 5 per cent less than the density at absolute zero. Furthermore, the heat of sublimation of the high-temperature form is still about 4500 cal. per mole at its melting point, which is much larger than the value of 1689 estimated above as maximum dipole action at 0°K.

In the case of HF, the situation is entirely different. Here the dipole forces play a predominant role. This is evidenced by the relatively high boiling point, which will be discussd below, and by the tendency of HF to polymerize in the gas phase. Simons and Hildebrand[2] measured the vapor density of HF from −39 to 88°C., and found that at the lowest temperature the apparent molecular weight of saturated vapor became 87.4 (molecular weight of the monomer is 20). The other hydrogen halides show no such tendency to polymerize in the gas phase, and since the van der Waals forces would be expected to be less for HF, it is obvious that some other stronger force must be holding the molecules together. Examination by electron diffraction[3] shows that the HF molecules stick together in short kinked chains, with the negatively charged part of one molecule near the positively charged part of the next.[4] The angle between two adjacent links[5] of the chain is about 142°. It seems a little strange that these chains are not straight, as that would give the lowest electrostatic energy; however, it is probable that the angle of 142° is not rigidly fixed but is, rather, an average, the bond between different HF molecules being weak enough so that the chains bend easily under thermal agitation. The distance between two

[1] HCl has a transition point at about 98°K., such as described in Chap. XIII, which has been ascribed to the setting in of rotation in the solid by Pauling, *Phys. Rev.*, **36,** 441 (1930).

[2] Simons and Hildebrand, *J. Am. Chem. Soc.*, **46,** 2183 (1924).

[3] Bauer, Beach, and Simons, *J. Am. Chem. Soc.*, **61,** 19 (1939).

[4] It has been suggested by Pauling that in the ion HF_2^-, which exists in aqueous solution, the binding is ionic, the two negative fluorine ions being held together by a positively charged hydrogen ion between them. It does not seem likely that this occurs in the polymerized HF, in gas phase, however.

[5] A "link" can be taken as the line joining one fluorine atom with the next.

adjacent fluorine atoms in the chain[1] is 2.55Å., and from this the energy of interaction is found to be 5900 cal. per bond mole. (In making this calculation, it is assumed that the HF molecules are oriented parallel to the chain, as shown in Fig. 66, which is not quite the assumption made by Bauer, Beach, and Simons.) This bond energy is, therefore, quite appreciable, and may well be expected to have a marked influence on the properties of the substance. It is considerably larger than would be expected for the van der Waals potential between two HF molecules, which (from comparison with HCl, see Table 39 and footnote 2, page 357) should not exceed 500 cal. per pair mole. The permanent dipole-induced dipole potential, calculated from Eq. (21)

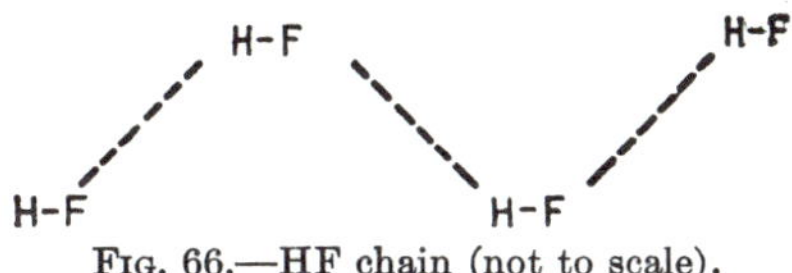

Fig. 66.—HF chain (not to scale).

of Appendix III (a reasonable estimate of the polarizability of HF being made), is also small.

A comparison between the HF and water molecules will be of interest. The heat of sublimation of ice is approximately 12,000 cal. per mole. As shown in Chap. XIX, ice has a tetrahedral structure, each molecule being surrounded by four others. Since each contact is a contact between *two* molecules, the number of "bonds" between molecules is twice the number of molecules. Therefore (if the effect of more distant molecules is neglected) the mutual energy of two water molecules that are in contact is about 6000 cal. per pair mole, which is just about the same as calculated for one pair of HF molecules. But even this energy is not sufficient to cause association of water in the vapor phase. The marked association of HF would seem to indicate that the binding energy of a pair is even greater than this. However, in view of the approximate nature of the calculations, the discrepancy indicated does not seem serious.

[1] It may be of interest to compare some other observed F-H-F (fluorine to fluorine) distances: $NaHF_2$, 2.51; KHF_2, 2.25; NH_4HF_2, 2.37. See Table III of Huggins, *J. Organic Chem.*, **1**, 407 (1936). In these cases we most probably have to do with an ionic binding between H and F, as suggested in footnote 4, p. 360.

17.3. The Experimental Material.—It is not usually possible to make so complete an analysis as in the case of the hydrogen halides. Usually not so many data are available, but a great deal can be learned simply from the boiling points of a series of substances.[1] If the molecules are held together by van der Waals forces in the solid state, the same forces are operative in the liquid state, and the heat of vaporization is a good measure of the strength of these forces. But according to Trouton's rule, the heat of vaporization in calories per mole of a normal nonassociated liquid is about twenty-one times the absolute boiling point. This rule is only an approximation which is not very good at low temperatures, but it is not necessary to go into these details; for the rest of our discussion the boiling point may be taken as a sufficiently good indication of the magnitude of the van der Waals forces, or other cohesive forces, such as dipole forces if they are operative. If there are special directed forces like dipole forces, the result is always that the heat of vaporization becomes more than twenty-one times the boiling point. So an abnormally high boiling point will, in general, mean a still greater abnormality in the heat of vaporization. Since van der Waals are in general the weakest type of force operative in solids and liquids, they ordinarily result in substances that boil at lower temperatures than is the case when other forces are operative.

The hydrogen halides, whose properties have already been extensively discussed, will illustrate the preceding statements. In the accompanying table are given their boiling points T_b and their molal heats of vaporization at the boiling point ΔH_b, as well as the dipole moments M for comparison.[2] It is seen that the boiling point and heat of vaporization increase in the order HCl, HBr, HI, as expected for van der Waals forces. On the other hand, HF has an abnormally high boiling point.

Similar results are found with other hydrogen compounds, the dipole forces being appreciable only when the central atom is in

[1] The boiling point has long been used as a criterion in the classification of chemical compounds. See footnote 1, p. 205.

[2] The heats of vaporization are taken from Bichowsky and Rossini, "Thermochemistry of the Chemical Substances," Reinhold Publishing Corporation, 1936. Boiling points for the chapter are, in general, from "International Critical Tables" or Landolt-Börnstein, "Tabellen." The dipole moments are the same as in Table 15. That for HF is the estimate of Smallwood, *Zeits. physik. Chem.*, **B19**, 253 (1932).

TABLE 40.—PROPERTIES OF HYDROGEN HALIDES

	T_b, °K.	ΔH_b, kg.-cal. per mole	$M \times 10^{18}$, e.s.u.
HF	292.5	....	2.0
HCl	188.1	3.86	1.03
HBr	206.1	4.21	0.78
HI	237.7	4.72	0.38

the first row of the periodic table. This is illustrated by the boiling points of the series of hydrogen compounds of the elements at the right of the periodic system, some of which are given in Table 46, of Chap. XIX, which is presented in connection with a further discussion of the intermolecular forces of polar liquids. It is interesting to observe that when the molecule is symmetrical, so that there is no resultant dipole moment (methane), the boiling point is low even for a hydride of an element of the first row.

It will be seen from this discussion and from that of Sec. 17.2 that hydrogen is capable of forming what may be called a "bridge" between two negative atoms. When it does this, it is sometimes said to form a "hydrogen bond,"[1] though the term "bridge" is probably more descriptive. Such hydrogen bridges are responsible for the polymerization of many organic compounds. For example, formic acid and other organic acids occur as dimers even in the gas phase. The structure of the dimer of formic acid is as follows:[2]

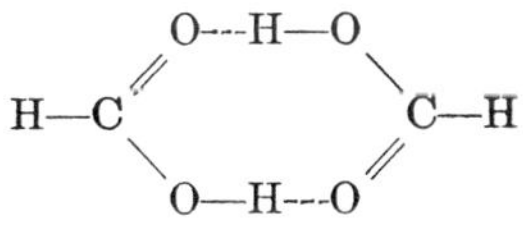

```
       O---H—O
     //       \
H—C             C—H
     \        //
       O—H---O
```

a dotted line being used to indicate the "bonds" between a hydrogen and an oxygen on the other molecule. It is thinkable in such a case as this, where the binding of the hydrogen tends toward the polar type, that the hydrogen is symmetrically situated between the two oxygens, rather than being definitely

[1] LATIMER and RODEBUSH, *J. Am. Chem. Soc.*, **42**, 1419 (1920). For an extended discussion and references, see Huggins, reference 1, p. 361; see also Rodebush, *Chem. Rev.*, **19**, 59 (1936).

[2] PAULING and BROCKWAY, *Proc. Nat. Acad. Sci.*, **20**, 336 (1934).

attached to one of them. Actually, however, it is probable the hydrogen remains attached to a particular molecule. In the case of water, which is discussed in detail in Chap. XIX, a strong hydrogen bridge is formed between water molecules, causing the relatively high boiling point of water, but the hydrogen forming the bridge undoubtedly belongs to one of the water molecules rather than the other.

Hydrogen bridges also occur in inorganic crystals, such as $NaHCO_3$. We have already discussed one such case, that of NH_4F, in Sec. 14.10.

The distance between oxygens in various O—H—O linkages varies from 2.55 to 2.8Å. It is obvious that the smaller the distance the more likely it will be that the hydrogen will be placed symmetrically with respect to the atoms it is bridging. However, the smallest of these distances is considerably greater than twice the normal O—H distance of 0.955Å in water.

Huggins has estimated the energies of hydrogen bridges in various compounds, involving F—H—F, O—H—O, C—H—N, N—H—F, C—H—O, and N—H—O linkages. (His estimate for F—H—F, however, rested on data now known to be inapplicable.) They are almost all around 6 kg.-cal. (only three varying appreciably from this figure, the total range being from 4 to 10.5). In view of the different electronegatives of the atoms and the different properties of the compounds involved, this figure seems to be remarkably constant. The N—H—N bridge formed in solid or liquid ammonia is apparently considerably weaker than those considered by Huggins. It should be noted that Huggins' estimates of hydrogen bridge strengths do not make any allowance for van der Waals forces.

It seems probable that only in compounds containing hydrogen are the dipole forces of considerable importance. Thus we may consider the series NF_3, boiling at 144°K., PF_3 boiling at 178°K., and AsF_3 boiling at 336°K. These compounds are all undoubtedly unsymmetrical (pyramidal in structure); one would expect the dipole forces, if they exist, to be more effective the smaller the molecule, but it is seen that although they may cause the boiling point of NF_3 to be higher than it otherwise would be, they are not sufficient to upset the order. On the other hand, it is probable (judging from the other halides of this group) that the dipole moments actually increase with the molecular weight, on account of the increasingly positive character of the central atom; this might be the cause of the relatively high boiling point of AsF_3, but it seems rather unlikely.

The halides of boron also are examples of molecular compounds. One might, perhaps, reasonably expect dipole forces to play a significant role in the case of BF_3, in which there is a relatively electropositive element combined with the extremely negative fluorine. However, BF_3 has the form of a triangle with the boron in the plane of the fluorines.[1] It is thus quite symmetrical, with no resultant dipole moment, and is extremely volatile. (The boiling points of the halides of boron are BF_3, 172°K.; BCl_3, 286°K.; BBr_3, 364°K.; and BI_3, 483°K., showing no indication of dipole-dipole forces.) This is, perhaps, a little surprising in view of the fact that BH_3 does not exist, but is always polymerized to the form B_2H_6, and that the halides of boron have a very considerable tendency to add an atom or group that can furnish an electron pair. Thus compounds like BCl_3PH_3 and BF_4^- are known,[2] and the latter has a tetrahedral structure. It is to be noted, however, that in these compounds the BF_3 or BCl_3 group really acts like a negative element, sharing the electrons that are furnished by another element. It may thus be the very fact that fluorine is strongly electronegative that prevents an electron pair from a fluorine in BF_3 from being shared with another BF_3 molecule. The negatively charged ion, F^-, may well be less electronegative, and so more prone to share electrons with a BF_3 molecule, giving BF_4^- (see also Secs. 16.9 and 16.13).

The aluminum halides furnish an interesting series, inasmuch as a sudden change of type takes place as one goes from the fluoride to the chloride. Aluminum fluoride has a high boiling point (sublimation point[3] = 1530°K.), indicating that the forces are too large to be merely van der Waals forces; in fact, one might well conclude that they are also too great to be dipole forces. X-ray analysis of the crystal[4] shows that each aluminum is surrounded by six fluorines, three at 1.70Å. and three at 1.89Å,

[1] Braune and Pinnow, *Zeits. physik. Chem.*, **B35**, 251 (1937); Linke and Rohrmann, *ibid.*, **B35**, 256 (1937).

[2] See, *e.g.*, Mellor, "Comprehensive Treatise on Inorganic and Theoretical Chemistry," vol. 5, pp. 132, 123, Longmans, Green & Company, 1924.

[3] Ruff and LeBoucher, *Zeits. anorg. allgem. Chem.*, **219**, 380 (1934).

[4] For crystal structures and references see the "Strukturbericht."

There is no indication that molecules are present in the lattice, and it seems rather probable that the forces are largely ionic in type, though they may be partly covalent. In the case of aluminum chloride, on the other hand, the situation is decidedly different. The boiling point is relatively low, approximately 433°K. Vapor-density measurements show that the formula is Al_2Cl_6, indicating that the crystal is probably composed of molecules of this formula. The higher halides resemble Al_2Cl_6 in their properties.[1] X-ray measurements of solid Al_2Cl_6 and electron-diffraction measurements[2] of the aluminum halide gases

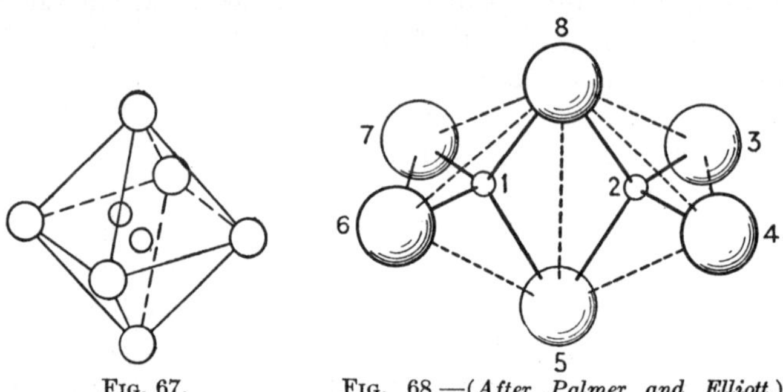

Fig. 67. Fig. 68.—(*After Palmer and Elliott.*)

have been made, but they are not in agreement. The crystal measurements of Al_2Cl_6 indicate a structure of the type shown in Fig. 67, consisting of six chlorines in the form of an octahedron with two aluminums inside, very close together (0.63Å.). The electron-diffraction measurements, on the other hand, indicate a molecule like that shown in Fig. 68, with the chlorines forming two tetrahedra with a common edge, one aluminum being inside each tetrahedron (Al-Al distance 3.4Å.). This arrangement would correspond to an electron structure like this:

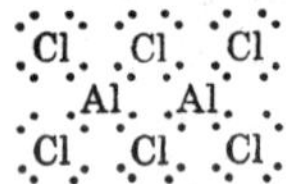

[1] An investigation of the thermodynamic properties of the aluminum halides has been made by Fischer and Rahlfs, *Zeits. anorg. Chem.*, **205,** 1 (1932). At high temperatures Al_2Cl_6 dissociates appreciably, and the other halides dissociate more easily.

[2] Palmer and Elliott, *J. Am. Chem. Soc.*, **60,** 1852 (1938).

It seems very unlikely that the structure of the molecule can be as different as this in the solid and gas. The distance between an aluminum atom and an adjacent halogen atom, according to the electron-diffraction measurements, is 2.06Å. if the halogen is one of the outside ones (No. 3 in Fig. 68, for example) or 2.21Å. if it is one of the inside ones (No. 8 in Fig. 68, for example). The calculated distance from Table 33 is 2.25Å. For Al_2Br_6, the distances are 2.21 and 2.33, calculated 2.40Å.; for Al_2I_6 they are 2.53 and 2.58, calculated 2.59. The binding is probably intermediate between ionic and covalent. It may be that the higher aluminum halides form molecular compounds, in contrast to AlF_3, because the bonds are more covalent, so that each aluminum is bound to special halogen atoms, which belong to it in particular and are not bound to any other aluminum. On the other hand, the higher halides may form molecular compounds simply because with a larger halogen the coordination number of aluminum is smaller. If the coordination number of aluminum is smaller, the coordination number of halogen (with respect to aluminum) must be smaller also. Thus halogens cannot form bridges between aluminum atoms to give a continuous network through the crystal.

In sharp contrast with the properties of aluminum chloride are those of the oxide and nitride. The former has a boiling point of 2480°K., and the forces may be either ionic or covalent, or something intermediate, but there are certainly no molecules formed. The nitride also has an extremely high boiling point, and the forces are almost certainly predominantly covalent.

The properties of scandium trichloride are also of interest. This compound is not molecular, as the boiling point is high (melting point 1212°K.). There is thus another abrupt transition from molecular to nonmolecular binding going from aluminum chloride to the corresponding scandium salt. Again, it cannot be stated with certainty whether this is due to $ScCl_3$ being more ionic than Al_2Cl_6, or whether it is because the larger scandium atom or ion has a larger coordination number than aluminum. Crystal-structure investigations of $ScCl_3$ should prove very interesting in this connection.

Halides in which there are four or more halogen atoms surrounding a central atom generally have fairly low boiling points; this is especially true of many of the fluorides. These compounds may thus be considered to be molecular compounds. Thus MoF_6 has a boiling point of 308°K., and WF_6 has a boiling point of 291°K. There are some exceptions, however. For example, SnF_4 boils at 978°K. as contrasted with 387°K. for $SnCl_4$. It is probable that the binding in SnF_4 has considerable ionic character. In the case of thorium, which in view of its high atomic number may be expected to be quite electropositive, it seems probable that the chloride $ThCl_4$ which boils at 1195°K. is largely ionic, and even the bromide $ThBr_4$ which sublimes at 883°K. is also probably largely ionic.

It is interesting that where two or more halides with different formula exist, the compound that contains the greater number of halogen atoms is often the lower boiling. Thus $SnCl_2$ boils at 896°K., whereas $SnCl_4$ boils at 387°K.; SeF_4 boils at 373°K., whereas SeF_6 boils at 224°K.; and OsF_6 boils at 476°K., whereas OsF_8 boils at 320°K. It seems reasonable to suppose that in these cases the compound with fewer halogen atoms has a more open structure, allowing close approach of electrically polarized groups and thus giving a better opportunity for the play of dipole forces; the forces may even verge toward the ionic type in $SnCl_2$. When the number of halogens is larger, there is less room about the central atom, and the close approach of groups having electric moments is prevented by steric hindrance. On the other hand, in a case like that presented by the phosphorus fluorides, it is probable that the dipole forces are unimportant, due to the negative character of the central atom, as is evidenced by the very low boiling points (PF_3, 178°K.; PF_5, 198°K.). In such a case, the van der Waals forces are greater the greater the number of atoms in the compound, and the compound with the greater number of halogens boils at the higher temperature. As is to be expected if the forces in the fluorides are predominantly of the van der Waals type, the chlorides boil at higher temperatures.[1]

[1] In a general way, van der Waals forces increase with the molecular weight of the compound, regardless of the atomic weight of the atoms forming the compound. This is well illustrated by the hydrocarbons, with their increasing boiling points in a homologous series. In a case like this, we can think of each of the atoms of the molecule having its own van der Waals (or in some cases dipole) forces operating between it and surrounding molecules, the forces on a given molecule being the sum of the forces on its parts. A large molecule is thus held in the liquid at many points, and this makes it difficult for such a molecule to leave the liquid phase.

Oxygen has less tendency than the halogens to form molecular compounds as is evidenced by the small number of volatile oxygen compounds. With a few exceptions (*e.g.*, OsO_4), volatile compounds are confined to the oxides of the halogens, the sulfur group, nitrogen, phosphorus, arsenic, and carbon. Particularly remarkable is the great difference between CO_2, on the one hand, and SiO_2 and B_2O_3 on the other.[1] Individual molecules do not exist in the latter crystals, probably because Si and B are electropositive enough to form oxides with bonds of predominantly ionic character, whereas this is not so with CO_2. However, oxygen being less negative than fluorine, the fact that it has, in general, less tendency to form molecular crystals cannot be explained by assuming that it forms more ionic compounds. It seems rather to be connected with the fact that it has a valence of two instead of one, so that with a given element in a given valence state there will be only half as many atoms of oxygen per atom of the other element in the oxide as there are atoms of fluorine in the fluoride. Thus it is easier for an oxygen to come into contact with, and be bound to, more than one atom of another element, and so it is easier for valence bonds to run from atom to atom throughout the crystal in an oxide than in a fluoride. It may be added that the small size of carbon may be part of the reason that this element prefers to form double bonds with oxygen rather than have a coordination number of four. Thus the difference between SiO_2 and CO_2 may be somewhat similar in nature to the difference between oxides and fluorides.

Summarizing, then, we may say that molecular crystals tend to be formed

1. When the strong bonds tend to be covalent rather than ionic, because the former are more definitely localized than the latter.

2. When one kind of atom tends to predominate in number in the crystal, especially if it is univalent.

Exercise

1. Argon has a face-centered cubic structure. Assuming as an approximation that each atom is affected only by its nearest neighbors, calculate the heat of sublimation from the density and value of α given in Table 39.

[1] In this connection, note the discussion in Sec. 16.7.

CHAPTER XVIII

METALLIC CRYSTALS

18.1. The Alkali Metals and Alkaline Earth Metals.—It was indicated in Chap. XIII that the typical metal (*e.g.*, an alkali metal) can be thought of as consisting of a series of positive ions immersed in a sea of free electrons. The stabilizing influence is the electrostatic attraction between the positive ions and the negative electrons. The origin of the balancing repulsive force is of a rather unusual character; it is due to the Pauli exclusion principle (Sec. 7.2) as applied to the *free* electrons in the metal.[1] These "free" electrons are actually moving in a sort of "box," similar to that described in Sec. 4.2. There is quantization of the energy levels in this box, and but one electron can occupy a given quantum state. Just as this property works out in the case of an atom to prevent it from collapsing, so it prevents the metal as a whole from collapsing.

These ideas can readily be worked out quantitatively and applied to the alkali metals.[2] The attractive force in an alkali metal may be considered to be very similar to that prevailing in an ionic uni-univalent crystal. It is true that in the crystal the negative ion has a practically fixed position, whereas in the metal the electron does not; however, it will be reasonable to write the expression for the attractive potential of the metal in the form $-\frac{Ae^2}{r}$, where here r is the average distance between electron and positive ion, A a constant, and e the electronic charge. Assuming that this average distance is proportional to the distance

[1] FRENKEL, *Zeits. Physik*, **49**, 31 (1928); FRENKEL, "Wave Mechanics, Elementary Theory," pp. 221*ff*., Oxford University Press, 1932. For the origins of the modern electron theory of metals see Pauli, *Zeits. Physik*, **41**, 81 (1927); Sommerfeld, Houston, and Eckart, *ibid.*, **47**, 1–60 (1928); and, also, Fermi, *ibid.*, **36**, 902 (1926).

[2] RICE, *J. Chem. Phys.*, **1**, 649 (1933). GOMBÁS, *Zeits. Physik*, **94**, 473 (1935); **95**, 687 (1935); and later papers in *Zeits. Physik*, has worked out a more elaborate theory of the same general type.

between adjacent positive ions, and hence to $V^{1/3}$, where V is the atomic volume of the metal, we may write it as $-\frac{A_1 e^2}{V^{1/3}}$, where A_1 is a new constant. A similar constant can be defined for ionic crystals. Sherman[1] has tabulated values of A_1 for the uni-univalent crystals and (what amounts to the same thing) $A_1/4$ for the bi-bivalent crystals whose Madelung constants are given in Table 18 of Chap. XIV; the range in A_1 (or $A_1/4$) is from 2.0 to 2.4.

Let us now return to a consideration of the repulsive force. As noted above, this may be referred to the quantization of the free electrons in the "box" provided by the metal. The box, however, will not be a simple box because of the presence of the positive ions. These ions consist of nuclei surrounded by shells of electrons occupying the inner quantum states which may be considered to be substantially unaffected by the presence of other positive ions. Since these inner states are already occupied, the free electron can enter the region near the nucleus only as a penetrating electron; as such, it can stay there only a short time. The presence of the ions thus effectively shuts out a certain portion of the space in the metal from the free electrons. Let us make the assumption now that the kinetic energy of the electrons can be quantized independently of their potential energy (which assumption may be said to be the quantitative characterization of the assumption that the electrons are "free") and that the effect of the ions can be taken into account by subtracting a volume V_i from V to get the volume per ion in which the electrons are free to roam. It will turn out later that V_i must be assigned values about five times as great as the volume that would be calculated from the crystal radii for the alkali ions; however, it seems not unlikely that the penetration of the electron begins already outside this arbitrarily defined distance, so that the electron will in general spend a relatively smaller time in regions near the ion than in regions farther away and this, together with the boundary conditions which the electron waves must meet at the points where the ions are located, may combine in the responsibility for the relatively large values of V_i.

If there are N ions in the portion of metal considered, the total volume available to them will be $N(V - V_i)$. The number of

[1] Sherman, J., *Chem. Rev.*, **11**, 107 (1932).

quantum states n_0 having kinetic energies less than, or equal to, T_0 will, from Eq. (21) of Chap. IV, be given by

$$n_0 = 2 \times \tfrac{4}{3}\pi(2T_0m)^{3/2}(V - V_i)Nh^{-3}, \tag{1}$$

where m is the mass of the electron; the factor 2 allows for the fact that the number is doubled owing to the two possible states of the electron spin. Electrons going into these quantum states will fill up the lowest energy levels first. Suppose that all the quantum states are filled up whose energy is, say, $T_{0,\text{max}}$ or less; then, since the total number of electrons is equal to the number of ions N, the value of $T_{0,\text{max}}$ will be found by setting $n_0 = N$ and solving for T_0. This gives

$$T_{0,\text{max}} = \frac{h^2}{2m}\left(\frac{3}{8\pi(V - V_i)}\right)^{2/3}.$$

Actually it is true only at absolute zero that *all* the states with energy less than $T_{0,\text{max}}$ are filled and *all* those with greater energy are empty, but even at room temperature this is true to a good approximation.

From Eq. (1) it is seen that the number of electrons having energy less than T_0 ($T_0 < T_{0,\text{max}}$) is proportional to $T_0^{3/2}$. Likewise the number having energy less than $T_0 + dT_0$, where dT_0 is a small increment, will be proportional to $(T_0 + dT_0)^{3/2}$; so the number in the range of energies between T_0 and $T_0 + dT_0$ will be proportional to

$$(T_0 + dT_0)^{3/2} - T_0^{3/2} = dT_0^{3/2} = \tfrac{3}{2}T_0^{1/2}\,dT_0,$$

hence will be equal to $aT_0^{1/2}\,dT_0$, where a is a constant. The average energy of all the electrons will be

$$T_{0,\text{av}} = \frac{\int_0^{T_{0,\text{max}}} T_0 \times aT_0^{1/2}\,dT_0}{\int_0^{T_{0,\text{max}}} aT_0^{1/2}\,dT_0} = \tfrac{3}{5}\,T_{0,\text{max}}$$

$$= \frac{3}{10}\frac{h^2}{m}\left(\frac{3}{8\pi(V - V_i)}\right)^{2/3}. \tag{2}$$

It will be observed that this average energy increases as the volume V decreases. It may be treated as a potential due to a repulsive force. The total energy E of the metal is thus given as a function of V, by the expression

$$E = -\frac{e^2 A_1}{V^{1/3}} + \frac{3}{10}\frac{h^2}{m}\left(\frac{3}{8\pi(V - V_i)}\right)^{2/3}. \tag{3}$$

This, of course, neglects all zero-point energy and thermal agitation of the ions. At absolute zero, however, the latter factor is nonexistent, and in any event, Eq. (3) should be a good approximation.

The equilibrium volume of the metal, the observed value of V at absolute zero, which we shall call V_e, will be obtained in terms of V_i, by finding the minimum of E by differentiation of Eq. (3). Since V_e is observed and V_i is unknown, the latter instead may be determined in terms of the former. Substituting the resulting value of V_i back into Eq. (3), we get an expression for E_e, the energy of formation of the metal at absolute zero (neglecting zero-point energy) from free positive ions and electrons. This should be the negative of the sum of the energy of sublimation S and the ionization potential I, since the decomposition of the metal into gaseous ions and electrons can be thought of as taking place in steps, first sublimation, then ionization.

The final theoretical evaluation of E_e depends upon the assignment of a value for A_1. The value that has been used is the entirely reasonable one of 2.08, for all the alkali metals. A comparison of observed and calculated values of E_e is given in Table 41.

TABLE 41.—BINDING ENERGIES OF ALKALI AND ALKALINE EARTH METALS
(Energies in electron volts, volumes in cu.Å per molecule)

Metal	V_e	V_i	I	S	$-E_e = I + S$	$-E_e$ (calc.)
Li..........	20.71	6.5	5.36	1.69	7.05	7.13
Na..........	37.60	14.9	5.11	1.12	6.23	6.18
K..........	71.6	33.3	4.32	0.86	5.18	5.25
Rb..........	87.6	42.6	4.16	0.82	4.98	4.98
Cs..........	108.7	55.3	3.87	0.82	4.69	4.70
Mg..........	23.5	8.	22.57	1.57	24.14	21.96
Ca..........	43.0	18.	17.91	2.07	19.98	18.98
Sr..........	55.4	25.	16.65	2.04	18.69	17.80
Ba..........	62.9	29.	15.14	2.12	17.26	17.23

The experimental values of S are for room temperature, but these are practically the same at absolute zero. Values of S from Bichowsky and Rossini, "Thermochemistry of the Chemical Substances" Reinhold Publishing Corporation, 1936; for source of V_e see Rice, reference 2, p. 370.

A similar calculation for the alkaline earth metals can be made, if we take into account the fact that there are two electrons per positive ion and that A_1 must be assigned another value in order that $-\frac{A_1e^2}{V^{1/3}}$ may give the correct value of the potential energy per positive ion. An alkaline earth metal may be compared with a uni-bivalent crystal. For a number of different types of such crystals, A_1 has been found by Sherman[1] to range from 6.5 to 8.0. The results of some calculations using $A_1 = 6.6$ are given in Table 41. In the case of the alkaline earth metals, I stands for the sum of the first two ionization potentials.

It will be observed that the agreement between the experimental and theoretical values is reasonably close, especially in the case of the alkali metals. For these, the agreement is so close that the method may be considered a theoretical way of calculating S, this quantity being found as the difference between E_e calculated and I observed. Though S is a small difference between these two relatively large quantities, the results obtained are very good. In the case of the alkaline earth metals, the agreement is not sufficiently good to use the results in this way. It might be imagined that this poorer agreement in the case of the alkaline earth metals is due to variation in A_1 from metal to metal. However, it seems more likely that it is due to deviation from the pure metallic type of binding setting in for the lighter alkaline earth metals. That this would occur is to be inferred from work by Wigner, Seitz, Slater, and others,[2] who have considered the binding forces in metals by actually finding approximate solutions for the wave equation for the free electrons. These calculations show (in verification of earlier less exact work) that instead of there being a completely continuous series of close-spaced energy levels for the free electrons, these occur in bands, there being certain fairly wide energy regions in which no allowed energy levels exist. That this is to be expected is seen if we consider the process of expanding a metal, allowing it to go

[1] Sherman, *Chem. Rev.*, **11**, 107 (1932).

[2] Based on earlier work of Bloch and Brillouin. For reviews see Slater, *Rev. Mod. Phys.*, **6**, 209 (1934); Mott and Jones, "Properties of Metals and Alloys," Oxford University Press, 1936. Numerous applications have appeared in the recent literature, the method having even been extended to substances which are far from being metallic.

over to neutral atoms (the state it naturally would assume if expanded indefinitely) instead of ions and electrons. At the end of the process, the free electrons have become bound electrons in definite energy levels. In the actual metal, we might well expect that some intermediate condition would exist. The energy levels would not be reduced to the uniformity of levels in an empty box, such as considered in Chap. IV, but would retain some of the characteristics proper to atoms, thus giving rise to the banded structure. This phenomenon will be further discussed in Sec. 18.4. If the energies for which no levels occur are higher energies than those normally possessed by any of the electrons, that is to say, if deviations from the uniform empty-box distribution occur only for levels not actually occupied, then the electrons will act like free electrons. This is the situation presumably existing in the alkali metals, but apparently not in any others, except perhaps barium. This accounts for the good agreement obtained with the alkali metals, using the very rough picture presented above. It may be remarked that it does not work any better for the copper group than it does for the lighter members of the alkaline-earth-metal group, and in the copper group, also, the electrons are probably not "entirely free."

Just as in the case of ionic crystals it was possible to get an expression for the compressibility by two-fold differentiation of the energy expression [Eq. (10), Chap. XIV], so this should be possible here. When this is attempted, very poor agreement with experimental values is obtained. This may be in part due to the fact that the experimental value is obtained at room temperature, and there are some indications that the values thus obtained may differ rather widely from the values at absolute zero. But, in any event, it appears that relatively insignificant changes in the energy expression may make a great difference in the calculated compressibility, so that the latter has but little significance from the point of view of the calculation outlined. When corrections are made so as to make the compressibilities fit, the value of A_1 necessary to make the theoretical energies agree with the experimental is slightly changed, and the agreement as between the different alkali metals is made slightly less exact. Nevertheless, the general picture and the assumption that A_1 has a reasonable value practically the same for all the alkali metals seem to work very well. Slight discrepancies show themselves in the case of the alkaline earth metals, and with metals of higher valence the deviation from metallic character increases until finally the metallic binding merges into atomic or covalent binding. Some of the elements just to the right of the center of the periodic table, having low valences, exhibit more typically metallic binding than some of those to their left. It must be said, however, that there is no proof that these elements of lower valence have more free electrons per atom than the less metallic metals of higher valence; in fact, since the latter have more valence electrons, the contrary is probably often true, but in any event, the elements of higher valence have fewer free electrons per valence electron.

18.2. The Transition between Metallic and Covalent Binding. From the preceding discussion, it appears that the deviation from metallic type binding is related to the existence of bands of energy levels. The properties of substances will be largely determined by the number of energy levels available per molecule in each band, for this will determine the distribution of valence electrons within the bands, and the disposition of the valence electrons, in turn, largely fixes the properties of the substance. It seems probable that a strictly covalent binding is marked by a structure in which the bands are exactly filled. When the bands are almost filled, or just fail to take care of all the electrons, the substance is predominantly of covalent type, but has some residual metallic properties, such as electrical conductivity, though the conductivity will be relatively small. Such a case is bismuth. The study of energy bands may be considered as analogous to the Hund-Mulliken point of view in the study of molecules (see Chap. XI). In Chaps. XV and XVI, we have rather stressed the Heitler-London-Pauling-Slater point of view, and for many purposes this is simpler, but with its localized bonds it is not applicable to metallic binding.

The role played by the number of valence electrons in the transition from metallic to covalent binding has been brought out by Hume-Rothery[1] in a study of the crystal structure of the elements toward the right of the periodic table. Although, as

[1] Hume-Rothery, "The Metallic State," Oxford University Press, 1931, pp. 306*ff*. This book and Hume-Rothery's more recent work, "The Structure of Metals and Alloys," Institute of Metals, London, 1936, contain much material on the metallic state.

will appear in Sec. 18.4, these considerations are not wholly independent of the energy bands, they emphasize a rather different aspect of the matter. The tendency of atoms to share electrons so as to complete shells of eight would, if carried to its logical conclusion, cause valence electrons to be shared between more than two atoms. In the case of arsenic, for example, this is not necessary, for three valence electrons from each atom can be shared with three other atoms. As a result every atom has three shared pairs which with the unshared pair complete the shell of eight. In the case of germanium, the arrangement is of the tetrahedral type in which each atom has four shared pairs. On the other hand, if the shell of eight is to be completed in the case of gallium, it will require that an atom share five electrons from other atoms, and it will then have to share its three electrons with five atoms; there are not enough electrons to go around, and each electron must be shared among more than two atoms. This may be regarded as an approach to the metallic state in which all electrons are shared among all atoms. This discussion suggests that each gallium should be surrounded by five nearest neighbors, and at first this was thought to be the case, but more recent evidence indicates that the number is seven.[1] This is a very peculiar type of structure, and though it shows that some of the atomic character of the binding remains, it also shows that the shift to metallic binding does not occur smoothly, but that various complications enter.

The number of valence electrons undoubtedly plays an important part in determining whether a substance is metallic or not, but it is by no means the only factor; in fact, the number of elements to be counted as metallic in any given row in the periodic table depends, as we know, upon the location of that row. Thus in the first row, probably only lithium and beryllium would be classified as metallic, whereas in later rows such elements as arsenic, antimony, and bismuth, and even tellurium are often considered to be metallic, the first three mentioned, especially, having some of the physical properties of metals. As indicated above, this does not, by any means, signify that all valence electrons are free, and in fact, in these very negative elements it

[1] For details of crystal structure and references see the "Strukturbericht." See also Dehlinger, "Gitteraufbau metallischer Systeme," Akademische Verlagsgesellschaft, Leipzig, 1935.

TABLE 42.—CRYSTAL STRUCTURES OF CERTAIN ELEMENTS

Element	Number neighbors	Distance	Remarks
Cu	12	2.55	Close-packed spheres (face-centered cubic,—see Appendix IV)
Zn	6 6	2.65 2.94	Structure almost close-packed
Ga	1 6	2.44 2.71–2.80	A slight displacement would change coordination number from 7 to 5
Ge	4	2.43	Tetrahedral, like diamond
As	3 3	2.51 3.15	Angle between bonds 97° (see p. 269) (Other forms exist—some amorphous)
Se	2	2.32	Spiral chains (allotropic forms exist—some amorphous)
Br	1	2.27	Molecular crystal
Ag	12	2.88	Like Cu (Possibly an allotropic form exists)
Cd	6 6	2.97 3.30	Like Zn
In	4 8	3.24 3.37	Slightly distorted close-packed
White Sn	4 2	3.02 3.15	Much distorted tetrahedral
Gray Sn	4	2.80	Tetrahedral, like diamond
Sb	3 3	2.87 3.37	Angle between bonds 96° (see p. 269) (Less stable amorphous forms exist)
Te	2	2.86	Spiral chains (allotropic forms probably exist)
I	1	2.70	Molecular crystal
Au	12	2.88	Like Cu
Hg	6 6	3.00 3.47	Rhombohedral (distorted close-packed)
Tl	12	3.45	Hexagonal close-packing (see Appendix IV). An allotropic form occurs which is also close-packed—face-centered cubic.
Pb	12	3.49	Close-packed
Bi	3 3	3.10 3.47	Angle between bonds 94° (see p. 269) (Amorphous forms may exist)

There seems to be no reason to suppose that in amorphous forms the immediate surroundings of any single atom are very different than in the crystalline form.

seems unlikely that there are as many free electrons as would correspond to one per atom. It does not take many free electrons (which may be defined as electrons in incompletely filled energy bands) to lend some of the physical metallic properties to a substance, and in such cases in which there are four or more valence electrons, *e.g.*, bismuth and silicon, the valence forces are presumably largely covalent. These substances are different from most metals in that they are brittle, and their electrical conductivity is relatively small. The conductivity of most metals decreases steadily with the temperature, the result of the thermal agitation of the lattice. In the case of a number of borderline metals, germanium, tellurium, and silicon, the conductivity increases with increasing temperature, but only at low temperatures.[1] This is probably due to an increase in the number of free electrons with increasing temperature, which is counteracted by other effects at higher temperatures.[2] Lead appears to be fairly metallic, but in the case of tin there are two forms known, one of which, gray tin, has a tetrahedral structure and does not appear to have metallic properties. White tin is metallic, and the tetrahedral arrangement is very much distorted, though signs of special structure, which would not be expected with purely metallic binding, are present. Such signs of structure are also marked in zinc, in which each atom has six neighbors at a distance of 2.65Å. and six at 2.94Å., in cadmium with a similar structure, and in mercury, though these elements are generally conceded to be metallic. In the more definitely metallic substances, the structure is close-packed, or nearly so, each atom having twelve or eight nearest neighbors. In Table 42, we summarize some of the pertinent properties of the elements in this region of the periodic table.

18.3. Intermetallic Compounds.—Grimm[3] has given a classification of binary compounds into various types, corresponding in the main to the types of binding considered in Chap. XIII. His tables show a great majority of all known and inferred compounds to be metallic compounds. His classification is for the most part based upon inference from the physical properties of

[1] The conductivity of silicon continues to increase up to a high temperature, where phase transitions occur.

[2] Hume-Rothery, "The Metallic State," p. 309.

[3] See Grimm, *Angewandte Chemie*, **47**, 53 (1934).

the compounds, but is undoubtedly accurate in the main. Metals and metallic compounds can be recognized by their relatively high electrical conductivity, due to the free electrons, by their metallic glance, arising from their high reflective power, which is also due, as optical theory indicates, to the free electrons they contain, and by such physical properties, as ductility, which depend upon easy slipping of the atoms past each other, which occurs with the metallic bond, but not with the localized covalent bond (because in this case it would require breaking of bonds) nor in the ionic bond (presumably because it could not occur without bringing ions of like charge close to each other).

The great preponderance of intermetallic compounds is of course to be referred back to the preponderance of metallic elements. Atoms with a relatively small number of valence electrons tend to lose electrons or, as Hume-Rothery has noted, share them with a *number* of other atoms, while elements whose number of valence electrons approaches eight tend to complete the shell of eight by adding electrons. There are far more elements with less than four valence electrons than with more than four valence electrons, which, as pointed out by Grimm, accounts for the large number of metallic elements and compounds. The elements that form true intermetallic compounds probably do not extend beyond the trivalent elements on the right of the periodic table (*e.g.*, gallium).

The so-called intermetallic compounds range from compounds, such as Na_4Sn, in which the binding probably has some of the properties of ionic, covalent, and metallic bindings, and Na_2Te (often called an intermetallic compound though probably largely covalent or ionic) to compounds in which the metallic character is largely preserved, as evidenced by such physical properties as electrical conductivity, metallic luster, and ductility.[1] Even these more metallic compounds often appear to be less metallic than the elements, at least as far as one of the typical metallic properties, electrical conductivity, is concerned. This is illustrated by Table 43, which is taken from a table given by

[1] For a general account, see Dehlinger, *Ergebnisse der exakten Naturwiss.*, **10,** 325 (1931). For a more recent general discussion with some attempt at classification with respect to composition and crystal structure, see Dehlinger, *Naturwiss.*, **24,** 391 (1936), and "Gitteraufbau metallischer Systeme."

TABLE 43.—SPECIFIC CONDUCTANCE (μ) OF METALLIC COMPOUNDS

Compound	Mg_2Sn	$MgCu_2$	Mg_2Cu	$MgZn_2$	Mg_3Bi_2	MgAl	Mg_3Al_2
$\mu \times 10^{-4}$	23.0	23.0	23.0	23.0	23.0	23.0	23.0
	0.092	19.4	8.38	6.3	0.76	2.63	4.53
	8.60	64.1	64.1	17.4	0.84	35.1	35.1
Compound	$MnAl_3$	$FeAl_3$	$NiAl_3$	Ag_3Al_2	Ag_3Al	AgMg	$AgMg_3$
$\mu \times 10^{-4}$	22.7	11.0	8.51	68.1	68.1	68.1	68.1
	0.20	0.71	3.47	3.85	2.75	20.52	6.16
	35.1	35.1	35.1	35.1	35.1	23.0	23.0
Compound	Ag_3Sb	Te_2Sb_2	TeSn	Te_3Bi_2	Cu_3As		
$\mu \times 10^{-4}$	68.1	0.017	0.017	0.017	64.1		
	0.93	0.48	0.97	0.045	1.70		
	2.56	2.56	8.60	0.84	2.85		

First row gives specific conductance for first element appearing in formula of compound; second row gives it for the compound; third row gives it for second element appearing in formula.

TABLE 44.—HEATS OF FORMATION OF METALLIC COMPOUNDS
(In kilogram-calories per formula weight)

SnBi	0.37	Fe_3Si	−20.	Mg_4Al_3	49.0	$NaSn_2$	20.
$SnBi_2$	−0.17	AlCu	68.	Mg_3Ce	17.0	Na_2Sn	21.
$SnBi_5$	−0.19	$AlCu_3$	23.	CaSi	87	Na_4Sn	34.
Sn_2Bi	−0.12	Al_2Cu	84.	$CaSi_2$	220	Na_4Sn_3	56.
Sn_5Bi	−0.78	Al_5Fe	25.0	$CaSn_3$	52	$NaCd_2$	8.5
CdSb	2.7	AlCo	32.	$CaZn_4$	29.5	$NaCd_5$	12.5
Cd_3Sb_2	4.0	Al_5Co	86.	$CaZn_{10}$	48.	Na_3Hg	11.1
$HgPb_2$	−0.05	$CeHg_4$	23.2	Ca_2Zn_3	40.	Na_3Hg_2	22.2
Hg_5Tl_2	2.50	$CeZn_4$	49.	Ca_4Zn	32.	NaHg	11.0
$HgCd_3$	0.74	$CeAl_4$	22.	$CaCd_3$	30.	$NaHg_2$	18.5
HgCd	1.96	Ce_3Al	39.	$CaAl_3$	51	$NaHg_4$	22.2
Hg_3Cd	3.99	$LaAl_4$	20.	Ca_3Mg_4	43.	KHg	11.0
Cu_3Sb	2.5	Mg_2Sn	59.	LiHg	20.8	KHg_3	26.0
Cu_3Sn	8.0	$MgZn_2$	13.1	$LiHg_2$	25.0	NaK	2.1
Cu_2Zn_3	16.	MgCd	9.2	$LiHg_3$	26.8	NaK_2	5.3
Cu_2Cd_3	3.0	$MgHg_4$	17.3	NaSn	16	NaK_3	5.6
Ag_3Hg_4	0.7					Na_2K	0.4

From Bichowsky and Rossini, "Thermochemistry of the Chemical Substances." The table gives heats evolved.

Kraus.[1] That many of these intermetallic combinations are real compounds of some stability is indicated by the heats of formation tabulated in Table 44. The heats of formation appear to follow no fixed rule except that they are low for compounds of two transition elements or two alkali metals.

Even more striking is the irregularity in the formulas themselves. With the exception of a few like Na_4Sn which involve relatively electronegative metals, or substances which have not been treated in this book as metals at all, the intermetallic compounds in general exhibit an almost complete breakdown of the valence rules. Even with elements that do form compounds of normal valence, abnormal compounds are likely to occur. Thus sodium forms a whole series of compounds with tin, Na_2Sn, Na_4Sn_3, NaSn, $NaSn_2$. It has been suggested by Kraus[1] that in this case there are complex tin ions, somewhat analogous to the well-known tri-iodide ion I_3^- and the various sulfide ions SS^{--}, S_2S^{--}, S_3S^{--}, S_4S^{--}, S_5S^{--} which are known to exist in aqueous solution. Kraus and his coworkers have also demonstrated the existence of complex tellurium ions $TeTe^{--}$ and Te_3Te^{--} in ammoniacal solution.

Incidentally, it may be mentioned that liquid ammonia has proved a most useful tool for the study of metallic substances. In general, metals are not easily volatile, nor are they soluble in most other nonmetallic solvents. Practically the only opportunity of studying metals in a dispersed or dilute condition has been afforded by liquid ammonia solutions. Metallic substances dissolved in ammonia act, in dilute solution, like electrolytes. A dilute solution of sodium, for example, acts as though it contains Na^+ and NH_3^- ions, the sodium apparently giving up an electron which then becomes more or less associated with a solvent molecule. At higher concentrations, however, this solution appears to contain free electrons as indicated by its high electrical conductivity, its metallic luster (it is a deep blue), and also by its magnetic properties. The tendency of metals to act like electrolytes in liquid ammonia solutions brings about the possibility of some rather queer replacement reactions. Thus KCl is soluble in liquid ammonia, but $CaCl_2$ is not, which results in the following reaction taking place:

$$2KCl + Ca \rightarrow CaCl_2 \text{ (ppt)} + 2K$$

[1] Kraus, *J. Am. Chem. Soc.*, **44**, 1216 (1922).

Many, perhaps most, intermetallic compounds do not have an exact, definitely defined composition, but can vary between certain limits, the possible leeway differing in different cases. In this respect, they resemble solid solutions, though they may have rather different properties from the constituent elementary substances. As stated above, they do not obey the ordinary valence rules, but certain interesting regularities do appear.

Some of the more striking of these regularities appear in the alloys of copper, silver, and gold.[1] These metals will dissolve another metal to some extent without change in crystal structure, the foreign atoms taking places on the lattice, forming a so-called α-phase. Addition of more of the foreign material causes a change in crystal structure to a body-centered cubic type known as the β-phase (usually stable only at fairly high temperatures). Addition of still more results in a more complicated though still cubic structure, the γ-phase, and still more may result in an ϵ-phase, which is hexagonal close-packed (see Appendix IV). The latter two types of alloy are hard and brittle, and so have lost some of their metallic characteristics. Each type of phase exists over a small range of compositions, but there is usually a gap in compositions between the various phases (compositions in these regions giving heterogeneous mixtures). Complications can occur, and phases may be stable only at high temperatures. Though the β-, γ-, and ϵ-phases have possible ranges of compositions, they can be described roughly by a stoichiometrical formula. Examples of the various kinds of phases are shown in Table 45. In a few cases, *e.g.*, CuZn, the composition in which the alloy exists actually does not quite reach the composition indicated by the formula. In $AuZn_3$ and perhaps $AuCd_3$ the structure is cubic; an ϵ-type structure occurs with a larger percentage of Zn. The crystal structure of Cu_5Si is different from that of other β-phases listed, but similar to several others not given.

Each of these types of phases appears to be characterized by a certain ratio of the number of valence electrons to the number of atoms, known from its discoverer as the "Hume-Rothery ratio."

[1] For more detailed accounts of the material described in this paragraph, see Hume-Rothery, "The Metallic State," pp. 328*ff.*, "The Structure of Metals and Alloys," pp. 98*ff.*; Dehlinger, *Ergebnisse der exakten Naturwiss.*, **10**, 325 (1935), and "Gitteraufbau metallischer Systeme," pp. 98*ff.*; and Westgren and Phragmén, *Trans. Faraday Soc.*, **25**, 379 (1929).

TABLE 45.—COMPOSITIONS OF ALLOYS

β-phases	γ-phases	ϵ-phases
Cu_5Si	Cu_5Zn_8	$CuZn_3$
Cu_5Sn	Ag_5Cd_8	Cu_3Sn
CuZn	Cu_9Al_4	Cu_3Sb
CuBe	$Cu_{31}Sn_8$	$CuBe_3$
AgZn	Cu_5Cd_8	Cu_3Ge
AuZn	Ag_5Zn_8	$AgZn_3$
AgCd	Ag_5Hg_8	$AgCd_3$
AuCd	Au_5Zn_8	Ag_5Al_3
Cu_3Al	Au_5Cd_8	Ag_3In
AgMg		Ag_3Sn
		Ag_3Sb
		$AuZn_3$
		$AuCd_3$
		Au_5Al_3
		Au_3Hg

Consider, for example, Cu_5Si. The five copper atoms contribute each a valence electron, the silicon atom contributes four; the total, nine, is divided among six atoms, giving a ratio 3:2. This ratio holds for all the β-phases listed. For the γ-phases the ratio is 21:13, for the ϵ-phases[1] it is 7:4. The metals of the iron and platinum groups also form β- and γ-alloys, for example, CoAl and Fe_5Zn_{21}. The electron-atom ratio follows the Hume-Rothery rule if the iron or platinum metal is assumed to contribute no valence electrons. Some attempts have been made to explain these rules, based on quantum states in the energy bands in the alloys (see Sec. 18.4).

In some alloys, each kind of atom takes its place in the lattice in a regular way, so that one could consider that the crystal is composed of two perfectly regular but interpenetrating superlattices. In other cases, the atoms of various kinds are arranged at random on a single lattice and the superlattice structure does not exist. Sometimes a transition from the more ordered super-

[1] It should be noted, however, that according to Westgren and Phragmén the homogeneity range of some ϵ-phases (Ag-Sn and Ag-Sb alloys) is so great that it covers the range which would be expected for the β- and γ-phases, also. In these cases β- and γ-phases have not been reported. In the case of Cu_3Sb, Ag_3In and Ag_3Sb, which are reproduced in our table as given by Dehlinger, it will be noted that the formulas given do not correspond to the 7:4 electron-atom ratio, but rather to the normal valence formula.

lattice arrangement to the less ordered random arrangement can be followed. Such order-disorder transitions have recently been rather extensively discussed[1] and present a number of points of interest. It will be recognized that if an ordered state tends to exist at all it must be a state of lower energy and will therefore exist in the lower temperature range. As the temperature is raised, the thermal agitation tends to produce the disordered condition. As the amount of disorder increases, the amount of energy necessary to still further increase the disorder by a given amount decreases; in other words, a certain amount of disorder already present tends to increase the ease with which further disorder is produced. The result is that there is a fairly sharp transition temperature above which the disorder suddenly becomes practically complete. As the crystal is heated, energy must be supplied to effect the order-disorder transition. The specific heat therefore rises, first slowly as disorder begins, then rapidly (since the *rate* of increase of disorder itself increases, as we have seen, which more than counterbalances the decrease in the amount of energy necessary to further increase the disorder), finally reaches a sharp maximum, and then drops suddenly at the transition temperature.[2] Other properties also, such as electrical resistance, show an abrupt bend at the critical temperature. By suddenly cooling an alloy in its disordered state (quenching), it is usually possible to "freeze" it in this state, as the transition to the ordered state takes place slowly at low temperatures. Properties such as electrical resistance and particularly hardness and malleability depend greatly on whether a state of order or dis-

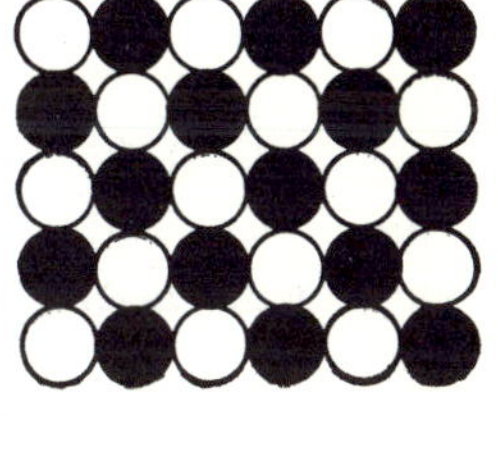

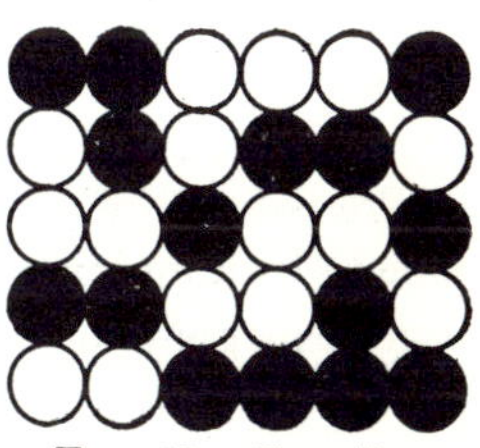

FIG. 69.—Two-dimensional illustration of order and disorder.

[1] See FOWLER, "Statistical Mechanics," 2d ed., Chap. XXI, The Macmillan Company, 1936, and NIX and SHOCKLEY, *Rev. Mod. Phys.*, **10**, 1 (1938) for reviews.

[2] In some cases latent heats of transition are found at a definite temperature, as in melting. If there is a latent heat, properties like electrical resistance should show actual discontinuities at the transition point.

order exists in an alloy, and so, when the alloy has finally been brought to room temperature, its physical properties depend largely on its history—whether it has been allowed to come to equilibrium or not.

An example of an alloy that exhibits a transition between ordered and disordered conditions is β-brass[1] which has (very approximately) the formula CuZn. The lattice is a body-centered cube, and in the ordered state the copper atoms occupy the positions at the center of cubes formed by zinc atoms, and the zinc atoms occupy similar positions with respect to the copper atoms. In the disordered state, the zinc and copper atoms are randomly placed in the body-centered lattice. The transition temperature is 470°C., and the transition begins around 200°C. In this case, the transition has not actually been followed by X rays, but the existence of the ordered superlattice at room temperature has been demonstrated,[2] and the specific heat curve has the expected form, described above. In this case, although quenching is possible, it is easy to get complete equilibrium at all temperatures in the interesting range.

Transitions of the type considered may also have very complicated connections with changes in crystal structure. This occurs in gold-copper alloys,[3] and the situation is further complicated experimentally by the fact, due to difficulty in securing equilibrium, that the results depend greatly on the heat treatment the alloy has received.

Another type of alloy is the interstitial[4] "compound." Such an alloy, having some metallic physical properties, is formed between a metal in or near the transition region of the periodic table and an element like boron, carbon, or nitrogen. In these alloys, the small atom finds a place in the spaces left between the metal atoms. If the size of the metal atom is sufficiently large compared with the other, this takes place with very slight change in the metal lattice, and phases that range around the compositions M_4X, M_2X, MX, and MX_2 (M = metal, X = boron, carbon, or nitrogen) are usual. If the metal atom is not

[1] BRAGG and WILLIAMS, *Proc. Roy. Soc. (London)*, **A151**, 540 (1935).

[2] JONES and SYKES, *ibid.*, **A161**, 440 (1937).

[3] See JOHANSSON and LINDE, *Ann. Physik*, **25**, 1 (1936).

[4] HÄGG, *Zeits. physik. Chem.*, **B12**, 33 (1931); DEHLINGER, *Ergebnisse der exakten Naturwiss.*, **10**, 357*ff*. (1931).

quite so large, then the concentration of the light atoms in the alloy cannot usually become so great, and other formulas appear; for example, cementite, which is of importance in the metallurgy of iron, is Fe_3C. When the percentage of carbon is as large as this, it is not in the spaces between iron atoms, but is substituted, resulting in a complicated lattice.

Hydrogen is strongly occluded by some transition metals; it is well known, for example, that palladium, in particular, can "dissolve" large quantities of hydrogen. In general, we may expect that in cases of this sort the molecular structure of the hydrogen is broken down. It has been suggested that the hydrogen plays the role of a metal in these occlusions, so they may be considered to be alloys.[1] Experiments on the transference of hydrogen in palladium under the influence of an electric field seem to indicate that the hydrogen exists at least in part in the form of positive ions. The released electrons would then join the other free electrons from the palladium. After a small amount of hydrogen has been added to palladium, there is a change of phase. There is evidence that this phase has the definite composition Pd_2H, at least above 80°C.

18.4. Energy Bands in Metals and Alloys.—Though a detailed discussion of this matter is beyond the scope of this book, some indication may be given of the methods used in considering the energy states of electrons in metals.[2] The energy state of a free or partly free electron can be characterized by three quantum numbers, which describe the state of its motion in the x-, y-, and z-directions, respectively. These quantum numbers are very closely associated with the wave number (*i.e.*, the reciprocal of the wave length) of the wave associated with the electron, as will be evident from the considerations of Chap. IV, in which we treated the motion of an electron in a box. In fact, it will be seen from Eq. (3c) of Chap. IV that the wave numbers for the successive allowed energy levels in the case of a one-dimensional motion will be in the ratios 1:2:3:4: . . . If the motion is three-dimensional and the wave under consideration has a direction that is not parallel to one of the axes, its wave number k

[1] See Coehn and Jürgens, *Zeits. Physik*, **71**, 179 (1931); Gillespie and Hall, *J. Am. Chem. Soc.*, **48**, 1207 (1926); and Ubbelohde, *Proc. Roy. Soc.* (*London*), **A159**, 295 (1937).

[2] For references see footnote 2, p. 374.

is related to the components of the wave number k_x, k_y, k_z in the x-, y-, and z-directions by the relation

$$k^2 = k_x^2 + k_y^2 + k_z^2.$$

The exact relations involved will be clear from the two-dimensional case illustrated in Fig. 9, page 38. The wave is proceeding in an arbitrary direction with wave length λ. Its "wave length" in the x-direction is $\lambda_x = \frac{\lambda}{\cos \alpha}$, and in the y-direction it is $\lambda_y = \frac{\lambda}{\sin \alpha}$. Hence $k_x = k \cos \alpha$ and $k_y = k \sin \alpha$, and since $\cos^2 \alpha + \sin^2 \alpha = 1$, it is seen that $k^2 = k_x^2 + k_y^2$, and the relation is easily extended to three dimensions. Giving the values of k_x, k_y, k_z will determine both k and the direction of the wave. We shall denote the vector with components k_x, k_y, and k_z by the symbol $\mathbf{k}$. It points in the direction of motion of the wave front.

In an actual metal, in which there are centers of force acting upon the "free" electrons, the situation is, of course, more complicated than in the empty box of Chap. IV. It is still possible, however, to assign wave numbers k_x, k_y, k_z, which may be considered as equivalent to quantum numbers for these electrons. If the metal were an empty box, there would be a very closely and regularly spaced set of energy levels as in the three-dimensional case of Sec. 4.2. In the real metal, the energy is still determined by k_x, k_y, k_z, but the functional relationship is more complex, and at certain quantum states jumps in the energy occur. That is, the energies of the lower quantum states are fairly regularly spaced, but finally we come to a certain quantum state whose energy is considerably higher than that of the next lower quantum state. This accounts for the energy bands and gaps in the energy-level system mentioned earlier in this chapter.

If electrons are allowed to bombard a crystal, they are reflected according to the same laws as X rays [see Eq. (1) of Chap. III]. Remembering that $k = 1/\lambda$, we can write this equation in the form

$$k \sin \theta = \frac{n}{2d}. \tag{4}$$

It is to be noted (see Fig. 7, page 27) that $k \sin \theta$ is just the projection of the vector **k** along the normal to the plane of atoms in the crystal from which the reflection takes place. The condition that Bragg reflection of a beam of electrons should take place is just that the projection of the vector **k** be given by Eq. (4). Now the wave mechanical theory of the motion of the electrons in a crystal leads to the remarkable but still very reasonable result that the gaps in the energy occur for combinations of the quantum numbers k_x, k_y, and k_z, which fulfill the condition given by Eq. (4).

Suppose now we set up a system of axes, such as shown in Fig. 70, and draw a line OP normal to one of the planes of atoms of the crystal. On this line, we mark off a point Q such that the length of OQ is equal to $n/2d$. If we lay off our vector **k** from the origin, it will be clear that every vector ending in a plane though Q perpendicular to OP (*i.e.*, parallel to the original plane of atoms in the crystal) will satisfy Eq. (4). This can be done for any plane in the crystal, for any value of n; so we find a series of planes in our diagram which lay off regions of space such that the jump in the energy occurs on crossing over from a k_x, k_y, k_z that lies on one side of one of these planes to a k_x, k_y, k_z that lies just across it, in the next region. In particular, there will be a closed region around the origin for which k_x, k_y, and k_z have their lowest values and for which the energies are low. A quantum state to which corresponds a vector that extends just beyond the boundary of this region has a considerably higher energy. If there are just enough electrons available to fill all the low energy levels, the energy of a metal will be relatively low. However, if there are too many electrons for the low energy levels, then the energy will be relatively high.

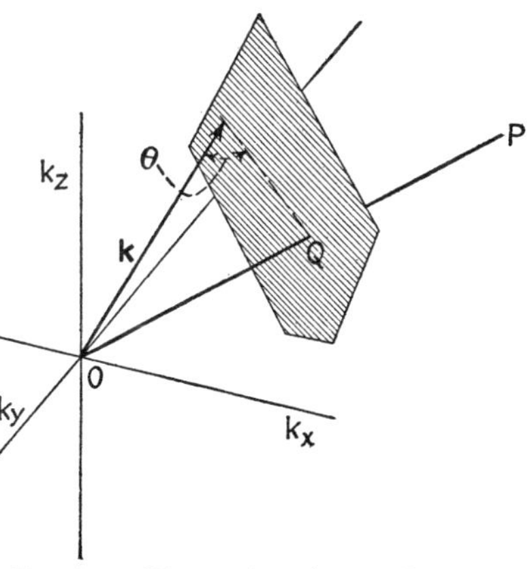

FIG. 70.—Illustrating the condition implied by Eq. (4). The plane passing through Q, parallel to the plane of atoms and perpendicular to OP, upon which the vector **k** ends, is shown shaded.

The number of energy levels per atom will depend on the arrangement of the atoms in a crystal, *i.e.*, on the type of lattice involved, but it will not depend upon the distance between atoms provided their relative positions are held fixed. This is understood by reference to Eq. (3c) of Chap. IV, which shows that the *number* of different values which $k_x = 1/\lambda_x$ may take before it reaches any given value is proportional to a, the linear dimension of the box. In general, it is true that the number of different values which k_x, k_y, and k_z may take before reaching any assigned value is proportional to the linear dimensions of the crystal. If, therefore, a crystal is expanded or contracted, the number of quantum states will be accordingly increased or decreased. On the other hand, by Eq. (4) above we see that the values of $k \sin \theta$ for which the gaps in energy occur are *inversely* proportional to the distance between planes in the crystal. The net result is that expansion or contraction of a crystal leaves the number of energy levels below the first energy gap unchanged. A crystal with too many electrons to fill the low-energy levels cannot, therefore, greatly increase its stability by expanding or contracting, but it may be able to revert to another crystal form which has more low-energy levels. In general, a completely filled energy band is quite stable, but one with a few extra electrons is disproportionately unstable. Although it is clear that a structure with a few electrons more than are necessary to fill the first band of energy levels is unstable, it may yet be that the *most* stable structure is one in which the band is not completely full. A band always contains a few electrons of relatively high energy corresponding to the states whose k_x, k_y, k_z lies near one of the corners of the polygonal figure bounding the first energy band. Electrons filling these high-energy states cause a certain instability, which may result in the alloy crystallizing in some other form. However, these energy states in the corners will not be included in a sphere[1] inscribed inside the polygonal figure, and if only enough electrons are present to fill the states represented by k_x, k_y, k_z values inside the inscribed sphere, the structure may be stable. In the case of the β-phases described above, the first energy band contains 2 quantum states per atom, but

[1] Sometimes a sphere cannot be inscribed in such a way as to nearly fill the polygonal figure, in which case some other similar, but less symmetrical, figure would be used.

the inscribed sphere contains[1] only 1.48, which corresponds very well with the Hume-Rothery ratio of 3:2. In the case of the γ-phases, the first band contains[1] 1.73 states per atom and the inscribed sphere 1.54. The Hume-Rothery ratio in this case is 21:13 = 1.61. In the case of the ϵ-phase, the number of states in the first band is 1.746 per atom,[1] which corresponds practically exactly to the Hume-Rothery ratio of 7:4.

The fact that the β-phases are metallic is probably connected with the fact that the first zone of energy levels is not filled, whereas in the case of the γ- and ϵ-phases some of the metallic properties have been lost because the first energy band is nearly filled. In the γ-phases the electrical conductivity is said to have its lowest value when the composition is such that the Hume-Rothery ratio is exactly fulfilled.

Exercises

1. Calculate the heat of sublimation of copper at room temperature, using its density of 8.93 g. per cc., and compare with the experimental value of 81 kg.-cal. per mole.

2. Write a Lewis electron structure formula for crystalline arsenic.

[1] See Mott and Jones, "The Properties of Metals and Alloys," Chap. V; Dehlinger, *Zeits. Physik*, **94**, 231 (1935).

CHAPTER XIX

THE STRUCTURE OF WATER, HYDRATES, AND AQUEOUS SOLUTIONS

A great part of inorganic chemistry consists in the study of reactions of substances dissolved in water. A consideration of the structure of water and the nature of aqueous solutions, being necessary for the understanding of these processes, is therefore of preeminent importance.

19.1. The Structure of Glasses.—As a prelude to the study of the structure of water and aqueous solutions, it will be appropriate to discuss the structure of certain glasses. A glass is a liquid that has been supercooled without freezing to a temperature so low that it has lost its fluidity. Many of its properties are still those of an ordinary liquid rather than a crystal.

Warren[1] has investigated vitreous SiO_2 by means of X rays. In all the various forms of SiO_2, each silicon is surrounded by four oxygens at the vertices of a regular tetrahedron, and each oxygen shares two silicons, acting as a connecting link between them. These different crystalline forms differ in the relative orientation of the tetrahedra about the two silicons joined by a common oxygen bond, the tetrahedra themselves being all alike.[2] In any given crystal type, there is a perfectly definite and regular arrangement. It was found, however, that the X-ray analysis of the glass indicated a *random* relative orientation of the tetrahedra with a common oxygen. The glass thus has the same arrangement as a crystal, provided one looks at a sufficiently small portion of it; the difference consists in the irregular arrangement

[1] Warren and Hill, *Zeits. Krist.*, **89**, 481 (1934); Warren, *Phys. Rev.*, **45**, 657 (1934). For a general discussion of glasses, see Zachariasen, *J. Am. Chem. Soc.*, **54**, 3841 (1932); Randall, "Diffraction of X-rays and Electrons by Amorphous Solids, Liquids and Gases," pp. 173*ff.*, John Wiley & Sons, Inc., 1934, Hägg, *J. Chem. Phys.*, **3**, 42 (1935); Zachariasen, *ibid.*, **3**, 162 (1935).

[2] They also differ in the angle between the two Si—O bonds to a given oxygen atom (see footnote 1, p. 312, Chap. XVI).

of the glass when one looks at the whole thing. The glasses GeO_2 and BeF_2 are similar to SiO_2.

19.2. The Structure of Water.—A detailed theory of the structure of liquid water has been developed by Bernal and Fowler[1] and has recently been further discussed by Katzoff and by Morgan and Warren.[2] This theory is based upon the structure

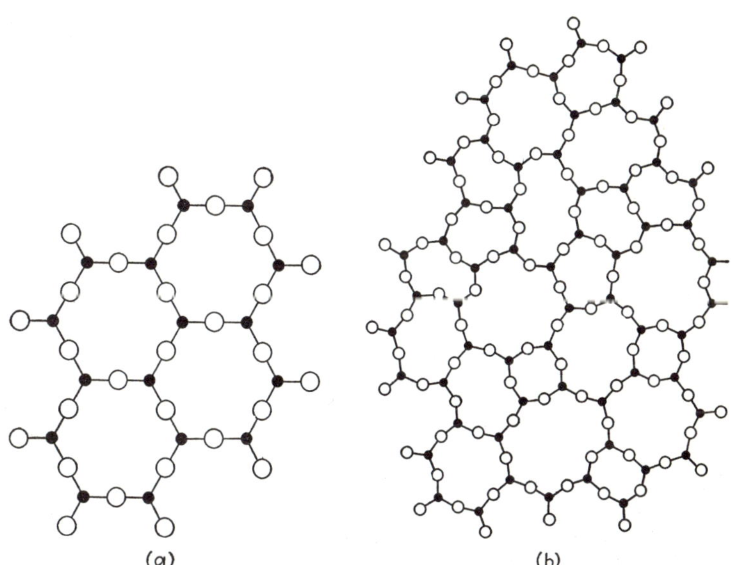

Fig. 71.—Illustrating by a hypothetical two-dimensional case the difference between (*a*) a crystalline structure, and (*b*) a glass with the same composition. It will be observed that the arrangement about any one of the black atoms is the same in the glass as in the crystal. [*After Zachariasen, J. Am. Chem. Soc.*, **54,** 3845*f.* (1932).]

of ice, X-ray examination of water itself, and the known properties of the water molecule, in particular the fact that the angle between the two hydrogen-oxygen bonds is nearly the tetrahedral angle. It seems reasonable to assume that the electronic structure of water is of the sp^3 tetrahedral type[3] (Sec. 15.2), and that

[1] Bernal and Fowler, *J. Chem. Phys.*, **1,** 515 (1933).

[2] Katzoff, *J. Chem. Phys.*, **2,** 841 (1934); Morgan and Warren, *ibid.*, **6,** 666 (1938).

[3] As pointed out on pp. 266 and 269, in the case of the water molecule it is not possible to state unequivocally whether the bond-eigenfunctions involve only *p*-states, or whether they are of the sp^3 type. There does seem

the electron wave functions are such as to cause a considerable density of negative electricity in directions corresponding to the vertices of a tetrahedron. At two of these positions of large density of negative electricity, however, are located the positive hydrogen nuclei, and since hydrogen is less electronegative than oxygen, these positions may be expected to constitute centers of net positive charge, while the other two tetrahedral positions constitute centers of negative charge. In liquid water, the positive centers of one molecule will attract the negative centers of another molecule. In this way, each oxygen atom tends to be surrounded by four hydrogens at the corners of a more or less regular tetrahedron, two of the hydrogens belonging to the same molecule, while the other two hydrogens belong each to a different molecule. (It is, of course, possible for the two hydrogens of a single molecule to be attracted to the negative spots of a given molecule, but it seems probable that usually the two negative spots will attract hydrogens from two different molecules.) The hydrogens act as links between the oxygens in much the same way that the oxygens act as links between silicons in SiO_2, with the difference that in SiO_2 the oxygen is equally shared between the surrounding silicons. Ordinary ice has essentially the same crystal structure as the form of SiO_2 known as tridymite.

X-ray investigation of water indicates that this tetrahedral arrangement persists to a considerable extent in the liquid.[1] The arrangement of molecules in the liquid probably resembles the arrangement of the silicons in vitreous SiO_2, but is even more indefinite, in that the links between adjoining water molecules are undoubtedly not at all permanent, but are continually breaking and reforming, so that any given water molecule is on the average definitely attached to less than four neighbors. Water, then, has what has been termed a broken-down ice structure. This shows up in the X-ray diffraction pattern for water in that instead of getting definite angles of diffraction, the diffraction merely shows a broad maxima at certain angles, indicating that the intermolecular distances are variable, but giving, neverthe-

---

to be, however, a tendency for the water molecule to have a tetrahedral coordination, which is easily understood if the bond-eigenfunctions are of the sp^3 type.

[1] See Katzoff, and Morgan and Warren, *loc. cit.*

less, good evidence of the essentially tetrahedral character of the arrangement.

Further evidence on the structure of water is furnished by a consideration of its density. It will be clear from a consideration of Figs. 83 and 84 of Appendix IV that a tetrahedral arrangement is a very open one, in which the space available is far from being filled. The X-ray analysis shows that the distance between nearest neighbor molecules in ice is 2.76Å. If ice were composed of spheres of this diameter in a close-packed arrangement with twelve nearest neighbors instead of four, it would have a molal volume of 9.0 cc. Actually, the molal volume of ice is 19.6 cc. In water, the X-ray analysis shows that the average distance between nearest neighbors is 2.90Å. at 1.5°C.; yet we know melting actually causes a contraction, the molal volume of water being 18.0 cc. This is in good accord with the picture of water as a broken-down ice structure. The decrease in volume is due to the partial breaking down of the tetrahedral structure, with filling in of some of the spaces. It should be noted that the decrease of the volume of water is by no means so great as to indicate a very extensive breaking down of the tetrahedral structure. Close-packed spheres of effective diameter 2.90Å. would give a molal volume of 10.5 cc. If the arrangement in ice were unchanged, but the volume increased by the amount indicated by the increase in intermolecular distance, the volume would be 21.8 cc. The actual molal volume of water is considerably nearer the latter figure.

When water is heated, there are two opposing tendencies operative, the normal tendency for neighboring molecules to be spaced farther apart, on the one hand, and the tendency toward further breakdown of the tetrahedral structure and filling in of the spaces, on the other. These tendencies just balance at 4°C., where the molal volume has a minimum.

Recently, Cross, Burnham, and Leighton[1] have investigated the Raman spectrum of water, and have thus obtained information about the vibrational frequencies of liquid water and ice. They have calculated the vibrational frequencies of liquid water on the assumption that the natural vibrations of any molecule are relatively slightly changed by its neighbors and, in particular, have discussed how much the frequency would be changed if one,

[1] Cross, Burnham, and Leighton, *J. Am. Chem. Soc.*, **59**, 1134 (1937).

two, three, or four neighboring molecules bound to the given water molecule by hydrogen bonds were causing the perturbations. Interpreting their experimental results with the aid of these theoretical calculations, they have come to the conclusion that in liquid water a given water molecule is, on the average, acted upon by only a little more than two of its neighbors with sufficient intensity to cause a shift in its vibrational frequencies. If their interpretation is correct, it would seem to indicate that the intermolecular binding in water is even less definite than indicated above. In any event, however, the binding is definitely enough oriented and sufficiently powerful to support the open tetrahedral structure of the liquid and prevent a collapse to a close-packed structure, which is the main point of importance for our purpose.

19.3. Comparison of Water with Other Liquids.—In the light of the foregoing account of the structure of water, it will be of interest to compare the boiling point of water with that of some related liquids, as seen in the accompanying table. It will be

TABLE 46.—ABSOLUTE BOILING POINTS OF VARIOUS SUBSTANCES

CH_4	112	SiH_4	161	H_3CCH_3	185	H_2NNH_2	387
NH_3	240	PH_3	186	H_3CNH_2	267	H_2NOH	330
OH_2	373	SH_2	213	H_3COH	338	HOOH	425
FH	292	ClH	188	H_3CF	195	$H_2C{=}CH_2$	169
						H_2CO	252

noted that water has a much higher boiling point than any of the substances CH_4, NH_3, and HF that have the same number of valence electrons. It seems most probable that this is due to the fact that in the water molecule there are two positive regions and two negative regions, which makes possible an interlocking network of connections through the liquid, each water molecule having a coordination number of four. In the case of NH_3, if nitrogen is assumed to have the tetrahedral electronic structure, there are three positive spots and one negative spot. This makes possible the formation of chains in which the negative end of each molecule is attracted to one of the positive spots of an adjacent one. It is impossible, however, to have a network, however irregular, running through the liquid with coordination number of four, for this requires an equal number of positive and negative places. A similar remark holds for HF if the fluorine is tetrahedral; if it is not (as is probably the case), no remark is necessary. Methane, in which the whole outside of the molecule is positive, has a very low boiling point, as is to be expected.

In the next row of the periodic table, hydrogen sulfide, corresponding to water, has the highest boiling point. The effect is rather slight, however, presumably because with these molecules van der Waals forces predominate.

The table shows how the substitution of a methyl radical for a hydrogen affects the boiling point. Methyl alcohol has a high boiling point, but not so high as water, probably because forces around the larger methyl cannot be so large as in the neighborhood of a hydrogen. CH_3F has a rather surprisingly low boiling point. Hydrazine has a high boiling point, 387°K., and this is true also of hydroxylamine and hydrogen peroxide.[1]

19.4. The Dielectric Constant of Water.—One of the most important properties of water, which it is necessary to consider in connection with the properties of solutions of electrolytes, is its dielectric constant. As remarked in earlier discussions, the water molecule has a dipole moment. Such a property is naturally to be expected from our description of the structure of the water molecule, involving, as it does, localization of positive charge on one side of the molecule and negative charge on the other. The electric dipole will tend to be oriented in an electric field in such a way as to decrease the energy of the system, and the electron shells of the molecules will also tend to be distorted because of their mutual polarizability. Suppose now that we have two electric charges of magnitude e_1 and e_2. Suppose further that the charges are sufficiently far apart so that as far as their effects on each other are concerned they may be treated as point charges. If they were in vacuum, they would have a mutual potential energy e_1e_2/r, where r is the distance between them. But we have seen in Sec. 12.9 that the introduction of a dielectric medium between two condenser plates will cut down the potential between them by a factor equal to D, the dielectric constant. In exactly the same way, if the two charges are immersed in a dielectric medium like water, the potential energy

[1] In this connection see Randall, *Proc. Roy. Soc. (London)*, **A159,** 83 (1937).

is decreased by a factor D, so that it becomes[1] e_1e_2/rD. The dielectric constant D is actually constant, independent of e_1, e_2, and r, provided the electric field does not become too large. If the field at any point becomes so great, however, that the dipoles are already practically completely lined up, corresponding to a state of saturation of the dielectric, then it is obvious that a further increase in the field can cause no further change in the orientation of the dipoles, and they will be without effect on the energy of the system. D thus appears to decrease as the dielectric approaches saturation. Such a situation is likely to arise if the bodies carrying the charges e_1 and e_2 are very small (of the order of molecular dimensions, as in the case of ions), for in that case the fields in the neighborhood of the charges become very great. Furthermore, other irregularities may appear because of the molecular structure of the dielectric.

The dielectric constant of a liquid ordinarily decreases with temperature, because the increasing thermal agitation of the molecules makes it harder to line them up. In the case of water, it is obvious that there is another force tending to prevent the orientation of water molecules in an electric field; this is the tendency for them to be lined up with respect to each other due to the forces considered in Sec. 19.2. This makes small displacements of individual water molecules more difficult but does not prevent them from lining up in such a way that, while still preserving their tetrahedral coordination, a preponderance of water molecules have their negative ends pointing toward the external positive charge. The dielectric constant of water at room temperatures has the unusually large value of approximately 80.

19.5. Ionization and the Properties of Ions in Aqueous Solution.—When a polar substance dissolves in water, it is generally ionized. The reactions of ions are of particular importance to the inorganic chemist. Since an understanding of these reactions must be based on a knowledge of the nature of the ions themselves, some space will be devoted to a consideration of this question.

We have seen that the formation of the gaseous ions from a solid salt is a process requiring a considerable amount of energy,

[1] In the term "potential energy" is included also the energy of the medium in which the charges are located (but see also Sec. 19.6).

and in no known case is there any appreciable tendency for such a reaction to proceed spontaneously. On the other hand, the formation of aqueous ions from a solid salt is one that frequently takes place very readily. It must, therefore, require much less energy than the formation of gaseous ions. The reason for this is readily understood if we remember the high dielectric constant of water. If the dielectric were really a nonmolecular continuum, we should have to break the process of solution into two steps, the first being the introduction of dielectric between the ions of the solid, which would result in a lowering in the energy of the system, and the second being the dispersion of the ions in the dielectric. The latter requires energy but, because of the reduction of the potential between ions by the dielectric, only $\frac{1}{80}$ as much as would be required in vacuum, and possibly not enough to balance the energy in the first step, so that energy might even be expected to be evolved on solution. In any event, the heat of solution would be small. When the molecular structure of water is considered, the situation appears more complicated. Saturation effects cause the effective dielectric constant to be lower than the measured value, as has already been explained. Actually, of course, the molecular structure also prevents the occurrence of the first step described above, the interpenetration of solvent between the ions of the solute. It may still be, however, that the arrangement of dielectric molecules around an ion in solution is more compact than the arrangement of other ions about it in the solid. A compact arrangement of the dipoles of the solvent about an ion results in a relatively low energy. In view of the various complicated factors involved, it is not surprising that solution of the alkali halides sometimes results in evolution and sometimes in absorption of heat; this may be seen from Table 47, where L is the heat evolved when one mole of salt is dissolved to form an infinitely dilute solution.[1] The heat evolved or absorbed is always small in magnitude, however, compared with the lattice energy (compare $-U_0$, Table 47). (Regarding energy units see page 461.)

[1] Fajans and Schwartz, *Zeits. physik. Chem.*, *Bodenstein Festband*, **717** (1931), have tabulated L for various salts and discussed its significance. In Table 47, we have used the very slightly different values of Bichowsky and Rossini, "Thermochemistry of the Chemical Substances," Reinhold Publishing Corporation, 1936.

TABLE 47.—HEATS OF SOLUTION OF ALKALI HALIDES AT ROOM TEMPERATURE
(In kilogram-calories per mole)

Salt	$-U_0$	L	$L^+ + L^-$
LiF	245.1	−0.7	244.4
LiCl	201.1	8.7	209.8
LiBr	189.9	11.5	201.4
LiI	176.2	14.9	191.1
NaF	216.4	−0.3	216.1
NaCl	184.0	−1.2	182.8
NaBr	175.9	−0.6	175.3
NaI	164.4	1.6	166.0
KF	193.2	4.0	197.2
KCl	168.3	−4.4	163.9
KBr	161.5	−5.1	156.4
KI	152.5	−5.2	147.3
RbF	183.4	6.0	189.4
RbCl	162.1	−4.4	157.7
RbBr	156.1	−6.1	150.0
RbI	147.9	−6.6	141.3
CsF	175.9	8.6	184.5
CsCl	153.2	−4.6	148.6
CsBr	149.6	−6.9	142.7
CsI	142.4	−8.5	133.9

In order to gain a more detailed understanding of the process of solution and the meaning of the values of L, the matter may be considered from a different point of view. The process of solution to form aqueous ions will again be thought of as broken up into two steps, but different from those considered in the preceding paragraph. It is assumed that the substance is first ionized into gaseous ions, in which process the lattice energy $-U_0$ is absorbed. From the considerations of Chap. XIV, the energy of this process is known for many salts. In the second step, the gaseous ions are dissolved in the water, the energy evolved in this process being the heat of solution of the gaseous ions. Let the energy evolved on solution of the gaseous positive ion be L^+, and that of the negative ion be L^-. The heat L

evolved on solution of the solid salt is then given by

$$L = L^{+} + L^{-} + U_0. \tag{1}$$

We may now make an attempt at a rough theoretical evaluation of L^+ and L^-. According to electrostatic theory, when an electrically charged body of radius r, with a spherically symmetrical distribution of electricity of total charge e within this distance r from the center, is introduced into a dielectric with dielectric constant D, an amount of energy equal to $\frac{e^2}{2r}\left(1 - \frac{1}{D}\right)$ is evolved. This gives a basis of comparison with the experimental results. The use of this relation in this connection was first suggested by Born,[1] and it was early applied by Latimer.[2]

In Table 47, we include values of $L^+ + L^-$ obtained by adding the quantities in the two preceding columns. It is impossible to separate the sum experimentally into the values L^+ and L^- for the individual ions. However, differences between the individual values can readily be found. For example, the difference between the heat of solution of $K^+ + Cl^-$ and $K^+ + Br^-$ will be the difference between Cl^- and Br^-. It should be possible to substitute any other alkali ion for potassium and (assuming the values of U_0 to be correct) still find the same difference between the chloride and the bromide. This is, of course, an immediate consequence of the ionic theory of Arrhenius, and the result can be expected to hold only if the heats of solution at infinite dilution are considered, for only at infinite dilution can we measure the interaction of an ion with the solvent, unaffected by interaction with other ions. In the tables, all values of heats of solution are those obtained by extrapolation to infinite dilution, and the result obtained is independent of the choice of the common ion within, at worst, about ± 2 kg-cal., which is within the limit of error of, and may be considered a further check on, the calculation of the lattice energy.

The average value for the difference Δ between the heats of solution of adjacent halogen and alkali ions is given in Table 48 (column marked "obs."). For comparison, there is also given the difference to be expected theoretically from Born's formula[3]

[1] Born, *Zeits. Physik*, **1,** 45 (1920).

[2] Latimer, *J. Am. Chem. Soc.*, **48,** 1234 (1926).

[3] Born's formula gives ΔE's of solution, and the L's are ΔH's (see Appendix II), but the difference is negligible.

TABLE 48.—HEATS OF SOLUTION OF ALKALI AND HALOGEN IONS AT ROOM TEMPERATURE
(In kilogram-calories per gram ion)

Ion	Radius, Å.	L^+ or L^- (calc.)	Δ (calc.)	Δ (obs.)	L^+ or L^- (used)
F^-	1.36	119.9	29.9	33.8	125.9
Cl^-	1.81	90.0	6.4	7.4	92.1
Br^-	1.95	83.6	8.1	9.2	84.7
I^-	2.16	75.5			75.5
Li^+	0.59	276	104	26.6	117.2
Na^+	0.95	172	49	18.9	90.6
K^+	1.33	123	13	6.6	71.7
Rb^+	1.48	110.0	13.5	7.2	65.1
Cs^+	1.69	96.5			57.9

given above. This is done simply by inserting the ionic radius from Table 16 in the formula. No allowance is made for the effect of coordination number (see Sec. 14.6), which is equivalent to assuming that the coordination number is six and that the radius of the ion is the same as that of a uni-univalent electrolyte. Furthermore, in the use of Born's formula, we continue to neglect the molecular character of the dielectric. The results can, therefore, be taken only as the roughest sort of an approximation; nevertheless, it will be seen that the "experimental" and theoretical differences agree pretty well for the negative ions, though there is no agreement at all for the positive ions. This suggests that the energy of solution of at least the largest negative ion, I^-, should be fairly close to the theoretical value, and if it is assumed that it does actually have this value, then from it and the data of Table 47 values for the heat of solution of all the other ions can be calculated[1]; these are also given in Table 48. The

[1] It must be said that this method of dividing the heat of solution between the positive and negative ions is open to some objections, and other methods of making this division have been proposed. Bernal and Fowler, Ref. 1, page 393, have made a calculation that yields considerably different values for the energies of solution of the positive and negative ions. Klein and Lange, *Zeits. Elektrochem.*, **43**, 570 (1937), have recently estimated energies of solution from absolute electromotive force measurements, getting values for the alkali ions about 15 kg.-cal. greater than those of Table 48. However, the older method of estimating the absolute electromotive force of cells, based on electrocapillary measurements, gives values for the alkali

values obtained for the positive ions are all much less than the theoretical values. If our method for dividing the heat of solution between positive and negative ions is correct, it indicates some asymmetry in a water molecule which makes it easier for a negative than a positive ion to orient the water molecule. It seems, indeed, very likely that the centers of positive charge are very close to the surface of the water molecule. This might account for the apparent readiness with which a negative ion effects the orientation of a water molecule, as compared to a positive ion. The readiness with which hydrogen bridges are formed is evidence in favor of the supposition that the positive charge is close to the surface, not only in water but in many polar groups in which hydrogen is present. An apparent asymmetry of the water molecule shows up also in other properties which will be discussed below.[1] It should also be remarked that

TABLE 49.—HEATS OF SOLUTION OF VARIOUS IONS AT ROOM TEMPERATURE
(In kilogram-calories per gram ion)

Ion	Radius, Å.	L^+ (obs.)	L^+ (calc.)
H^+	....	253	
Ag^+	1.26	106	129
Tl^+	....	68	
Mg^{++}	0.82	446	795
Ca^{++}	1.18	369	552
Sr^{++}	1.32	339	494
Ba^{++}	1.53	305	426
Fe^{++}	....	446	
Al^{+++}	0.72	1071	2040

the special tetrahedral structure of the water molecule would have to be taken into account in any attempt at a quantitative discussion of the process of hydration; it is an oversimplification to treat it as a simple dipole.

ions only 5 or 6 kg.-cal. greater than those of Table 48. See Latimer, "The Oxidation States of the Elements and their Potentials in Aqueous Solutions," Prentice-Hall, Inc., 1938, pp. 21–22. Latimer, Pitzer, and Slansky, *J. Chem. Phys.*, **7**, 108 (1939), have suggested a method of dividing the heat of solution, similar to that used here, which gives even better agreement with the results based on electrocapillary measurements.

[1] See LATIMER, *Chem. Rev.*, **18**, 354–356 (1936), and LATIMER, PITZER, and SLANSKY, *J. Chem. Phys.*, **7**, 108 (1939).

In Table 49, we give some values for the energy of solution of some polyvalent positive gaseous ions. These are to be compared with the theoretical values obtained from the radii of Table 16, appropriate modifications being made to allow for the double or triple charge. These corrections, however, do not include a correction, like that in Sec. 14.6, for the effect of the charge on the apparent radius. It is observed that, as is to be expected, the discrepancy between the theoretical and experimental values is considerable, and this discrepancy would be increased by the correction for the effect of charge on the radius.

The experimental values were obtained, as in the case of the univalent ions, by the use of heats of solution and lattice energies. The latter, however, were not calculated directly, but were calculated from the heats of formation of the crystal, and other thermal data,[1] and the already determined electron affinities of the halogens, using the appropriate modification of Eq. (16) of Chap. XIV. Excellent checks were obtained with different halides of the same metal.

Although the heat of solution is of paramount importance in determining the solubility of salts, it is not possible to get a complete understanding of the phenomenon from the heat of solution alone. Even if it were possible to assume that the ions formed a perfect solution, and the interionic attraction were completely negligible, it would be necessary to take into account the effect of the entropy change on solution. As shown in Appendix II, the entropy is essentially a measure of the freedom of motion of the system, and a system tends to assume that state in which its motion is as unrestricted as possible, provided the energy of this state is not too great. For example, when a substance evaporates, its entropy increases because the gas molecules enjoy greater freedom of motion, and as far as this factor is concerned evaporation is favored, but the energy factor favors the condensed state. When the volume of a gas increases, the entropy likewise increases; in fact, in an isothermal process the increase of entropy per mole is given by the gas constant times the change in the natural logarithm of the volume occupied by

[1] The energies of formation and solution were taken mostly from Bichowsky and Rossini, "Thermochemistry of the Chemical Substances." A few heats of solution are from Latimer, Schutz, and Hicks, *J. Chem. Phys.*, **2**, 82 (1934).

one mole. In the case of a solution, the change of entropy of a mole of solute with concentration is likewise given by the gas constant times the change in the natural logarithm of the volume containing one mole, if the solution is dilute enough. The entropy of the mole of solute is, however, by no means the same as if it existed in the same volume of empty space. In the first place, the molecules of solvent occupy a large fraction of the space, and the volume actually available to the solute molecules is therefore by no means equal to the total volume. Besides this, the solute molecules may well have an effect on the solvent molecules, and this effect is best included as part of the entropy of the solute. If the solute is an electrically charged ion in aqueous solution, it tends to orient the water molecules in its neighborhood; this decreases the freedom of motion of the water molecules and results in the entropy being lower than it would otherwise be.[1]

Solution of a solid salt is a process in many respects resembling vaporization, and so would be expected to result in an increase of entropy, in spite of the fact that the dissolved ions are free to move only in a fraction of the solvent volume. However, the effect of the orientation of solvent molecules may be so great as to actually cause a decrease in entropy on solution. In any event, the effect of orientation is to cause there to be less tendency for the substance to dissolve. The solubility product of a salt is the equilibrium constant for the process of solution and so is a measure of the solubility of the salt. The dependence of such an equilibrium constant on the energy and the entropy of solution is given in Appendix II [Eq. (6) and footnote 1, page 460]. The relation, which applies strictly only to perfect solutions, in which forces between solute molecules can be neglected, is as follows:

$$RT \ln K = L + T\Delta S^0. \tag{2}$$

Here R is the gas constant, T the absolute temperature, K the solubility product, L the heat evolved on solution per mole (this includes any work done against the constant pressure of the

[1] For a more detailed discussion of the significance of the entropy of solution see Latimer and Kasper, *J. Am. Chem. Soc.*, **51,** 2293 (1929), or Latimer, *Chem. Rev.*, **18,** 354–356 (1936). See also Eley and Evans, *Trans. Faraday Soc.*, **34,** 1093 (1938).

atmosphere: L is the negative of the ΔH of the process), and ΔS^0 is the hypothetical change of entropy when one mole of solid salt dissolves to form a solution whose concentration is one mole per liter. By hypothetical change of entropy is meant the change that would occur were the solution a perfect solution, *i.e.*, if it were possible to neglect the interaction of the ions on each other while in solution. (We call such a hypothetical perfect solution with a concentration of one mole per liter a hypothetical molal solution.) Since salt solutions are far from perfect, Eq. (2) can be only a rough approximation, but it should suffice to give the solubility of slightly soluble salts (for in such cases we never deal with large concentrations and may neglect ionic interactions) and, in any case, will give a fair approximation. The fact that a concentration of one mole per liter may not be attainable does not prevent us from calculating what the entropy of such a solution would be if it could be realized.

The value of ΔS^0 is of considerable importance in determining and understanding the solubility of various salts, and we shall now concern ourselves with the evaluation of this quantity. As may be seen by Appendix II, ΔS^0 is given by the heat absorbed when one mole of the salt dissolves to give a hypothetical molal solution, divided by T, provided the process is carried out reversibly, but ΔS° is usually obtained by calculation from other thermodynamic quantities. The values for a number of solid salts are given in Table 50.

To understand the values of ΔS^0, we again find it convenient to think of the process of solution as divided into two steps, first the formation of gaseous ions with hypothetical concentration of one mole per liter, and secondly, the solution of the gaseous ions to form a solution, also with the hypothetical molal concentration.[1]

[1] The method of obtaining these entropies is briefly as follows. We can first get an absolute value for the entropy of a solid salt by use of the third law of thermodynamics, which states that the entropy of any substance at absolute zero is zero. To find the entropy at some other temperature, use is made of the law stated in Appendix II, that the change in entropy on changing the temperature is equal to the heat absorbed divided by the temperature. The heat absorbed when the temperature is increased by an increment dT is $C_p\,dT$, where C_p is the specific heat at constant pressure, and the corresponding entropy change is $C_p\,dT/T$. The entropy at some temperature, say T_1, is, therefore, given by $\int_0^{T_1} (C_p/T)\,dT$, which may be

We write

$$\Delta S^0 = \Delta S_1{}^0 + \Delta S_2{}^0,$$

where $\Delta S_1{}^0$ and $\Delta S_2{}^0$ refer, respectively, to the two steps just described. Values of $\Delta S_1{}^0$ for a number of solid salts are given in Table 51. $\Delta S_2{}^0$ is the entropy of solution of a pair of ions and may be written

$$\Delta S_2{}^0 = \Delta S^+ + \Delta S^-$$

where ΔS^+ and ΔS^- refer to the positive and negative ions, respectively. By a study of electrolytic cells in which the two

TABLE 50.—ENTROPIES OF SOLUTION OF SOLID SALTS TO FORM HYPOTHETICAL MOLAL SOLUTIONS AT 25°C.
(In calories per mole per degree Centigrade)

	Li^+	Na^+	K^+	Ag^+	Mg^{++}	Ca^{++}	Ba^{++}	Zn^{++}	Cd^{++}	Hg^{++}
F^-	....	−1	6	....	....	−32	−25			
Cl^-	3	10	18	8	....	....	......	−23	−21	−9
Br^-	....	14	21	12	....	....	......	−19	−11	
I^-	....	17	25	15	....	....	......	−14	−4	
NO_3^-	....	21	27	....	....	....	21			
SO_4^{--}	−26	−3	8	−8	....	−33	−25	−52	−46	
CO_3^{--}	....	−17	..	−18	−60	−46	−38	−58	−68	

electrodes are at different temperatures, it is theoretically possible to get the entropies of the separate ions in solution.[1] In this way, it has been possible to get the values of ΔS^+ and ΔS^- given in Table 52.

---

evaluated if C_p is known as a function of T. In this way the absolute entropy of the salt can be obtained; the entropy of dissolved ions can be obtained from the entropy of the solid salt and the entropy change on solution, and the entropy of the gaseous ions can be obtained by statistical mechanical calculations. This gives all the data necessary for finding the quantities given in the tables. Actual data in the tables are taken from Landolt-Börnstein, "Tabellen," Latimer, Pitzer, and Smith, *J. Am. Chem. Soc.*, **60,** 1829 (1938), and Latimer, "The Oxidation States of the Elements and Their Potentials in Aqueous Solutions," pp. 328–333.

[1] Knowing the absolute entropy of only one ion of course makes it possible to determine it for all others. For this purpose we use the value for Cl^-, obtained by Eastman and Young (quoted by Latimer, *Chem. Rev.*, **18,** 355 (1936).

From Table 51, it is seen that the values of $\Delta S_1{}^0$ are all accounted for if the entropy of vaporization to form gaseous ions is in all cases about 20 cal. per mole per degree per ion. The great variation in the values of ΔS^0 is to be ascribed to variation in $\Delta S_2{}^0$, and this is understood by considering the values of ΔS^+

TABLE 51.—ENTROPIES OF IONIZATION OF SOLID SALTS TO FORM GASEOUS IONS AT HYPOTHETICAL MOLAL CONCENTRATIONS, AT 25°C.
(In calories per mole per degree Centigrade)

	Li^+	Na^+	K^+	Ag^+	Ca^{++}	Zn^{++}	Cd^{++}	Hg^{++}
F^-	..	45	43	..	71			
Cl^-	41	42	41	41	..	68	63	66
Br^-	..	42	41	41	..	65	65	
I^-	..	41	41	40	..	62	63	

TABLE 52.—ENTROPIES OF SOLUTION OF GASEOUS IONS, AT 25°C.
(In calories per mole per degree Centigrade)

H^+	−24.3	Ca^{++}	−51.2	Fe^{++}	(−69.)
Li^+	−25.3	Ba^{++}	−41.2	Fe^{+++}	(−112.)
Na^+	−19.6	Cu^{++}	(−69.)	Al^{+++}	−119.3
K^+	−11.0	Zn^{++}	−67.0	F^-	−26.1
Rb^+	−8.8	Cd^{++}	−59.3	Cl^-	−12.2
Ag^+	−20.7	Hg^{++}	−51.1	Br^-	−8.4
Tl^+	−9.6	Sn^{++}	−48.0	I^-	−4.2
Mg^{++}	−70.0	Pb^{++}	−40.8	S^{--}	−26.3

Doubtful values in parentheses.

and ΔS^-. It will be seen that the smaller the ion (of given sign and magnitude of charge) and the larger the charge, the more negative is the value of ΔS^+ or ΔS^-. This is exactly what is to be expected, for the greater the force the ion exerts on the surrounding water molecules, the more the water molecules should be tied up, with consequent reduction of their freedom of motion. This results in the lowering of the entropy of the system. It will be seen that this effect is great enough to make the molal entropy of solution (Table 50) of most of the salts involving polyvalent ions negative, even in the cases where there are three ions per molecule, in spite of the large values of $\Delta S_1{}^0$ when three ions are present. Where complex ions are involved, we have not attempted to break down the entropy of solution into component parts, but roughly speaking, the entropy of solution is of similar magnitude to that of molecules involving simple ions of the same valence type.

It is of interest to note that in the entropies, as well as in the energies of solution of gaseous ions, there is a lack of symmetry between positive and negative ions.[1] We may, for example, compare K^+ and F^- which have nearly the same ionic radii. The value of L^+ for K^+ has been taken as 71.7 kg.-cal., whereas the value of L^- for F^- is 125.9 kg.-cal. This much greater heat of solution would indicate that F^- has a greater effect on the solvent than K^+, and this is borne out by the much more negative value of the entropy of solution of F^-, indicating that it restricts the motion of the water molecules much more than does K^+. In the case of I^-, the entropy of solution is so low that it is of the right order of magnitude to be accounted for on the supposition that the only effect of the solvent is to occupy space, so that a dissolved I^- does not have so much room to move around in the solution as in the gas phase.[2] This is the sort of situation in which a solvent would act as though it were a continuous dielectric in which the molecular structure were unimportant, and may offer some justification for our procedure in calculating the energy of solution of I^- on that basis.

Another property that is related to the entropy of solution of an ion is the apparent volume of the ion when in solution. This property has been discussed especially by Bernal and Fowler,[3] but our discussion will differ somewhat from theirs. The apparent volume of an ion in solution is defined as the increase in volume[4] per ion when a salt containing that ion is dissolved. As in the case of other properties, it is not possible without further hypothesis to separate the change of volume due to the positive and that due to the negative ions. However, at infinite dilution the difference in the apparent volumes of two halogen ions, let

[1] See LATIMER, *Chem. Rev.*, **18**, 354–356 (1936), and LATIMER, PITZER, and SLANSKY, *J. Chem. Phys.*, **7**, 108 (1939).

[2] The ion is probably free to move in a space equal to only $\frac{1}{10}$ to $\frac{1}{100}$ of that occupied by the solvent [see Horiuti, *Zeits. Elektrochem.*, **39**, 22 (1933); Evans and Polanyi, *Trans. Faraday Soc.*, **31**, 891 (1935); Rice, *J. Chem. Phys.*, **5**, 353 (1937); Eyring and Hirschfelder, *J. Phys. Chem.*, **41**, 249 (1937)]. This would indicate an entropy of solution of $-R \ln 10$ to $-R \ln 100$, or -4.6 to -9.2 calories per mole per degree.

[3] BERNAL and FOWLER, *J. Chem. Phys.*, **1**, 531 (1933).

[4] This increase in volume is the difference in volume between solution and pure solvent. The volume of the solid salt is not added to the latter.

us say, combined with a particular positive ion, should be independent of which positive ion this is. This is found to be the case, presumably within experimental error. Slightly modifying the procedure of Bernal and Fowler, we have attempted to get the apparent volume in solution of the separate ions by assuming that in CsI, which has large ions that presumably have the smallest effect on the water molecules, the volume is divided between Cs^+ and I^- in the ratio of the cubes of the ionic radii. The apparent volumes of other ions can then be obtained, and these are given (together with the volumes of spheres with radii equal to the ionic radii of Table 16) in[1] Table 53. It will be seen that

TABLE 53.—APPARENT IONIC VOLUMES IN SOLUTION
In Å³.

	Experimental, °C.			Calc. from Table 16
	0	25	50	
H^+	......	−3.		
Li^+	−4.7	−5.7	−6.7	0.9
Na^+	−9.8	−6.6	−5.1	3.6
K^+	7.8	10.3	11.3	9.9
Rb^+	16.3	18.7	19.8	13.6
Cs^+	28.2	30.9	32.1	20.2
Ag^+	......	*ca.* −10	......	8.4
Mg^{++}	−43.6	−44.0	−45.2	2.3
Ca^{++}	−39.1	−36.3	−37.2	6.9
Sr^{++}	−42.0	−37.8	−36.8	9.6
Ba^{++}	−34.2	−28.8	−26.9	15.0
F^-	......	1.9[a]	......	10.5
Cl^-	30.6	33.6	34.5	24.8
Br^-	41.4	45.3	46.6	31.1
I^-	58.8	64.4	67.1	42.2
NO_3^-	......	53.		

[a] 18°C.

[1] Values for alkali and halide ions (except F^-) from tabulation of differences of partial molal volumes, Landolt-Börnstein "Tabellen," III. Ergänzungsbd., p. 382. Value for NO_3^- from $RbNO_3$ after H. Smith, in Landolt-Börnstein, "Tabellen," III., p. 383. Values for F^- and Ag^+ from data in "International Critical Tables." Values for alkaline earth ions from data and calculations kindly furnished by W. C. Root (Thesis, Harvard University); for H^+ from calculations by W. C. Root based on "International Critical Tables."

for a number of the large ions the apparent ionic volumes are slightly larger than the calculated values and parallel them rather well. The small ions, however, which exert large forces on the water molecules, sufficiently increase the density of the water so that the apparent ionic volume becomes small. Negative values presumably mean that the electrostatic attraction between ion and water is so great that the openwork structure of the water is broken down, and a number of water molecules coordinate with the ion, forming a complex ion.

There is, apparently, a limit to the negative value of the apparent volume of a univalent ion; in fact, at 25° those of Li^+ and H^+ are less negative than that of Na^+. Lithium and hydrogen are smaller than the holes in the water structure and so are probably not centered in the holes.[1] This seems to diminish their effectiveness in breaking down the water structure, at least at the lower temperatures.

In the case of polyvalent ions, the forces are much greater, and large negative volumes occur.

It will be seen from a comparison of the tables that there is a relation between the entropy values and the volumes. The parallelism between calculated volume and observed apparent volume breaks down badly in those cases in which there is a large negative value of the entropy of solution of the gaseous ion.

19.6. Relation between Energy and Free Energy of Solution of Ions.—In view of the facts brought out in the preceding section, it may be profitable to attempt a more exact interpretation of Born's expression for the energy of solution of an ion

$$\frac{e^2}{2r}\left(1 - \frac{1}{D}\right).$$

This is the actual reversible work necessary, according to electrostatic theory, to take an ion bodily out of solution. It thus resembles a free energy of solution rather than an energy of solution, because removing the ion from solution also changes the condition of the molecules of the solvent, and this may be reflected in absorption or evolution of heat, which has to be taken into account in computing the energy change, but not in computing the free-energy change, which is given by the reversible mechanical work done on the system. From thermodynamics,

[1] Compare GIBSON, *Sci. Monthly*, **46**, 115 (1938).

this reversible work ΔF is equal to $\Delta H - T\Delta S$, where ΔH is the change of the heat content on taking a mole out of solution, in this case equal to L^+ or L^-, while ΔS is the change of entropy[1] (see Appendix II). This suggests that it would be better to compare the expression[2] $\frac{Ne^2}{2r}\left(1 - \frac{1}{D}\right)$ with $L^\pm + T\Delta S^\pm$, where $\Delta S^\pm$ is taken from Table 52, rather than to compare it with $L^\pm$ itself. However, the theory visualizes a process in which each individual ion is handled and moved separately by some external agency, but in which, of course, no such control is maintained over the solvent molecules, which are free to move under the influence of the field of the ion. This means that the ΔS term used to calculate ΔF should not contain any contribution from the change in the freedom of motion of the ions themselves on solution, but only from that part of $\Delta S^\pm$ that is due to the effect of the ions on the solvent. Now it was indicated above that it is reasonable to assume that the entropy of solution of I^- ions, -4.2 cal. per mole per degree, is entirely due to the change of the freedom of motion of the ions; therefore, in calculating ΔF for comparison with the theoretical expression, ΔS will be set equal to zero, so $\Delta F = L^-$, for I^-. It is then clear that the set of values of L^+ and L^- obtained by assuming L^- to be given correctly by the theoretical expression in the case of iodide ion is consistent with the present considerations. In all other cases, ΔF will be different from L^+ or L^-. If it is assumed that the change of the freedom of motion of all the halide ions is the same as for I^- (and this is certainly a reasonable assumption as the correction is practically negligible in any event), then the change of entropy due to the effect on the solvent is obtained by subtracting -4.2 from ΔS^-, giving

$$\Delta F = L^- + T(\Delta S^- + 4.2).$$

Values of ΔF at room temperature, obtained in this way for the

[1] Care must be taken to keep the signs straight. ΔH is the change of heat content on taking an ion *out* of solution, hence it is positive if heat is *added* to the system when an ion is *removed.* $L^\pm$ is positive if heat is *lost* by the system when an ion is *put in.* Hence $\Delta H = +L^\pm$. Here ΔS is the change of entropy on removing an ion from solution, whereas $\Delta S^\pm$ is the change of entropy when an ion is dissolved. So if these are equated, it must be with opposite signs.

[2] N = Avogadro's number (see note on page 461).

several halide ions, are as follows: F^-, 119.5; Cl^-, 89.8; Br^-, 83.5. It is observed that these are in remarkable agreement with the calculated values of L^- given in Table 48; this may, however, be to some extent fortuitous.

If a similar correction is attempted for the positive ions, the disagreement between calculated and observed values of L^+ is greater.[1]

It seems worth while to consider the matter under discussion from another point of view. It is known from thermodynamics[2] that $\Delta S = -(\partial\, \Delta F/\partial T)_p$ (the subscript indicates that the differentiation is to be performed at constant pressure). So the equation $\Delta F = \Delta H - T\Delta S$ may be written

$$\Delta F = \Delta H + T\left(\frac{\partial\, \Delta F}{\partial T}\right)_p.$$

If we set $\Delta F = \dfrac{Ne^2}{2r}\left(1 - \dfrac{1}{D}\right)$, it is found that

$$T\left(\frac{\partial\, \Delta F}{\partial T}\right)_p = \frac{Ne^2}{2r}\,\frac{T}{D^2}\left(\frac{\partial D}{\partial T}\right)_p$$

which is always small compared with ΔF if D is set equal to the dielectric constant of water; $-T(\partial \Delta F/\partial T)_p$ turns out to be approximately $0.018Ne^2/2r$.

The calculated value of $-T(\partial\Delta F/\partial T)_p$ or $T\Delta S$ should, of course, be compared, for the halide ions, with the experimental value of[3]

$$-T(\Delta S^- + 4.2)$$

[1] LATIMER, PITZER, and SLANSKY, *J. Chem. Phys.*, **7**, 108 (1939), have recently made a calculation which is similar to the one given in this section. They found that by adding 0.85Å. to the radii of the univalent positive ions, they could be brought into line also. This is connected with the asymmetry of the water molecules discussed above, it being assumed that the center of the electrical dipole cannot get closer than 0.85Å. if the dipole has the negative end nearest the ion. Latimer, Pitzer, and Slansky also believe that better results are obtained if 0.10Å. be added to the radii of the negative ions. This leads, it is true, as noted in the footnote, p. 402, to a set of values for the energy of hydration which are in slightly better agreement than ours with those based on electrocapillary measurements.

[2] See LEWIS and RANDALL, "Thermodynamics and the Free Energy of Chemical Substances," p. 172, McGraw-Hill Book Company, Inc., 1923.

[3] With regard to the signs, see the footnote on the preceding page.

rather than with $-T\Delta S^-$, itself. But our apparently reasonable assumptions have made $T(\Delta S^- + 4.2)$ equal to zero for I^-, and it is indeed doubtful whether I^- should tie up the water molecules more than they are tied up in pure water. It is true that $-T\Delta S^-$ for I^- checks approximately with the theoretical value of $0.018Ne^2/2r$, but it is questionable whether this is significant as neither $-T\Delta S^-$ nor $-T(\Delta S^- + 4.2)$ checks at all well for the other negative ions, not to mention the positive ions. These difficulties are all connected with the rough approximation of treating a molecular medium as a continuous dielectric with unvarying dielectric constant, as Born's formula does. The validity of applying Born's formula for the calculation of ΔF may, then, also be questioned. The difficulties involved in this case should not be too serious, however, if the entropy effect is not too great, since the term involving D has a relatively small effect on the calculated value of ΔF.

19.7. The Factors Affecting Solubility. Illustrative Examples. All of the alkali halides are very readily soluble except lithium fluoride, which is soluble only to the extent of 5×10^{-2} mole per liter. The heat absorbed on solution of lithium fluoride is, however, less than is the case with a number of the other alkali halides. The insolubility of lithium fluoride must therefore be referred to the entropy of solution. It is interesting to compare lithium fluoride with potassium bromide, which absorbs more heat on solution, but is very soluble. The contributions to the quantity on the right-hand side of Eq. (2), page 405, may be summarized as follows (with all entropies expressed in kilogram-calories per mole per degree to be comparable with energy term expressed in kilogram-calories per mole):

	L	ΔS_1^0	ΔS_2^0	ΔS^0	$L + T\Delta S^0$ (T = 298° abs.).
LiF.....	−0.7	0.044	−0.051	−0.007	−2.8
KBr....	−5.1	0.041	−0.019	0.022	1.5

The value of ΔS_1^0 for LiF has been obtained from Table 51 by extrapolation. The value of the solubility product calculated for LiF from Eq. (2), page 405, is about 8.7×10^{-3}, corresponding to a solubility of 9×10^{-2} mole per liter, which is in good accord with the observed value. It is of interest to note that the low solubility as compared with potassium bromide may be traced to the large negative value of ΔS_2^0, which is ultimately due to the fact that the small lithium and fluoride ions exert great forces on the water molecules and so restrict the freedom of motion of the latter. Although it is hardly correct to say that any one factor is responsible for the solubility or insolubility of a salt, it may, at least, be said that the large forces exerted by the ions on water and consequent entropy effect are an important contributing factor in making LiF only slightly soluble.

The silver halides show a very different behavior from the alkali halides, silver fluoride being soluble, whereas silver chloride, bromide, and iodide are practically insoluble, the solubility decreasing in that order. But the entropies of solution of the silver salts are almost identical with those of the corresponding sodium salts. It is, therefore, obvious that the difference in solubility of sodium chloride and silver chloride has to do principally with the heat of solution, and it is, indeed, true that when silver chloride is dissolved 16 kg.-cal. per mole are absorbed, whereas in the case of sodium chloride only 1 kg.-cal. per mole is absorbed. In order to analyze this difference, let us turn to Eq. (1), page 401. L^- is the same for the two salts. L^+ is 15 kg.-cal. greater for Ag^+ than for Na^+. This tends to favor the solubility of the silver salts; however, it is more than counterbalanced by the fact that $-U_0$ is 30 kg.-cal. more for AgCl than for NaCl. Since these salts have almost the same interionic distance, it is seen that this difference must be mostly due to the difference in the type of binding in the respective lattices. Indeed, we have seen (Chap. XIV, Table 21) that van der Waals forces contribute 29 kg.-cal. and covalent forces 11 kg.-cal. to the lattice energy of AgCl. These forces, then, are very largely responsible for the insolubility of AgCl.

In Chap. XVI, we noted that the distance between silver and chlorine atoms in silver chloride is smaller than expected theoretically for ionic compounds (see Table 38, page 335). This is, presumably, another indication of the partially covalent character of the bond. The theoretical radius of the silver ion (Table 16) is closer to that of potassium than to that of sodium ion. In calculating the lattice energy of silver chloride, the actual instead of the theoretical atomic distance is used. Had the latter been used, the discrepancy between the actual and theoretical lattice energies would have appeared greater, which would seem to reinforce the opinion that the covalent character of the bond in AgCl is in large part responsible for its insolubility. In the case of silver fluoride the interionic distance coincides fairly closely with the theoretical. In this salt, the forces are probably largely ionic. The salt might be expected to resemble somewhat potassium fluoride. The silver ion in solution, however, does not differ from that produced by silver chloride, and its high heat of solution L^+ would tend to increase the solubility as compared with potassium fluoride. On the other hand, the entropy of solution of silver fluoride is lower than that of potassium fluoride, but not enough lower to counteract the energy effect.

There is thus an interesting difference between silver chloride and silver fluoride, resulting largely from the nonionic contribution to the binding in the former. The increasing insolubility of the bromide and iodide is related to the increasing amount of covalency in their binding. The solubility of various silver salts may be compared with the solubility of corresponding sodium salts, and a general view of the behavior of the silver salts thereby obtained.[1] The solubility of most sodium salts is rather large, and it may be assumed that the solubility product of a series of such salts will not vary except by a factor that is small compared with the variation of the solubility products of the corresponding silver salts. The solubility product of the sodium salts, therefore, may be roughly taken as constant. Furthermore, it is seen from Table 50 that the entropies of solution of corresponding sodium and silver salts are much alike. Therefore, from Eqs. (1) and (2) we get for any pair of salts AgX and NaX the following equation[2]:

$$RT \ln K_{\text{AgX}} - RT \ln K_{\text{NaX}} = L^+_{\text{Ag}} - L^+_{\text{Na}} + U_{\text{AgX}} - U_{\text{NaX}}.$$

Since $RT \ln K_{\text{NaX}}$ is assumed to be constant and L^+_{Ag} and L^+_{Na} are also constants, the value of $\ln K_{\text{AgX}}$ for different salts[3] (different X's) should give an approximately straight line when plotted against $U_{\text{AgX}} - U_{\text{NaX}}$. That this is actually the case is shown in

[1] See FAJANS, *Zeits. Krist.*, **66**, 343*ff*. (1928).

[2] For the sake of simplicity, we drop the subscript, 0, on the U's.

[3] We can include on the same diagram salts of the type Ag_2X, where X is a bivalent radical, if for these salts we plot $\frac{1}{2} \ln K_{\text{Ag}_2\text{X}}$ as ordinate and $U_{\text{Ag}_2\text{X}} - U_{\text{Na}_2\text{X}}$ as abscissa, where $U_{\text{Ag}_2\text{X}}$ and $U_{\text{Na}_2\text{X}}$ are the lattice energies for that amount of salt containing one gram atom of silver and sodium

Fig. 72. The straight line does not appear to have the right slope; but this is due to the fact that the solubility products of the sodium salts are actually not all exactly the same.[1] The interesting point which it is desired to bring out here is the

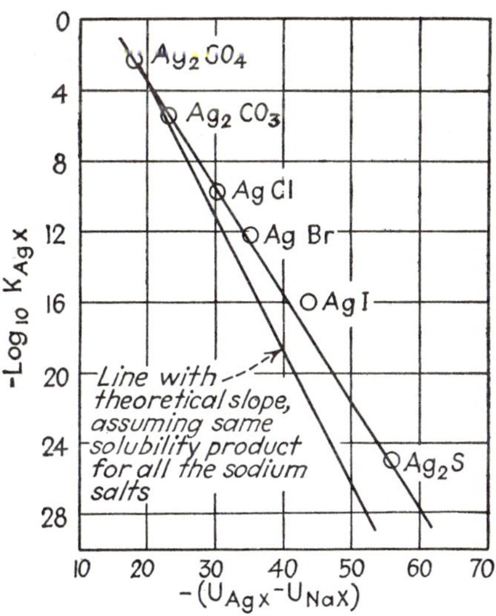

FIG. 72.—Solubility products and lattice energies. Theoretical slope for 18°C. Data for the carbonate and halides are for 25°C., but difference is within experimental error. Value for Ag_2S very rough. (*Data from Landolt-Börnstein, "Tabellen."*)

dependence of the solubility product on the character of the binding of the salt. Roughly speaking, assuming that the sodium

---

respectively. It should be noted that the solubility product is the equilibrium constant for the reaction

$$AgX\ (solid) \rightleftarrows Ag^+(aq) + X^-(aq)$$

or

$$Ag_2X\ (solid) \rightleftarrows 2Ag^+(aq) + X^-(aq)$$

and the actual solubility of the substance may be greater than that calculated from this equilibrium because of the effects of other reactions, namely, hydrolysis and complex formation.

[1] The discrepancies are practically entirely accounted for by this variation in the solubility products of the sodium salts, as appears from a tabulation in Chap. XXI of Latimer, reference 1, p. 406. (These solubility products are corrected for deviations from the perfect solution laws, and so can be used directly in the equation—actually Latimer gives free energies of solution from which corrected solubility products are calculated directly.)

salts are ionic, $U_{\text{AgX}} - U_{\text{NaX}}$ is a measure of the deviation from the pure ionic type of binding in the silver salt, and it is seen that a large deviation means, on account of the logarithmic relation, an extremely small solubility.

A similar analysis could undoubtedly be made of the solubilities of many other salts, but none will be attempted here. Various complications may, of course, arise. In some cases, the ions formed react with water, so that the ions present are not those which would be expected offhand from the formula of the salt. Not all salts in which the binding is purely ionic are soluble. For example, this is the case with $BaSO_4$ and $BaCO_3$. In such cases, ions of higher valence are involved, and the solubility depends on the relation between the ionic forces in the lattice and those in solution.

In no case will a stable atomic compound dissolve to give neutral atoms in solution. It is only the possibility of ionization that makes solution of a nonmolecular solid possible. On the other hand, there is the possibility of the molecules of molecular compounds dissolving in water, depending on the relative strengths of the attraction between molecules of the substance involved and the attraction between these molecules and water. These matters have been discussed in detail by Hildebrand,[1] and will not be considered here. Molecular compounds may also react with the water, perhaps forming hydrated ions, as in the case of Al_2Cl_6 (see Sec. 19.9).

19.8. Ionization of Halides of the Transition Metals.—In the account of the complex salts given in Chap. XVI, we have noted several cases in which a halogen ion is attached to a central ion and is not ionized in aqueous solution. Whether such a halide is ionized depends upon the mode of attachment of the halogen; if water or ammonia molecules, for example, are directly attached to the central ion, filling up its coordination sphere, then the halide is ionized, but if the halogen itself is in the coordination sphere it is not. This behavior may seem very different from that exhibited by the salts of the more electropositive metals, but in reality the contrast in behavior is not so great as it may seem. For since the ions of the more positive metals also have an affinity for water, when the salts ionize in aqueous solution, we

[1] HILDEBRAND, "Solubility," 2d ed., (Reinhold Publishing Corporation, 1936).

may, to all intents and purposes, consider that these ions have formed a compound with several molecules of water, *i.e.*, they are hydrated. This is especially true if the charge on the ion is greater than 1. These ions are thus essentially complex ions in solution, as has, indeed, been indicated already.

The question as to whether a halide will be ionized in aqueous solution is largely a question as to whether the positive ion has a greater attraction (under the conditions existing in the solution) for the halide ion or a water molecule. This question is always decided in favor of the water molecule in the case of the alkali and alkaline earth metals, but the matter is more doubtful in the case of a transition metal. The situation is further complicated in the case of the latter by the fact that equilibria are frequently not established rapidly. Thus we have seen in Sec. 16.2 that the equilibrium in aqueous solution between the various aquochromichlorides is slowly established.

In the case of the typical weak electrolyte $HgCl_2$, which ionizes only very slightly in aqueous solution,[1] the equilibrium must establish itself rapidly, for it undergoes the usual ionic reactions, such as precipitation of mercuric iodide by potassium iodide.

The equilibria involved in the interchange of halogen ions and water in complex ions have not been found at all simple, nor have the conditions under which such interchange takes place been systematically discussed. It is thus scarcely safe to make any definite quantitative statements with regard to the difference between the alkali and alkaline earth metals on the one hand and the transition metals on the other. Qualitatively, it appears that the force between the transition ions and chlorine, bromine, or iodine is relatively greater (as compared with the force between the ions and water molecules) than is the case with the more metallic ions. As seen in Sec. 19.7 by comparing sodium and silver salts, it seems probable that this is an expression of a greater tendency of the bond to be covalent and of a greater importance of van der Waals forces when a transition metal is involved. The water molecule is so much smaller than a chloride ion that one would expect the force between the water

[1] $HgCl_2$ is perhaps $\frac{1}{500}$ ionized in saturated solution at 25°. See Mellor, "Comprehensive Treatise on Inorganic and Theoretical Chemistry," vol. IV, p. 822, Longmans, Green & Company, 1923.

and the metallic ion (though only an ion-dipole force as contrasted to an ion-ion force) to be greater than that between the metallic ion and a chloride ion, or at least great enough so that in the presence of the large excess of water in aqueous solutions the positive ion would be hydrated without question, as always occurs with an alkali or alkaline earth metal, and with the metals of the aluminum group.[1] However, the very fact that the chloride ion is fairly large renders it liable to distortion in the field of a small positive ion; *i.e.*, the bond tends toward covalency instead of being electrostatic.

Furthermore, as we have seen, hydration may, if the ion is small, involve a large decrease of entropy due to the tying up of the water molecules; this discourages ionization, and makes it possible for weak electrolytes to exist (especially when the positive ion is at least divalent) if the bond is sufficiently covalent. The transition ions are very small, and it seems likely that the energy involved in the distortion of the halogen is what turns the balance so that the complex ions containing chlorine, bromine, or iodine become more or less stable in aqueous solution. The magnetic criterion applied to the salt $(NH_4)_2\,[FeCl_5H_2O]$ does indeed indicate that the binding is not covalent, but in this case the complex ion is undoubtedly not extremely stable.

On the other hand, the binding in most complex ions containing fluorine is in all probability ionic. An example is FeF_6^{---}, and the statement would certainly be true for ScF_6^{---}. The latter ion is quite stable in aqueous solution,[2] and it is probable that FeF_6^{---} is also—at least it is known that Fe^{+++} and F^- have a strong tendency to combine in aqueous solution,[3] but this is not surprising, for the fluoride ion is so small that the force between it and the central ion is quite large. One could not in such a case predict offhand whether the complex ion should be stable or not,

[1] This happens despite the possibly covalent character of the binding in aluminum chloride discussed in Sec. 17.3, and the fact that its crystal contains molecules of the formula Al_2Cl_6 (see Sec. 19.9). This bespeaks an extremely strong bond between aluminum and the oxygen of the water, in agreement with the amphoteric character of aluminum hydroxide (see Sec. 19.10).

[2] See WEINLAND, "Chemie der anorganischen Komplexverbindungen," p. 149, Ferdinand Enke Verlag, Stuttgart, 1924.

[3] See ABEGG, "Handbuch der anorganischen Chemie," 4. Band. 3. Abt., 2. Teil, p. B158, Verlag von S. Hirzel, Leipzig, 1935.

and no attempt at theoretical calculations have yet been made, as far as the author is aware.

It must be stated that we have made little attempt in the foregoing discussion to evaluate the possible influence of van der Waals forces. This is hardly possible in the present state of knowledge, but Sec. 19.7 indicates that it may not be inappreciable.

19.9. Hydrates.—Since many salts readily dissolve in water, it is not at all surprising that in many cases hydrated crystals are formed. These may be considered to be in a sense intermediate between aqueous solutions and anhydrous crystals.

Bernal and Fowler[1] suggest a classification of hydrates into three types. (1) There are hydrates in which the water merely occupies holes in the crystal. (2) There are hydrates in which the water appears to be definitely attached to a specific ion, surrounding it in what may be called a coordination sphere. (3) The crystal may resemble ice in which the ions are dissolved, the crystal melting, on heating, in its own water of hydration; such a crystal would be $Na_2SO_4 \cdot 10H_2O$, for example. It is not certain, however, that crystals of the first two classes will always lose water by evaporation rather than melting in their own water of hydration,[2] nor it is certain that the reverse will be true of the latter class of crystals. Furthermore, it may often be difficult to decide, even when the structure of a crystal is known, to which class it may most logically be assigned.

One thing is fairly certain, however, and that is that the more or less tetrahedral structure of the water molecule is often of considerable importance in determining the structure of a crystal. An example of this is furnished by the hexahydrated magnesium halides.[3] In these salts, the magnesium ion is surrounded by

[1] Bernal and Fowler, *J. Chem. Phys.*, **1**, 533 (1933).

[2] It might be supposed that water molecules which merely fill a hole in a crystal would be easy to remove, but Fajans (private discussion) has suggested that they will be relatively hard to remove. If the space for the water molecules is already present it will not be necessary to do any work in order to make the holes in which they are to rest. On the other hand, if it is necessary to make the holes before putting in the water molecules, the hydrate will have a relatively high energy and hence be unstable. $AlF_3 \cdot \frac{1}{2}H_2O$ is an example of an extremely stable hydrate in which the water occupies holes in the crystal structure.

[3] For crystal structures and references see the *Strukturbericht*. In

six water molecules in the form of a regular octahedron. Two halogens are close to this group and near the center of opposite faces of the octahedron. At each of the six other sides, there is a halogen which is displaced from the center of the face toward one of the water molecules. Although it is not possible by means of X-ray determinations to locate the hydrogens in a crystal, it seems probable that each of the two halogens that are near the centers of opposite faces has a hydrogen from each of the three waters of the face directed toward it. The other hydrogen from each H_2O molecule is directed toward one of the other halogens. Although the tetrahedral structure of the water molecule is not too directly in evidence, it appears that each water molecule has its negative end (or ends) directed toward the central magnesium ion and is in contact through hydrogen bonds with two halogen ions.[1] Each of the latter is joined to several waters through hydrogen bonds. This peculiar arrangement of the chlorines about the water tetrahedron naturally results in a rather complicated crystal structure. It is in contrast to the situation that exists in the case of several ammonia complexes of similar structure, *e.g.*, $Ni(NH_3)_6Cl_2$, $Ni(NH_3)_6Br_2$, $Ni(NH_3)_6I_2$. In these compounds, which have the fluorite structure, the $Ni(NH_3)_6^{++}$ ion is likewise surrounded by eight halogen ions, but each of the latter is opposite the center of a face of the ammonia octahedron, thus making possible a more regular crystal arrangement. Each halogen is surrounded by four $Ni(NH_3)_6^{++}$ groups at the corners of a regular tetrahedron.

Another interesting example of the influence of the structure of the water molecule is furnished by $NiSO_4 \cdot 7H_2O$. In this case, each Ni^{++} has six waters coordinated about it, and there is one

---

most cases we give only those details of the crystal structure which are necessary for the understanding of the role of the water in the crystal. However, concerning the crystal discussed in this paragraph see Appendix IV.

[1] The H—O—H angle appears to be greater than the tetrahedral angle. It is suggested by Andress and Gundermann, *Zeits. Krist.*, **87**, 366 (1934), who determined the structure of these crystals, that this is connected with the strong force the magnesium exerts on the surrounding oxygens. This loosens the hydrogens (as evidenced by the tendency of the salt to hydrolyze) (see Sec. 19.10) and may well cause a change in the valence angle. For, when the OH bond is made more polar, the positive charges on the hydrogens are less neutralized, and tend to push away from each other.

remaining water that does not come into contact with an Ni^{++}. There are two types of coordinated water molecule. (1) Of the six water molecules about a given nickel ion, four have three bonds in a plane (*i.e.*, there is what might be called a double bond to the nickel, and the two hydrogens make contact with oxygens, which may be from the sulfate group, the negative ends of an uncoordinated water molecule, or the negative end of one of the other type of water of coordination belonging to a different nickel ion). (2) The other two waters of the coordination group are tetrahedral, one of the negative spots making a contact with the nickel, and the other with a hydrogen of an uncoordinated water, or a hydrogen of a water of coordination from another group, while the two positive spots each contact a sulfate oxygen, or the negative end of an uncoordinated water. The uncoordinated water is also tetrahedral. In $NiSO_4 \cdot 6H_2O$, the nickel is again six-coordinated, and the general structure of the coordination group is the same as before, *i.e.*, four of the waters have three bonds and two are tetrahedral.

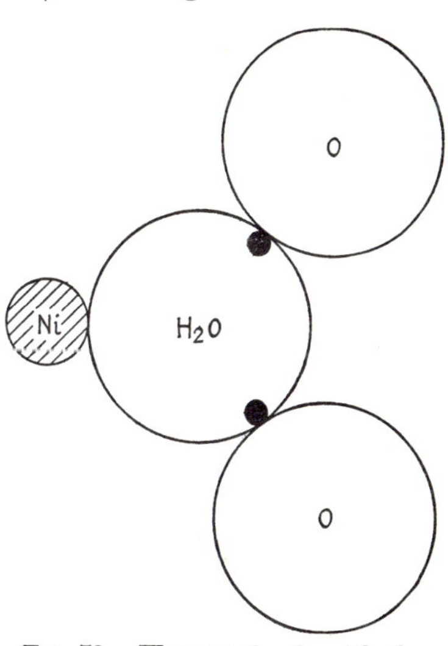

Fig. 73.—Water molecule with three bonds in a plane in hydrates of $NiSO_4$. The black circles indicate the location of the hydrogens in the water molecule.

In $Al(H_2O)_6Cl_3$ and $Cr(H_2O)_6Cl_3$, there are six water molecules arranged in the form of a regular octahedron about the metal ion. Each water is combined with two Cl^-'s, presumably through hydrogen bonds, the angle between these bonds being about 105°, nearly the tetrahedral angle. The Cl^-'s form bridges between the complex ions. These salts have a tendency to split off HCl even at room temperature (hydrolysis), indicating the great affinity of the metal ion for oxygen and the force with which it displaces the electrons away from the hydrogen.[1]

[1] See footnote 1, p. 420, and Andress and Carpenter, *Zeits. Krist.*, **87,** 462 (1934).

It will be noted that in the aluminum and chromium salts each water has only two bonds to atoms other than the central metal, while in the nickel salts described above two of the waters exhibit three "external" bonds. It has been suggested by Beevers and Schwartz[1] that this difference is characteristic of the metals, Al^{+++} and Cr^{+++} being able to satisfy all twelve negative bonds from the six water molecules, whereas Ni^{++} can satisfy only ten. The exact reason for this is not clear, though it might reside in the greater positive charge of the trivalent ions. If there really are ten or twelve bonds of a covalent nature, they are obviously a different type of bond than those considered in Sec. 15.2, and as a matter of fact measurements of magnetic susceptibility indicate that the bonds in $NiSO_4 \cdot 7H_2O$ are ionic or semicovalent.

The great contrast between the structure of anhydrous aluminum chloride (which, as seen in Sec. 17.3, is a molecular compound) and that of the hydrated salt is of interest. Aluminum chloride in sufficiently dilute solution appears to be completely ionized,[2] though it is somewhat hydrolyzed. The ion in aqueous solution undoubtedly has a structure like that in the hydrated salt. It is this possibility of reversion to a stable hydrated form that makes aluminum chloride soluble. The large magnitude of the forces operating between aluminum and oxygen may, then, be said to be chiefly responsible for the solubility of the salt, but the possibility of hydration of chloride ion also plays a necessary role. Even though the binding in solid anhydrous aluminum chloride is not ionic, we may consider the process to be broken up into the usual steps (s = solid, g = gas, aq = aqueous):

$$\tfrac{1}{2}Al_2Cl_6\ (s) \rightarrow Al^{+++}\ (g) + 3Cl^-\ (g)$$
$$Al^{+++}\ (g) \rightarrow Al^{+++}\ (aq)$$
$$3Cl^-\ (g) \rightarrow 3Cl^-\ (aq)$$

The first step requires an energy of 1270 kg.-cal., and as may be

[1] Beevers and Schwartz, *Zeits. Krist.*, **91**, 168 (1935).

[2] This must be the case, for the value obtained for the heat of solution of Al^{+++} appears to be the same regardless of whether data for Al_2Cl_6 or another halide are used to calculate it. Thus the state of Al^{+++} in solution is independent of the particular aluminum salt. It should be remarked that this use of the heat of solution is an illustration of a general method. It may often prove a valuable source of information regarding the state of an ion in solution.

seen from Tables 49 and 48, the latter two evolve energies of 1071 and $3 \times 92.1 = 276$ kg.-cal., respectively. The affinity of aluminum ion for liquid water, given by the second reaction, is not quite so great as its affinity for gaseous chloride ion, given by the first reaction; but the additional energy of hydration of the chloride ion more than balances the difference. Crystalline $Al(H_2O)_6Cl_3$ gives out only about 13 kg.-cal. on dissolving.

It is of interest that the aluminum fluoride crystal does not give evidence of molecular structure. The structure is rather complex and irregular, but each aluminum ion is nestled in among fluoride ions, and the structure is probably close to being truly ionic. It is not close-packed, and hydrates are known, ranging from $AlF_3 \cdot 3\frac{1}{2}H_2O$ to $AlF_3 \cdot \frac{1}{2}H_2O$. In $AlF_3 \cdot \frac{1}{2}H_2O$, which has a structure practically identical with the anhydrous salt, the water is probably in interstices in the crystal. It is quite evident that in this crystal the structure of the water molecule cannot play the same role that it does with the other crystals we have discussed, for it is close only to the negative fluorine ions and does not come near to the positive ions at all. It is, nevertheless, very difficult to drive the last half molecule of water out of the crystal requiring strong heating[1] [the heat absorbed in the reaction $2AlF_3 \cdot H_2O\ (s) \rightarrow 2AlF_3\ (s) + H_2O\ (g)$ is 35 kg.-cal.]. It seems very likely that there is at least one hydrogen bond between the oxygen of the water and a fluoride ion.

It is of some interest to note that the dehydration of AlF_3 is irreversible. The anhydrous salt does not readily take up water because of the difficulty of opening up the structure so that the water molecules can enter. This difficulty must also appear when the water molecules leave. Some energy is required to open the structure, which is later regained when the structure again closes after the water molecules have left. In other words, the dehydration requires an activation energy.

An example of a salt containing only one molecule of water of hydration is $Li_2SO_4 \cdot H_2O$. In this case, each lithium is surrounded by a tetrahedron of four oxygens. In half these tetrahedra, each of these oxygens also belongs to a sulfate tetrahedron; in the other half, however, one of them belongs to a water molecule. This last oxygen forms hydrogen bonds to neighboring sulfate oxygens, the

[1] See footnote 2, p. 421.

angle between the two hydrogen bonds of a given water molecule being the tetrahedral angle.

It seems safe to draw the following general conclusions about hydrated compounds: (1) There is a tendency for the water molecules to be definitely coordinated [through the oxygen, as, for example, in $Al(H_2O)_6^{+++}$] with a metal ion. This is undoubtedly due to the marked displacement of negative charge toward the oxygen in the water molecule, together with its small size. The oxygen thus can displace negative ions (except fluoride ion) from the neighborhood of the positive ion.[1] It is seen that this tendency is ultimately due to the outstandingly electronegative character of oxygen. (2) The structure of the water, except in a few cases (where the adjustment is probably difficult because of geometrical reasons), plays an important role in the determination of the crystal structure.

19.10. Acids and Bases.—A large number of substances, when dissolved in water, give rise to either H^+ ions or OH^- ions. These ions, of course, exert strong forces on the water molecules, and the H^+, in particular, is probably pretty firmly attached to one of the negative spots of a water molecule, so that what is present is actually an H_3O^+ ion. The question as to whether a substance is a strong acid or not will depend on the force with which the ionized residue holds a proton as compared with the attraction of a water molecule for a proton, *in aqueous solution.*[2] If the latter predominates, the substance is a strong acid.

Most of the commoner acids may be classified into two groups: (1) the hydroacids, such as HCl, in which the hydrogen to be ionized is attached directly to some negative ion which may be considered to be the central atom of the molecule; (2) the oxygen acids, in which the hydrogen is attached to an oxygen atom which is in turn attached to a central atom. Other acids are known in which the hydrogen is attached to some negative element, other than oxygen, which is in turn attached to a central atom, such as chloroplatinic acid H_2PtCl_6.

[1] This, of course, does not occur with the stable complex ions of the transition elements, in which, as we have already marked, the forces are, at least to a considerable extent, covalent.

[2] By adding this phrase we automatically include any effect due to hydration.

In the case of the hydroacids, the strength of the acid increases with molecular weight. Thus the strength of the acids H_2O, H_2S, H_2Se, H_2Te increases in that order. The halogen acids HCl, HBr, and HI are all strong acids, and may be considered to be completely dissociated in water, but HF is a weak acid. This trend of the acid strength is merely a manifestation of the fact that chemical forces in general become less when atomic or ionic sizes increase. It is easier to split a hydrogen ion away from a large ion than from a small ion.[1] On going across the periodic table, the strength of these acids decreases as the electronegativity of the element decreases. An electronegative element draws electrons to it; these electrons, then, are less likely to be shared with the hydrogen, which is thus more readily split off as an ion. In the case of the hydroacids, however, the electronegativity of negative elements is not always important. Thus, though fluorine is the most electronegative element known, it does not draw the shared electrons away from the hydrogen sufficiently to make HF a strong acid. The small size of the fluorine ion causes the forces to be so strong that the acid is weak (in fact, it is probable that H_2F_2 as well as HF_2^- can exist in solution). However, the greater electronegativity of fluorine as compared with oxygen is sufficient to make HF a stronger acid than water, and water is in turn a stronger acid than ammonia. Ammonia, in fact, does not behave as an acid at all, but as a base, as the tendency of the nitrogen to share its extra pair of

[1] This statement, due to Kossel, *Ann. Physik*, **49**, 276 (1916), holds as a first approximation, when we consider the effective forces between ions in solution as well as when we consider the forces between gaseous ions. As Fajans, *Naturwiss.*, **9**, 729 (1921), has pointed out, the process of ionization should really be broken up into steps, for example,

$$\mathrm{HX}\ (aq) \rightarrow \mathrm{HX}\ (g) \qquad (a)$$

$$\mathrm{HX}\ (g) \rightarrow \mathrm{H}^+\ (g) + \mathrm{X}^-\ (g) \qquad (b)$$

$$\mathrm{H}^+\ (g) + \mathrm{X}^-\ (g) \rightarrow \mathrm{H}^+\ (aq) + \mathrm{X}^-\ (aq) \qquad (c)$$

The energies involved in processes (*b*) and (*c*) are large and opposite in sign. The overall ionization energy is thus a small difference between large quantities. However, in systems as simple as these it seems possible to suppose that the results, obtained by assuming that the ionization in solution resembles ionization in a continuous medium of constant dielectric constant, are roughly correct. Thus if one arranges a series of reactions with different X's in the order of the energies of ionization the result will be the same for gas phase and for solution.

electrons is sufficient to enable ammonia in aqueous solution to rob some of the water molecules of a hydrogen ion.[1] The tendency of phosphorus in phosphine to share its electrons is also present, but the forces are so much weaker on account of the large size of the phosphorus ion that phosphine is not appreciably basic in aqueous solution. However, phosphonium compounds are known; these salts decompose in water, giving the acid and phosphine. The heavier members of the nitrogen group of the periodic system do not form this type of compound.

In the case of oxygen acids, the strength of the acid is determined to a large extent by the electronegativity of the central element. With acids of similar structure, such as H_2SO_3 and H_2SeO_3, the one containing the more electronegative element, in this case H_2SO_3, is the stronger. With extremely electropositive elements, the hydroxide is a base instead of an acid, and in intermediate cases it is amphoteric.

An oxygen acid is made stronger by the substitution of a more electronegative element for some atom in the molecule. A well-known example of this effect of a negative atom is furnished by the comparison of chloracetic acid with acetic acid, the former being much the stronger. In general, the greater the number of negative atoms present in an acid, other things being equal, the stronger the acid. This results in a hydroxide, in which the central atom has a high valence, being more acidic than one in which the same central atom has a lower valence. Thus sulfuric acid is a stronger acid than sulfurous. Many elements in the transition region of the periodic table exhibit a series of hydroxides or oxides which range from those having basic properties, through those which are amphoteric, to compounds which are acidic in nature. Manganese is an excellent example of such an element.[2]

[1] The reader should compare the reactions (in aqueous solution and hence under comparable conditions) $NH_3 + H_2O \rightarrow NH_4^+ + OH^-$ and $H_2O + H_2O \rightarrow H_3O^+ + OH^-$. The first has far the greater tendency to go in the direction written.

[2] The acids of phosphorus constitute an exception to the general rule. Phosphorous acid, H_3PO_3 is stronger than phosphoric and hypophosphorous, H_3PO_2 is stronger still. It is found, however, that only two hydrogens ionize in H_3PO_3 and only one in H_3PO_2. The ease of ionization of those hydrogens that do ionize and the fact that the others do not has been explained by assuming that the latter hydrogens play the role of a negative

A qualitative explanation of the above described properties of the oxygen acids can be given.[1] The more electronegative the central atom, the more strongly it attracts electrons, and the more the electrons on the oxygen are pulled toward the central atom. This makes it less easy for the oxygen to share electrons with the hydrogen, which then becomes more loosely bound and will instead become attached to a water molecule. However, if the central atom is sufficiently electropositive, the electrons on the oxygen tend to be displaced toward the hydrogen. Thus the hydroxyl radical tends to break off. In other words, a polar character of the oxygen-hydrogen bond promotes the splitting off of hydrogen ion in aqueous solution, while a polar character of the oxygen-metal bond tends to cause the splitting off of hydroxyl ion. The strengthening of the acid by the introduction of a negative atom occurs because the latter draws electrons toward it, and thus, by a shift transmitted throughout the molecule, away from the hydrogen. The effect is naturally smaller, the farther the negative atom is from the acid hydrogen.

A different form of theory can be given, from which quantitative results can be obtained. A number of attempts have been made to calculate ionization constants of organic acids[2]; here, only the recent work of Kossiakoff and Harker[3] on inorganic oxygen

---

element, being attached directly to the phosphorus, so that the electronic structures of these compounds may be written as follows:

```
      H
      ..
     :O:           ..
      .. ..       :O:
   H:P:O:H        .. ..
      .. ..     H:P:O:H
     :O:          .. ..
      ..          H
```

See Table 31, Chap. XVI, and Latimer and Hildebrand, "Reference Book of Inorganic Chemistry," p. 176, The Macmillan Company, 1929. This structure of H_3PO_3 has been confirmed by the calculation of Kossiakoff and Harker (see p. 437).

A very interesting account of some of the properties of complex hydrates and acids has been given by Hall, *Chem. Rev.*, **19**, 89 (1936).

[1] See Lewis, "Valence and the Structure of Atoms and Molecules," p. 138, Reinhold Publishing Corporation, 1923.

[2] See Smallwood, *J. Am. Chem. Soc.*, **54**, 3048 (1932); Eucken, *Angewandte Chemie*, **45**, 203 (1932); Kirkwood and Westheimer, *J. Chem. Phys.*, **6**, 506, 513 (1938).

[3] Kossiakoff and Harker, *J. Am. Chem. Soc.*, **60**, 2047 (1938). Their account has been altered here in some details, but the results remain the same.

acids will be considered. These authors consider the actual strength of the O—H electron-pair bond, which is broken to give an H^+, to be the same in all acids, and calculate the differences in ionization on the basis of differences in electrostatic forces due to effective charges on the various atoms in the acid radical. In calculating these charges, a shared electron pair is supposed to contribute one electronic charge to one of the atoms and one to the other. Thus consider the two reactions:

$$HSO_3^- \rightarrow SO_3^{--} + H^+$$

and

$$HSO_4^- \rightarrow SO_4^{--} + H^+$$

It is necessary to compare the electron structures of the SO_3^{--} and SO_4^{--} ions, the entities which are attracting the H^+ ions. These electron structures are as follows:

$$\begin{array}{l} :\ddot{O}: \\ :\ddot{\underset{\cdot\cdot}{S}}:\underset{\cdot\cdot}{\ddot{O}}: \\ :\underset{\cdot\cdot}{\ddot{O}}: \end{array} \quad \text{and} \quad \begin{array}{c} :\ddot{O}: \\ :\underset{\cdot\cdot}{\ddot{O}}:\underset{\cdot\cdot}{\ddot{S}}:\underset{\cdot\cdot}{\ddot{O}}: \\ :\underset{\cdot\cdot}{\ddot{O}}: \end{array}$$

In SO_3^{--}, there are three shared pairs, so the sulfur has five valence electrons, which gives it a net charge of $+1$, and each oxygen with seven electrons has a net charge of -1 electronic charge. In SO_4^{--}, the sulfur has a net charge of $+2$ and each oxygen a net charge of -1. A hydrogen attached to an oxygen in SO_4^{--} thus comes under the influence of a greater positive charge on the sulfur than a hydrogen attached to an oxygen on SO_3^{--}. This tends to make the hydrogen on SO_4^{--} ionize more readily. On the other hand, there is an extra negatively charged oxygen on the SO_4^{--}, but this is farther away from the attached hydrogen than is the sulfur, hence has less effect on it. The net result is that HSO_4^- is a stronger acid than HSO_3^-. It is seen from this description that, as in the older theory, the effect is due to the tendency of the extra oxygen in SO_4^{--} to take one of the sulfur electrons for itself, leaving the sulfur more positively charged than in SO_3^{--}. Instead of supposing, however, that this effect is transmitted through displacement of electrons away from the hydrogen, it is supposed in Kossiakoff and Harker's theory that it is transmitted directly electrostatically. In the older theory, the larger charge on the sulfur in SO_4^{--} is partly counteracted through the sulfur pulling electrons in from the oxygens, which of course leaves a smaller negative charge on the oxygens, and in particular the oxygen to which the H^+ is attached. So the repulsion of the sulfur and the attraction of the oxygens are both less in the older theory than in Kossiakoff and Harker's theory. If the differences in the attractive and repulsive forces should just about cancel each other, the simple electrostatic calculation would give correct results, even though there were some truth in the older theory, and the electrostatic theory does give excellent results, as we shall see.

Before proceeding to the calculations, it may be well to indicate why the first hydrogen ion of a polyvalent acid is always most readily lost. Compare, for example, the reactions

$$H_2SO_4 \rightarrow HSO_4^- + H^+$$

and

$$HSO_4^- \rightarrow SO_4^{--} + H^+$$

To understand why the first tends to go more readily than the second, it is necessary to compare SO_4^{--} and HSO_4^-. The latter has the electronic structure

```
      ..
     :O:
  ..  ..  ..
 :O:S:O:H
  ..  ..  ..
     :O:
      ..
```

In Kossiakoff and Harker's method of calculating charges on the various atoms, the hydrogen is assumed to share an electron pair; so there is a sulfur with charge of $+2$, two oxygens with charge of -1, one oxygen with charge of 0, and one hydrogen with charge of 0. There is, therefore, one less negative oxygen than there is in SO_4^{--}, with the result that a hydrogen is more readily ionized from H_2SO_4 than from HSO_4^-.

It is very probable that a hydrogen ion in aqueous solution is rather definitely attached to a water molecule, forming a hydronium ion, H_3O^+. This ion will be hydrated by other more loosely bound water molecules. The process of ionization may be supposed to consist of the formation of a hydronium ion. In the neighborhood of any OH group in an acid, a number of water molecules will be oriented; in particular there will usually be one water molecule oriented so that one of the negative spots on the oxygen of the water will be near the hydrogen of the OH group. Consider the case of HClO, for example. In solution, there will exist a configuration like this:

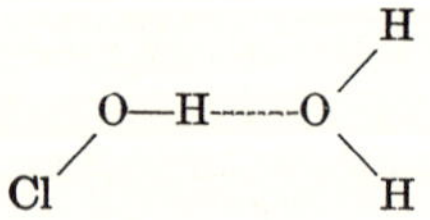

```
                H
               /
   O—H-------O
  /            \
Cl              H
```

Of course there will also be other oriented water molecules in the neighborhood. The first step in the ionization will be the formation of the following configuration:

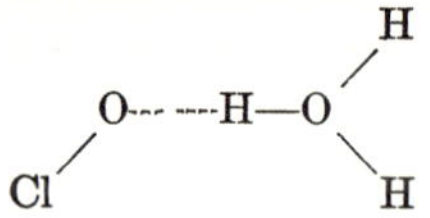

```
                H
               /
   O------H—O
  /            \
Cl              H
```

This will require the breaking of an O—H bond and the formation of an O—H bond; the energies involved will presumably about cancel. But in addition, since the ClO^- group has a net negative charge and H_2O is neutral, it will require a certain amount of electrostatic energy to remove the proton from the one position to the other; let this be E_1 per mole. The next step in the ionization will be the breaking away of one of the *other* hydrogens on the hydronium ion, leaving the configuration

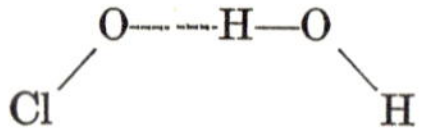

```
   O-------H—O
  /            \
Cl              H
```

This second step will again involve the breaking and formation of a hydrogen bond, and the electrostatic energy, due to the attraction of the ClO^- ion, will be small because of the greater distance of the other hydrogen. Eventually, however, there will exist in the solution a hydronium ion with oriented water molecules about it. The net energy effect, after the first step, will consist, at least approximately, in the difference in energy of the hydration of a hydronium ion and the hydration of a water molecule. Let this energy (*i.e.*, the energy *absorbed* on ionization due to the difference in hydration) be E_2. It is supposed that E_2 is the same no matter what acid is used. The total energy of ionization ΔE is, therefore, given by

$$\Delta E = E_1 + E_2,$$

where only E_1 depends upon the nature of the acid.

In calculating E_1, it is assumed that no change in the orientation of water molecules about the acid ion takes place. For it will be seen, though the tetrahedral water molecule was inadequately represented in a two-dimensional diagram, that the water molecule which was left near the ClO^- ion after the hydronium ion broke up was already properly oriented. Therefore E_1 can be taken simply as the electrostatic energy required to move the proton from the ClO^- to the adjacent water molecule. If all distances involved are known, this can be calculated, provided an estimate can be made for the effective dielectric constant of the liquid. The large dielectric constant of water is due to the possibility of orientation of the water molecules in an electric field. If there is to be no change in the orientation, however, there will be no contribution from this factor; only the polarizability of the electrons in the molecules will contribute. Measurements of the index of refraction indicate that with orientation excluded the dielectric constant is about 1.77. The value that has actually been used is[1] 3. This may be supposed to allow for the possibility of displacement of protons in the electric field, without orientation, or for a slight amount of orientation. We assume that any entropy change involved in this process is negligible.

The distances of the various atoms composing the oxygen ion are known, or may be readily inferred, from X-ray data on crystals. Only the O—H bond distance and the distance between the oxygens of the acid and of the water need to be considered. Kossiakoff and Harker take the O—H distance to be 0.95Å., which is consistent with Table 29. The distance between oxygens is taken as 2.70Å., which is slightly less than the distance between water molecules in ice. Using these values, it is seen that the proton must be moved along the line joining the two oxygens by a distance of 0.80Å., on going from the acid to the water molecule. With these data, then, and the assumption already made as to the charge to be assigned to each

[1] See Debye, "Polar Molecules," Chap. VI, Reinhold Publishing Corporation, 1929.

atom, E_1 is readily calculated as the difference in electrostatic potential of the proton at the two positions.

The ionization constant K is the equilibrium constant for the reaction

$$HAc + H_2O \rightarrow H_3O^+ + Ac^-$$

where Ac^- stands for the acid radical. It is related to the standard free-energy change of this reaction (see Appendix II) by the usual formula

$$RT \ln K = -\Delta F^0$$

where $\Delta F^0 = \Delta H - T\Delta S^0$. Since there is little change in volume on ionization, ΔH is practically equal to $\Delta E = E_1 + E_2$, but the change in entropy on ionization needs to be considered. There are two reasons for a change of entropy in the foregoing chemical reaction. In the first place, it will be realized that there is a difference in the volume through which a proton is free to move when it is attached to Ac^- and when it is attached to[1] H_2O. This is not because of any particular difference in the binding of the proton in the two cases, but simply because ΔS^0 is the change in entropy that occurs on reaction when the concentration of HAc is at its standard state, 1 mole per liter, and when water is at its standard state, 55.5 moles per liter (see footnote 1, page 460; in dilute solution, the solvent is treated as a pure liquid). If there were one oxygen atom per acid molecule, the effective volume for H^+ would be about 55.5 times as great after the reaction had occurred as before, since there are 55.5 times as many water molecules as acid radicals for it to rest on. This would mean an increase of entropy on ionization given by

$$R \ln 55.5 = 8.0 \text{ cal. per degree.}$$

If there is more than one oxygen (or more than one hydrogen) in the acid, the entropy change is very slightly different.

There is one other entropy effect, a decrease on ionization due to the hydration of the H_3O^+, which should not be far different from the entropy of solution of H^+, or about -24.3 cal. per deg. per gram ion. The value of ΔS^0 should therefore be

[1] The method of making such a calculation was indicated by Harker in a private communication.

about $-24.3 + 8.0 = -16.3$. In a few cases, there are enough data available to find ΔS^0 experimentally.[1] These are as follows:

$$
\begin{array}{lll}
HSO_4^- & \rightarrow H^+ + SO_4^{--} & \Delta S^0 = -26.2 \\
HSO_3^- & \rightarrow H^+ + SO_3^{--} & \Delta S^0 = -29.6 \\
H_2SO_3 & \rightarrow H^+ + HSO_3^- & \Delta S^0 = -22.1 \\
HCO_3^- & \rightarrow H^+ + CO_3^{--} & \Delta S^0 = -35.2 \\
H_2CO_3(aq) & \rightarrow H^+ + HCO_3^- & \Delta S^0 = -22.9 \\
HPO_4^{--} & \rightarrow H^+ + PO_4^{---} & \Delta S^0 = -43 \\
H_2PO_4^- & \rightarrow H^+ + HPO_4^{--} & \Delta S^0 = -30.3 \\
H_3PO_4(aq) & \rightarrow H^+ + H_2PO_4^- & \Delta S^0 = -16.0
\end{array}
$$

It is seen that the ΔS^0 values are more negative than calculated, and the entropies of ionization of HCO_3^- and HPO_4^{--} vary rather considerably from the expected result. This can be due only to a change in the orientation of water molecules about the ionized anion residue. If this occurs, however, E_2 is also affected, for in assuming that E_2 had a common value for all acids, any such reorientation about the ionized residue was neglected. If such a reorientation gives a negative contribution to $T\Delta S^0$, it must give a *greater* negative contribution to E_2. Otherwise no reorientation would occur, for a system does not revert to a state of low entropy unless this condition is more than compensated by being a state of low energy, the general thermodynamic condition for equilibrium being that the free energy should be a minimum. Since the variation of $T\Delta S^0$ at room temperature from acid to acid is not extremely large, amounting to only a few kilogram-calories, it seems reasonable to suppose that the variation in $E_2 - T\Delta S^0$ is still less. We therefore set as an approximation

$$\Delta F^0 = E_1 + C,$$

where C is a constant and $C \cong E_2 - T\Delta S^0$. Using $C = -43.0$ kg.-cal. and making the small correction for the number of available oxygens per proton in the acid radical, Kossiakoff and Harker have calculated values of $\log_{10} K$ for a series of oxygen acids. In Fig. 74, $\log_{10} K$, observed, is plotted against $\log_{10} K$, calculated. If the calculations were always exactly correct, all points would lie on the diagonal line. The result may be said

[1] See Latimer, "The Oxidation States of the Elements and their Potentials in Aqueous Solutions," pp. 328*ff*., Prentice Hall, Inc., 1938; Latimer, Pitzer, and Smith, *J. Am. Chem. Soc.*, **60**, 1829 (1938).

to be a remarkable check, considering the approximations and arbitrary assumptions involved. In spite of this, it seems best to reserve judgment as to how literally the theory should be taken, but, in any event, it promises to be a useful working hypothesis.

Kossiakoff and Harker have found that agreement was obtained only for the correct structure of the acid; the assignment

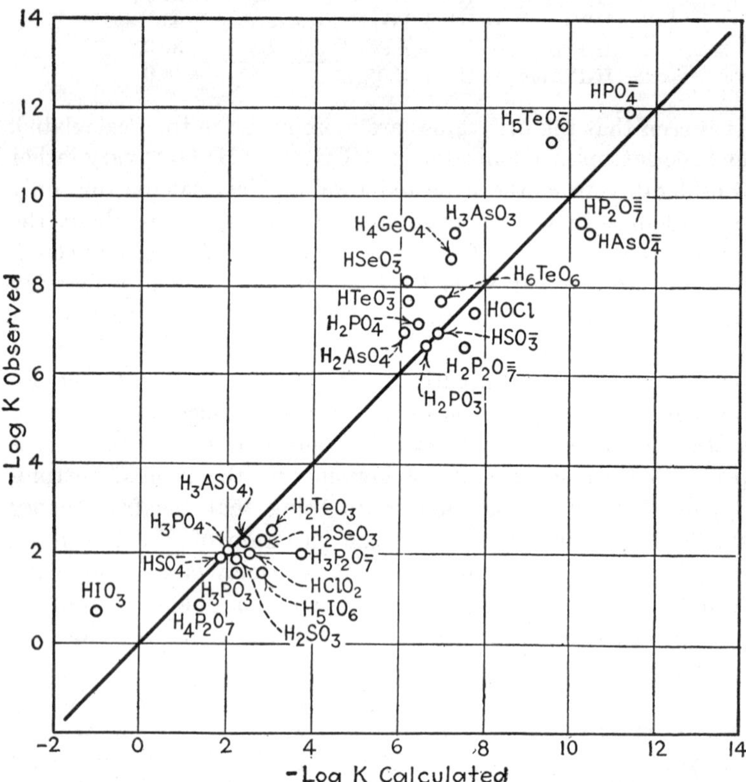

FIG. 74.—[*After Kossiakoff and Harker, J. Am. Chem. Soc.*, **60**, 2053 (1938).]

of other structures destroyed the agreement. For example, in the acid H_5IO_6, iodine has a coordination number of six, and from this structure the correct value of $\log_{10} K$ was obtained from the calculation; assuming that the acid is $HIO_4 \cdot 2H_2O$, with a coordination number of four for iodine, gives no agreement between experimental and calculated values of $\log_{10} K$. It was possible, then, to predict the structure in the case of germanic acid, where it was not previously determined. It was found necessary to assume that germanium has a coordination number of four, the proper formula being $Ge(OH)_4$. For H_3PO_3 the structure given on page 429, above, was used.

Finally, it should be noted that the value of C of -43.0 kg.-cal. is very reasonable. Taking $T\Delta S^0$ as approximately 7 kg. cal., this makes $-E_2 = 50$ kg.-cal. This may be compared with the heat of solution L^+ of H^+, which is 253 kg.-cal. The heat of solution includes the heat of formation of the O—H bond in H_3O^+ (the proton affinity of water) and the heat of hydration of the hydronium ion, whereas $-E_2$ consists principally of the latter.[1] This interpretation yields a value of about 200 kg.-cal. for the proton affinity of water. This is about the same as the value for the proton affinity of ammonia found in Sec. 14.10, which is very reasonable, especially as the calculations are rough.

In special cases, it is possible for special factors to affect the strengths of acids. For example, it is possible that, after removal of the proton, the electronic configuration of the acid radical might in some cases revert to some other arrangement. This would cause the ion to be more stable than it otherwise would be, and thus make the ionization take place more readily. As an illustration, consider the ionization of HCO_3^-. This leaves a CO_3^{--} ion, which very likely has the structure

```
       ..
      :O:
   ..  ..  ..
  :O: C::O
   ..      ..
```

in which there is an oxygen with a double bond. If this is the case, the double bond resonates between the three C O linkages, as explained in Sec. 16.8, giving the carbonate ion a symmetrical structure. But if a proton is attached to one of the oxygens, the attraction of the positive charge on the electrons will make it less likely that that oxygen will share its electrons to give a double bond. Only two C—O linkages, then, participate (at least to the full extent) in the resonance. Thus the electron structure of the acid radical changes, becoming more symmetrical, after ionization. Naturally this change would not take place unless it made the CO_3^{--} ion more stable. It has, then, a tendency to shift

[1] From each of these should be subtracted the heat of hydration of water, but this cancels.

the equilibrium

$$HCO_3^- \rightleftarrows CO_3^{--} + H^+$$

to the right. The calculations of Kossiakoff and Harker, using the double-bonded structure and neglecting resonance, might be expected to give too low a value for the ionization constant, and this is true in the cases of H_2CO_3, HCO_3^-, HNO_3, HNO_2, HCOOH, and CH_3COOH, though the difference between calculated and observed values is within the limits of error in H_2CO_3 and HCO_3^-. In the case of H_3BO_3, a plane single-bonded structure $[B(OH)_3]$ gave a calculated value of the ionization constant that agreed with the experimental value within the limits of error. This is of interest in connection with the discussion of boron compounds given on pages 322–324.

19.11. Oxidation and Reduction Reactions.—Among the most important reactions occurring in solution are those involving oxidation and reduction. These are often exceedingly complicated, and it is possible to offer but few generalizations as to the strength of oxidizing and reducing agents. In many cases, various solid substances are involved, and in practically all cases the hydration of various substances plays a significant role. Often the ions of water are involved, so that the strength of an oxidation or reduction agent depends upon the concentration of these ions in solution.

One important series of reactions of this type consists of those in which one metal displaces another in solution. In the displacement series, we are accustomed to thinking of a metal that displaces another metal as being more electropositive than the one displaced. In a general way, this is true, but it is by no means always true. The ionization potential being a rough measure of the electropositiveness of a metal, we might thus expect cesium to displace lithium ion from solution (if this could be carried out without reaction with the water itself). Actually, lithium ion tends to displace cesium. Chief factor in this reversal of expectation is the fact that the energy of hydration[1] of Li^+ is so much greater than that of Cs^+, making Li^+ ion in solution very stable. In other cases, the stability of the solid metal may play a predominating part.

[1] The energy of hydration is here the same as L^+, Table 48, the convention as to sign being the same as there.

Sometimes no solids are involved, but there are likely to be enough other complications to make predictions and explanations difficult. Suppose, for example, it is desired to compare the following, entirely analogous, reactions:

$$ClO_3^- + 6H^+ + 6E^- \rightarrow Cl^- + 3H_2O$$
$$BrO_3^- + 6H^+ + 6E^- \rightarrow Br^- + 3H_2O$$

and

$$IO_3^- + 6H^+ + 6E^- \rightarrow I^- + 3H_2O$$

all taking place in aqueous solution (E^- is used as the chemical symbol for electrons). The relative tendency of the reactions to go depends upon the stability and condition of hydration of Cl^-, Br^-, and I^-, and ClO_3^-, BrO_3^-, and IO_3^- In such cases as these, it is often of assistance to analyze the reactions into steps. Thus for the reactions just considered, we set down the

TABLE 54.—ANALYSIS OF OXIDATION REACTIONS, I

Partial reaction	Heat absorbed, kg-cal.		
	If X = Cl	If X = Br	If X = I
$XO_3^-\ (aq) + H^+\ (aq) \rightarrow HXO_3\ (s)$ $HXO_3\ (s) \rightarrow \frac{1}{2}H_2\ (g) + X\ (g) + \frac{3}{2}O_2\ (g)$	49.7	38.1	−2.3 82.4
$X\ (g) + E^- \rightarrow X^-\ (g)$	−87.3	−81.8	−74.3
$X^-\ (g) \rightarrow X^-\ (aq)$	−92.1	−84.7	−75.5
$5H^+\ (aq) + \frac{1}{2}H_2\ (g) + \frac{3}{2}O_2\ (g) + 5E^- \rightarrow 3H_2O\ (liq)$	−768.2	−768.2	−768.2
$XO_3^-\ (aq) + 6H^+\ (aq) + 6E^- \rightarrow X^-\ (aq) + 3H_2O\ (liq)$	−897.9	−896.6	−837.9

(E^- = electron, *s* = solid, *g* = gas)

partial reactions[1] shown in Table 54. From the results of this table, we may expect ClO_3^- and BrO_3^- to be much stronger oxidizing agents than IO_3^-. It is not possible to say that this is due to any one cause. Of considerable importance, however, is the stability of the ions ClO_3^-, BrO_3^-, and IO_3^-. As bromine is a

[1] Data from Bichowsky and Rossini, "Thermochemistry of the Chemical Substances." It should be borne in mind that the overall reaction is a half-reaction. When it is considered together with the half reaction of the reducing agent the extremely large negative values for the energy of reaction will not be so much in evidence. Only the *differences* in the energy of reaction are of significance in determining the *relative* power of the oxidizing agents.

larger atom than chlorine, it is not surprising that BrO_3^- is less stable than ClO_3^-. However, it is rather surprising that IO_3^- is more stable in aqueous solution than either of the others. This may be due to the normally larger coordination number of iodine, which enables it to be more completely hydrated. Thus in solution, IO_3^- probably really becomes $I(OH)_6^-$, whereas BrO_3^- and ClO_3^- take on at most one molecule of water. Since the size of the central atom thus has a double effect, influencing the bond energy directly and also affecting the possibility of hydration, it is not easy to predict the result in any given case. Thus, with respect to reduction to the halide ion, IO_4^- is only about 24 kg.-cal. more stable than ClO_4^-, while BrO_4^- is apparently unknown and hence undoubtedly quite unstable. The reactions of the ions SO_3^{--}, SeO_3^{--}, and TeO_3^{--} as oxidizing agents to give S, Se, and Te, respectively, are analyzed in Table 55. The higher oxygen ions of bromine or selenium are invariably relatively unstable.

TABLE 55.—ANALYSIS OF OXIDATION REACTIONS, II

Partial reaction	Heat absorbed, kg.-cal.		
	If X = S	If X = Se	If X = Te
XO_3^{--} (*aq*) + $2H^+$ (*aq*) → H_2 (*g*) + X (*g*) + $\frac{3}{2}O_2$ (*g*)	201	174	180
X (*g*) → X (*s*)	−53	−51	−39
$4H^+$ (*aq*) + H_2 (*g*) + $\frac{3}{2}O_2$ (*g*) + $4E^-$ → $3H_2O$ (*liq*)	−656	−656	−656
XO_3^{--} (*aq*) + $6H^+$ (*aq*) + $4E^-$ → X (*s*) + $3H_2O$ (*liq*)	−508	−533	−515

(E^- = electron, *s* = solid, *g* = gas)

Various other oxidation and reduction reactions could be studied in the same way. It must be remembered, however, that the entropy of the various substances involved also has an influence on the reaction, though it is less important than the energy. Furthermore, the actual course of many reactions is determined not so much by the products and reactants as by the rate of reaction. Of course, if a reaction is to proceed at all, the products must be more stable than the reactants, but the rate of reaction may be so slow that, even though this condition is fulfilled, the reaction for all practical purposes will not go. Sometimes there is more than one set of possible products, and the actual products obtained will depend very largely upon the relative rates of reaction. Thus, just how much the valence of an oxidizing agent will be reduced by a given reducing agent may depend upon the rate of the respective reactions that give the various valence states. These considerations greatly complicate the discussion of oxidation and reduction reactions, and indeed, of all reactions. On the other hand, equilibria can often be measured, and even in the cases where they are not the controlling factors, their study is of importance as a general guide.

Although it is usually unsafe to make generalizations about oxidizing and reducing agents, there is one statement that can, apparently, safely be made concerning the valence states of the more metallic transition elements. In any given column of the periodic table to the left of the iron, palladium, and platinum metals, the elements of greater atomic weight tend to exist in the higher valence states; the compounds in which the element exhibits the high valences are therefore less powerful oxidizing agents with the heavier elements. Thus the hexavalent compounds of molybdenum, tungsten, and uranium are not strong oxidizing agents.

Exercises

(s = solid; g = gas)

1. From the heat of formation of aqueous $ZnCl_2$ from its elements, 115.3 kg.-cal. (evolved); the heat of vaporization of zinc, 27.4 kg.-cal; and other quantities obtainable from the tables in the book, calculate the heat of solution of gaseous Zn^{++}. Repeat, using ZnI_2 (heat of formation of aqueous ZnI_2, 61.4 kg.-cal.) Calculate the heat of solution theoretically, and compare.

2. From the heat of formation of aqueous $InCl_3$, 145.4 kg.-cal. (evolved); the heat of vaporization of indium, 52 kg.-cal.; and other quantities obtainable from tables in the book, calculate the heat of solution of gaseous In^{+++}. Calculate the heat of solution theoretically, and compare.

3. HCl is ionized in aqueous solution, but liquid HCl itself is not ionized. Remembering that the reactions to be compared are

$$HCl + H_2O \rightarrow H_3O^+ + Cl^-$$

and

$$HCl + HCl \rightarrow H_2Cl^+ + Cl^-$$

and that solvation must play an important role in the phenomena, explain this difference in behavior.

4. Analyze the reaction Li (s) + Cs^+ (aq) → Li^+ (aq) + Cs (s) into steps whose energies may be found in the tables in the book, and verify the statement that the chief reason for the tendency of lithium to displace cesium is the large energy of hydration of lithium. What will be the effect of the entropy of hydration?

5. Similarly analyze the reaction Zn (s) + Fe^{++} (aq) → Zn^{++} (aq) + Fe (s), and indicate why zinc replaces iron. (See Exercise 1.)

6. From Tables 16 and 49, make a rough estimate of the heat of solution of Au^+, and compare with Tl^+.

7. From Table 49, make a rough estimate of the heat of solution of Al^{++}. Then, using data to be found in the book, estimate the energy of the reaction

$$Al^{+++}\ (aq) + Cl^-\ (aq) \rightarrow Al^{++}\ (aq) + \tfrac{1}{2}Cl_2\ (g).$$

APPENDIX I

SOME DEFINITIONS AND THEOREMS OF CLASSICAL MECHANICS

Since considerable use is made, in various parts of the text, of the principles of the dynamics of a particle and of a system of particles, it seems desirable to collect the formulas used in an appendix. It is assumed that the definitions of such terms as force, energy, mass, and momentum, as well as the process of resolving forces and other directed quantities along axes, are familiar to the student. In the more elementary cases, the appendix is intended to provide a summary rather than an explanation of the various laws and equations. For more detailed accounts, the student may consult any good text on general physics or mechanics.

The equation of motion of a single particle may be written in various ways, the most important of which may be summarized as follows:[1]

$$f = ma = m\dot{v} = \dot{M}, \tag{1}$$

where f = force exerted on the particle.
m = mass of the particle.
$a = \dot{v}$ = acceleration produced by the force.
v = velocity.
M = momentum = mv.

f, a, v, M, and $\dot{M}$ are all directed or vector quantities; a, and $\dot{M}$ are in the same direction as f. These may all be resolved along x-, y-, and z-axes, and there is a set of equations for the component along each axis, as follows:

$$\begin{aligned} f &= ma_x = m\ddot{x} = m\dot{v}_x = \dot{M}_x \\ f &= ma_y = m\ddot{y} = m\dot{v}_y = \dot{M}_y \\ f &= ma_z = m\ddot{z} = m\dot{v}_z = \dot{M}_z. \end{aligned} \tag{2}$$

[1] The dot is used throughout to denote differentiation with respect to time. The double dot denotes a second derivative.

According to Eq. (1), a force operating for a definite time produces a change in the momentum of the body it acts upon equal to the force times the time. If it operates through a given distance, it changes the kinetic energy (which is, of course, equal to $\frac{1}{2}mv^2$) by an amount equal to the force times the distance through which it operates. A particle that is placed in a field, such as an electrical, magnetic, or gravitational field, which exerts a force on that particle, is said to have a potential energy due to the field, which depends upon the position of the particle in the field. The potential energy is equal to the work needed to be done by some external agency (which is supposed to be capable of grasping and holding the particle) in order to bring the particle from a position outside the field (in general, an infinite distance away) to a point in the field, without giving the particle any kinetic energy. If the field repels the particle, the potential energy is positive. If the field attracts the particle, the work done by the external agency is negative (*i.e.*, actually the field does work on the external agency) so the potential energy is negative. The total energy of the particle moving in the field and without other constraints or forces is equal to the algebraic sum of the kinetic and potential energies.

Motion of a Particle Constrained to Move a Fixed Distance from a Given Point.—A particle so constrained, say by a rigid rod of length r, will, if no external force is exerted upon it, rotate with constant velocity in a circle with radius r and center at the given point. As such circular motion involves a constant acceleration of the particle toward the fixed center (otherwise the particle would move along a straight line), the rod must exert a force on it. This is the centripetal force and is equal to mv^2/r, where m and v are mass and velocity of the particle. The equal and opposite force exerted by the rotating particle on the rod is called the "centrifugal force."

Motion of a Particle That Is Attracted to a Fixed Point.—If a particle moves under the influence of a force that is directed along the line joining it to a fixed center, its motion is best described by means of a system of polar coordinates, with origin at the center of force. The position of the particle is designated by giving the coordinates r, θ, and ϕ, where r is the distance to the center of force and θ and ϕ are the usual polar angles. This is illustrated in Fig. 75, where the point P is the position of the particle. In this figure, r is equal to $\overline{\mathrm{OP}}$, θ is the angle made by the line $\overline{\mathrm{OP}}$ with the polar axis $\overline{\mathrm{QO}}$, and ϕ is the angle made by the intersection of the plane QPO and the plane Π (which is perpendicular to the polar axis at the origin) with an arbitrary line $\overline{\mathrm{OR}}$ in Π. This line, however, is taken as the intersection of the plane of motion with Π, as this is convenient for later calculations, and involves no loss of generality. Suppose the particle is moving with a velocity v in some direction. The line $\overline{\mathrm{OP}}$ and the vector representation of the velocity v determine a plane. Since the force acting on the particle is in this plane, the particle will have no tendency to move out of the plane, and will continue in it indefinitely. This plane of motion is shown as the plane POR in Fig. 75. The angle POR will be designated as χ.

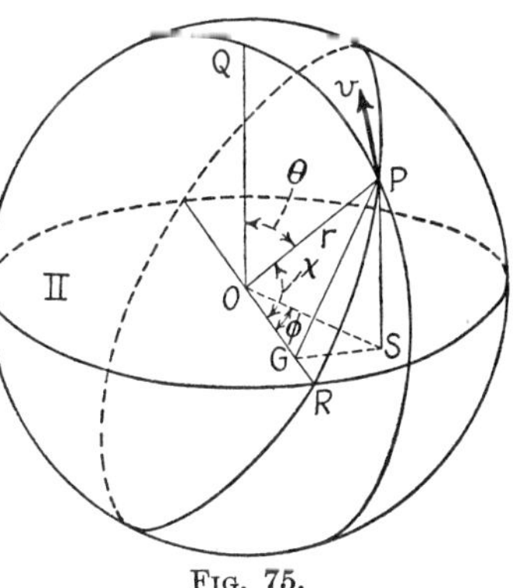

Fig. 75.

The angular momentum p_χ of the particle with respect to the point O is defined as the product of (1) the length $\overline{\mathrm{OP}} = r$, (2) the projection of the velocity v along a line in the plane of motion perpendicular to $\overline{\mathrm{OP}}$, and (3) the mass of the particle. This projection or component of the velocity will be equal to $r\dot{\chi}$; we may, therefore, write

$$p_\chi = mr^2\dot{\chi}. \tag{3}$$

p_χ is a very important dynamical quantity. Its importance arises, in part, from the fact that for central motion, *i.e.*, motion in which the force is directed toward (or away from) a fixed point, it is constant. This may be readily proved, following the original discussion of Newton, in an intuitionally obvious manner. First it is convenient to note that the conservation of angular momentum is equivalent to the statement that for a central field of force the areas swept out by the radius vector in equal times are equal (see Fig. 76). (Applied to the motions of the planets about the sun, this is Kepler's second law.)

In order to prove this statement, we think of the central force as being replaced by an intermittent force, which acts at intervals and for a very short time only. The shorter the time force acts, however, the greater the force itself is supposed to be. It is assumed that the integral of the force over any appreciable time

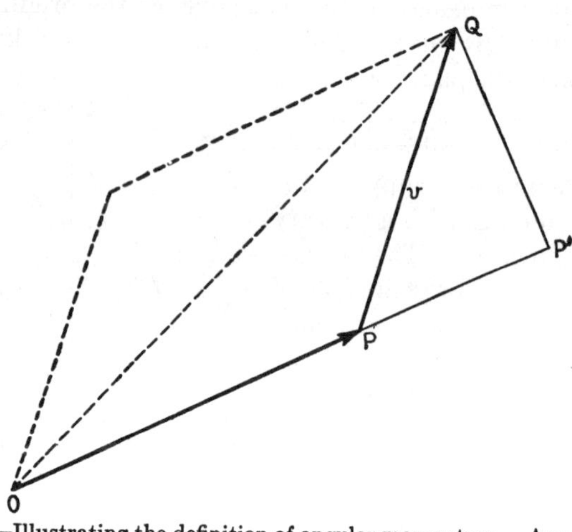

FIG. 76.—Illustrating the definition of angular momentum. Angular momentum is equal to the mass, times the distance OP, times the perpendicular projection of v, here represented by QP′. It is noted that the area of the triangle OPQ is equal to $\frac{1}{2}$OP × QP′. Since QP $= v$ is the distance moved by the particle in unit time, the triangle OPQ is the area swept out by the radius vector (line joining the particle to O) in unit time. Hence angular momentum = 2 × mass × area swept out in unit time.

interval, $\int_{t_1}^{t_2} f\,dt$, is the same for the true force as for the force that is assumed to replace it. Let the moving body start at the point P (Fig. 77) and move to the point Q, at which time it is acted upon by the intermittent force, which is directed toward the center O. The body moves in a straight line from P to Q, since no force is acting upon it in that interval, and it sweeps out an area equal to $\frac{1}{2}\overline{OP'} \times \overline{PQ}$, where $\overline{OP'}$ is the perpendicular distance from O to the line $\overline{PQ}$. If no force had acted at Q, the body would have moved to R, where $\overline{QR} = \overline{PQ}$, in a time equal to that necessary to go from P to Q, and would have swept out an area OQR obviously equal to the area OPQ. However, if a force does act on the body at Q, it acquires there a component of velocity parallel to $\overline{OQ}$; therefore, instead of moving to

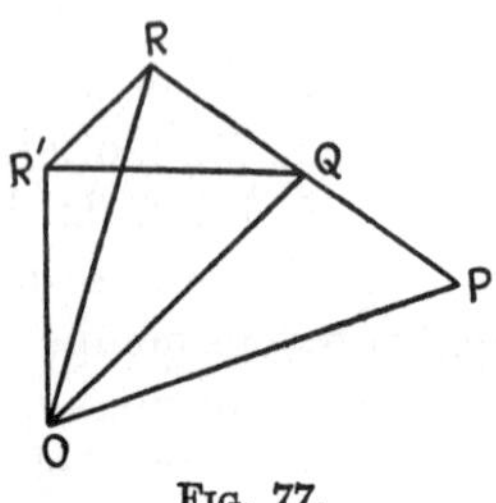

FIG. 77.

R, it will move to a point R′ such that $\overline{RR'}$ is parallel to $\overline{OQ}$, and will thus sweep out an area OQR′. But since $\overline{RR'}$ is parallel to $\overline{OQ}$, the altitude of the triangle OQR is the same as that of the triangle OQR′, $\overline{OQ}$ being considered as the common base. Therefore the two triangles have the same area, and it is seen that after the force has acted the same area is swept out, per unit time, as before. As the force acts intermittently, this will continue to be true indefinitely. But if we consider a smaller and smaller force acting more and more often, we approach the limiting case, the law of equal description of areas continuing to hold good as this limit is approached. Thus angular momentum is conserved as a body moves under the influence of a central field.

In considering the motion of a particle attracted to a fixed point, it is often desirable to express its kinetic energy in terms of the polar coordinates and the corresponding velocities. It has already been noted that the component of the velocity of the particle perpendicular to the line $\overline{OP}$ joining it to the center of force (see Fig. 75) is given by $r\dot{\chi}$; the velocity along $\overline{OP}$ is equal to $\dot{r}$. The velocity component $r\dot{\chi}$ can be resolved further into two components that are mutually perpendicular and also perpendicular to $\overline{OP}$, namely, a component parallel to the plane II and a component tangent to the great circle QP. Since these components are perpendicular to the component $\dot{r}$, they are entirely independent of the magnitude of $\dot{r}$. The component along QP will, obviously, be equal to $r\dot{\theta}$. The component parallel to plane II will be the angular velocity, $\dot{\phi}$, times the perpendicular distance of P to the polar axis, which is $r \sin \theta$. Thus, this component of velocity is $\dot{\phi} r \sin \theta$. Since $r\dot{\theta}$ and $\dot{\phi} r \sin \theta$ are components of $r\dot{\chi}$, and all three of these quantities are components of v, we may write

$$r^2\dot{\chi}^2 = r^2\dot{\theta}^2 + \dot{\phi}^2 r^2 \sin^2 \theta \tag{4}$$

and

$$v^2 = \dot{r}^2 + r^2\dot{\chi}^2 = \dot{r}^2 + r^2\dot{\theta}^2 + \dot{\phi}^2 r^2 \sin^2 \theta. \tag{5}$$

Generalized Definition of Momentum.—Before proceeding further, it will be necessary to formulate a general definition of momentum.[1] So far we have considered ordinary momentum and angular momentum. The former is of particular value when the position of a particle is described in terms of rectangular coordinates. The component of momentum $m\dot{x}$ is said to be conjugate to the variable x, $m\dot{y}$ is conjugate to y, etc. In like manner, the angular momentum $mr^2\dot{\chi}$ is conjugate to the variable χ. In general, no matter what system of coordinates is used, a set of conjugate momenta may be found.

Usually, in the case of a single particle, there will be three independent coordinates with their conjugate momenta. If the motion is constrained in some way, there may be fewer independent coordinates (as, for example, in the case of the rotator constrained to move at a fixed distance from a given point). When there is more than one particle, there will be more than three variables. Suppose there are N independent variables $q_1, q_2, \ldots, q_N$. It will always be possible to express the kinetic energy of the system in terms of these coordinates and their corresponding velocities $\dot{q}_1, \dot{q}_2, \ldots, \dot{q}_N$. Thus the kinetic energy may be expressed as a function $T(q_1, q_2, \ldots, q_N, \dot{q}_1, \dot{q}_2, \ldots, \dot{q}_N)$ of all these variables. Then the momenta $p_1, p_2, \ldots, p_N$ conjugate to the coordinates $q_1, q_2, \ldots, q_N$ are defined by the following set of equations:

$$\begin{aligned} p_1 &= \frac{\partial T(q_1, q_2, \cdots, q_N, \dot{q}_1, \dot{q}_2, \cdots, \dot{q}_N)}{\partial \dot{q}_1} \\ p_2 &= \frac{\partial T(q_1, q_2, \cdots, q_N, \dot{q}_1, \dot{q}_2, \cdots, \dot{q}_N)}{\partial \dot{q}_2} \\ &\cdots\cdots\cdots\cdots \\ p_N &= \frac{\partial T(q_1, q_2, \cdots, q_N, \dot{q}_1, \dot{q}_2, \cdots, \dot{q}_N)}{\partial \dot{q}_N}. \end{aligned} \tag{6}$$

In the partial differentiations,[2] all variables are held constant except the one indicated. Thus, in the case of p_1, the expression for the kinetic energy is differentiated as though $q_1, q_2, \ldots, q_N, \dot{q}_2, \ldots, \dot{q}_N$ (all except $\dot{q}_1$) were constants.

The momenta already used are examples of the definition just given. Thus suppose the kinetic energy is expressed in terms

[1] For a more extended treatment see, *e.g.*, Byerly, "Generalized Coördinates," Ginn and Company, 1916.

[2] See Daniels, "Mathematical Preparation for Physical Chemistry," Chap. XVII, McGraw-Hill Book Company, Inc., 1928.

of the Cartesian components of velocity $\dot{x}$, $\dot{y}$, and $\dot{z}$,

$$T = \frac{1}{2}m(\dot{x}^2 + \dot{y}^2 + \dot{z}^2).$$

The x-component of momentum is given by

$$\frac{\partial T}{\partial \dot{x}} = m\dot{x}$$

which is, of course, the usual expression.

In the case of a body rotating in a plane at a fixed distance from a center of rotation,

$$T = \frac{1}{2}mv^2 = \frac{1}{2}mr^2\dot{\chi}^2.$$

So

$$\frac{\partial T}{\partial \dot{\chi}} = mr^2\dot{\chi},$$

which is the usual expression for the angular momentum.

In the case of a body moving under the influence of a central force, the kinetic energy can be written from Eq. (5)

$$T = \frac{1}{2}m\dot{r}^2 + \frac{1}{2}mr^2\dot{\theta}^2 + \frac{1}{2}mr^2\dot{\phi}^2 \sin^2 \theta$$

The momenta conjugate to r, θ, and ϕ, are, respectively,

$$p_r = \frac{\partial T}{\partial \dot{r}} = m\dot{r} \quad (7a)$$

$$p_\theta = \frac{\partial T}{\partial \dot{\theta}} = mr^2\dot{\theta} \quad (7b)$$

$$p_\phi = \frac{\partial T}{\partial \dot{\phi}} = mr^2\dot{\phi} \sin^2 \theta. \quad (7c)$$

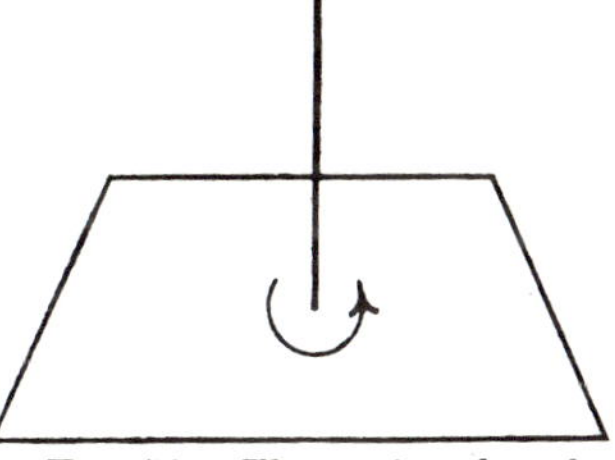

FIG. 78.—Illustrating the relation between direction of the angular momentum vector and the direction of motion.

Components of Angular Momentum.—The angular momentum may be considered to be a *vector*, rather than a scalar, the direction of the vector being perpendicular to the plane of motion, and such that if the direction of rotation of the rotator appears counterclockwise to the observer the angular momentum vector points toward the observer (see Fig. 78).

The component of the angular-momentum vector along the polar axis is of particular importance. Let the angular momentum be p_χ, and let α be the angle made by the angular-momentum vector and the polar axis. This component is then $p_\chi \cos \alpha$, and it may be shown that it is equal to p_ϕ of Eq. (7c). To prove this statement, consider Fig. 75. The particle is supposed to be at point P and moving in the plane POR. The point P has the coordinates r (= $\overline{\text{OP}}$), θ (= angle QOP), and ϕ (= angle ROS). χ is the angle ROP. The plane PGS is constructed perpendicular to $\overline{\text{OR}}$. Therefore the angle PGS is the dihedral angle between Π and plane POR; this is equal to the angle between the perpendiculars to Π and POR (*i.e.*, the polar axis and the angular momentum vector, respectively); α is defined as the latter angle, so angle PGS = α. It is now desired to prove that $p_\chi \cos \alpha = p_\phi$, or, by Eqs. (3) and (7c), $\dot{\chi} \cos \alpha = \dot{\phi} \sin^2 \theta$. We proceed by first getting a number of relationships between the various angles and distances. From Fig. 75, the following relationships are seen to be true.

$$\overline{\text{OS}} = r \sin \theta \tag{8a}$$
$$\overline{\text{PS}} = r \cos \theta \tag{8b}$$
$$\overline{\text{GS}} = \overline{\text{GP}} \cos \alpha \tag{8c}$$
$$\overline{\text{PS}} = \overline{\text{GP}} \sin \alpha \tag{8d}$$
$$\overline{\text{GS}} = \overline{\text{OS}} \sin \phi \tag{8e}$$
$$\overline{\text{OG}} = \overline{\text{OS}} \cos \phi \tag{8f}$$
$$\overline{\text{GP}} = r \sin \chi \tag{8g}$$
$$\overline{\text{OG}} = r \cos \chi. \tag{8h}$$

Various relations between the angles follow from these equations. Thus from (8f), (8h), and (8a),

$$\cos \chi = \cos \phi \sin \theta. \tag{9a}$$

From (8b), (8d), and (8g),

$$\cos \theta = \sin \alpha \sin \chi. \tag{9b}$$

From (8c), (8e), (8g), and (8a),

$$\cos \alpha \sin \chi = \sin \phi \sin \theta. \tag{9c}$$

If now we differentiate (9c) with respect to time, we get

$$\dot{\chi} \cos \alpha \cos \chi = \dot{\phi} \cos \phi \sin \theta + \dot{\theta} \sin \phi \cos \theta. \tag{10}$$

But from (9b),

$$\dot{\theta} \sin \theta = -\dot{\chi} \sin \alpha \cos \chi. \tag{11}$$

From (10) and (11),

$$\dot{\chi} \cos \chi \left(\cos \alpha + \sin \alpha \sin \phi \frac{\cos \theta}{\sin \theta} \right) = \dot{\phi} \cos \phi \sin \theta.$$

On taking note of (9a), this reduces to

$$\dot{\chi} \left(\cos \alpha + \sin \alpha \sin \phi \frac{\cos \theta}{\sin \theta} \right) = \dot{\phi}.$$

Applying (9c) and multiplying both sides by $\sin^2 \theta$, we get

$$\dot{\chi} \cos \alpha \, (\sin^2 \theta + \sin \chi \sin \alpha \cos \theta) = \dot{\phi} \sin^2 \theta. \tag{12}$$

But now it is seen from (9b) that the parenthesis on the left-hand side of Eq. (12) is equal to $\sin^2 \theta + \cos^2 \theta = 1$, so that the relation,

$$\dot{\chi} \cos \alpha = \dot{\phi} \sin^2 \theta, \tag{13}$$

which was to be proved, follows.

Systems of Two Particles.—There is often occasion to consider a system composed of two particles that attract or repel one another. Let the Cartesian coordinates of the first particle, referred to some fixed set of axes, be x_1, y_1, and z_1, and let those of the second particle, referred to the same set of axes, be x_2, y_2, and z_2. Let the masses of two particles be m_1 and m_2, respectively. As in the case of a single particle,

$$\begin{aligned} f_{x_1} &= m_1\ddot{x}_1 \qquad & f_{x_2} &= m_2\ddot{x}_2 \\ f_{y_1} &= m_1\ddot{y}_1 \qquad & f_{y_2} &= m_2\ddot{y}_2 \\ f_{z_1} &= m_1\ddot{z}_1 \qquad & f_{z_2} &= m_2\ddot{z}_2. \end{aligned} \tag{14}$$

If there is no other force than that which the particles exert on each other, then since the forces on the two particles are equal and opposite

$$\begin{aligned} m_1\ddot{x}_1 &= -m_2\ddot{x}_2 \quad \text{or} \quad & m_1\ddot{x}_1 + m_2\ddot{x}_2 &= 0 \\ m_1\ddot{y}_1 &= -m_2\ddot{y}_2 \quad \text{or} \quad & m_1\ddot{y}_1 + m_2\ddot{y}_2 &= 0 \\ m_1\ddot{z}_1 &= -m_2\ddot{z}_2 \quad \text{or} \quad & m_1\ddot{z}_1 + m_2\ddot{z}_2 &= 0. \end{aligned} \tag{15}$$

Integrating twice with respect to time gives

$$\begin{aligned} m_1x_1 + m_2x_2 &= \alpha_x t + \beta_x \\ m_1y_1 + m_2y_2 &= \alpha_y t + \beta_y \\ m_1z_1 + m_2z_2 &= \alpha_z t + \beta_z \end{aligned} \tag{16}$$

where the α's and β's are constants of integration.

Let us now introduce quantities, X, Y, Z, the coordinates of the center of gravity of the pair of particles, which are defined by the equations

$$\begin{aligned}(m_1 + m_2)X &= m_1x_1 + m_2x_2\\(m_1 + m_2)Y &= m_1y_1 + m_2y_2\\(m_1 + m_2)Z &= m_1z_1 + m_2z_2.\end{aligned} \tag{17}$$

It is seen from Eqs. (17) and (16) that the center of gravity of two particles which exert forces on each other, but on which no other forces are acting, moves with a uniform velocity. We also introduce three other new variables x, y, and z, such that

$$\begin{aligned}x &= x_2 - x_1\\y &= y_2 - y_1\\z &= z_2 - z_1.\end{aligned} \tag{18}$$

x, y, and z are the components of the distance between the particles, and their changes in the course of time determine the relative motion of the two particles. They can, of course, be expressed in terms of polar coordinates. From Eqs. (18) and (15),

$$\begin{aligned}\ddot{x} &= \ddot{x}_2 - \ddot{x}_1 = \ddot{x}_2\left(1 + \frac{m_2}{m_1}\right)\\\ddot{y} &= \ddot{y}_2 - \ddot{y}_1 = \ddot{y}_2\left(1 + \frac{m_2}{m_1}\right)\\\ddot{z} &= \ddot{z}_2 - \ddot{z}_1 = \ddot{z}_2\left(1 + \frac{m_2}{m_1}\right).\end{aligned} \tag{19}$$

From Eqs. (14) and (19),

$$\begin{aligned}f_x &= \ddot{x}\mu\\f_y &= \ddot{y}\mu\\f_z &= \ddot{z}\mu\end{aligned} \tag{20}$$

where $\mu = m_1m_2/(m_1 + m_2)$ and is called the reduced mass, and where $f_x = f_{x_2} = -f_{x_1}$, etc.

These are the same as the equations of motion of a particle of mass μ upon which a force with components f_x, f_y, and f_z is acting. The relative motion of a pair of particles can therefore be treated in the same way as the motion of a single particle with mass μ and coordinates x, y, and z. As the force between the particles depends on the distance between them, f_x, f_y, and f_z are definite functions of x, y, and z.

The kinetic energy of the system of two particles can readily be separated into two parts, (1) the kinetic energy of the motion of the center of gravity, which is equal to

$$\tfrac{1}{2}(m_1 + m_2)(\dot{X}^2 + \dot{Y}^2 + \dot{Z}^2) = \tfrac{1}{2}(m_1 + m_2)V^2, \tag{21}$$

where V is the velocity of the center of gravity, and (2) the relative kinetic energy,

$$\tfrac{1}{2}\mu(\dot{x}^2 + \dot{y}^2 + \dot{z}^2) = \tfrac{1}{2}\mu v^2, \tag{22}$$

where v is the relative velocity of the two particles. These two parts of the kinetic energy are quite independent of each other and may be treated separately. By differentiation of (17) and (18), solving for $\dot{X}$, $\dot{Y}$, $\dot{Z}$, $\dot{x}$, $\dot{y}$, and $\dot{z}$ in terms of $\dot{x}_1$, $\dot{y}_1$, $\dot{z}_1$, $\dot{x}_2$, $\dot{y}_2$, $\dot{z}_2$, substituting in Eqs. (21) and (22), and adding, it may readily be shown that the total kinetic energy of the system

$$\tfrac{1}{2}m_1(\dot{x}_1{}^2 + \dot{y}_1{}^2 + \dot{z}_1{}^2) + \tfrac{1}{2}m_2(\dot{x}_2{}^2 + \dot{y}_2{}^2 + \dot{z}_2{}^2)$$

is equal to the sum of the contributions of (21) and (22). In Eq. (21) or (22), as in the other equations, the rectilinear coordinates can be replaced in the usual way by polar coordinates.

Many of the preceding results can be generalized to the case of n particles. In this case there are $3n$ equations like Eqs. (14). In general the components of the forces will be functions of the positions of all the particles. Theoretically, the equations can be integrated, and there will be $6n$ constants of integration, because there are $3n$ differential equations of the second order. The case of the two particles just considered illustrates this general rule, as there are 12 constants, the 6 of Eqs. (16) and 6 more that will come from the integration of Eqs. (20).

The generalized definition of the center of gravity is given by the equations

$$\begin{aligned}
(m_1 + m_2 + \cdots + m_n)X &= m_1x_1 + m_2x_2 + \cdots + m_nx_n \\
(m_1 + m_2 + \cdots + m_n)Y &- m_1y_1 = m_2y_2 + \cdots + m_ny_n \\
(m_1 + m_2 + \cdots + m_n)Z &= m_1z_1 = m_2z_2 + \cdots + m_nz_n.
\end{aligned}$$

If there are no forces other than those exerted by the particles on each other, the center of gravity again moves with constant velocity.

There will be no generalization strictly analogous to Eqs. (20).

APPENDIX II

THE PRINCIPLES OF EQUILIBRIUM

It is believed that there will be found in this book but few applications of thermodynamics with which a student who has had a course in physical chemistry is not familiar. It has, nevertheless, seemed desirable to incorporate some of the principles of chemical equilibrium in a form which, although making no claim to rigor, it is hoped may be readily visualized.

A state of chemical equilibrium results from a balance of two tendencies (1) the tendency of a system to spread out over as great a space (strictly, phase space—see Sec. 4.6) as possible, (2) the tendency for the individual elements of a system to have a low energy. The first tendency predominates at high temperatures, the second at low.

Without going too deeply into statistical mechanical theory, we shall now attempt to formulate these ideas. We shall first consider the tendency of a system to spread out in ordinary space. For our purposes, this will be sufficient, and the tendency of a system to spread out in phase space, including momentum space, will be mentioned only incidentally.

Let us now look at the behavior of a gas. If the gas is perfect, there being no forces between the molecules and all positions being equivalent as far as potential energy is concerned, the tendency to increase in volume is unopposed by any internal forces. The gas exerts a pressure on the vessel containing it, and if the vessel is enlarged, for example by withdrawing a piston in a cylinder, the gas does work on it. Suppose there are N molecules of gas in a vessel of volume V_1, and suppose this volume is increased to V_2; then the amount of work done by the gas is $\int_{V_1}^{V_2} P\,dV$, where $P = NkT/V$ is the pressure (k is the gas constant per molecule, T the absolute temperature). We find $\int_{V_1}^{V_2} P\,dV = NkT \ln \frac{V_2}{V_1}$. If the temperature of the gas is to be kept constant, an amount of heat q equal to $NkT \ln \frac{V_2}{V_1}$ must be absorbed by the gas. It is thus seen that q/NT is in a certain sense the measure of the relative molecular volume change. It is a quantity that is independent of the amount of the gas, the temperature, and the nature of the gas. This quantity—the amount of heat per molecule, added when a system undergoes any change which is effected in a mechanically reversible manner (*i.e.*, in such a way that if the process is carried in the reverse direction the amount of mechanical energy required to do this is the same as the amount furnished by the system in the direct process), divided by the temperature—is called the molecular entropy change in the process. Multiplying it by Avogadro's number gives the molal entropy change designated by the symbol ΔS.

This may be generalized to include cases in which the momentum space (see Sec. 4.6) is also involved. Thus any increase in the phase space available to a molecule of the system is reflected in an increase of the entropy. If the temperature of a perfect gas is increased without changing the volume occupied by the gas, the entropy increases because a greater region in momentum space becomes available to the molecule. This is measured again by the heat absorbed per molecule divided by the temperature, as may be shown by statistical mechanics. The molecular entropy of a system, being a measure of the phase space available to the particles of the system, depends only on the state of the system, and its relation to the heat absorbed in any mechanically reversible process is the essence of the second law of thermodynamics.

The entropy change connected with any change in a thermodynamic system thus has, in a certain sense, a double aspect, being related, on the one hand, to the heat absorbed if the change in the system is carried out reversibly, and on the other hand, directly to the change in the freedom of motion of the molecules. Let us return to the case of the expanding gas, and consider the second aspect. We see that the entropy change in expanding from volume V_1 to volume V_2 is equal to

$$\Delta S = Nk \ln (V_2/V_1).$$

Or, as we see by allowing V_2 to approach V_1, we may write

$$\frac{\partial S}{\partial V} = Nk\frac{d \ln V}{dV}.$$

Integrating this equation, we may write for the entropy of the gas as a function of the volume

$$S = Nk \ln V + C_1,$$

where C_1 is a constant, which depends on the nature of the gas, and also, since these quantities are held constant in the process considered, on N and T. If now we set $C_1 = -Nk \ln N + NC$, where C is independent of N, but dependent on temperature, we get

$$S = Nk \ln \frac{V}{N} + NC,$$

and it will be seen that assuming this dependence of C_1 on N gives S the proper dependence on N. For example, if N is doubled and V is also doubled, so that we have twice as much gas at the same pressure, the entropy will be doubled. In the foregoing expression, V/N is a measure of the freedom of motion of the gas in ordinary space; and its freedom of motion in momentum space, which depends only on the temperature and the nature of the gas, is contained in C.

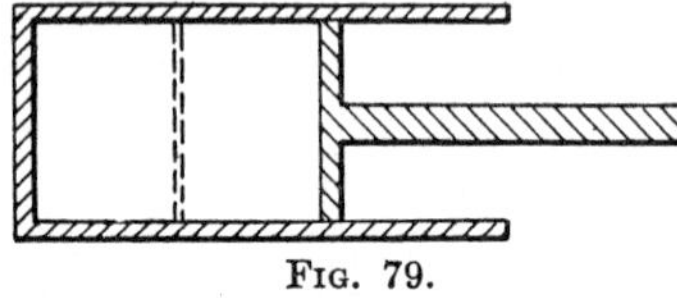

FIG. 79.

In order to illustrate the principles involved in equilibrium problems, we shall consider a special example. Let us suppose that a gas is enclosed in a cylinder with a piston as shown in the figure. It is supposed that between the two dotted lines a strong force acts on the molecules, so that those in the left part of the cylinder have a much lower energy than those in the right part. Let the volume of the left part be V_l, that of the right part V_r, let the difference in energy per molecule be $\Delta\epsilon$, and suppose equilibrium is established in all parts of the cylinder. The entropy of the system will be the sum of the entropies of the left- and right-hand sides, and will have the form

$$N_l k \ln \frac{V_l}{N_l} + N_l C + N_r k \ln \frac{V_r}{N_r} + N_r C$$
$$= N_l k \ln \frac{V_l}{N_l} + N_r k \ln \frac{V_r}{N_r} + NC,$$

if $N_l + N_r = N$. Now let the piston move to the right a small amount, V_r increasing by δV_r. Assume this to take place so slowly that equilibrium is maintained, so that the whole process is reversible. Then, if the process occurs isothermally, the change of entropy will be (since N and C are constant in the isothermal process)

$$\begin{aligned}
\delta S &= \delta\left(kN_l \ln \frac{V_l}{N_l} + kN_r \ln \frac{V_r}{N_r}\right) \\
&= k\left(\ln \frac{V_l}{N_l} - 1\right) \delta N_l + k\left(\ln \frac{V_r}{N_r} - 1\right) \delta N_r + k\left(\frac{N_r}{V_r}\right) \delta V_r \\
&= k\left(\ln \frac{V_r}{N_r} - \ln \frac{V_l}{N_l}\right) \delta N_r + k\frac{N_r}{V_r} \delta V_r \\
&= k\left(\ln \frac{V_r}{V_l} - \ln \frac{N_r}{N_l}\right) \delta N_r + k\frac{N_r}{V_r} \delta V_r. \qquad (1)
\end{aligned}$$

(The simplification comes from noting that, since N is constant, $\delta N_l = -\delta N_r$.) But this must also be equal to the heat absorbed in the process divided by the temperature. To maintain constant temperature, it will be necessary to absorb an amount of energy equal to $\Delta\epsilon\, \delta N_r$ due to transfer of molecules from the low- to the high-energy region and another amount equal to

$$P\delta V_r = kTN_r \frac{\delta V_r}{V_r}$$

due to the work the gas does on the piston in expanding. This latter term, when divided by T, cancels the last term in (1), and we can write

$$\frac{\Delta\epsilon}{T} = k\left(\ln \frac{V_r}{V_l} - \ln \frac{N_r}{N_l}\right)$$

or

$$\frac{N_r}{N_l} = \frac{V_r}{V_l} e^{-\frac{\Delta\epsilon}{kT}}. \qquad (2)$$

In other words, the relative probability per unit volume that a molecule find itself in the right-hand part is $e^{-\frac{\Delta\epsilon}{kT}}$. This is a result that holds in general for phase space. Since a quantum state occupies a definite volume in phase space, the relative probability of two quantum states with energies ϵ_1 and ϵ_2, respectively, is given by $e^{-\frac{\epsilon_1-\epsilon_2}{kT}}$.

A very similar treatment can be given a system in which an equilibrium of two atoms forming a diatomic molecule

$$\mathrm{A} + \mathrm{B} \rightleftarrows \mathrm{AB}$$

is established. Combination of A and B restricts their motion and is very much like forcing one of them to move in a small volume. Suppose we let the volume in which the gases (considered perfect) are free to move be designated as V and suppose we let the effective volume for the *relative* motion of a pair of combined atoms be V_0. Then, by analogy with the first expression in Eq. (1), we can write for the change in entropy when N_A, N_B, and N_{AB}, the numbers of the various types of molecules present, are changed,[1]

$$\begin{aligned}\delta S &= \delta\left(kN_A \ln \frac{V}{N_A} + kN_B \ln \frac{V}{N_B} + kN_{AB} \ln \frac{V}{N_{AB}} + kN_{AB} \ln V_0\right) \\ &= k\left(\ln \frac{V}{N_A} - 1\right) \delta N_A + k\left(\ln \frac{V}{N_B} - 1\right) \delta N_B \\ &\qquad + k\left(\ln V_0 + \ln \frac{V}{N_{AB}} - 1\right) \delta N_{AB}. \end{aligned} \tag{3}$$

(The quantity V_0 is not divided by N_{AB} because it is the effective volume for *one* pair.) In the expression above

$$\delta N_A = \delta N_B = -\delta N_{AB},$$

since the total number of atoms of A and B remain always constant if both those combined and uncombined are counted. If $\Delta\epsilon$ is the energy of dissociation of AB (*i.e.*, the energy necessary to break an AB of *average* energy into an A and a B of *average* energy), then the energy absorbed is $\Delta\epsilon\, \delta N_A$, and if

[1] The entropy of the whole is the entropy of A plus the entropy of B plus the entropy of AB, each calculated as though the other gases were not there. The complete expression for the entropy is

$$kN_A \ln \frac{V}{N_A} + kN_B \ln \frac{V}{N_B} + kN_{AB} \ln \frac{V}{N_{AB}} + N_A C_A + N_B C_B + N_{AB} C_{AB}$$

It will be seen by comparing with Eq. (3) that, since $\delta N_A = \delta N_B = -\delta N_{AB}$, we have the relation $k \ln V_0 = C_{AB} - C_A - C_B$.

this is divided by T it must be equal to the foregoing expression, provided the process is carried out reversibly.[1] If we let $n_A = N_A/V$, *i.e.*, the concentration of the atoms A, etc., then the resulting expression may be written

$$\frac{n_A n_B}{n_{AB}} = (eV_0)^{-1} e^{-\frac{\Delta\epsilon}{kT}}, \tag{4}$$

where the quantity on the left is the equilibrium constant κ expressed in terms of concentrations. Since V_0 does not differ greatly for reactions of this type, the value of $\Delta\epsilon$ may be considered a measure of the equilibrium constant at any given temperature. This is the basis of the usage of the text in comparing energy changes for similar reactions.

From Eq. (3) it is seen that $\Delta\sigma^0 = -k(\ln V_0 + 1)$ is the change of entropy per molecule of AB dissociating when the concentrations are unity,[2] and Eq. (4) may be written in the form

$$kT \ln \kappa = -\Delta\epsilon + T\,\Delta\sigma^0. \tag{5}$$

If we let K be the equilibrium constant expressed in terms of partial pressures, *i.e.*, $K = P_{AB}/P_A P_B$, and let ΔE be the energy change per mole and ΔS^0 the entropy change per mole when $P_A = P_B = P_{AB} = 1$ (and with *pressure* constant instead of volume), then it may be readily shown that Eq. (5) is equivalent to

$$RT \ln K = -\Delta H + T\Delta S^0 = -\Delta F^0, \tag{6}$$

where ΔH, which is called the change in heat content, in this case equals $\Delta E + RT$ and in general is equal to $\Delta E + \Delta(PV)$ and ΔF^0 is defined by the equation. Equations (5) and (6) can be applied in general to cases of equilibrium, if more than two kinds

[1] This is the same thing as saying that A, B, and AB are in such concentrations that they are in equilibrium. If the rate of reaction were so slow that A, B, and AB could exist together without equilibrium being established, Eq. (3) would still give the entropy change when N_A, N_B, and N_{AB} were changed. But then we could not set the entropy change equal to $\frac{\Delta\epsilon\, \delta N_A}{T}$. This equality is the *condition* for equilibrium.

[2] Of course when all concentrations are unity, the system will not in general be at equilibrium, but as we have noted, Eq. (3) holds anyhow.

of atoms are involved, and Eq. (6) is especially useful when a process takes place at constant pressure.[1]

The quantity ΔF^0 is called the *standard* free-energy change of dissociation, the term "standard" referring to the fact that the process takes place at partial pressures equal to unity. In general, for an isothermal process $\Delta F = \Delta H - T\Delta S$, and at constant pressure as well as constant temperature

$$\Delta F = \Delta E + P\Delta V - T\Delta S.$$

Now ΔE is the total change of energy of the system, and $P\Delta V$ is the work done against a piston or other device (for example, the atmosphere) used to keep the pressure constant.[2] If the change is carried out in a mechanically reversible manner, then $T\Delta S$ is the heat absorbed by the system. ΔF must then, under these circumstances, be the reversible work done on the system by any mechanical device operating upon it other than the piston or the atmosphere which maintains constant pressure. The term "mechanical device" includes also electrically operating instruments.

From Eq. (5), we may readily obtain an important equation that gives the rate of change of $\ln \kappa$ with temperature. Dividing through by T and differentiating, we have

$$k\frac{d \ln \kappa}{dT} = -\frac{1}{T}\frac{d\Delta\epsilon}{dT} + \frac{\Delta\epsilon}{T^2} + \frac{d\Delta\sigma^0}{dT}.$$

However, since the entropy change in a mechanically reversible process is the heat absorbed divided by the temperature, the following equation must hold for any of the substances, A, B

[1] Equation (6) also applies when solid and liquid substances are involved. The solid and liquid states normally existing are themselves taken as the standard states and it may be shown that if this is done no pressures corresponding to the solid or liquid substances appear in K, but only pressures of substances not present in the solid or liquid state. Substances in solution may also be present. The normal state is the state where the concentration is 1 molal, and concentrations in solution appear in K in lieu of pressures. For further details, the student should refer to some standard work on thermodynamics, such as Lewis and Randall, "Thermodynamics and the Free Energy of Chemical Substances," McGraw-Hill Book Company, Inc., 1923.

[2] It is seen that ΔH differs from ΔE simply in that it includes the change in energy of the piston or the atmosphere.

or AB, (or any other substance which may be involved in the equilibrium) if the temperature is changed while holding the volume constant:[1]

$$\frac{1}{T}\frac{d\epsilon}{dT} = \frac{d\sigma}{dT}.$$

Hence we must have[2]

$$-\frac{1}{T}\frac{d\Delta\epsilon}{dT} + \frac{d\Delta\sigma^0}{dT} = 0,$$

and

$$\frac{d \ln \kappa}{dT} = \frac{\Delta\epsilon}{kT^2}. \tag{7}$$

In a similar manner, if the pressure is held constant, it is readily shown from Eq. (6) that

$$\frac{d \ln K}{dT} = \frac{\Delta H}{RT^2}. \tag{8}$$

Note on Energy Units.—In this appendix we have distinguished between *molecular* and *molal* quantities. In general, however, symbols representing energies are, throughout the text (except as otherwise noted and in Chap. XIX), so defined as to refer to one molecule, or one atom, or one bond, etc. An effort has been made to follow this usage consistently in the equations, and the gas constant per molecule k (the Boltzmann constant) has been used rather than R. In the case of chemical reactions the energy change is that which occurs when the number of molecules indicated in the equation react.

However, it is often convenient to express energies numerically as calories or kilogram-calories per mole, or, in the case of reactions, as kilogram-calories change upon reaction of the number of moles indicated by the chemical equation. When the term "calories" or "kilogram-calories" is used alone, it refers to such molal quantities. This should cause no confusion, for the molal energies are obtained from molecular energies simply by multiplication by Avogadro's number.

In Chap. XIX, except when otherwise stated, energies and entropies have been defined to refer to one mole of substance.

[1] This is a mechanically reversible process since there is no mechanical change.

[2] Since the gases are perfect, we can, without affecting the energy, take any particular volume, say unit volume, and hence deal with standard entropies.

APPENDIX III

ELECTRICAL FORCES

All the types of forces between atoms are ultimately electrical in origin. Those which are operative in covalent or nonpolar bonds are, as seen in the text, rather complex in character, involving electron exchange and the wave properties of the electron in a rather fundamental way. Other forces, such as those operative in ionic crystals, molecular crystals, and metals are of somewhat simpler character. In this appendix, a summary will be given of the various types of electrical forces (in vacuum).

Before proceeding, it will be well to recall some fundamental definitions. The *mutual potential* (potential energy) of any two bodies containing electrical charges is the work it is necessary to do on those bodies in order to bring them from positions in which they are an infinite distance apart to the positions in which their potential is desired. Let f be the force between two bodies, and suppose it is a function of the distance r between them; it will be convenient to write $f(r)$ for the force when the distance is r. The force is assumed to be positive if it is repulsive. We shall designate the mutual potential of two bodies when at a distance r by $\phi(r)$. To increase the distance between the charges by an amount dr, an amount of work equal to $-f(r)dr$ must be done on them. Hence by the definition of $\phi(r)$ we can write

$$\phi(r_0) = -\int_{\infty}^{r_0} f(r)dr = \int_{r_0}^{\infty} f(r)\, dr \tag{1}$$

and

$$f(r) = -\frac{d\phi(r)}{dr}. \tag{2}$$

The *electric field* $E(r)$ due to some body containing electric charge is the force exerted on unit positive charge ($E(r)$ is positive if this force is directed away from the body producing the field). The force $f(r)$ on a charge e is given by

$$f(r) = eE(r). \tag{3}$$

The most elementary type of electrical force is that between two charged bodies (assumed to be points) with charges e_1 and e_2, let us say (and we may assume that e_1 and e_2 are positive or negative according as the corresponding charge is positive or negative). In this case,

$$\phi(r) = \frac{e_1 e_2}{r} \tag{4}$$

and

$$f(r) = \frac{e_1 e_2}{r^2}. \tag{5}$$

These are negative if the charges are of opposite sign, corresponding to attraction. By Eqs. (3) and (5), it is seen that the field at the charge e_1, due to the charge e_2, is equal to e_2/r^2.

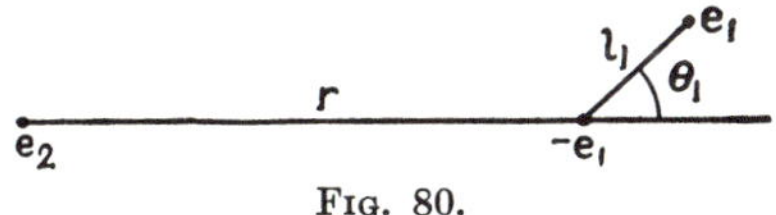

FIG. 80.

The electrostatic unit of electrical charge is defined in such a way that if r is in centimeters, $f(r)$ is in dynes. Thus two unit charges 1 cm. apart exert on each other a force of 1 dyne.

Ion-Dipole Forces.—As seen in Sec. 12.4, two equal and opposite charges, say e_1 and $-e_1$, separated by a distance l_1, constitute a dipole moment

$$M_1 = e_1 l_1. \tag{6}$$

The potential between such a dipole and another charge e_2 depends upon the distance r between the charge and the dipole (we assume that r is so large compared with l_1 that we can speak roughly in this manner of distance between the charge and the dipole) and the orientation of the dipole with respect to the line joining it and the charge. Let us consider Fig. 80. For definiteness, let us call the distance between e_2 and $-e_1$ simply r; the distance between e_2 and e_1 will then be $r + l_1 \cos \theta_1$ to a very good approximation, since $r \gg l_1$. The potential between e_2 and $-e_1$ will be $-\frac{e_1 e_2}{r}$, and the potential between e_2 and e_1 will be $e_2 e_1/(r + l_1 \cos \theta_1)$. The total potential between charge and dipole will be the sum of these

$$e_2e_1\left(\frac{1}{r + l_1 \cos \theta_1} - \frac{1}{r}\right).$$

We may expand the first fraction in the parentheses by the binomial theorem and drop terms containing l_1^2 and higher powers, obtaining

$$\phi(r) = -\frac{e_2e_1l_1 \cos \theta_1}{r^2} = -\frac{e_2M_1 \cos \theta_1}{r^2} \tag{7}$$

and

$$f(r) = -\frac{2e_2M_1 \cos \theta_1}{r^3}. \tag{8}$$

This force, of course, does not include any torque tending to turn the dipole, but is only the force directed along the line joining e_1 and e_2. The dipole will have turning forces on it, and there is also a torque on the system as a whole, *i.e.*, equal and opposite forces tending to push the ion and the dipole, respectively, perpendicularly to the line joining them, in the plane e_2, $-e_1$, e_1. If the dipole is held fixed in position and the ion moves a distance dy upward in Fig. 80, the angle θ_1 will change by $d\theta_1 = dy/r$. So the component of force is

$$f_1(r) = -\frac{1}{r}\frac{d\phi(r)}{d\theta_1} = -\frac{e_2M_1 \sin \theta_1}{r^3} \tag{8a}$$

A force of the ion-dipole type will in general be much less than the force between two charges, unless r becomes of the order of l_1, in which case the expression given for the potential is no longer valid. An ion-dipole force is the type of force operative between ions in aqueous solution and the surrounding water molecules, and is responsible for most solvation effects, though as noted in Sec. 19.5, this is a somewhat oversimplified statement.

Dipole-Dipole Forces.—The potential between two dipoles is readily calculated in a very similar manner. Let the first dipole be composed of charges e_1 and $-e_1$ at a distance l_1, so that $e_1l_1 = M_1$, and let the second dipole be composed of charges e_2 and $-e_2$ at a distance l_2, so that $e_2l_2 = M_2$ (see Fig. 81). We again assume that the distance r between the two dipoles is large compared with l_1 and l_2. Now for definiteness let the distance from $-e_2$ to $-e_1$ be r. In calculating the other distances it will not be allowable this time to make approximations. Let the distance between $-e_2$ and e_1 be r_1; it is given by the relation $r_1^2 = (r + l_1 \cos\theta_1)^2 + (l_1 \sin\theta_1)^2$. Similarly the distance r_2 between $-e_1$ and e_2 is given by the relation $r_2^2 = (r + l_2 \cos\theta_2)^2 + (l_2 \sin\theta_2)^2$. Finally the relation $r_{12}^2 = (r + l_1 \cos\theta_1 + l_2 \cos\theta_2)^2 + (l_1 \sin\theta_1 - l_2 \sin\theta_2)^2$ holds for the distance r_{12} between e_1 and e_2, for the special case that both dipoles lie in the same plane. The total potential energy is $\phi(r) = e_1e_2(r^{-1} - r_1^{-1} - r_2^{-1} + r_{12}^{-1})$. Expanding r_1^{-1}, r_2^{-1}, and r_{12}^{-1}, which involve square roots, by the binomial theorem, and discarding all terms of the third degree or higher in l_1 and l_2, we obtain

$$\phi(r) = \frac{2M_1M_2 \cos\theta_1 \cos\theta_2}{r^3} + \frac{M_1M_2 \sin\theta_1 \sin\theta_2}{r^3} \tag{9}$$

and, for the force component along r,

$$f(r) = \frac{6M_1M_2 \cos\theta_1 \cos\theta_2}{r^4} + \frac{3M_1M_2 \sin\theta_1 \sin\theta_2}{r^4}. \tag{10}$$

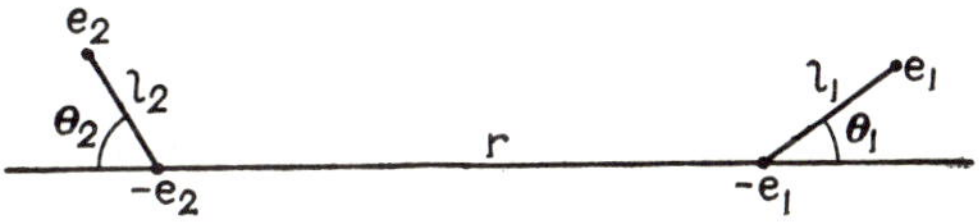

FIG. 81.

This type of force is the kind designated as a dipole force in the discussion of molecular crystals.

We could also consider forces on quadripoles, *i.e.*, equal positive and negative charges arranged in the form of a square $\begin{smallmatrix} + & - \\ - & + \end{smallmatrix}$; but although they play a role in certain phenomena they will not be important for our purposes.

Ion-Induced Dipole Forces.—If a body, composed of electrical charges but in which the positive and negative charges balance, is placed in an electric field it was seen in Sec. 12.4 that there is a displacement of electric charge in the body, resulting in the separation of positive and negative charge, and producing a dipole whose moment (at least if the field is not too large) will be proportional to the field; the constant of proportionality is called the "polarizability" and is designated by the symbol α. Thus a body with a charge e_1 at a distance r from a neutral body will produce a field equal in magnitude to e_1/r^2 at the neutral body, causing a dipole moment equal to $\alpha e_1/r^2$. The dipole moment is oriented along the line joining the charge and the body, in such a direction that attraction results between the charge and the polarized body. On applying the expression for an ion-dipole potential, the mutual potential between the charges and the polarized body is given by

$$\phi_m(r) = -\frac{\alpha e_1^2}{r^4}. \tag{11}$$

If, however, we are interested in the potential energy of the whole system, as is generally the case, there is another potential involved that must be considered. The creation of a dipole in a polarizable body requires the expenditure of energy, and this self-potential of the induced dipole must be added to $\phi_m(r)$. Since the strength of the induced dipole depends upon the distance of the charged body that induces it, the self-potential will also be a function of r, and we can write

$$\phi(r) = \phi_m(r) + \phi_s(r), \tag{12}$$

where $\phi_s(r)$ is the self-potential. We shall now proceed to calculate $\phi_s(r)$.

Suppose the induced dipole to consist of a charge e and an equal and opposite charge $-e$, separated by a distance l. The electric field E producing this separation exerts a force eE in the direction of the field on the positive charge e, and an equal and opposite force $-eE$ on the negative charge. It is this force which is necessary to effect the separation by the amount l. By the definition of α, we have

$$el = M = \alpha E. \tag{13}$$

Setting the force eE equal to $g(l)$, since it is a function of l, we have by Eq. (13)

$$g(l) = \frac{e^2 l}{\alpha}. \tag{14}$$

The work, then, which is necessary to change l from a value of zero to its final value l is obtained by integrating g from 0 to l;

$$\int_0^l g(x)\,dx = \int_0^l \frac{e^2 x}{\alpha}\,dx = \frac{e^2 l^2}{2\alpha} = \frac{M^2}{2\alpha}. \tag{15}$$

This is just the value of the self-potential, and putting in the value of M, namely, $\alpha e_1/r^2$ gives

$$\phi_s(r) = \frac{\alpha e_1^2}{2r^4}. \tag{16}$$

This represents a positive potential, since work is done on the system to separate the charges e and $-e$ from their normal positions of coincidence. We get, therefore,

$$\phi(r) = -\frac{\alpha e_1^2}{2r^4} \tag{17}$$

and

$$f(r) = \frac{-2\alpha e_1^2}{r^5}. \tag{18}$$

Dipole-Induced Dipole Forces.—A dipole also can induce another dipole in a polarizable body. By Eqs. (8), (8*a*), and (3), a dipole of moment M_1 produces a field at a distance r, whose components along and perpendicular to the line joining the dipoles are, respectively,

$$E(r) = -\frac{2M_1 \cos\theta_1}{r^3}; \qquad E_1(r) = -\frac{M_1 \sin\theta_1}{r^3}. \tag{19}$$

A body at distance r will therefore have induced in it a moment with components $(2\alpha M_1 \cos\theta_1)/r^3$ and $(\alpha M_1 \sin\theta_1)/r^3$, respectively. Each component can be treated as independent of the other. Each one will be in such a direction as to lower the total potential energy of the system. By Eq. (9) the respective electrostatic potential energies due to interaction with the original dipole will be

$$\phi_m(r) = -\frac{4\alpha M_1^2 \cos^2\theta_1}{r^6}; \qquad \phi_{m1}(r) = -\frac{\alpha M_1^2 \sin^2\theta_1}{r^6}. \tag{20}$$

But, by Eq. (15), there will be self-potentials of just half these amounts and opposite in sign, so the total potential energy will be

$$\phi(r) = -\frac{2\alpha M_1^2 \cos^2\theta_1}{r^6} - \frac{\alpha M_1^2 \sin^2\theta_1}{2r^6} \tag{21}$$

and the force component along r will be

$$f(r) = -\frac{12\alpha M_1^2 \cos^2\theta_1}{r^7} - \frac{3\alpha M_1^2 \sin^2\theta_1}{r^7}. \tag{22}$$

It should be noticed that these equations, as well as Eqs. (17) and (18), break down at small distances for two reasons. In the first place, they cannot be valid under conditions under which Eqs. (7) and (9) break down, and in the second place, if the inducing body is too close to the one in which a dipole is induced, the field over the latter will not be constant, so that Eq. (13) will not hold.

It is forces of the type we have just been considering that produce the van der Waals attraction. The inducing dipole does not need to be a permanent dipole; the temporary displacement of charge resulting from the motion of the electrons in their orbits is quite sufficient. Thus any two molecules or atoms, due to their momentary dipoles, are always inducing in each other secondary dipoles. The primary dipoles may be in such a direction as to either attract or repel each other; they continually change in direction, and on the average neither attraction nor repulsion is experienced. But the secondary induced dipoles are always in such a direction as to attract the primary dipoles. There is thus a net attraction, the magnitude of which is discussed in Chap. XVII.

It is seen that the forces considered present a series in which the dependence on r becomes more and more marked. In a general way, it may be said that the higher the power of r, the smaller the actual magnitude of the force. All the expressions for the potential break down when the distance r becomes of the order of the distances separating the charges within a given dipole. In most cases, however, the latter separations appear to be small compared with atomic distances that give the lower limit for r, so that the formulas can be used with a fair degree of approximation.

Exercises

1. Calculate the force and the potential between two electrical charges equal to the charge on the electron at a distance of 2×10^{-8} cm. Calculate (approximately) the force and the potential between two dipoles, produced by displacing one electronic charge 0.1×10^{-8} cm., when placed 2×10^{-8} cm. apart in the relative positions shown:

$$\begin{array}{llll} (a) & +- \quad +- & (c) & \begin{matrix} + & + \\ - & - \end{matrix} \\ \\ (b) & +- \quad -+ & (d) & \begin{matrix} + & - \\ - & + \end{matrix} \end{array}$$

2. Derive Eq. (18) directly from Eq. (8) and Eq. (22) directly from Eq. (10).

APPENDIX IV

SOME REMARKS ON THE GEOMETRY OF MOLECULES AND CRYSTALS

The Geometry of the Tetrahedron and Tetrahedral Structures.—The geometry of the tetrahedron is best understood by considering its relation to the cube. Alternate corners of a cube are the vertices of a regular tetrahedron (Fig. 82). All the main diagonals of the cube go through its center of gravity O which is also the center of gravity of the tetrahedron. Since the ratios of the lengths of the edges, the face diagonals, and the main diagonals of a cube are readily deduced by use of the Pythagorean theorem, all the important relations of the tetrahedron are easily found. The "tetrahedral angle" is the angle made by the lines joining the center of gravity with two of the vertices, for example, the angle AOC. The relation of an atom having a coordination number of four (S in SO_4^{--} ion, for example) to the atoms that surround it tetrahedrally is that of the center of the tetrahedron O to its four vertices A, C, H, and F.

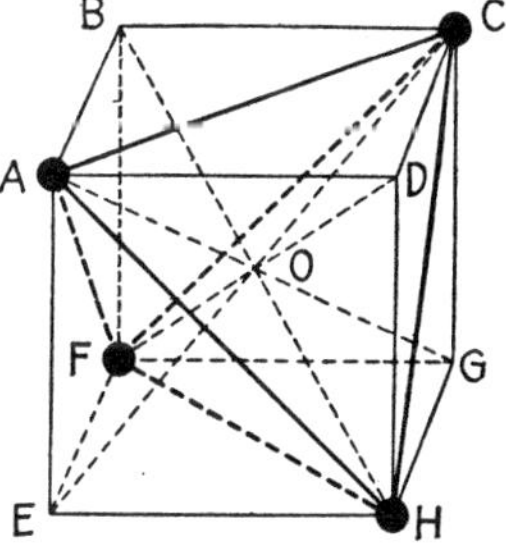

FIG. 82.—Geometry of the regular tetrahedron.

In view of the intimate relationship between a tetrahedron and a cube, it is to be expected that there will be a close relationship between a cesium chloride (body-centered cubic) lattice (Fig. 83) and a tetrahedral lattice. Indeed a tetrahedral lattice can be made out of a cesium chloride lattice by removing half of each kind of atom. This is shown in Fig. 84. Figure 84a shows the structure as mutually contacting spheres, and Fig. 84b shows the positions of the centers of gravity with the mutual contacts indicated by heavy lines going from one center to the other. This is the sphalerite or zincblende structure. It is clear from Fig. 84a that if the two types of atoms are of anywhere near equal size a tetrahedral arrangement will be a very open one in which the atoms do not come near to filling the space available, even though they are in contact.

If the white and black spheres shown in Fig. 84 are considered to represent the same kind of atom instead of different kinds, then the sphalerite structure becomes the diamond structure.

FIG. 83.—Cesium chloride lattice.

The lattice points of Fig. 84b are shown viewed from another angle in Fig. 85a. It is readily seen that the structure may be extended indefinitely in all directions, with the black centers always surrounded by white ones in the directions shown in Fig. 85b, and the white centers always surrounded by black ones in the manner shown by Fig. 85c. This is not the only type of extended tetrahedral structure. Another, the wurtzite structure, is shown in Fig. 85d. It will be seen that in this structure the black centers can be divided into alternate planes, all the centers in one set of planes being surrounded by white centers, as shown in Fig. 85b (these black centers labeled b in

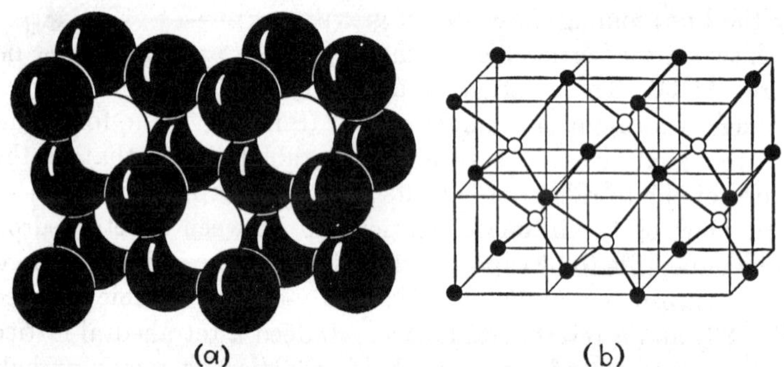

FIG. 84.—Sphalerite or zincblende structure.

the figure), whereas all those of the other planes are surrounded as shown in Fig. 85e (labeled e in the figure). Similarly, white centers are in alternate planes, having neighbors arranged as in Figs. 85c and 85f, respectively (labeled c and f, respectively). Figures 85a and 85d do not exhaust the possibilities of tetrahedral lattices. Several others, which differ in various details, are known, but in all, the immediate surroundings of any atom are the same (except for the directions in which the neighbors are to be found).

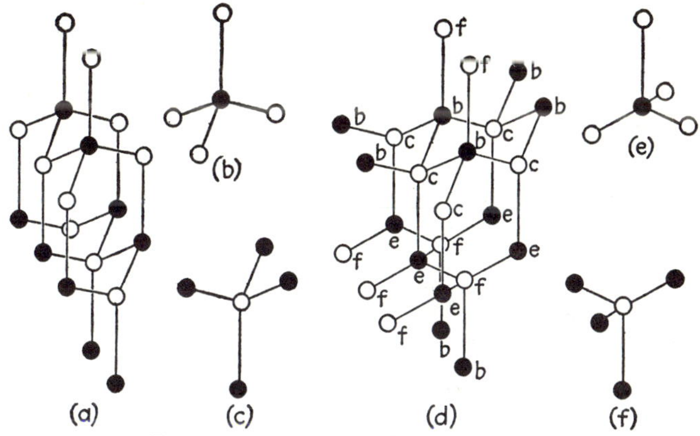

FIG. 85.—Sphalerite and wurtzite structures.

The structures of two of the common varieties of SiO_2 are easily understood from Figs. 85a and 85d. In β-cristobalite, the Si atoms form a diamond lattice and so may be represented by the black and white spheres of Fig. 85a, the oxygens being laid midway on the connecting lines.[1] Similarly, the black and white

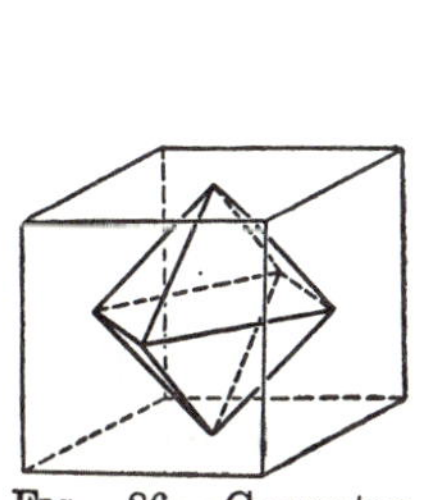

FIG. 86.—Geometry of the octahedron.

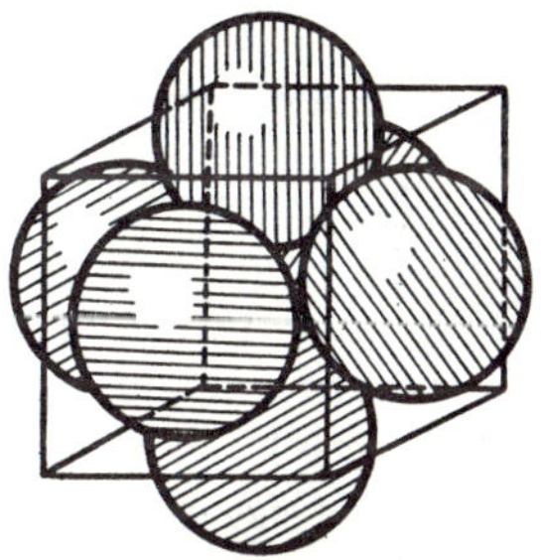

FIG. 87.—Octahedral arrangement of atoms.

spheres of Fig. 85d may be considered to represent the Si atoms of β-tridymite, with the oxygens laid midway on the connecting lines. Some other forms of SiO_2 differ in that the Si—O—Si angle is not 180°.

[1] "Strukturbericht," vol. I, p. 169; but see also vol. II, p. 261.

Geometry of the Octahedron and of Fluorite and Related Structures.—The geometry of the octahedron may be appreciated from Fig. 86. An octahedron is formed by joining the mid-points of the faces of a cube. Vertices of an octahedron correspond to faces of a cube and, conversely, faces of an octahedron correspond to vertices of a cube. The structure obtained by placing atoms in mutual contact at the corners of an octahedron may be visualized from Fig. 87. This may, for example, represent the six water molecules in $Al(H_2O)_6^{+++}$, the Al^{+++} being then in the space at the center.

FIG. 88.—Fluorite structure.

In the light of what has been said about octahedra and tetrahedra, the structure of fluorite, CaF_2, may be understood from Fig. 88. Here the large cubes represent Ca^{++} and the small ones F^-. The relative size and the cubical shape of the cubes have no physical significance, but are used to make the structure easy to visualize. It will be seen that the Ca^{++} have eight F^-'s surrounding each one. On the other hand, each F^- has only four Ca^{++}'s surrounding it. These are tetrahedrally arranged around the F^- ions, as will be seen from the fact that every other corner of the cubes representing F^- ions is connected to a Ca^{++} cube. A great many crystals have the fluorite structure, including many with complex anions and cations. Thus with

$Ni(NH_3)_6Cl_2$ the $Ni(NH_3)_6^{++}$ ions are represented by the large cubes, the Cl^- ions by the small cubes, in Fig. 88. Each Cl^- is thus in contact with a face of the $Ni(NH_3)_6^{++}$ octahedron and joins four such faces belonging to different octahedra. In $Ni(NH_3)_6(ClO_4)_2$, the tetrahedral ClO_4^- replaces Cl^-. The corners (*i.e.*, the oxygens) of the ClO_4^- tetrahedra are undoubtedly directed toward the $Ni(NH_3)_6^{++}$ ions which they connect.

The structure of the hexahydrated magnesium halides, discussed in Sec. 19.9, is closely related to the fluorite structure. Although there is no molecule formation evident in Fig. 88, the whole crystal forming a giant molecule, we can arbitrarily think of the crystal as divided into molecules with their long axes all

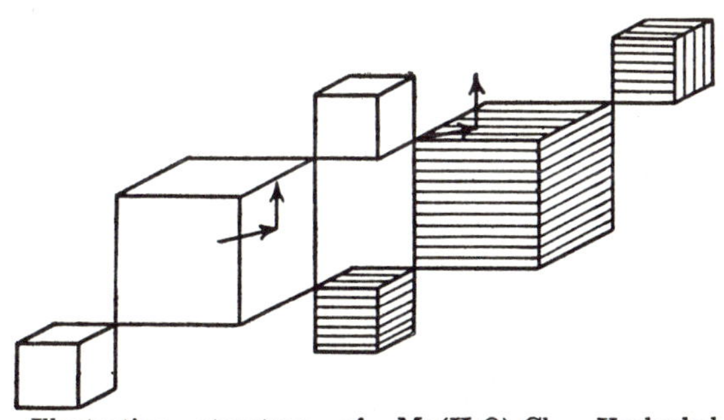

FIG. 89.—Illustrating structure of $Mg(H_2O)_6Cl_2$. Unshaded molecule in fluorite position with respect to shaded molecule. Displacement of unshaded molecule to get into position for $Mg(H_2O)_6Cl_2$ shown by arrows.

parallel. Two such molecules are shown in Fig. 89. Now with $Mg(H_2O)_6Cl_2$ the large cubes represent $Mg(H_2O)_6^{++}$ and the small ones represent Cl^-. The actual structure of $Mg(H_2O)_6Cl_2$ can be formed from the structure shown in Fig. 88 by displacements of the molecules with respect to each other. Such a displacement is shown in Fig. 89. The unshaded molecule moves with respect to the shaded molecule so that one of the chlorines of each molecule is nearer the water molecule at the corner of one of the vertices of the $Mg(H_2O)_6^{++}$ octahedron of the other molecule. The reason for the upward displacement, as well as the displacement along the long axis of the molecule, will be evident from Fig. 87. Since the water molecule rests at the center of the cube face, it is not possible for a Cl^- which moves nearer the center of the cube face to remain as close to the center of the cube as it was before. Other molecules surrounding the shaded molecule in Fig. 89 are displaced in other directions in forming the $Mg(H_2O)_6Cl_2$ structure. Some of the displacements are not quite so symmetrical as that shown in the figure.

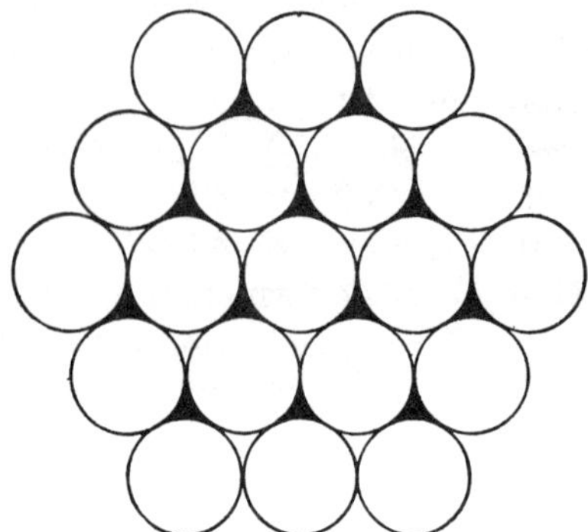

FIG. 90.—Illustrating close-packing.

Close-packed Structures.—A close-packed structure of spheres of equal size is composed of planes such as shown in Fig. 90. These rest one on top of the other, with the atoms of one plane

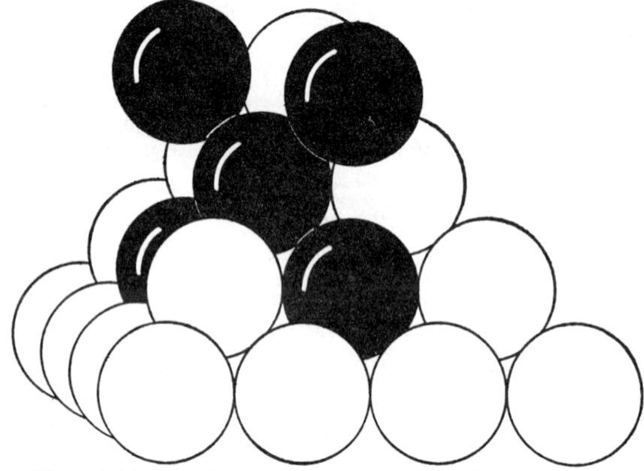

FIG. 91.—Illustrating cubic close-packing. The balls are piled in the form of a pyramid, with two extra ones added at the top. Five of the balls are shaded to bring out the fact that this is really a face-centered cubic arrangement. (*After model made by Dr. E. Mack, Jr.*)

just above half of the spaces between the atoms of the plane below. Let the plane shown in Fig. 90 be designated as B, the one just below it in the crystal as A, and the one just above it as C. Let us suppose that the atoms of C rest on those of B in such a way that their centers are just above the *black* spaces between the atoms of B. Then, if the atoms of A are just below the black spaces of B, we have hexagonal close-packing; on the other hand, if the atoms of A are below the white spaces of B, the arrangement is cubic close-packing, or face-centered cubic. The latter is illustrated in Fig. 91.

It will be observed that in a close-packed arrangement four adjacent atoms are tetrahedrally arranged about the space between them, and that any given atom in the close-packed structure has twelve closest neighbors.

Exercises

1. If the edge of a tetrahedron (octahedron) has the length l, find the distance from the center of gravity to a vertex.

2. Calculate the tetrahedral angle.

3. Calculate the anion-cation radius ratio which will just give anion-anion contact when the cation has a coordination number of four; of six.

4. From the distances given in Table 36 find the molal volume of CsCl; of CaF_2.

APPENDIX V

GENERAL REFERENCES

In order to avoid repeated references in the text, some books of general interest in connection with the subjects treated in this volume are listed herewith. This list makes no pretension of being complete; its object is merely to provide the student with a representative selection of books for supplementary reading. The books in the first group are elementary accounts of material related to that presented in the early part of the present volume. The books in the second group give greater detail or a different point of view on various special topics. In the third group are listed some tabulations of experimental material which have been found helpful in the preparation of the tables. In most cases we give, after the title of a book, the particular chapter or chapters of the present work to which the reference is most closely related.

I

H. T. Briscoe: "The Structure and Properties of Matter," McGraw-Hill Book Company, Inc., 1935; for Chaps. II, IV, VII, XI.

R. W. Gurney: "Elementary Quantum Mechanics," Cambridge University Press, 1934; for Chaps. III, IV, V, VI, IX, X, XI, XVIII.

G. F. Hull: "An Elementary Survey of Modern Physics," The Macmillan Company, 1936; for Chaps. II to IX.

N. F. Mott: "An Outline of Wave Mechanics," Cambridge University Press, 1930; for Chaps. III, IV, V, VI, X.

W. G. Penney: "Quantum Theory of Valency," Methuen & Co., Ltd., 1935; for Chaps. IX, X, XI.

"Outline of Atomic Physics," 2d ed., by the University of Pittsburgh Physics Staff, John Wiley & Sons, Inc., 1937; for Chaps. II to VII.

II

A. E. van Arkel and J. H. de Boer: "Chemische Bindung als elektrostatische Erscheinung," S. Hirzel, Leipzig, 1931; for Chap. XI and later chapters.

S. Dushman: "Elements of Quantum Mechanics," John Wiley & Sons, Inc., 1938; for Chaps. III, IV, V, VI, IX, X, XVII.

H. J. EMELÉUS and J. S. ANDERSON: "Modern Aspects of Inorganic. Chemistry," D. Van Nostrand Company, Inc., 1938; for Chap. XII, and later chapters.

K. FAJANS: "Radioelements and Isotopes: Chemical Forces and Optical Properties of Substances," McGraw-Hill Book Company, Inc., 1931; for Chap. XI and later chapters.

J. FRENKEL: "Wave Mechanics, Elementary Theory," Oxford University Press, 1932; for Chaps. III, IV, V, XVIII.

O. HASSEL: "Crystal Chemistry," William Heinemann, Ltd., 1935; for Chaps. XIV to XVIII.

L. PAULING: "The Nature of the Chemical Bond," Cornell, University Press, 1939; for Chap. XI and later chapters.

L. PAULING and E. B. WILSON, JR.: "Introduction to Quantum Mechanics," McGraw-Hill Book Co., Inc., 1935; for Chaps. III, IV, V, VI, IX, X.

E. RABINOWITSCH and E. THILO: "Periodisches System," Ferdinand Enke, Stuttgart, 1930; for Chaps. I, II, III, V, VII, VIII, XIII, XIV.

A. E. RUARK and H. C. UREY: "Atoms, Molecules and Quanta," McGraw-Hill Book Company, Inc., 1930 (Dover reprint); for Chaps. II to X.

N. V. SIDGWICK: "Electronic Theory of Valency," Oxford University Press, 1927; for Chaps. XI, XII, XIII, XVI, XVII, XIX

N. V. SIDGWICK: "Some Physical Properties of the Covalent Link in Chemistry," Cornell University Press, 1933; for Chap. XI and later chapters.

A. SOMMERFELD: "Atomic Structure and Spectral Lines, vol. I, 3d English ed., Methuen & Co., Ltd., 1934; for Chaps. II, IV to VIII.

H. SPONER: "Molekülspektren," vol. II, Julius Springer, Berlin, 1936; for Chaps. IV, VI, IX, X, XI, XIII.

C. W. STILLWELL: "Crystal Chemistry," McGraw-Hill Book Company, Inc., 1938; for Chaps. XIII to XVI.

H. E. WHITE: "Introduction to Atomic Spectra," McGraw-Hill Book Company, Inc., 1934; for Chaps. IV to VIII.

Handbuch der Physik, 2d ed., vol. XXIV, part 2, Julius Springer, Berlin, 1933.

III

F. R. BICHOWSKY and F. D. ROSSINI: "The Thermochemistry of the Chemical Substances," Reinhold Publishing Corporation, 1936; a tabulation of thermal data.

LANDOLT-BÖRNSTEIN: "Physikalisch-Chemische Tabellen," 5th ed., with three supplements, Julius Springer, Berlin, 1923–1936.

H. SPONER: "Molekülspektren," vol. I (supplementary tables in vol. II), Julius Springer, Berlin, 1935; results of band-spectrum analysis.

"International Critical Tables," McGraw-Hill Book Company, Inc., 1926–1930.

Strukturbericht, supplement to the *Zeitschrift für Kristallographie*, Akademische Verlagsgesellschaft, Leipzig, vol. I, 1931, vol. II, 1937, vol. III, 1937, vol. IV, 1938; a summary of crystallographic data.

AUTHOR INDEX

SUBJECT INDEX

See the Formula and Substance Index for reference to individual compounds mentioned in the text. For tables see this Subject Index, or the list of tables, pp. xiii–xiv. *n.* indicates reference to footnote; *ex.* indicates reference to exercise.

D

F

K

L

M

T

U

V

FORMULA AND SUBSTANCE INDEX

This index contains references to individual substances and certain groups of substances, mentioned in the text; individual substances, which appear in tables or lists only, or which are used as illustrations, without bringing out any particularly characteristic property of the substance, are not in general included, as these may be traced from the Subject Index or from the List of Tables, pp. xiii–xiv.

Alphabetization in this index is by the first symbol in the formula, then by the second symbol, etc. Numeral subscripts are disregarded in deciding between compounds containing different elements. Thus BN precedes Be, and Ag_2F follows AgCl and precedes AgI. Certain group names are used and alphabetized as if they were symbols. Thus H halides follows HF and precedes HI. The most important of these terms are the following: alkali, alkali-ions, alkali-metals, alkaline-earth, alkaline-earth-metals, complexes, Cu-group, Fe-group, halide-ions, halides, halogens, hydrides, oxides, Pt-group, rare earths, S-group, subhalides, sulfides, selenides, tellurides, Zn-group. A hyphenated word is alphabetized as a single word. Thus, Zn-group follows all Zn compounds, e.g., ZnS; and halides follows halide-ions.

The valence of an element is sometimes indicated by a Roman numeral in parentheses.

Contrary to the usual custom of always listing all compounds under the most positive elements, compounds are sometimes listed under a negative element or group, when the latter determines the characteristic properties, and many negative ions are listed.

n. indicates reference to footnote; *ex.* refers to exercises.

B

I

K

L

S

T

SOME DOVER SCIENCE BOOKS

THE RISE OF THE NEW PHYSICS (formerly THE DECLINE OF MECHANISM), *A. d'Abro*
This authoritative and comprehensive 2-volume exposition is unique in scientific publishing. Written for intelligent readers not familiar with higher mathematics, it is the only thorough explanation in non-technical language of modern mathematical-physical theory. Combining both history and exposition, it ranges from classical Newtonian concepts up through the electronic theories of Dirac and Heisenberg, the statistical mechanics of Fermi, and Einstein's relativity theories. "A must for anyone doing serious study in the physical sciences," *J. of Franklin Inst.* 97 illustrations. 991pp. 2 volumes.
T3, T4 Two volume set, paperbound $5.50

THE STRANGE STORY OF THE QUANTUM, AN ACCOUNT FOR THE GENERAL READER OF THE GROWTH OF IDEAS UNDERLYING OUR PRESENT ATOMIC KNOWLEDGE, *B. Hoffmann*
Presents lucidly and expertly, with barest amount of mathematics, the problems and theories which led to modern quantum physics. Dr. Hoffmann begins with the closing years of the 19th century, when certain trifling discrepancies were noticed, and with illuminating analogies and examples takes you through the brilliant concepts of Planck, Einstein, Pauli, de Broglie, Bohr, Schroedinger, Heisenberg, Dirac, Sommerfeld, Feynman, etc. This edition includes a new, long postscript carrying the story through 1958. "Of the books attempting an account of the history and contents of our modern atomic physics which have come to my attention, this is the best," H. Margenau, Yale University, in *American Journal of Physics*. 32 tables and line illustrations. Index. 275pp. 5⅜ x 8.
T518 Paperbound $2.00

GREAT IDEAS AND THEORIES OF MODERN COSMOLOGY, *Jagjit Singh*
The theories of Jeans, Eddington, Milne, Kant, Bondi, Gold, Newton, Einstein, Gamow, Hoyle, Dirac, Kuiper, Hubble, Weizsäcker and many others on such cosmological questions as the origin of the universe, space and time, planet formation, "continuous creation," the birth, life, and death of the stars, the origin of the galaxies, etc. By the author of the popular *Great Ideas of Modern Mathematics*. A gifted popularizer of science, he makes the most difficult abstractions crystal-clear even to the most non-mathematical reader. Index. xii + 276pp. 5⅜ x 8½ T925 Paperbound $2.00

GREAT IDEAS OF MODERN MATHEMATICS: THEIR NATURE AND USE, *Jagjit Singh*
Reader with only high school math will understand main mathematical ideas of modern physics, astronomy, genetics, psychology, evolution, etc., better than many who use them as tools, but comprehend little of their basic structure. Author uses his wide knowledge of non-mathematical fields in brilliant exposition of differential equations, matrices, group theory, logic, statistics, problems of mathematical foundations, imaginary numbers, vectors, etc. Original publications, appendices. indexes. 65 illustr. 322pp. 5⅜ x 8. T587 Paperbound $2.00

THE MATHEMATICS OF GREAT AMATEURS, *Julian L. Coolidge*
Great discoveries made by poets, theologians, philosophers, artists and other non-mathematicians: Omar Khayyam, Leonardo da Vinci, Albrecht Dürer, John Napier, Pascal, Diderot, Bolzano, etc. Surprising accounts of what can result from a non-professional preoccupation with the oldest of sciences. 56 figures. viii + 211pp. 5⅜ x 8½. S1009 Paperbound $2.00

A SOURCE BOOK IN MATHEMATICS,
D. E. Smith
Great discoveries in math, from Renaissance to end of 19th century, in English translation. Read announcements by Dedekind, Gauss, Delamain, Pascal, Fermat, Newton, Abel, Lobachevsky, Bolyai, Riemann, De Moivre, Legendre, Laplace, others of discoveries about imaginary numbers, number congruence, slide rule, equations, symbolism, cubic algebraic equations, non-Euclidean forms of geometry, calculus, function theory, quaternions, etc. Succinct selections from 125 different treatises, articles, most unavailable elsewhere in English. Each article preceded by biographical introduction. Vol. I: Fields of Number, Algebra. Index. 32 illus. 338pp. 5⅜ x 8. Vol. II: Fields of Geometry, Probability, Calculus, Functions, Quaternions. 83 illus. 432pp. 5⅜ x 8.
Vol. 1 Paperbound $2.00, Vol. 2 Paperbound $2.00,
The set $4.00

FOUNDATIONS OF PHYSICS,
R. B. Lindsay & H. Margenau
Excellent bridge between semi-popular works & technical treatises. A discussion of methods of physical description, construction of theory; valuable for physicist with elementary calculus who is interested in ideas that give meaning to data, tools of modern physics. Contents include symbolism; mathematical equations; space & time foundations of mechanics; probability; physics & continua; electron theory; special & general relativity; quantum mechanics; causality. "Thorough and yet not overdetailed. Unreservedly recommended," *Nature* (London). Unabridged, corrected edition. List of recommended readings. 35 illustrations. xi + 537pp. 5⅜ x 8. Paperbound $3.00

FUNDAMENTAL FORMULAS OF PHYSICS,
ed. by D. H. Menzel
High useful, full, inexpensive reference and study text, ranging from simple to highly sophisticated operations. Mathematics integrated into text—each chapter stands as short textbook of field represented. Vol. 1: Statistics, Physical Constants, Special Theory of Relativity, Hydrodynamics, Aerodynamics, Boundary Value Problems in Math, Physics, Viscosity, Electromagnetic Theory, etc. Vol. 2: Sound, Acoustics, Geometrical Optics, Electron Optics, High-Energy Phenomena, Magnetism, Biophysics, much more. Index. Total of 800pp. 5⅜ x 8.
Vol. 1 Paperbound $2.25, Vol. 2 Paperbound $2.25,
The set $4.50

THEORETICAL PHYSICS,
A. S. Kompaneyets
One of the very few thorough studies of the subject in this price range. Provides advanced students with a comprehensive theoretical background. Especially strong on recent experimentation and developments in quantum theory. Contents: Mechanics (Generalized Coordinates, Lagrange's Equation, Collision of Particles, etc.), Electrodynamics (Vector Analysis, Maxwell's equations, Transmission of Signals, Theory of Relativity, etc.), Quantum Mechanics (the Inadequacy of Classical Mechanics, the Wave Equation, Motion in a Central Field, Quantum Theory of Radiation, Quantum Theories of Dispersion and Scattering, etc.), and Statistical Physics (Equilibrium Distribution of Molecules in an Ideal Gas, Boltzmann Statistics, Bose and Fermi Distribution. Thermodynamic Quantities, etc.). Revised to 1961. Translated by George Yankovsky, authorized by Kompaneyets. 137 exercises. 56 figures. 529pp. 5⅜ x 8½.
Paperbound $2.50

CELESTIAL OBJECTS FOR COMMON TELESCOPES,
Rev. T. W. Webb
Classic handbook for the use and pleasure of the amateur astronomer. Of inestimable aid in locating and identifying thousands of celestial objects. Vol I, The Solar System: discussions of the principle and operation of the telescope, procedures of observations and telescope-photography, spectroscopy, etc., precise location information of sun, moon, planets, meteors. Vol. II, The Stars: alphabetical listing of constellations, information on double stars, clusters, stars with unusual spectra, variables, and nebulae, etc. Nearly 4,000 objects noted. Edited and extensively revised by Margaret W. Mayall, director of the American Assn. of Variable Star Observers. New Index by Mrs. Mayall giving the location of all objects mentioned in the text for Epoch 2000. New Precession Table added. New appendices on the planetary satellites, constellation names and abbreviations, and solar system data. Total of 46 illustrations. Total of xxxix + 606pp. 5⅜ x 8. Vol. 1 Paperbound $2.25, Vol. 2 Paperbound $2.25
The set $4.50

PLANETARY THEORY,
E. W. Brown and C. A. Shook
Provides a clear presentation of basic methods for calculating planetary orbits for today's astronomer. Begins with a careful exposition of specialized mathematical topics essential for handling perturbation theory and then goes on to indicate how most of the previous methods reduce ultimately to two general calculation methods: obtaining expressions either for the coordinates of planetary positions or for the elements which determine the perturbed paths. An example of each is given and worked in detail. Corrected edition. Preface. Appendix. Index. xii + 302pp. 5⅜ x 8½. Paperbound $2.25

STAR NAMES AND THEIR MEANINGS,
Richard Hinckley Allen
An unusual book documenting the various attributions of names to the individual stars over the centuries. Here is a treasure-house of information on a topic not normally delved into even by professional astronomers; provides a fascinating background to the stars in folk-lore, literary references, ancient writings, star catalogs and maps over the centuries. Constellation-by-constellation analysis covers hundreds of stars and other asterisms, including the Pleiades, Hyades, Andromedan Nebula, etc. Introduction. Indices. List of authors and authorities. xx + 563pp. 5⅜ x 8½. Paperbound $2.50

A SHORT HISTORY OF ASTRONOMY, *A. Berry*
Popular standard work for over 50 years, this thorough and accurate volume covers the science from primitive times to the end of the 19th century. After the Greeks and the Middle Ages, individual chapters analyze Copernicus, Brahe, Galileo, Kepler, and Newton, and the mixed reception of their discoveries. Post-Newtonian achievements are then discussed in unusual detail: Halley, Bradley, Lagrange, Laplace, Herschel, Bessel, etc. 2 Indexes. 104 illustrations, 9 portraits. xxxi + 440pp. 5⅜ x 8. Paperbound $2.75

SOME THEORY OF SAMPLING, *W. E. Deming*
The purpose of this book is to make sampling techniques understandable to and useable by social scientists, industrial managers, and natural scientists who are finding statistics increasingly part of their work. Over 200 exercises, plus dozens of actual applications. 61 tables. 90 figs. xix + 602pp. 5⅜ x 8½.
Paperbound $3.50

PRINCIPLES OF STRATIGRAPHY,
A. W. Grabau
Classic of 20th century geology, unmatched in scope and comprehensiveness. Nearly 600 pages cover the structure and origins of every kind of sedimentary, hydrogenic, oceanic, pyroclastic, atmoclastic, hydroclastic, marine hydroclastic, and bioclastic rock; metamorphism; erosion; etc. Includes also the constitution of the atmosphere; morphology of oceans, rivers, glaciers; volcanic activities; faults and earthquakes; and fundamental principles of paleontology (nearly 200 pages). New introduction by Prof. M. Kay, Columbia U. 1277 bibliographical entries. 264 diagrams. Tables, maps, etc. Two volume set. Total of xxxii + 1185pp. 5⅜ x 8. Vol. 1 Paperbound $2.50, Vol. 2 Paperbound $2.50, The set $5.00

SNOW CRYSTALS, *W. A. Bentley and W. J. Humphreys*
Over 200 pages of Bentley's famous microphotographs of snow flakes—the product of painstaking, methodical work at his Jericho, Vermont studio. The pictures, which also include plates of frost, glaze and dew on vegetation, spider webs, windowpanes; sleet; graupel or soft hail, were chosen both for their scientific interest and their aesthetic qualities. The wonder of nature's diversity is exhibited in the intricate, beautiful patterns of the snow flakes. Introductory text by W. J. Humphreys. Selected bibliography. 2,453 illustrations. 224pp. 8 x 10¼. Paperbound $3.25

THE BIRTH AND DEVELOPMENT OF THE GEOLOGICAL SCIENCES,
F. D. Adams
Most thorough history of the earth sciences ever written. Geological thought from earliest times to the end of the 19th century, covering over 300 early thinkers & systems: fossils & their explanation, vulcanists vs. neptunists, figured stones & paleontology, generation of stones, dozens of similar topics. 91 illustrations, including medieval, renaissance woodcuts, etc. Index. 632 footnotes, mostly bibliographical. 511pp. 5⅜ x 8. Paperbound $2.75

ORGANIC CHEMISTRY, *F. C. Whitmore*
The entire subject of organic chemistry for the practicing chemist and the advanced student. Storehouse of facts, theories, processes found elsewhere only in specialized journals. Covers aliphatic compounds (500 pages on the properties and synthetic preparation of hydrocarbons, halides, proteins, ketones, etc.), alicyclic compounds, aromatic compounds, heterocyclic compounds, organophosphorus and organometallic compounds. Methods of synthetic preparation analyzed critically throughout. Includes much of biochemical interest. "The scope of this volume is astonishing," *Industrial and Engineering Chemistry*. 12,000-reference index. 2387-item bibliography. Total of x + 1005pp. 5⅜ x 8. Two volume set, paperbound $4.50

THE PHASE RULE AND ITS APPLICATION,
Alexander Findlay
Covering chemical phenomena of 1, 2, 3, 4, and multiple component systems, this "standard work on the subject" (*Nature*, London), has been completely revised and brought up to date by A. N. Campbell and N. O. Smith. Brand new material has been added on such matters as binary, tertiary liquid equilibria, solid solutions in ternary systems, quinary systems of salts and water. Completely revised to triangular coordinates in ternary systems, clarified graphic representation, solid models, etc. 9th revised edition. Author, subject indexes. 236 figures. 505 footnotes, mostly bibliographic. xii + 494pp. 5⅜ x 8. Paperbound $2.75

A COURSE IN MATHEMATICAL ANALYSIS,
Edouard Goursat

Trans. by E. R. Hedrick, O. Dunkel, H. G. Bergmann. Classic study of fundamental material thoroughly treated. Extremely lucid exposition of wide range of subject matter for student with one year of calculus. Vol. 1: Derivatives and differentials, definite integrals, expansions in series, applications to geometry. 52 figures, 556pp. Paperbound $2.50. Vol. 2, Part 1: Functions of a complex variable, conformal representations, doubly periodic functions, natural boundaries, etc. 38 figures, 269pp. Paperbound $1.85. Vol. 2, Part 2: Differential equations, Cauchy-Lipschitz method, nonlinear differential equations, simultaneous equations, etc. 308pp. Paperbound $1.85. Vol. 3, Part 1: Variation of solutions, partial differential equations of the second order. 15 figures, 339pp. Paperbound $3.00. Vol. 3, Part 2: Integral equations, calculus of variations. 13 figures, 389pp. Paperbound $3.00

PLANETS, STARS AND GALAXIES,
A. E. Fanning

Descriptive astronomy for beginners: the solar system; neighboring galaxies; seasons; quasars; fly-by results from Mars, Venus, Moon; radio astronomy; etc. all simply explained. Revised up to 1966 by author and Prof. D. H. Menzel, former Director, Harvard College Observatory. 29 photos, 16 figures. 189pp. 5⅜ x 8½. Paperbound $1.50

GREAT IDEAS IN INFORMATION THEORY, LANGUAGE AND CYBERNETICS,
Jagjit Singh

Winner of Unesco's Kalinga Prize covers language, metalanguages, analog and digital computers, neural systems, work of McCulloch, Pitts, von Neumann, Turing, other important topics. No advanced mathematics needed, yet a full discussion without compromise or distortion. 118 figures. ix + 338pp. 5⅜ x 8½. Paperbound $2.00

GEOMETRIC EXERCISES IN PAPER FOLDING,
T. Sundara Row

Regular polygons, circles and other curves can be folded or pricked on paper, then used to demonstrate geometric propositions, work out proofs, set up well-known problems. 89 illustrations, photographs of actually folded sheets. xii + 148pp. 5⅜ x 8½. Paperbound $1.00

VISUAL ILLUSIONS, THEIR CAUSES, CHARACTERISTICS AND APPLICATIONS,
M. Luckiesh

The visual process, the structure of the eye, geometric, perspective illusions, influence of angles, illusions of depth and distance, color illusions, lighting effects, illusions in nature, special uses in painting, decoration, architecture, magic, camouflage. New introduction by W. H. Ittleson covers modern developments in this area. 100 illustrations. xxi + 252pp. 5⅜ x 8. Paperbound $1.50

ATOMS AND MOLECULES SIMPLY EXPLAINED,
B. C. Saunders and R. E. D. Clark

Introduction to chemical phenomena and their applications: cohesion, particles, crystals, tailoring big molecules, chemist as architect, with applications in radioactivity, color photography, synthetics, biochemistry, polymers, and many other important areas. Non technical. 95 figures. x + 299pp. 5⅜ x 8½. Paperbound $1.50

THE PRINCIPLES OF ELECTROCHEMISTRY,
D. A. MacInnes

Basic equations for almost every subfield of electrochemistry from first principles, referring at all times to the soundest and most recent theories and results; unusually useful as text or as reference. Covers coulometers and Faraday's Law, electrolytic conductance, the Debye-Hueckel method for the theoretical calculation of activity coefficients, concentration cells, standard electrode potentials, thermodynamic ionization constants, pH, potentiometric titrations, irreversible phenomena. Planck's equation, and much more. 2 indices. Appendix. 585-item bibliography. 137 figures. 94 tables. ii + 478pp. 5⅝ x 8⅜.
Paperbound $2.75

MATHEMATICS OF MODERN ENGINEERING,
E. G. Keller and R. E. Doherty

Written for the Advanced Course in Engineering of the General Electric Corporation, deals with the engineering use of determinants, tensors, the Heaviside operational calculus, dyadics, the calculus of variations, etc. Presents underlying principles fully, but emphasis is on the perennial engineering attack of set-up and solve. Indexes. Over 185 figures and tables. Hundreds of exercises, problems, and worked-out examples. References. Two volume set. Total of xxxiii + 623pp. 5⅜ x 8. Two volume set, paperbound $3.70

AERODYNAMIC THEORY: A GENERAL REVIEW OF PROGRESS,
William F. Durand, editor-in-chief

A monumental joint effort by the world's leading authorities prepared under a grant of the Guggenheim Fund for the Promotion of Aeronautics. Never equalled for breadth, depth, reliability. Contains discussions of special mathematical topics not usually taught in the engineering or technical courses. Also: an extended two-part treatise on Fluid Mechanics, discussions of aerodynamics of perfect fluids, analyses of experiments with wind tunnels, applied airfoil theory, the nonlifting system of the airplane, the air propeller, hydrodynamics of boats and floats, the aerodynamics of cooling, etc. Contributing experts include Munk, Giacomelli, Prandtl, Toussaint, Von Karman, Klemperer, among others. Unabridged republication. 6 volumes. Total of 1,012 figures, 12 plates, 2,186pp. Bibliographies. Notes. Indices. 5⅜ x 8½.
Six volume set, paperbound $13.50

FUNDAMENTALS OF HYDRO- AND AEROMECHANICS,
L. Prandtl and O. G. Tietjens

The well-known standard work based upon Prandtl's lectures at Goettingen. Wherever possible hydrodynamics theory is referred to practical considerations in hydraulics, with the view of unifying theory and experience. Presentation is extremely clear and though primarily physical, mathematical proofs are rigorous and use vector analysis to a considerable extent. An Engineering Society Monograph, 1934. 186 figures. Index. xvi + 270pp. 5⅜ x 8.
Paperbound $2.00

APPLIED HYDRO- AND AEROMECHANICS,
L. Prandtl and O. G. Tietjens

Presents for the most part methods which will be valuable to engineers. Covers flow in pipes, boundary layers, airfoil theory, entry conditions, turbulent flow in pipes, and the boundary layer, determining drag from measurements of pressure and velocity, etc. Unabridged, unaltered. An Engineering Society Monograph. 1934. Index. 226 figures, 28 photographic plates illustrating flow patterns. xvi + 311pp. 5⅜ x 8. Paperbound $2.00